P9-AZX-461

FOOD COLLOIDS

Special Publication No. 75

Food Colloids

The Proceedings of an International Symposium
Organized by the Food Chemistry Group of
The Royal Society of Chemistry

Unilever Research, Colworth, UK, 13th–15th April 1988

Edited by
R. D. Bee
Unilever Research, Sharnbrook

P. Richmond
AFRC, Norwich

J. Mingins
AFRC, Norwich

British Library Cataloguing in Publication Data

Food colloids.
1. Food. Constituents: colloids.
I. Bee, R. D. II. Richmond, P. III. Mingins, J.
IV. Royal Society of Chemistry, *Food Chemistry Group*
641.1

ISBN 0-85186-826-6

Published by The Royal Society of Chemistry,
Thomas Graham House, Cambridge CB4 4WF

Filmset by Bath Typesetting Ltd., Bath
and printed by J. W. Arrowsmith Ltd., Bristol

Made in Great Britain

Preface

Following the success of the international symposium on 'Food Emulsions and Foams' organised by the Food Chemistry Group of the Royal Society of Chemistry in 1986 in Leeds, it was decided to institute a bi-annual series of symposia on surface and colloidal aspects of foods. This book records the proceedings of the second symposium entitled 'Food Colloids' which was held on 13–15th April 1988 at Unilever Research, Colworth House, Sharnbrook, England.

Our focus on this topic is intended to display the importance of the colloidal state in foods; the excellent response in papers and posters, and the large international audience at the meeting illustrate that knowledge and control of these aspects is now recognized as fundamental to foods.

A major aim of these symposia is to bring together scientists from a wide range of disciplines representing both the acedemic and industrial communities. By engendering discussion on generic colloid problems facing the food industry, we hope to identify key scientific and technical stumbling blocks to progress; equally important, it provides a continuing forum to air new understanding of the colloidal behaviour of foods and the operations by which they are made.

The term 'Food Colloids' gave us a wide programme covering numerous examples of disperse phases. A major departure at this meeting was the presentation of some specifically invited papers in areas of physics, which it was hoped would stimulate new approaches to food-based problems. These physical models are put into context by expert papers from industrial laboratories on very practical problems in multicomponent, multiphase foods. Together with excellent contributions from academic laboratories, they show that food colloid research is flourishing and that we can look forward to a valuable meeting in Norwich in 1990.

It is a pleasure to be able to include in this volume the Royal Society of Chemistry Food Chemistry Group Junior Medal lecture given by Dr Mary Griffin. Her subject, 'Steric stabilisation and flow properties of concentrated casein micellar suspensions' amply illustrates the intriguing complexities of food colloids and her significant contributions to the science.

We should like to thank Dr Darling for his spontaneous review at the end of the meeting, the session chairmen and all the contributors to papers, posters, and discussions. We are also grateful to the many people involved with the local arrangements who ensured the smooth running of the meeting; also our sincere thanks to Unilever Research for providing the symposium facilities.

In particular, we thank June Knighton for her able work as symposium secretary and Wayne Morley for collecting the discussion material. Finally, our thanks go to the staff of the Royal Society of Chemistry for help in preparing this volume.

R. D. Bee, J. Mingins, P. Richmond
June 1989.

Contents

1	Edible Food Foams and Sponges *P. J. Lillford and F. J. Judge*	1
2	The Mechanical and Flow Properties of Foams and Highly Concentrated Emulsions *H. M. Princen*	14
3	The Physics of Froths and Foams *D. Weaire*	25
4	Contibution of Drainage, Coalescence, and Disproportionation to the Stability of Aerated Foodstuffs and the Consequences for the Bubble Size Distribution as Measured by a Newly Developed Optical Glass-fibre Technique *A. D. Ronteltap and A. Prins*	39
5	Beer Foams *C. W. Bamforth*	48
6	Solid Foams *A. C. Smith*	56
7	Competitive Adsorption Between Proteins and Small-molecule Surfactants in Food Emulsions *E. Dickenson and C. M. Woskett*	74
8	Electrostatic Interactions Between Proteins and their Effect on Foam Composition and Stability *D. C. Clarke, A. R. Mackie, L. J. Smith and D. R. Wilson*	97
9	Steric Stabilization and Flow Properties of Concentrated Casein Micellar Suspensions *M. C. Ambrose Griffin*	110
10	Effect of Adsorbed Proteins on Interactions Between Emulsion Droplets *L. R. Fisher and E. E. Mitchell*	123

11 Theoretical Studies of the Solid–Fluid Interface 138
G. Rickayzen

12 Electrochemical Approach to Studies of Binding and Electrostatic Interaction in Concentrated Food Dispersions 154
A. H. Clark and P. M. Hart

13 Lecithin-stabilized Silica Dispersions 173
A. C. Mackie and M. J. Hay

14 Suppression of Perceived Flavour and Taste by Food Hydrocolloids 184
Z. V. Baines and E. R. Morris

15 Effect of Surfactants and Non-aqueous Phases on the Kinetics of Reactions of Food Preservatives 193
L. Wedzicha and A. Zeb

16 Weak Particle Networks 206
T. Van Vliet and P. Walstra

17 Influence of the Emulsifier on the Sedimentation of Water-in-Oil Emulsions 218
F. van Voorst Vader and F. Groeneweg

18 Towards a Comprehensive Theory for Sedimentation in Colloidal Suspensions 230
G. L. Barker and M. J. Grimson

19 Mechanism of Fracture in Meat and Meat Products 246
P. Purslow

20 Continuous Sausage Processing 262
L. L. Borchert

21 Technological Problems in Margarine and Low-calorie Spreads 267
J. Madsen

22 The Role of Fat Crystals in Emulsion Stability 272
I. J. Campbell

23 The Colloid Chemistry of Black Tea 283
R. S. Harbron, R. H. Ottewill and R. D. Bee

24 Aspects of Stability in Milk and Milk Products 295
D. G. Dalgleish

25 Ultrasonic Measurements in Food Emulsions and Dispersions 306
M. J. W. Povey and D. J. McClements

26 The Packing and Movement of Particles 323
S. F. Edwards

27 Discussions 331

Abstracts of Posters

28 Studies on the Aggregation of Casein Micelles 355
D. S. Horne

29 Kinetics of the Partial Coalescence Process in Oil-in-Water Emulsions 360
K. Boode and P. Walstra

30 The α-Gel Phase of Glycerol Lactopalmitate in Whipped Emulsions 364
J. M. M. Westerbeck, A. Prins and K. Kussendrager

31 Importance of Milk Proteins to the Whipping Properties of 38% Fat Cream 368
E. C. Needs

32 Low Molecular Weight Surfactants and the Stability of Cream Liqueurs 372

33 Interfacial Competition Between α_{s1}-Casein and β-Casein in Oil-in-Water Emulsions 377
E. Dickinson and E. Rolfe

34 Non-intrusive Determination of Droplet Size Distribution in a Concentrated Oil-in-Water Emulsion from Sedimentation Profiles 382
A. M. Howe and M. M. Robins

35 Effect of Heat on the Emulsifying Properties of Gum Arabic 386
R. C. Randall, G. O. Phillips and P. A. Williams

36 Gel Formation after Heating of Oil-in-Water Emulsions Stabilized by Whey Protein Concentrates 391
G. Masson and R. Joot

37 Rheological Study of Interactions Among Wheat Flour, Milk Proteins, and Lipids of Béchamel Sauce 395
L. P. Martinez Padilla and J. Hardy

38 Casein Micelles, Polycondensation, and Fractals 400
D. S. Horne, T. G. Parker and D. G. Dalgleish

Edible Food Foams and Sponges

By P. J. Lillford and F. J. Judge

UNILEVER RESEARCH, COLWORTH HOUSE, SHARNBROOK, BEDFORD MK44 1LQ, UK

1 Introduction

It is not the intention that this paper should be definitive or report a comprehensive study of food foams. Instead, some of the factors of common interest will be identified in products which can be described as foams and sponges. First, what is the difference between a foam and a sponge? For the sake of this paper, a foam will be defined as a colloidal dispersion of gas in a liquid or solid-like structure, whereas a sponge has a degree of interconnectivity of the gaseous phase such that it is normally continuous in both the gaseous and the liquid or solid phases. This, however, introduces the first problem: how does one characterize the degree of connectivity or the degree of completeness of a gas dispersion? As all of the definitions refer to structures, the remainder of this paper will describe the types of structures so far observed and some of the methods applied in their examination.

2 Baked Products

Probably the largest category of foams and sponges is to be found in the area of baked foods derived from cereal flours. It is probably the oldest food technology, requiring flour, wild yeast, and fire—'Stone Age biotechnology.'

Figure 1 shows an X-ray projection photograph of a form of bread currently of high commercial value, *viz.* pizza base. Stereo shots reveal that this is both a foam and a sponge. Clearly, continuity of bubbles has occurred owing to coalescence, but some isolated bubbles still remain. How did this occur, and does it matter?

Figure 2 shows the surface structure of cut cake, and we have found a simple correlation between the subjective terms 'heavy' and 'light' and the observable phase volume of bubbles, *i.e.* the density.

At high magnification (Figure 3), we can see again that what was a foam has become a sponge owing to coalescence of the bubbles. Furthermore, observations at this degree of magnification allow us to pose the questions which are common to all of the subsequent structures. For example, how do we measure the phase volume of the air? How do we describe distribution of cell sizes and channelling? How do we obtain molecular details of the material providing the mechanical

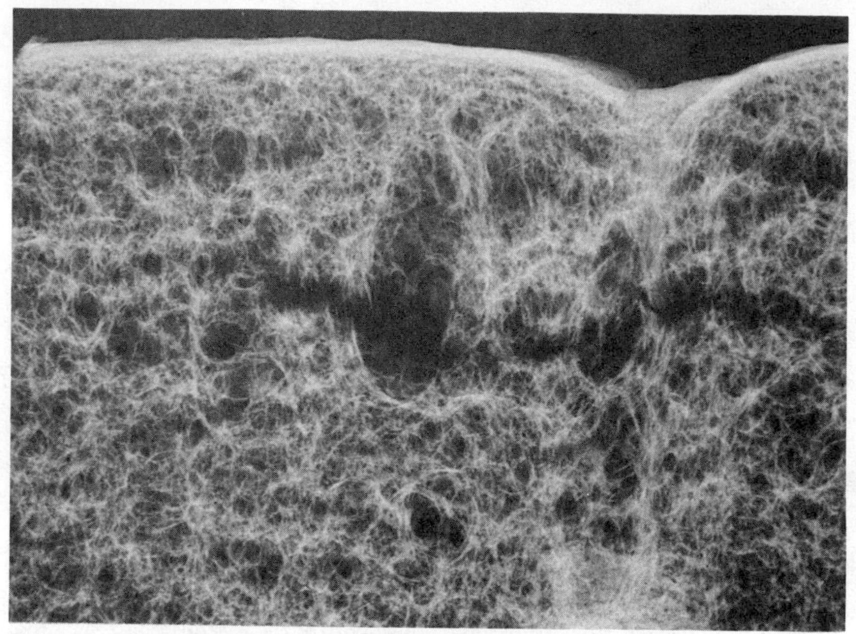

Figure 1 *Microfocal X-ray projection of pizza base*

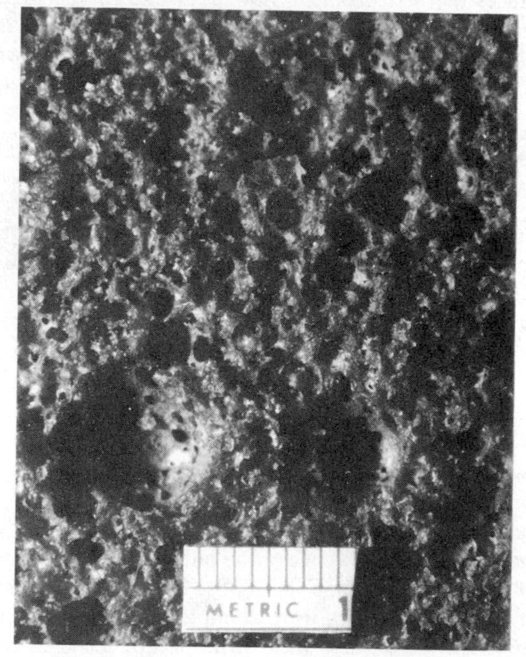

Figure 2 *Surface structure of a chocolate cake*

Figure 3 *Scanning electron micrograph of sponge cake*

structure? Finally, how does that molecular structure relate to the mechanical and therefore the perceived textural properties of the final product?

In this laboratory, recent work has focused on the formation and physical properties of baked structures.[1] For example, in Figure 4 the volume expansions of two cake batters are plotted against time in the oven. The two formulations differed in that the batter marked 1 had a lower viscosity throughout the baking process. This resulted in convective flow in the batter during cooking and the higher expansion was achieved. More important, however, both batters are seen to decrease in volume before baking is complete. By measurements of the permeability of these structures to a gas flow, we could show that this decrease in volume is primarily due to the air cells coalescing to form continuous channels. In other words, foams are becoming sponges. The formation of continuous air channels is found to be essential in maintaining the volume of the cake on cooling once removed from the oven. The unfortunate phenomenon of cake collapse appears to be due to the absence of channelling. When foams are removed from the oven, water vapour condenses in individual bubbles and the overpressure of the atmosphere is sufficient to cause a dramatic collapse of the structure if the gas cells remain intact. A simple calculation, assuming that as the steam condenses atmospheric pressure is applied, shows that the resultant overpressure on a 10 cm diameter sponge cake is equivalent to a 12-stone man standing on it! The collapse of such structures on removal from the oven is therefore not surprising. During baking, the viscosity of all batters increases as starch gelatinizes. Perhaps more important, however, is that in the later stages of baking water is driven off so that

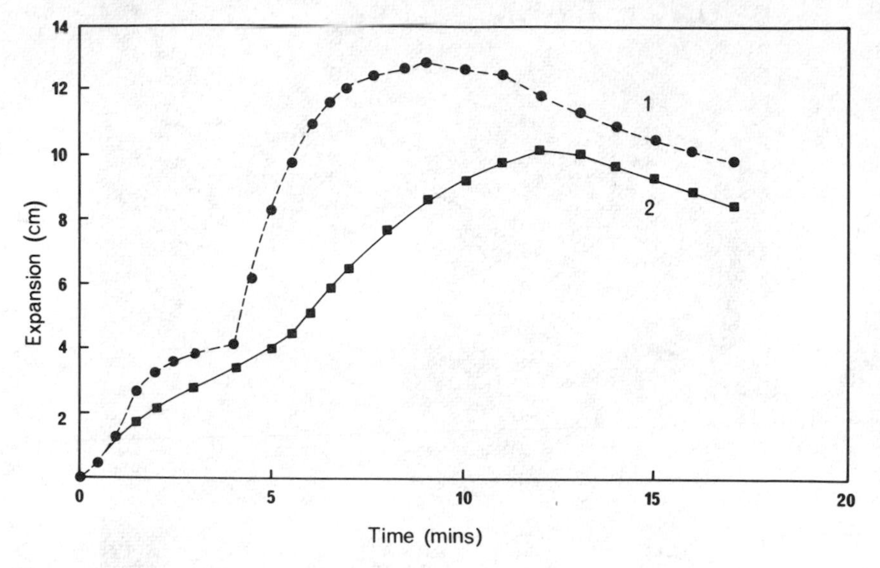

Figure 4 *Volumetric expansion of heated cake batters*
(Reproduced with permission from *Phys. Technol.*, 1988, **19**, 18)

a viscous matrix around air bubbles transforms to a brittle solid capable of cracking either under the increasing steam pressure, or alternatively when the overpressure is exerted during cooling. This allows further rupture of the air cells and increases permeability.

Mechanical methods of converting the foams to sponges are equally relevant. Thus, the common culinary practice of dropping cooked cakes while still in the baking tin serves not only to free the cake from the tin, but also to impart a rapid shock wave through the structure, which cracks further air cells and can indeed prevent subsequent shrinkage.

Many baked products can be considered as foams or sponges with brittle cell walls. Further examples of this kind of structure are meringues, and even puff pastry, where the laminar structure can be considered as a stacked set of extremely distorted holes. Fortunately, we can now borrow theories of the mechanical properties of such structures from other work. Pre-eminent in this field is the work of Ashby and Gibson[2] on brittle foams, where a simple theory was developed to explain the mechanics in terms of both modulus and failure properties of foams and sponges. The theory is oversimplified, as cell sizes of uniform dimensions are assumed and the polydispersity of foam cell sizes is not considered. Nonetheless, this theory allows us to examine first-order effects of matrix and air cell volume on the properties of the structures. A simple relationship between the product elastic modulus and the square of the bulk density is inferred. In Figure 5 results obtained for wafer structures are shown and a simple correlation obeying Ashby

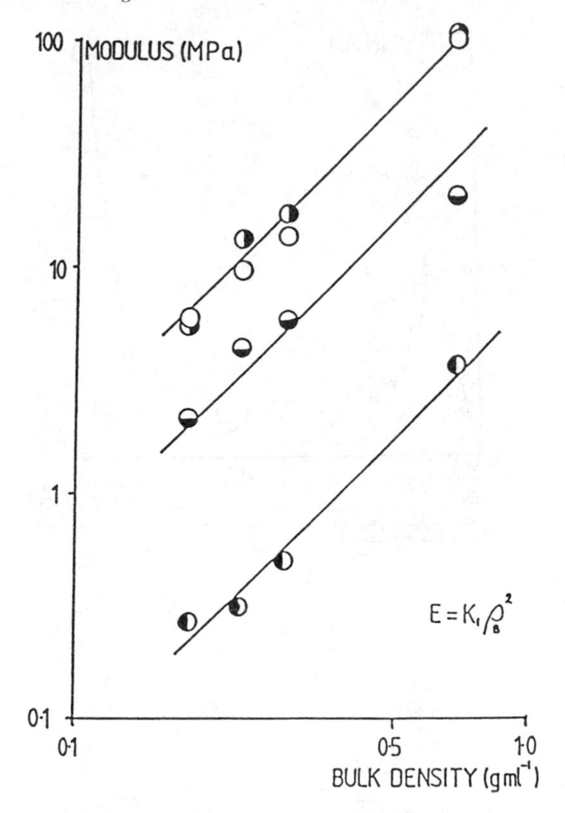

Figure 5 *Initial modulus of sponge cakes:* ○ *vacuum dried;* ◑, 33% *RH;* ◐, 57% *RH* ⊖, 75% *RH*
(Reproduced with permission from *J. Cereal Sci.*, 1989, **9**, 61)

and Gibson's model is observed.[3] In compression testing the stress and strain curves show the form given in Figure 6 and a critical stress to be related to the sensory properties of 'hardness', which is shown in Figure 7. A more complete use of Ashby and Gibson's models of brittle fracture in foodstuffs is described in 'Solid Foams', A. C. Smith in this volume.

There are many other confections which may also be amenable to this kind of examination, for example many familiar chocolate confectioneries now contain appreciable amounts of included air. Figures 8 and 9 show X-ray projections of an Aero Bar, having a low air phase volume but very large air cell size, and of a Wispa Bar having a similar air phase volume but at a much smaller air cell size. Textures of these products are now well recognized and noticeably different.

Sugar confections, on drying, also produce extremely brittle cell walls, presumably constructed from sugar glasses, and in Figure 10 the internal structure of a Crunchie Bar is shown. The brittle fracture properties of this structure are part of its appeal and its advertising claim.

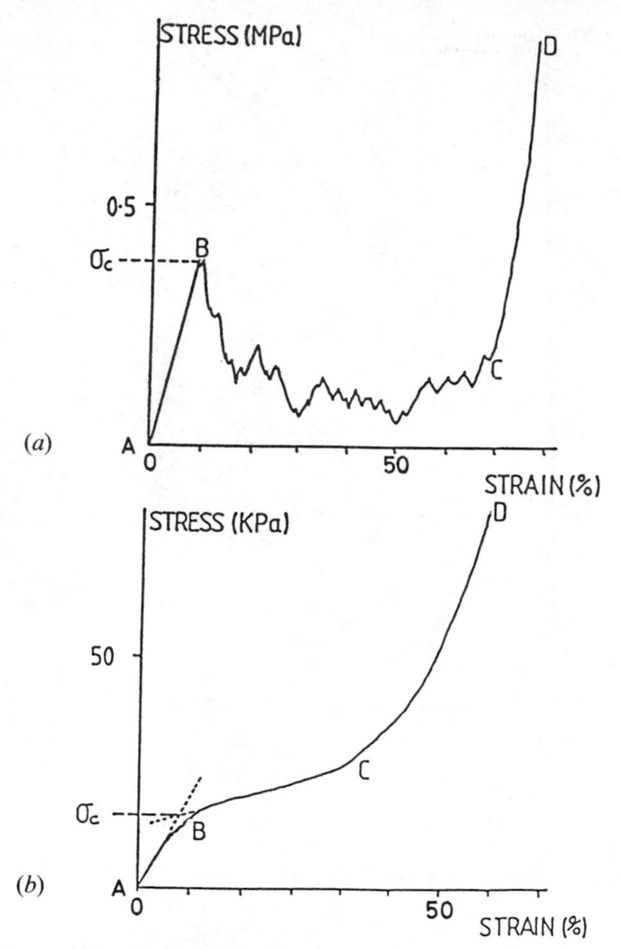

Figure 6 *Typical stress–strain curves for compression of sponge cakes: (a) vacuum dried or at* 33% *RH; (b)* 57% *or* 75% *RH*
(Reproduced with permission from *J. Cereal Sci.*, 1989, **9**, 61)

3 Dairy Products

Dairy-product foams and sponges are also well known, typical examples being whipped cream and ice cream. In these materials, cell walls are not necessarily required to be brittle and in fact fast fracture and crispness are not normally expected in such products. Examination of the structure of whipped cream (Figure 11) shows an appreciable distortion of the air cells, which implies a very high local phase volume and that the matrix surrounding air cells is sufficiently viscoelastic that rounding-off of the air cells does not readily occur. At higher magnification (Figure 12) we can see the unique and beautiful internal structure of

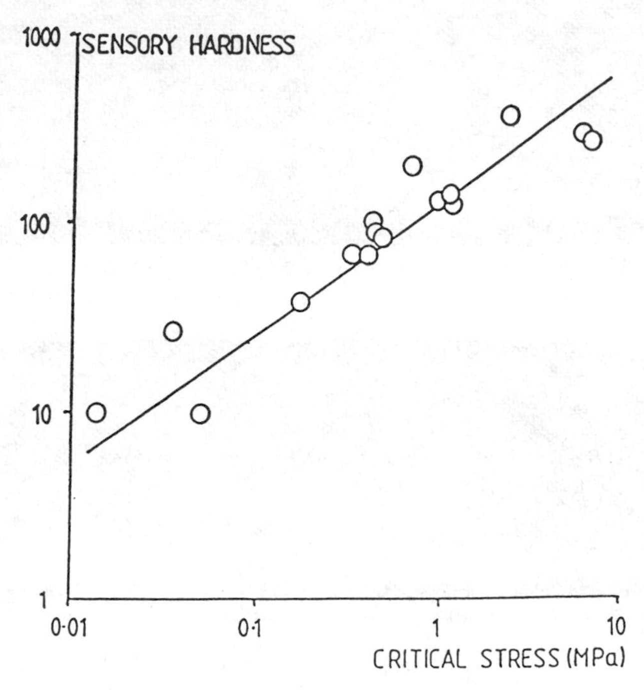

Figure 7 *Correlation of sensory 'hardness' with measured crushing stress. Correlation coefficient* r = 0.95
(Reproduced with permission from *J. Cereal Sci.*, 1989, **9**, 61)

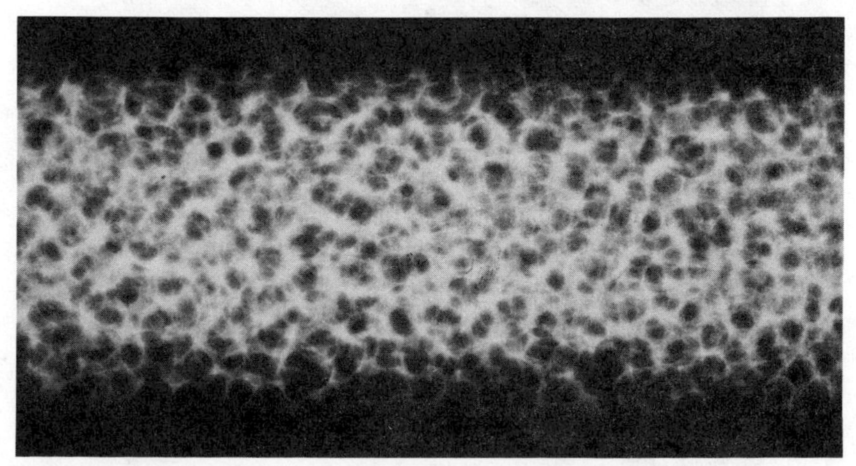

Figure 8 *Microfocal X-ray projection of an Aero chocolate bar*

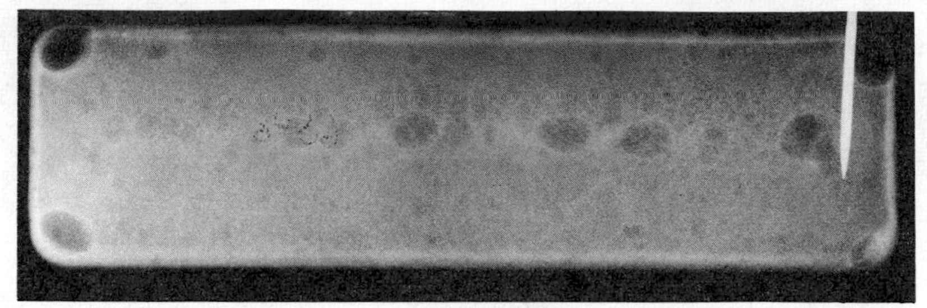

Figure 9 *Microfocal X-ray projection of a Cadbury's Wispa bar*

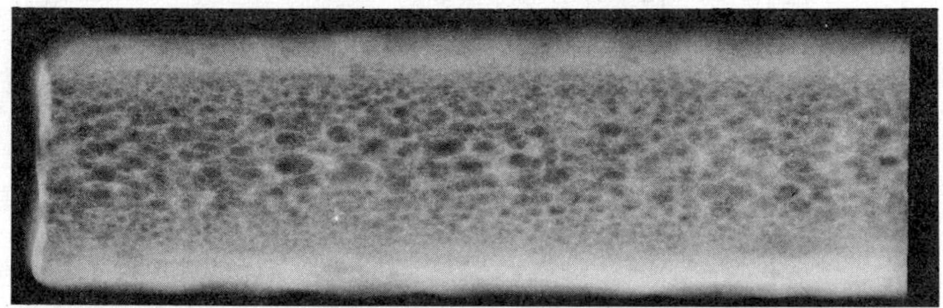

Figure 10 *Microfocal X-ray projection of a Cadbury's Crunchie bar*

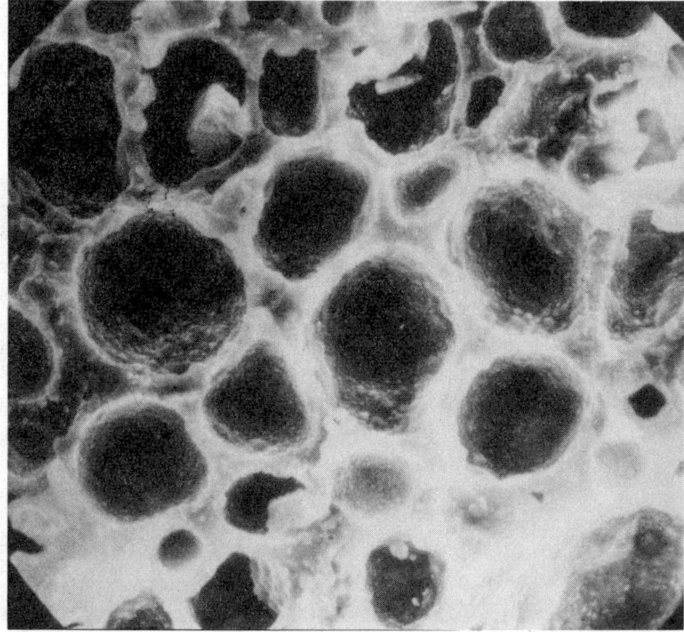

Figure 11 *Scanning electron micrograph of whipped cream*

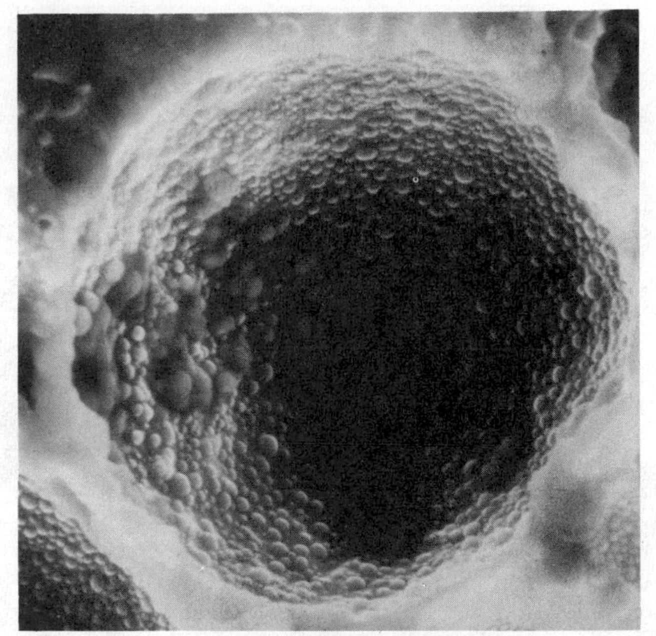

Figure 12 *Fat droplets in the air cell walls of whipped cream*

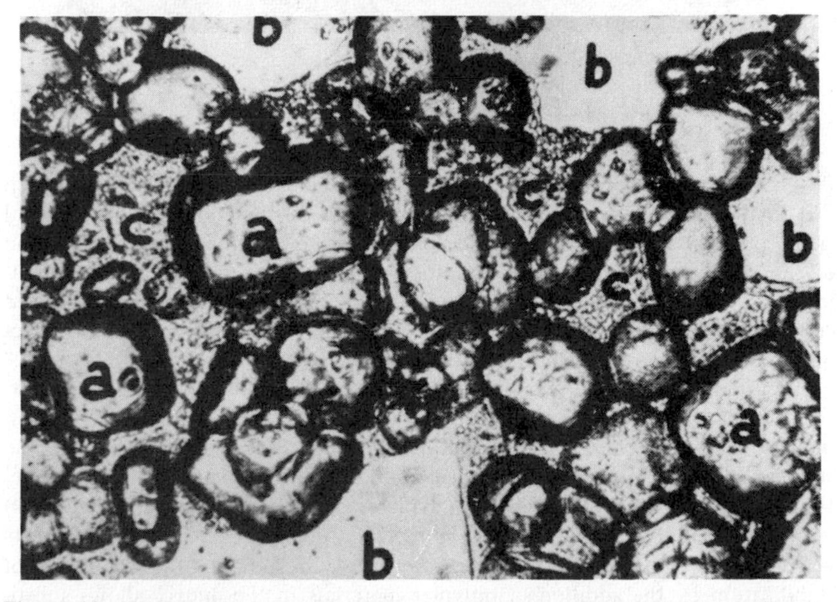

Figure 13 *Structural elements in ice cream: (a) ice crystals; (b) air cells; (c) unfrozen 'matrix'*
(Reproduced with permission from 'Ice Cream', 4th ed., 1986, AVI Publishing Corp. USA)

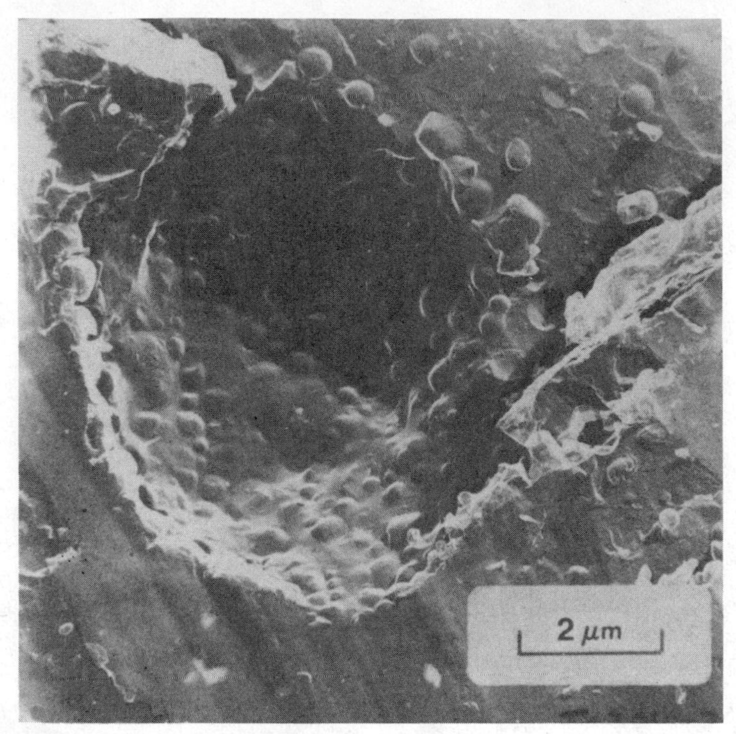

Figure 14 *Air cells in ice cream*
(Reproduced with permission from *Dairy Ind.*, 1972, **37**, 419)

the air cells themselves.[4] The surface of the cell is coated by small droplets, which presumably are spherical crystalline fats produced during cooling of the original cream prior to whipping. However, despite extensive research on the formulation of the properties of the cream and the whipping process, the role of small-molecule surface-active agents such as monoglycerides, lipoproteins, or added emulsifiers is not clear, and neither is the crystal network of the fats and its effect on the mechanism of interfacial stabilization.

High-quality ice cream is simply a frozen version of the creams seen so far. However, low-magnification photographs (Figure 13) show that the ice acts on the cell so as to cause considerable distortion.[4] Presumably, therefore, the stabilization mechanism must allow deformation of bubble walls without fracture. High-magnification photographs of a typical commercial ice cream (Figure 14) indicate that all the small air cells are spherical.[5] Also, the evidence of fat droplets heavily clustered at the interface is less apparent. Presumably, the technology of stabilization by the addition of polymer materials in the matrix allows substitution of fat stabilization, at least in part, by a concentrated viscoelastic matrix of polymers. Apart from the small intact air cells, numerous channels can be seen in most commercial ice creams, particularly after storage or temperature cycling

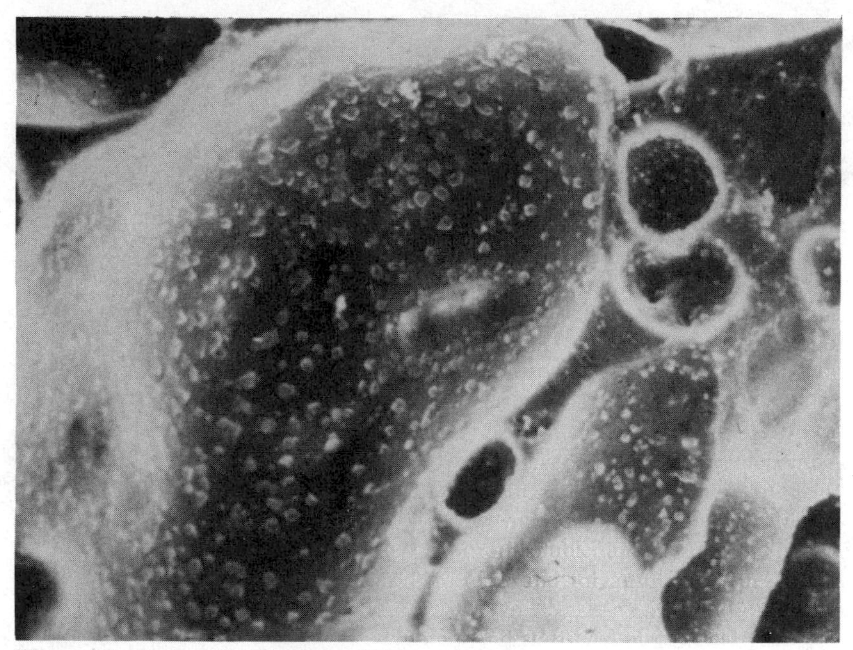

Figure 15 *Large air cells/channels in ice cream*

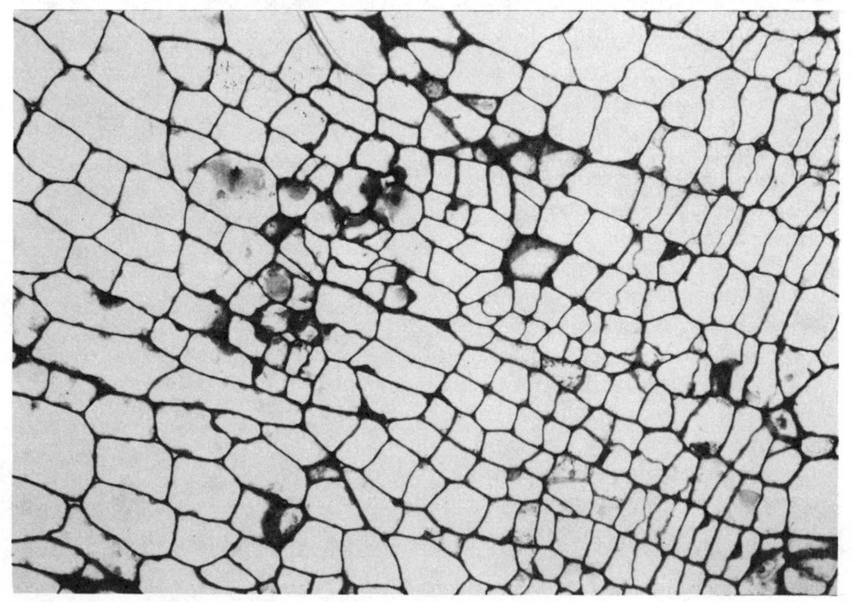

Figure 16 *Thin section of fresh carrot tissue*

(Figure 15). The mechanism by which these channels form is not clear, but their presence has been linked to the unfortunate phenomenon of shrinkage.

The matrix structure of ice cream is peculiarly complex. Obviously ice crystals are present, but also it is inevitable that at the low temperature of manufacture, storage, and consumption, the presence of ice will result in highly concentrated solutions of other small sugars and polymers in the matrix itself. The phase rules for these solutions at such high concentrations and low temperatures are not known. It is possible that the continuous matrix is itself a phase-separated composite of a complex mixture of biopolymers.

4 Liquid Filled Foams

All the previous examples considered are of classical sponges and foams in which air is the included phase. However, it is probably worth recognizing that many other foodstuffs contain liquids or solutions as the dispersed phase in a foam-like or cellular structure. Indeed, all commonly consumed vegetables can be considered as liquid-filled foams. Once again, whether the structure is a foam (*i.e.* with discrete inclusions) or a sponge (with continuity of liquid between the cells), it is vitally important to the mechanical properties and hence the texture of the foods. Figure 16 shows the structure of a piece of fresh carrot. As a result of the composition of the cytoplasm, turgor pressure is maintained in fresh vegetables and a typical crisp texture results. Breakdown of the cell membrane and the cell walls produces an entirely different flow mechanism of both the cell wall members

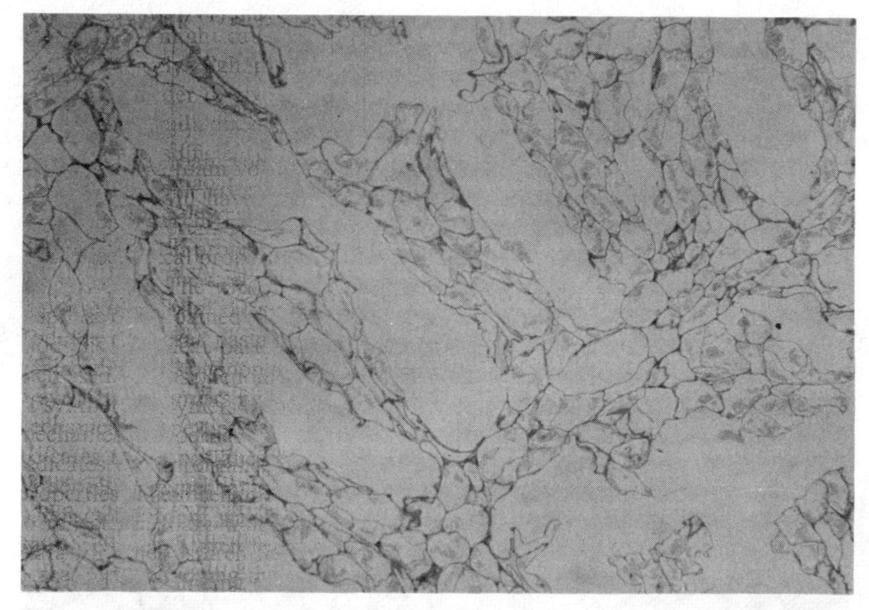

Figure 17 *Thin section of frozen/thawed carrot tissue. Large voids represent regions of thawed ice*

and the contents under mechanical deformation, and a very different texture is therefore detected. Figure 17 shows the effect of freezing damage in carrot tissue. It is obvious that massive damage to the cell walls, allowing channelling of the liquid throughout the structure, has been caused. Not surprisingly, the texture is considerably different and generally not preferred. Recently, attempts to modify Ashby and Gibson's theories to allow for fluid flow have been made,[6] but their value remains to be proved.

In conclusion, an attempt has been made to outline the enormous range and scope of sizes, shapes and mechanical properties of edible foams and sponges. Despite much work over many years, details relating molecular structure through mechanical properties to final eating texture are still not understood except in a very few cases. Much work remains to be done before the prediction of properties and proper control of processes can be achieved.

References

1. C. B. Holt, *Phys. Technol.*, 1988, **19**, 18.
2. M. F. Ashby and L. J. Gibson, *Metall. Trans. A*, 1983, **14**, 1755.
3. G. E. Attenburrow, F. M. Goodband, L. J. Taylor and P. J. Lillford, *J. Cereal Sci.*, 1989, **9**, 61.
4. W. S. Arbuckle, 'Ice Cream', 4th ed., 1986, AVI Publishing Corp. USA.
5. K. G. Berger, B. K. Bullimore, G. W. White and R. C. Wright, *Dairy Ind.*, 1972, **37**, 419.
6. M. Warner and S. F. Edwards, *Europhys. Lett.* 1988, **5**, 623.

The Mechanical and Flow Properties of Foams and Highly Concentrated Emulsions

By H. M. Princen*

GENERAL FOODS CORPORATION, TECHNICAL CENTRE, TARRYTOWN, NY 10591, USA

1 Introduction

Foams and concentrated emulsions play an important role in a variety of technologies, including foods (*e.g.* aerated desserts, mayonnaise, low-fat margarines, and salad dressings). The rheological properties of such food products are obviously crucial to their pourability, processing, mouth feel, etc. Other areas where such systems are used or contemplated include enhanced oil recovery, encapsulated fuels, fire fighting, and cosmetics.

Foams and highly concentrated emulsions are defined here as fluid–fluid dispersions (air-in-liquid or liquid-in-liquid) in which the (effective) volume fraction, φ, of the dispersed phase approaches or exceeds that of the close-packed sphere configuration, φ_0. When $\varphi > \varphi_0$, the bubbles or drops are deformed against their neighbours and acquire an increasingly pronounced polyhedral shape. They remain separated by thin films of continuous phase, which are stabilized against rupture by surfactants, polymers, or small solid particles. Because of this extreme crowding, these systems are plastic, *i.e.* they have an (elastic) shear modulus, a yield stress, and a shear-rate dependent viscosity.

In monodisperse systems, $\varphi_0 = 0.7405$. Perhaps contrary to intuition and popular belief, φ_0 is found to be slightly *smaller* (*ca.* 0.72) for typical unimodal polydisperse systems.[1,2] Although it is true that close-packed sphere systems can be envisaged in which φ_0 exceeds 0.74, and even approaches unity, this would require very specific, multimodal size distributions and highly ordered packing. Such dispersions will rarely, if ever, be encountered in practice.

Another and more prevalent reason why the peculiar rheological properties of such systems may be observed at nominal values of $\varphi < 0.74$ is that, particularly in foods, the dispersed bubbles or drops may be surrounded by thick adsorbed layers of stabilizing polymer, *e.g.* protein. For present purposes, these layers should be considered as part of the dispersed entities. It is readily shown[3,4] that the *effective* and nominal volume fractions, φ and φ_n are related through

$$\varphi^{-1/3} \simeq \varphi_n^{-1/3} - 1.10h/2R$$

* Present address: Mobil Research and Development Corp., Central Research Laboratory, P.O. Box 1025, Princeton, NJ 08543–1025, USA.

where h is the thickness of the interdroplet films (*i.e.* twice the thickness of the stabilizing adsorbed layer), and R is the radius of the core droplet. For example, when $2R = 1$ μm and $h = 0.1$ μm, we see that $\varphi = 0.7405$ already at $\varphi_n = 0.56$.

In spite of the technological importance of these classical systems, it is only in the past decade or so that considerable progress has been made toward a full understanding of their mechanical and rheological properties. Two-dimensional models have been extremely helpful in providing insight into the kinematics of deformation and flow and in establishing, at least qualitatively, how the various rheological properties depend on the system's physical parameters, *e.g.* volume fraction, drop size, and interfacial tension. The mathematical difficulties associated with the analysis of three-dimensional models, particularly when they involve polydispersity, are formidable. So far, the only way to extend the two-dimensional predictions to three dimensions has been through careful experimental work on well-characterized, real systems, especially concentrated oil-in-water emulsions.

For recent reviews on foam flow, see references 5 and 6.

2 Two-dimensional Modelling

Shear Modulus and Yield Stress—Even in two-dimensional analyses, polydispersity creates great difficulties. Therefore, with the notable exception of the work of

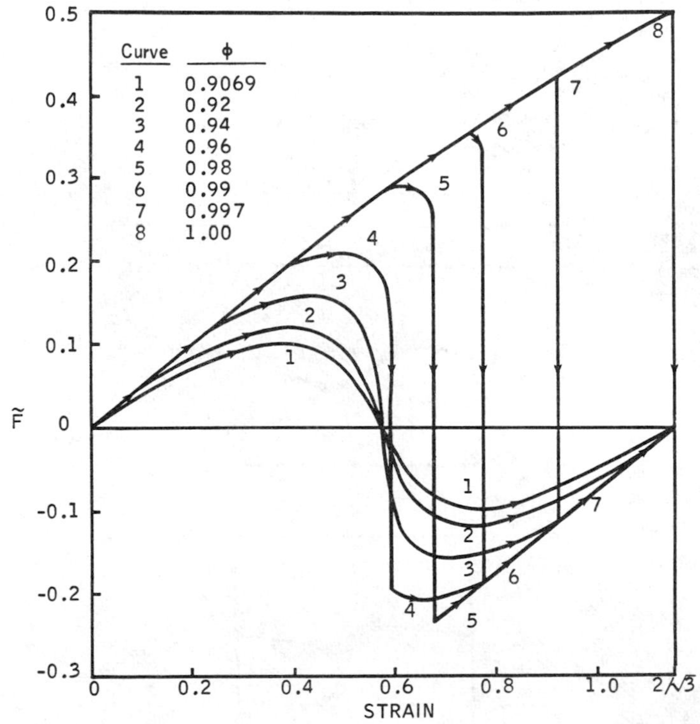

Figure 1 *Stress vs. strain curves for different volume fractions of the dispersed phase*

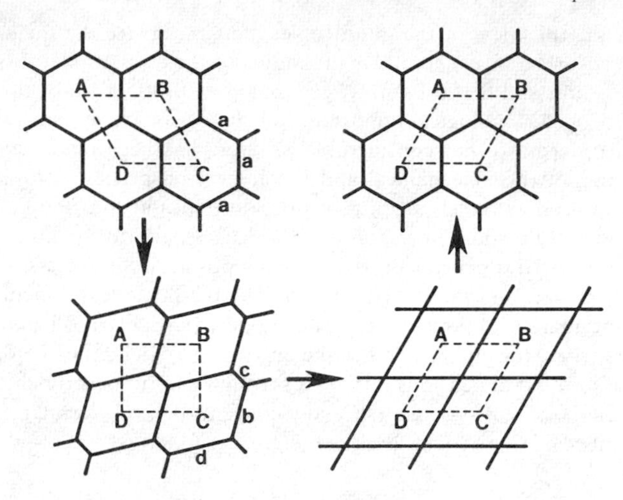

Figure 2 *Motion of four bubbles or drops surrounding a unit cell ABCD* ($\varphi = 1$)

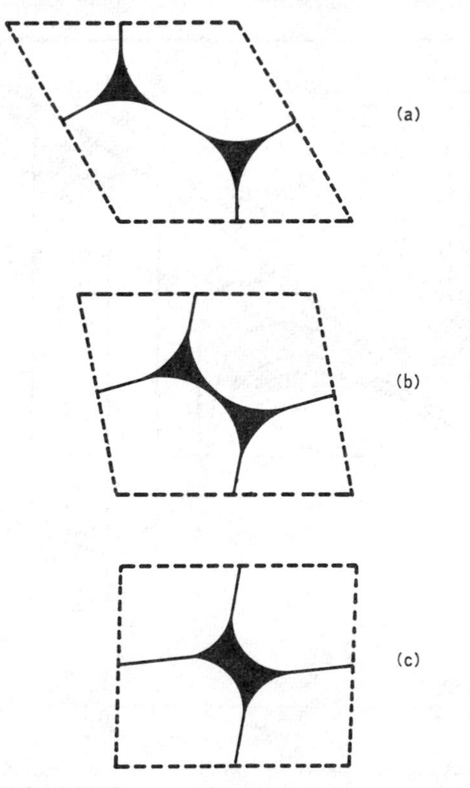

Figure 3 *Unit cell for* $0.9069 < \varphi < 1$

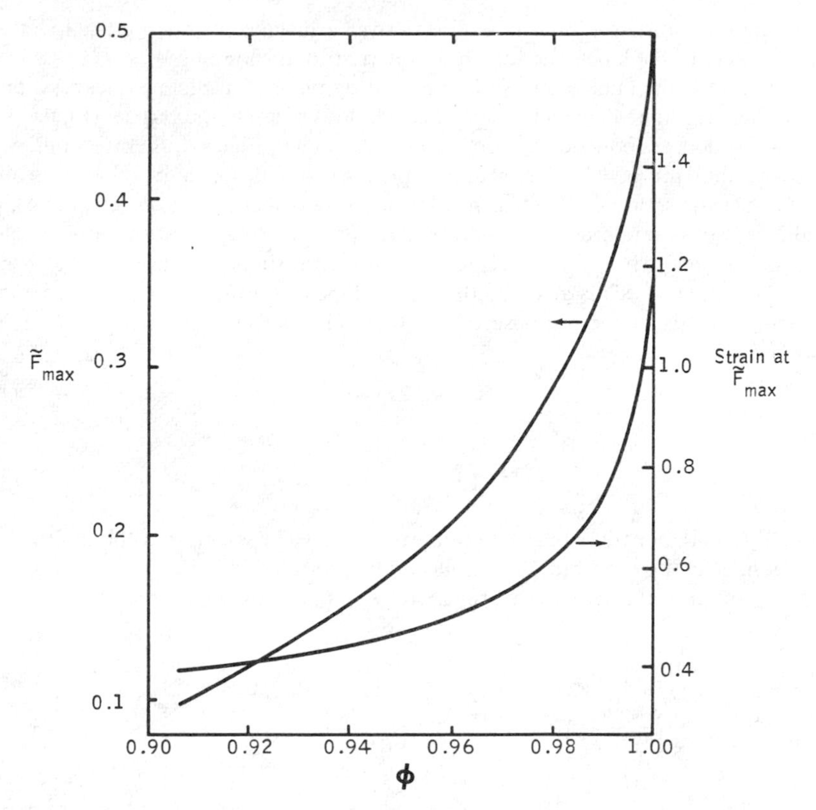

Figure 4 $\tilde{F}_{max}$ *(reduced yield stress) and yield strain as a function of* φ.

Weaire *et al.*,[7–9] all studies have dealt with monodisperse systems, whose behaviour may be derived from that of a single cylindrical drop or bubble (or unit cell) in a perfectly ordered lattice. For the case of simple shear deformation, Princen[10] analysed such a system in great detail, including the effect of volume fraction, $\varphi \geqslant \varphi_0$ (=0.9069), film thickness, and finite contact angle at the Plateau border–film junction. Independently, restricting himself to the limiting case of 'dry foams' ($\varphi = 1$), Prud'homme[11] advanced a similar aproach to the problem.

The most important results of Princen's study are contained in Figure 1, which shows the stress–strain relationship for a unit cell at different volume fractions (and zero contact angle). Curve 8 refers to the dry-foam case. The stress increases monotonically up to a maximum (the reduced yield stress/unit cell, $\tilde{F}_{max}$) and then drops to zero. Figure 2 shows the sequence of events in pictorial form for four bubbles around a unit cell ABCD. The maximum stress is reached when four films meet in a vertex, a configuration that is unstable (according to Plateau's first rule, stability requires that three, and not more than three, films meet at dihedral angles of 120°). The instability resolves itself by the formation of new film at the vertex and the system's return to the initial configuration. The cycle is accompanied by the relative translation of adjacent rows over a distance of one unit-cell width. This aspect of the kinematics has been referred to as 'hopping.'[11]

For $\varphi < 1$, the bubbles are still separated by films of negligible thickness, but now there are finite Plateau borders of continuous phase at the vertices (Figure 3). These borders are bounded by circular arcs. As seen in Figure 1, the stress initially follows that for $\varphi = 1$. This is the regime where the Plateau borders remain separated. As soon as two adjacent Plateau borders merge into one (Figure 3), a different stress–strain curve is followed. It passes through a maximum or yield stress per unit cell, $\tilde{F}_{max}$, that decreases with decreasing φ (Figure 4). The shear modulus per unit cell is given by the initial slope of the stress–strain curves.

On the basis of the response of the unit cell, one finds for the system as a whole[10]

$$G = 0.525 \frac{\sigma}{R} \varphi^{1/2} \tag{1}$$

and

$$\tau_0 = 1.050 \frac{\sigma}{R} \varphi^{1/2} \tilde{F}_{max}(\varphi) \tag{2}$$

where G is the (static) shear modulus, τ_0 is the yield stress, σ is the interfacial tension, R is the drop radius, *i.e.* the radius of a circle of the same area as the cross-section of the cylindrical drops, and $\tilde{F}_{max}(\varphi)$ is known (Figure 4).

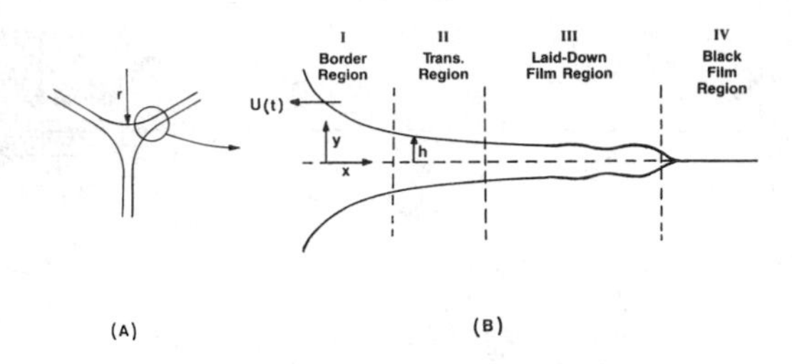

Figure 5 *Three films meeting in a Plateau border and detailed profile of a stretching or shrinking film*

Subsequently, Khan and Armstrong[12,13] and Kraynik and Hansen[14] considered the effect of the orientation of the unit cell, relative to the shear direction, for the dry-foam case. They found that the shear modulus is unaffected but that the yield stress is sensitive to the orientation. In addition, they considered planar extension deformation, as well as simple shear.

The cyclic character of the curves in Figure 1 is not observed in practical systems. It arises from the perfect order in the monodisperse model: instabilities occur simultaneously at all vertices. By the ingenious use of computer simulation, Weaire and co-workers[7–9] succeeded in introducing polydispersity into the two-dimensional model for $\varphi = 1$. Now, instabilities occur more randomly as 'flickering events' throughout the system. Apart from small perturbations, the result is a monotonically increasing stress with increasing strain, up to a plateau where the (yield) stress no longer depends on strain. This more closely reflects the behaviour of statically strained real systems. Because of the limited number of cells (64 or 100) in the repeating unit, the individual local instabilities are still seen as fluctuations in the stress. With increasing disorder, these are expected to decrease in size until a completely smooth curve is obtained.

Following Weaire and Fu,[9] one can introduce a reduced shear modulus $G^* = G\bar{R}/\sigma$, where $\bar{R}$ is a mean drop radius. Their results indicate that G^* decreases slightly with increasing polydispersity, if $\bar{R}$ is defined as

$$\bar{R} = \left(n^{-1} \sum_{i=1}^{n} R_i^2\right)^{1/2}$$

where R_i is the equivalent-circle radius of bubble i and n is the total number of bubbles. In other words, if R in equation (1) is replaced by $\bar{R}$, the value of the numerical coefficient (0.525) is expected to decrease slightly with increasing polydispersity. This dependence may well be reduced or eliminated with a different choice of the mean drop radius, namely the two-dimensional equivalent of the surface-volume or Sauter mean radius,

$$R_{21} = \sum_{i=1}^{n} R_i^2 \Big/ \sum_{i=1}^{n} R_i$$

which has been proposed, without rigorous proof but with some experimental justification,[1,2,4] as the appropriate mean radius in problems of this kind.

Viscous Effects—In the above analyses, deformations are imposed at a vanishingly small rate. The effects are purely elastic and are due exclusively to varying surface area at constant surface tension. Rate-dependent viscous effects are ignored. At finite deformation rates in, say, steady shear, one may write formally

$$\tau(\dot{\gamma}) = \tau_0 + \tau_s(\dot{\gamma}) \tag{3}$$

where $\dot{\gamma}$ is the shear rate and $\tau_s(\dot{\gamma})$ is a rate-dependent contribution to the stress, so that the effective viscosity is

$$\mu_e(\dot{\gamma}) \equiv \tau(\dot{\gamma}) / \dot{\gamma} = \tau_0/\dot{\gamma} + \tau_s(\dot{\gamma})/\dot{\gamma} \tag{4}$$

The first term, the 'elastic' or yield stress term, already accounts qualitatively for the well known shear-thinning behaviour of foams and concentrated emulsions.

To evaluate the second, viscous term, Khan and Armstrong[12,13,15] and Kraynik and Hansen[16] employed a two-dimensional, spatially periodic model, in which all the continuous phase is contained in the films (*i.e.* there are no Plateau borders), the films have mobile interfaces, and there is no exchange of fluid between the films. When the system as a whole is strained, the uniform films are stretched or compressed at constant volume. It turns out that this mechanism leads to viscous terms in equations (3) and (4) that are insignificant compared with the elastic terms up to extremely high shear rates that are unlikely to be encountered in practice ($\dot{\gamma} \approx 10^5\ \mathrm{s}^{-1}$ for typical systems). Experimentally, one finds a much more significant contribution.[4,17,18]

An alternative model has been advanced by Schwartz and Princen.[19] In many ways it assumes the exact opposites, *i.e.* the films are negligibly thin, so that all the continuous phase is contained in the Plateau borders, and the film surfaces are rigid. Moreover, hydrodynamic interaction between the films and the adjoining Plateau borders is considered to be crucial. This model, believed to be more realistic for common surfactant-stabilized foams and emulsions, was inspired by the work of Mysels *et al.*[20] on the dynamics of a planar, vertical soap film being pulled out of, or pushed into, a bulk solution via an intervening Plateau border. One important result of their analysis, commonly referred to as Frankel's law, relates the film thickness, $2h_\infty$, to the pulling velocity, U, and may be written in the form

$$H_\infty/r = 0.643(3Ca^*)^{2/3} \tag{5}$$

where $Ca^* = \mu U/\sigma$ is the capillary number, μ and σ are the viscosity and surface tension of the liquid, r is the radius of curvature of the Plateau border where it meets the film and is given by capillary hydrostatics, $r = (\sigma/2\rho g)^{1/2}$ where ρ is the density of the liquid, and g is the gravitational acceleration.

Frankel's law, which has its analogues in related problems,[21-23] has been verified experimentally[24,25] in the regime where the drawnout film thickness, $2h_\infty$, is sufficiently large for disjoining-pressure effects to be negligible. The interesting hydrodynamics and the associated viscous energy dissipation are confined to a transition region between the emerging, rigidly moving film and the macroscopic Plateau border. In this region the relative slope of the interfaces remains small, so that the lubrication version of the Stokes equations is justified.

Clearly, the same basic process operates in moving foams and emulsions, except that each Plateau border of radius r (set by drop size and volume fraction) is now shared by three films. At any given moment, one or two of the films will be drawn out of that border, while the other(s) is (are) pushed into it, at respective velocities U that are dictated by the macroscopic motion of the system (Figure 5). Schwartz and Princen[19] considered a periodic uniaxial, extensional strain motion of small frequency and amplitude, so that inertial effects are negligible, and complications due to merger of adjacent Plateau borders and capillary instabilities are avoided. They proceeded by calculating the instantaneous rate of energy dissipation in the transition regions of each of the three films associated with a Plateau border, and integrated the results over a complete cycle. When the effective strain rate is related to the frequency of the imposed motion, the result can be expressed as an effective viscosity which is given by*

$$\mu_e = 5.3\mu Ca^{-1/3} \tag{6}$$

where $Ca = \mu a\dot{\gamma}/\sigma$, a is the length of the sides of the hexagon that circumscribes a drop or bubble, and μ is the viscosity of the continuous phase. It was argued that, in the case of emulsions, the effect of the dispersed-phase viscosity is relatively

* The numerical coefficient of 5.3 was originally given a value of 6.7. The change results from the correction of some numerical errors in the original paper,[19] kindly pointed out to us by Drs. Kraynik and Reinelt.[12]

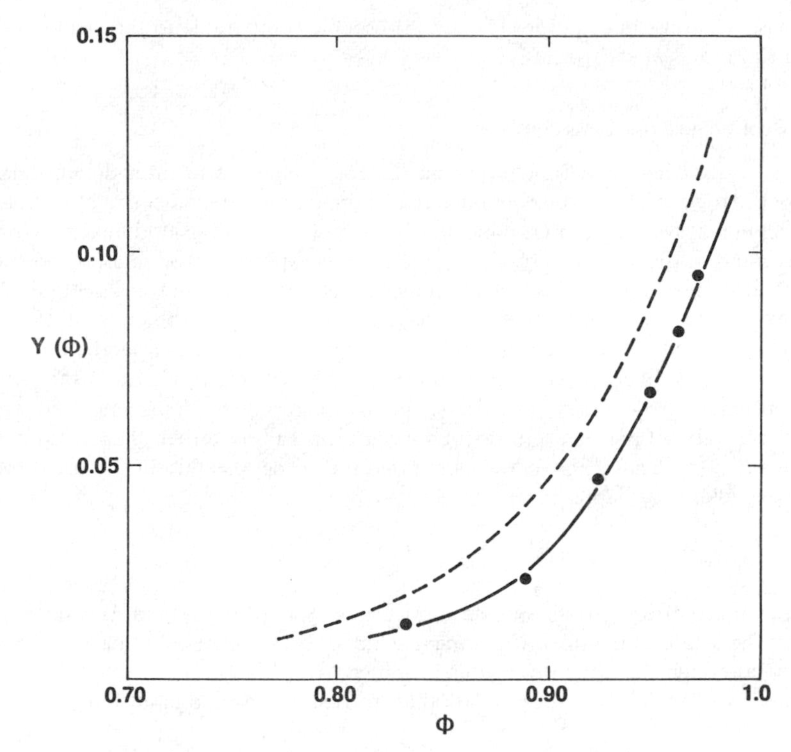

Figure 6 *Yield–stress function* Y(φ). *Solid line: most recent results.*[18] *Dashed line: results from older study,*[4] *prior to implementation of various experimental improvements*

insignificant, but this needs further analysis. Also, because of the small amplitude of the imposed motion, the result of equation (6) does not depend on the volume fraction.

Apart from a possible change in the numerical coefficient, equation (6) is expected to hold also for a periodic, small-amplitude *shearing* motion. However, in *steady shear*, the above process is periodically interrupted by rapid film motions associated with the capillary instabilities. Provided that Ca is sufficiently small, one expects the stored elastic energy at the instability to be completely dissipated, resulting in a contribution to μ_e of $\tau_0/\dot{\gamma}$. Further, as the strain at the instability depends on the volume fraction (Figures 1 and 4), the viscous term itself may become φ-dependent. Thus, for steady shear, one may anticipate

$$\mu_e = \tau_0/\dot{\gamma} + C_2(\varphi)\mu Ca^{-1/3} \tag{7}$$

or, for the shear *stress*,

$$\tau(\dot{\gamma}) = \tau_0 + C_2(\varphi)\frac{\sigma}{a} Ca^{2/3} \tag{8}$$

where $C_2(\varphi)$ is of the order of 1 and τ_0 is given by equation (2). For typical systems, the viscous terms in equations (7) and (8) become comparable to the yield stress terms at practical shear rates.

3 Real Foams and Emulsions

It has so far been impossible to extend the above analyses to three-dimensional models. However, the results obtained in two dimensions are indicative of what to expect in real systems. For example, the factor σ/R in the shear modulus and yield stress, and in the viscous term in equation (8), is expected to be retained on the basis of dimensional analysis, while the exponent of φ in equations (1) and (2) changes from 1/2 to 1/3 for geometric reasons. Also, the exponents of Ca in equations (7) and (8) are expected to be unaffected.[19] On the other hand, all static and dynamic properties will depend on φ in a different manner than in the two-dimensional case. Finally, practical systems are rarely monodisperse and are characterized by a mean equivalent-sphere drop radius rather than a unique radius R. As indicated above, we believe that the proper mean radius to use is the surface-volume or Sauter mean radius:

$$R_{32} = \sum_{i=1}^{n} R_i^3 \Big/ \sum_{i=1}^{n} R_i^2 \approx 3V/S$$

as it is linked directly to the specific surface area, S/V, of the system. When this is done, the details of the size *distribution* (*e.g.* its width) are believed to have a minor effect, although the evidence for this is limited.[4]

In view of the above, we speculate that, for real foams and emulsions,

$$G = \frac{\sigma}{R_{32}} \varphi^{1/3} E(\varphi) \tag{9}$$

$$\tau_0 = \frac{\sigma}{R_{32}} \varphi^{1/3} Y(\varphi) \tag{10}$$

whereas, for steady shear,

$$\tau = \tau_0 + C(\varphi)\frac{\sigma}{R_{32}} Ca^{2/3} \tag{11}$$

or

$$\mu_e = \tau_0/\dot{\gamma} + C(\varphi)\mu Ca^{-1/3} \tag{12}$$

where, for practical reasons, Ca has been redefined as

$$Ca \equiv \mu R_{32}\dot{\gamma}/\sigma \tag{13}$$

In the absence of further theoretical progress, the functions $E(\varphi)$, $Y(\varphi)$, and $C(\varphi)$ can be evaluated by experiment only. Similarly, the validity of the exponents of Ca in equations (11) and (12) needs to be confirmed experimentally. This requires foams or emulsions that are very well characterized in terms of σ, R_{32}, and φ, and careful rheological measurements that take account of complications such as wall slip and end effects.

We were unable to find published data that could be used for the above purpose. In a series of studies,[1,4,17,18] we generated our own data, using a modified concentric-cylinder viscometer and series of extremely stable oil-in-water emulsions, these having decided experimental advantages over foams. We refer to the original papers for experimental details. The following results were obtained for the static shear modulus and yield stress:

$$G = 1.77 \frac{\sigma}{R_{32}} \varphi^{1/3}(\varphi - 0.71) \tag{14}$$

$$\tau_0 = \frac{\sigma}{R_{32}} \varphi^{1/3} Y(\varphi) \tag{15}$$

where $Y(\varphi)$ is shown in Figure 6. Within the range considered ($0.83 < \varphi < 0.97$), Y is accurately represented by the empirical expression

$$Y(\varphi) = -0.080 - 0.114 \log (1 - \varphi)$$

For the steady-shear stress and viscosity we found

$$\tau = \tau_0 + 32 (\varphi - 0.73) \frac{\sigma}{R_{32}} Ca^{1/2} \tag{16}$$

and

$$\mu_e = \tau_0/\dot{\gamma} + 32(\varphi - 0.73)\mu Ca^{-1/2} \tag{17}$$

It is noted that the numerical constants of 0.71 in equation (14) and 0.73 in equations (16) and (17) are very close to φ_0.

The main difference between these experimental results and theoretical predictions is the value of the exponent of Ca. The reason for this deviation is not clear but may be linked to disjoining-pressure effects in the experimental systems. Typical values were $R_{32} = 10\ \mu m$, $\sigma = 5$ mN m^{-1}; $\mu = 1.5$ mPa s; $\dot{\gamma} = 20\ s^{-1}$ and $\varphi = 0.95$. With estimates of $U \approx \dot{\gamma} R_{32} \approx 200\ \mu m\ s^{-1}$ and $r \approx 2\ \mu m$, equation (5) would then predict a pulled-out film thickness $2h_\infty \approx 8$ nm(!), clearly in the range where disjoining-pressure effects are very important indeed. Hence our emulsions were not optimum for a true test of equations (11) and (12), and more experimental work remains to be done, while the theoretical model needs to be extended by incorporation of the disjoining pressure in the force balance in the transition region.

Nevertheless, much progress has been made and it is possible now to design foams and concentrated emulsions with certain desired rheological properties by a judicious combination of mean drop size, volume fraction, interfacial tension, and continuous-phase viscosity.

Acknowledgement. I thank Academic Press for granting permission to republish the figures, which first appeared in the *Journal of Colloid and Interfacial Science.*

References

1. H. M. Princen and A. D. Kiss, *J. Colloid Interface Sci., 1986*, **112**, 427.
2. H. M. Princen and A. D. Kiss, *Langmuir*, 1987, **3**, 36.

3. H. M. Princen, M. P. Aronson, and J. C. Moser, *J. Colloid Interface Sci.*, 1980, **75**, 246.
4. H. M. Princen, *J. Colloid Interface Sci.*, 1985, **105**, 150.
5. A. M. Kraynik, *Annu. Rev. Fluid Mech.*, 1988, **20**, 325.
6. J. P. Heller and M. S. Kuntamukkula, *Ind. Eng. Chem. Res.*, 1987, **26**, 318.
7. D. Weaire and J. P. Kermode, *Philos. Mag.*, 1984, **B50**, 379.
8. D. Weaire, T. L. Fu, and J. P. Kermode, *Philos. Mag.*, 1986, **B54**, L39.
9. D. Weaire and T. L. Fu, *J. Rheol.*, 1988, **32**, 271.
10. H. M. Princen, *J. Colloid Interface Sci.*, 1983, **91**, 160.
11. R. K. Prud'homme, paper presented at the 53rd Annual Society of Rheology Meeting, Louisville, KY, 1981.
12. S. A. Khan, *PhD Thesis*, MIT, Cambridge, MA, 1985.
13. S. A. Khan and R. C. Armstrong, *J. Non-Newtonian Fluid Mech.*, 1986, **22**, 1.
14. A. M. Kraynik and M. G. Hansen, *J. Rheol.*, 1986, **30**, 409.
15. S. A. Khan and R. C. Armstrong, *J. Non-Newtonian Fluid Mech.*, 1987, **25**, 61.
16. A. M. Kraynik and M. G. Hansen, *J. Rheol.*, 1987, **31**, 175.
17. A. Yoshimura, R. K. Prud'homme, H. M. Princen and A. D. Kiss, *J. Rheol.*, 1987, **31**, 699.
18. H. M. Princen and A. D. Kiss, *J. Colloid Interface Sci.*, 1989, **128**, 176.
19. L. W. Schwartz and H. M. Princen, *J. Colloid Interface Sci.*, 1987, **118**, 201.
20. K. J. Mysels, K. Shinoda, and S. Frankel, 'Soap Films: Studies of Their Thinning and a Bibliography,' Pergamon Press, New York, 1959.
21. L. Landau and B. Levich, *Acta Physicochim. URSS*, 1942, **17**, 42.
22. F. P. Bretherton, *J. Fluid Mech.*, 1961, **10**, 166.
23. L. W. Schwartz, H. M. Princen, and A. D. Kiss, *J. Fluid Mech.*, 1986, **172**, 259.
24. K. J. Mysels and M. C. Cox, *J. Colloid Interface Sci.*, 1962, **17**, 136.
25. J. Lyklema, P. C. Scholten and K. J. Mysels, *J. Phys. Chem.*, 1965, **69**, 116.

The Physics of Froths and Foams

By D. Weaire

DEPARTMENT OF PURE AND APPLIED PHYSICS, TRINITY COLLEGE, DUBLIN 2, IRELAND

1 Introduction

The structure of a froth* or foam* divides space into cells in such a way that the total area of their interfaces is minimized. Many natural structures conform to this pattern[1] for one reason or another; in the more straightforward cases it is a simple matter of the minimization of surface energy.

In this paper we shall examine such a structure and its properties, within the limitations of an idealized model. It should have wide relevance to froths and froth-like structures, but we principally have in mind the very simplest of these—made, for example, with a detergent solution.

We do not wish to underestimate the conceivable complexities of even this case. Ordinary soap films have presented many generations of scientists (Hooke, Newton, Rayleigh, Gibbs, *etc*) with a wealth of varied behaviour.[2–4] Within our 'simple' soap froth may lurk effects not yet suspected by those who seek to encapsulate their properties in a straightforward model; but we must begin somewhere, if we are to have a decent theory of froths, for which the relevant books published to date,[5,6] show a pressing need.

Our intention is to steer a prudent course between the Scylla of advanced mathematics[7] (some formidable treatises on surface area minimization) and the Charybdis of industrial chemistry.[5,6] A *direct* applicability to food science can hardly be asserted: from detergent foam to whipped cream is a small step in the kitchen, but a large one in terms of a fundamental theory. The author must confess that he has occasionally been distracted into biological areas by the inspiring prose of D'Arcy Wentworth Thompson,[8] but his standpoint remains essentially that of a physicist or materials scientist.

Like much else in research, the studies described here arose incidentally, as a by-product of the pursuit of other interests.[9] The complex three-dimensional atomic structures of amorphous solids (such as window glass) have remained controversial to this day. Attempts to understand better the topology and geometry of random structures have been hampered by the need to build or visualize large three-dimensional models. The metallurgist Cyril Stanley Smith, in the course of wide-ranging speculations on the structure of materials,[10] suggested

* For the purposes of this paper these terms are to be regarded as equivalent

a helpful analogy between amorphous solids and foam structures. Smith had, in turn, been led to this system by the closer analogy with the grain structure of metals.[11] He noted that foams can be realized as two-dimensional structures, which are much easier to study and analyse than their three-dimensional counterparts.

2 Two-dimensional Froth

Lest the idea of *two-dimensional* froth should seem totally abstract and artificial, a brief indication will be given of how such a froth may easily be made and observed.

Two glass plates, separated by a small gap, may be disposed vertically over a bath of suitable solution, in which bubbles are blown by means of a submerged tube, as in Figure 1. This arrangement, described by Boys,[2] and exhibited in a compelling display at the San Francisco Exploratorium, has been used by Glazier *et al.*[12] in a recent comprehensive experimental study.

Even more simply, a blob of detergent foam may be squashed between two glass slides, with a suitable spacer. Figure 2 is a photograph of a froth prepared in this way.

This is not to say that experimental measurements are not fraught with difficulties,[12] especially as regards mechanical properties, but the creation of the

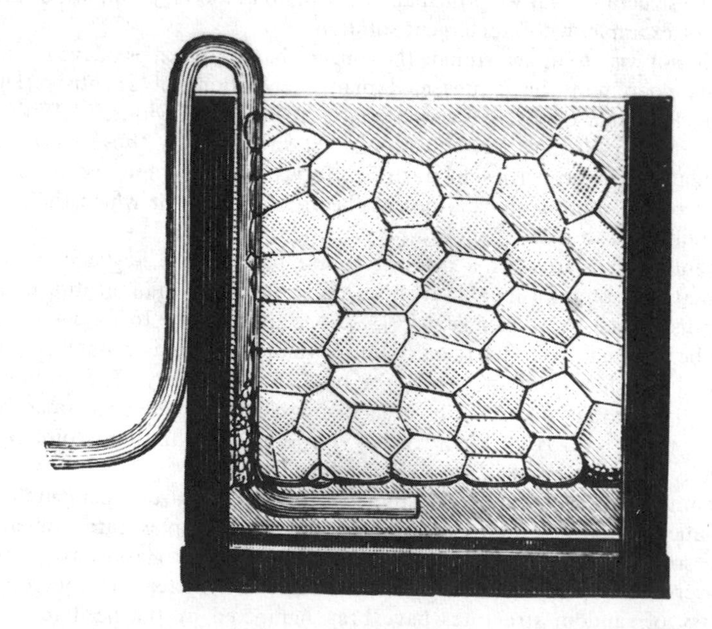

Figure 1 *C. V. Boys' suggestion[2] of a 'very simple experiment which you can easily try at home,' which he used in conjunction with projection, to demonstrate soap film equilibrium principles*

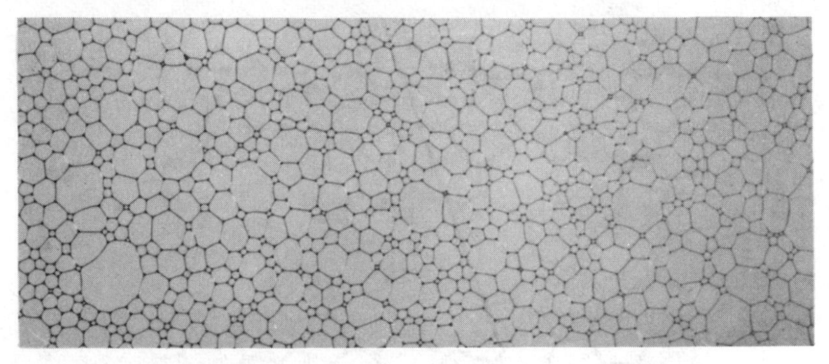

Figure 2 *Two-dimensional soap froth*

two-dimensional foam for qualitative demonstration purposes could hardly be easier.

3 Idealized Model

Our idealized model will be two-dimensional, for reasons outlined above. The essential definitions and principles may be generalized in an obvious way to three dimensions.

The froth consists of a system of *cells*, which contain a gas, and are separated by liquid cell walls. For simplicity, we define its basic properties as follows.

(1) The gas is incompressible.

(2) The walls are of negligible thickness and hence may be represented by lines which meet at points (vertices).

(3) The total energy of the system is taken to be the surface energy, and the energy per unit surface area (or surface tension) is a constant, σ. In the two-dimensional system the area in question is given by the product of cell wall length and the separation of the plates, together with a factor of 2, since the wall has two sides.

(4) The cell walls are permeable to the gas; diffusion between cells due to pressure differences obeys Fick's law (*i.e.* it is proportional to the pressure difference).

These definitions suffice for the discussion that follows, which concerns static or quasi-static properties. The dynamics of flow require further information regarding viscosity, boundary conditions, *etc.*[13] Such considerations would take us beyond what is contemplated here (but see Princen[14]).

4 Simulation

This idealized foam can be *simulated* by straightforward computational methods.[15,16] We shall not dwell on these, except to say that they consist of traditional numerical mathematics. The availability of graphical output such as Figure 3 is important. A failure of the algorithm soon produces something that looks wrong. In addition, graphical output can be transferred to video, and this is

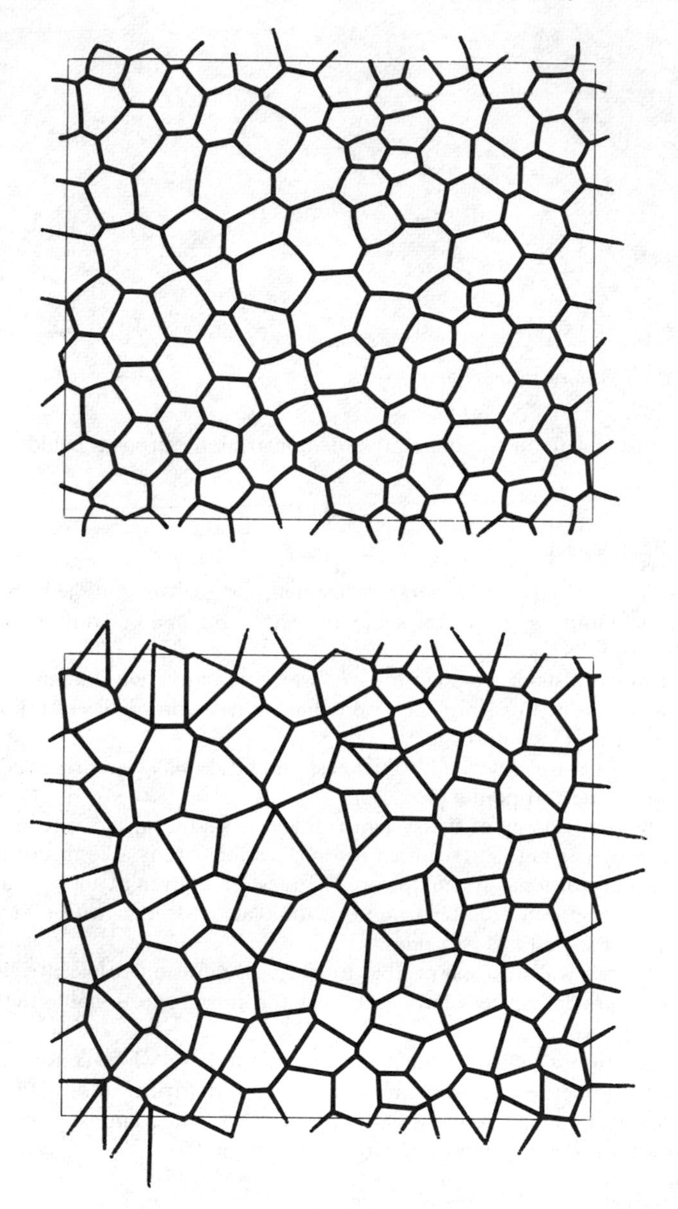

Figure 3 *(a) Typical simulated two dimensional soap froth. Note the periodic boundary conditions. (b) The sample was created by first performing this random geometrical construction, then equilibrating the structure, with fixed cell areas*

typical of the sort of computational simulation which will receive a stimulus from video techniques in the years to come.

One small technical point requires mention: it is convenient to use periodic boundary conditions throughout the simulation. This eliminates the effects associated with a free boundary, which would be very large for the small number of cells employed in such situations.

5 Properties of the Idealized Froth

The example which we have just seen (Figure 3) will serve to illustrate some general features. The cell walls are arcs of circles. Two equilibrium conditions must be met: (i) the walls meet at 120° at every vertex; and (ii) the curvatures are such that a pressure p can be assigned to every cell, pressure differences between neighbouring cells being related to radius of curvature by

$$\Delta p = 2\sigma/r \tag{1}$$

where σ is the surface tension (this is the type of relation, familiar from elementary physics courses, often invoked for bubbles, *etc.*). It must be remembered that the system adjusts to meet these requirements, while keeping cell areas at fixed values.

It is immediately evident that, in general, there will be intercellular diffusion, and this equilibrium can only be quasi-static. Hence the structure must slowly evolve under the constraint of changing cell areas. This smooth evolution is, however, punctuated by sudden local rearrangements, as shown in Figure 4. Moreover, individual cells may disappear, as in Figure 5. The scenario is already beginning to look complicated. A merciful simplification (available only in two dimensions) was pointed out by Von Neumann,[17] to the effect that

$$\text{Rate of increase of individual cell area} = \text{constant} \times (n - 6) \tag{2}$$

where n is the number of sides of the cell (the constant is a simple combination of diffusion constant and surface tension, both assumed to be independent of time).

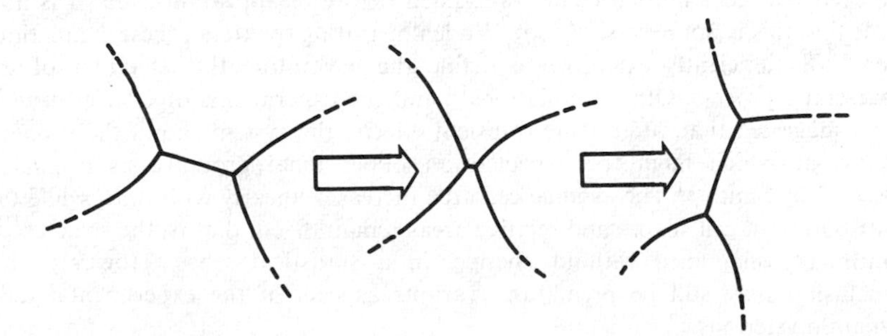

Figure 4 *Rearrangement of cells which takes place whenever they evolve in such a way that the length of a cell wall vanishes*

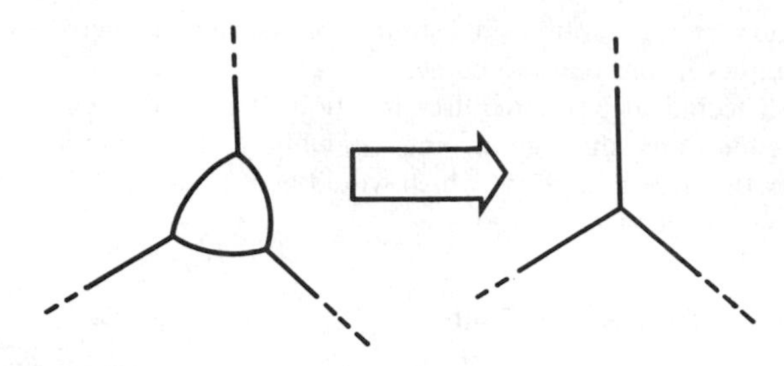

Figure 5 *Disappearance of a three-sided cell, due to diffusion. Four- and five-sided cells may disappear in a similar manner, immediately provoking rearrangements as in Figure 4, to maintain three-fold vertices*

However, although this is convenient for some purposes, it does *not* help much in the prediction of the complicated consequences of the cell area changes. Before we say any more about them, we should note that a structure made entirely of *hexagonal* cells is doubly stable—according to equation (2) it does not evolve at all. This is obvious when all cells are equal in area (Figure 6a), but remains true even when they are not, provided that they all have six sides. Cells of roughly equal area are made readily enough in practice.[12] This prompts the question of whether cells of equal area are necessarily hexagonal. The answer, as provided by our simulation, is *no*. It is just possible to rearrange cells of equal area in the manner in Figure 6. This illustrates, in passing, an important general feature of froth structures. There are many alternative (metastable) structures for given cell areas; this is all the more true when we make cell areas widely variable.

Returning to our defective hexagonal structure, we see that it contains 'the seed of its destruction' (the phrase is from Gibbs, in a related context[6]). The five-sided cells shrink; the seven-sided cells grow and hence, in a cancerous fashion, the whole structure is devoured. We shall have cause to return to this example shortly.

More typically, cells evolve throughout a disordered structure, as in Figure 7. The continual disappearance of cells (Figure 5) causes the structure to coarsen, that is, the average cell size increases steadily (incidentally, Smith seems to have believed that cells must become three-sided before disappearance, but it is now clear that this is not necessarily so). While illustrating the trend, these simulations were not sufficiently extensive to define the asymptomatic behaviour of the coarsening process. Other simulations[18] and the experiments of Glazier *et al.*[12] have suggested that, after some transient effects, the system shows the simplest behaviour which might be expected on dimensional grounds, as originally asserted by Smith.[11] The average cell area increases linearly with time, while the distribution of cell shapes and relative areas remain fixed; that is, the structure is continually magnified without change, in a statistical sense. However, this conclusion may still be premature; various aspects of the experimental data remain mysterious.[19]

The transient effects can be particularly large if we start from a relatively ordered structure, as we will have the 'cancerous' behaviour indicated above.[12]

For further details of the simulation of the 'life and death' of individual cells

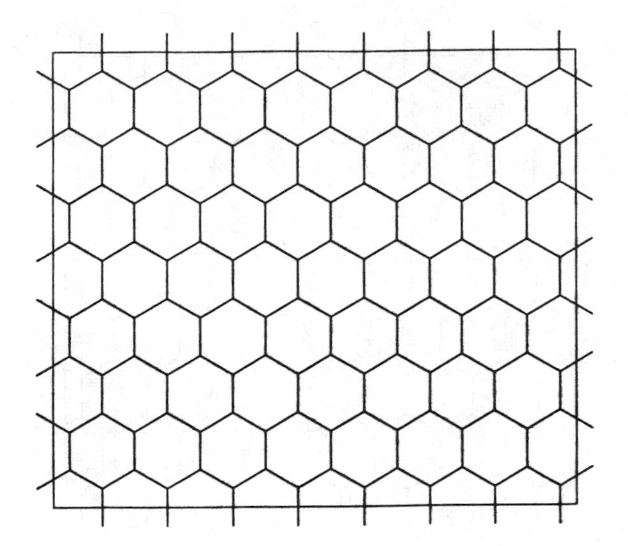

(*a*)

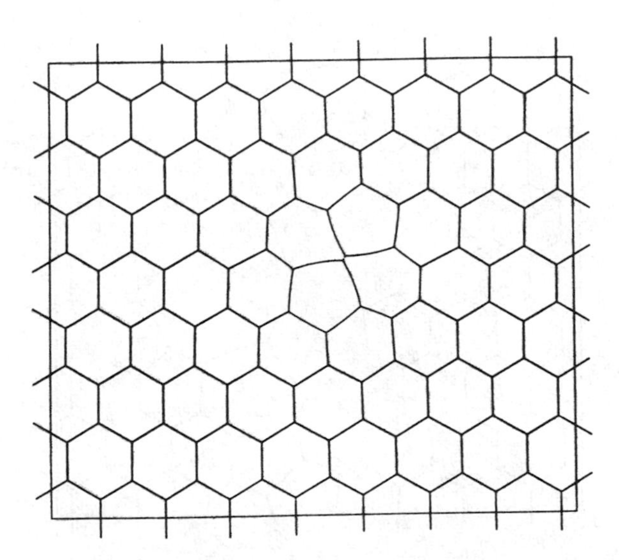

(*b*)

Figure 6 *The perfect honeycomb froth (a) is stable, but a single topological defect (b) is enough to render it unstable, even though the areas are initially unchanged. Later stages in the subsequent evolution are shown in (c) and (d)*

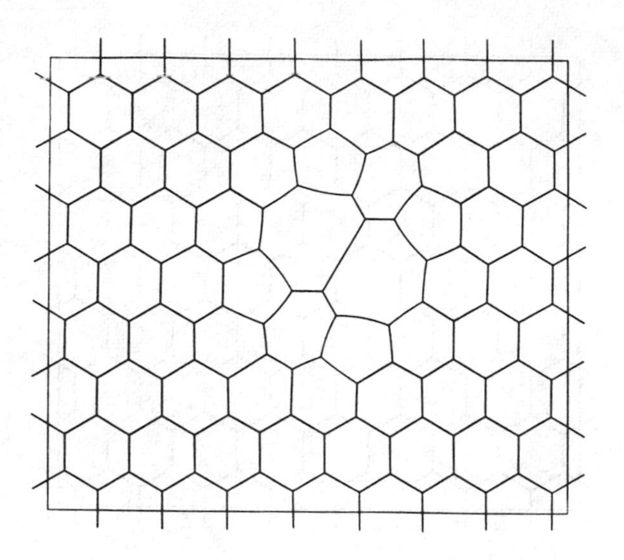

(*c*)

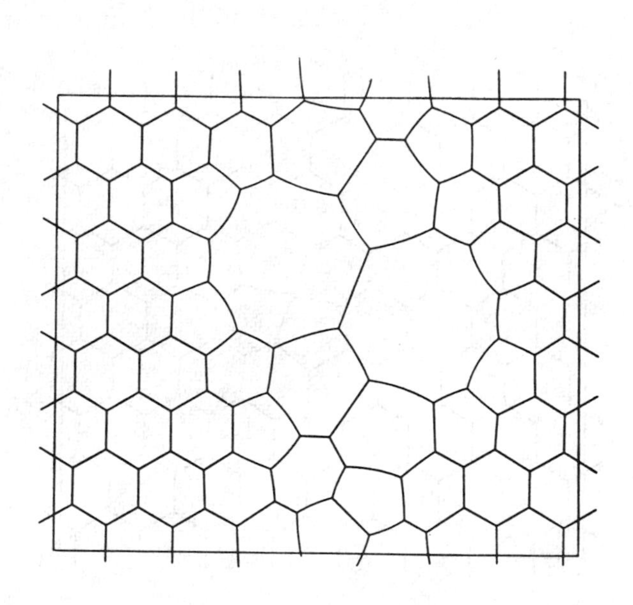

(*d*)

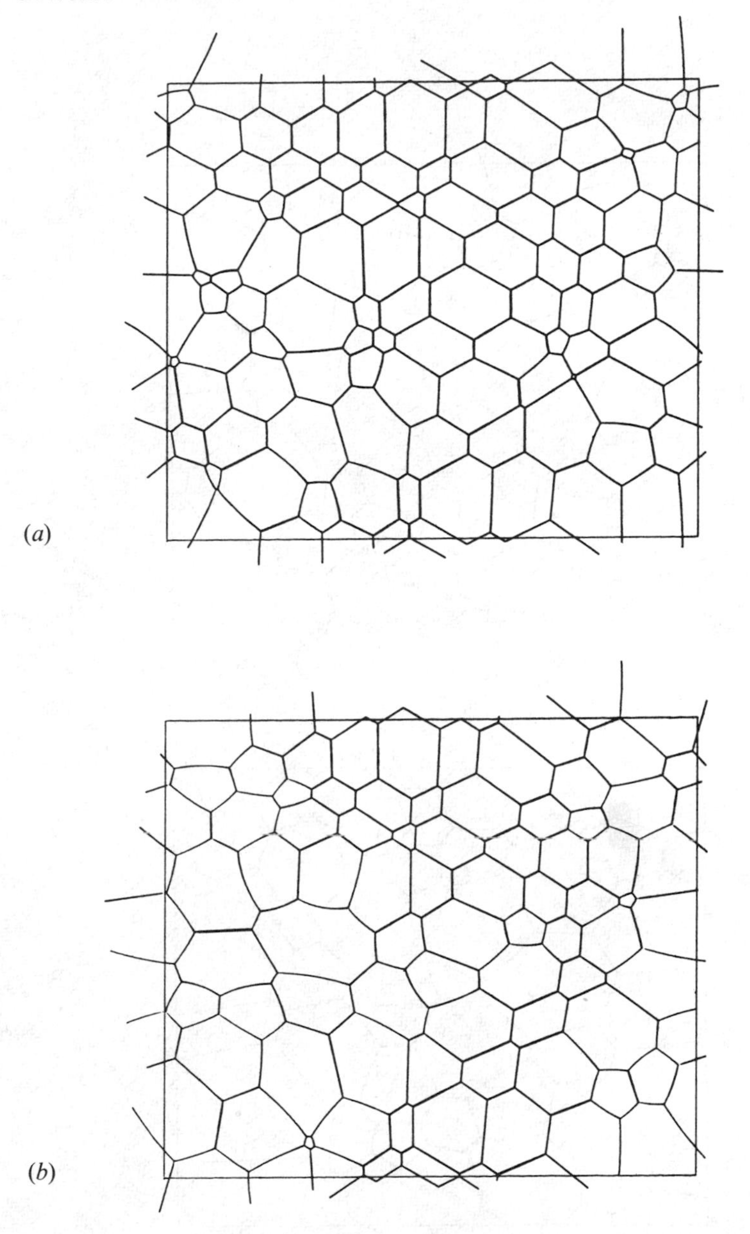

Figure 7 *Evolution of a typical froth structure. A sequence of configurations of the same froth is shown, (a)–(f), in chronological order*

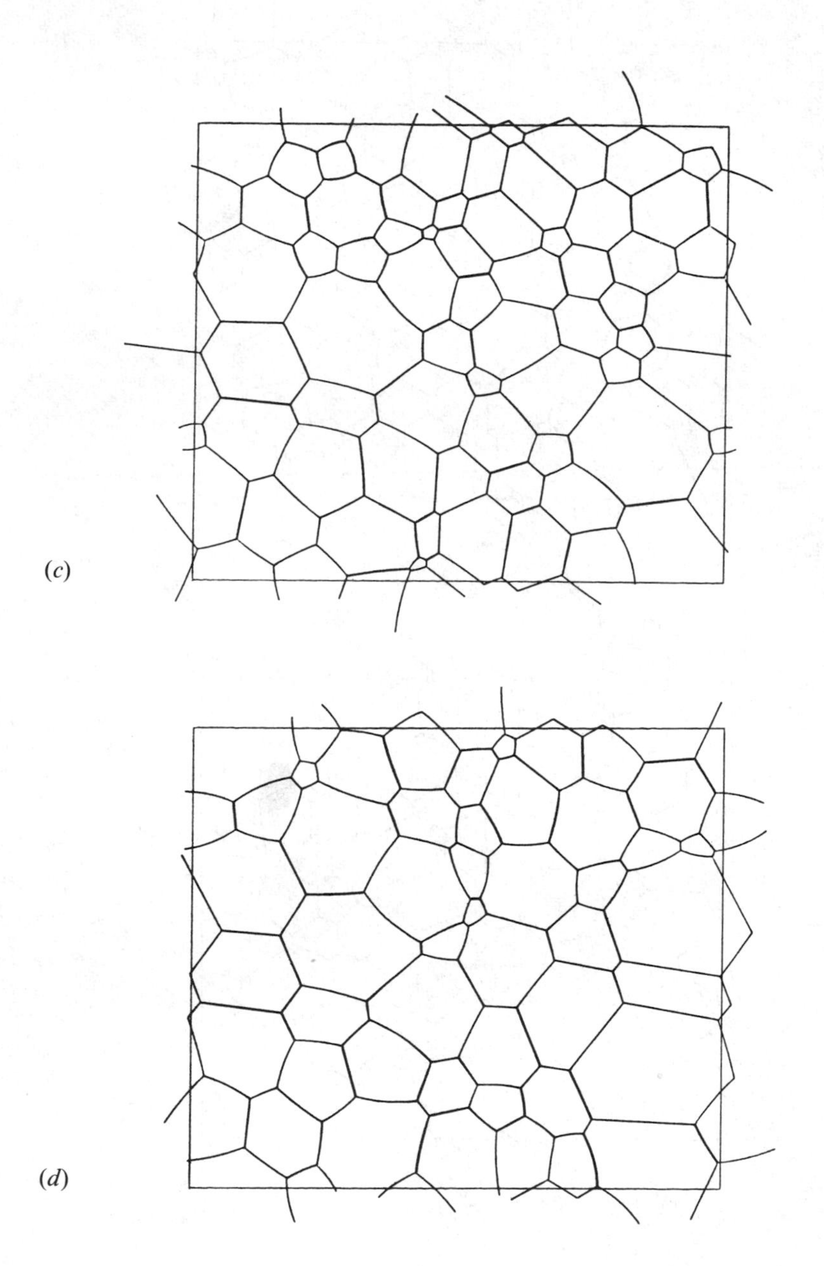
(c)
(d)

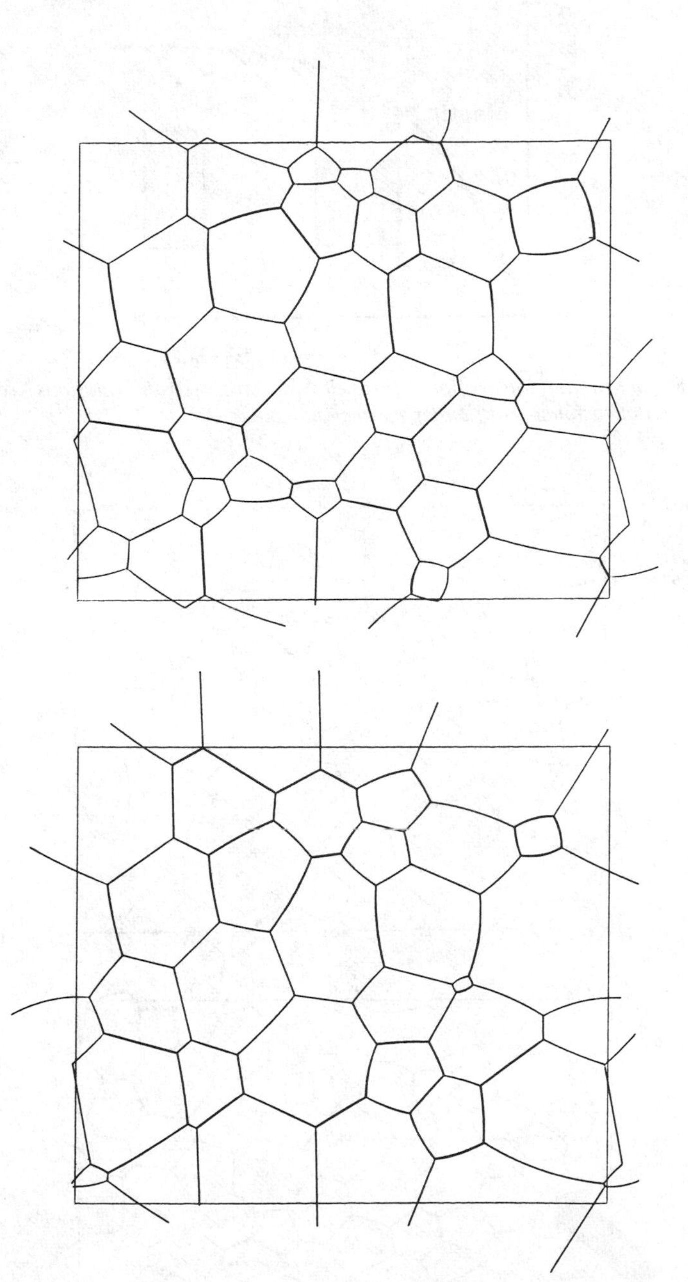
(e)
(f)

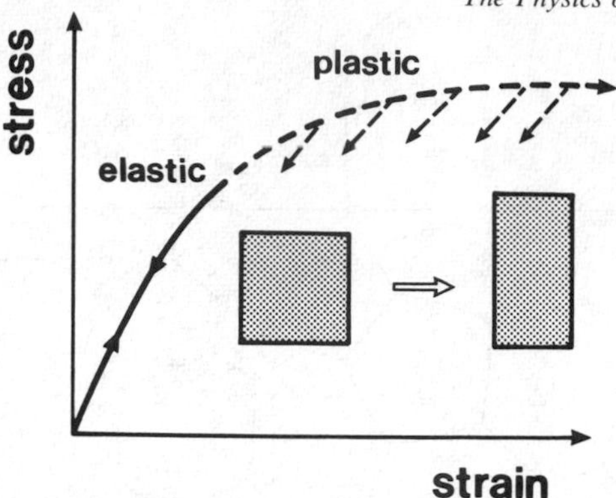

Figure 8 *Schematic illustration of results for stress–strain relations, for two-dimensional froth under extensional shear stress*

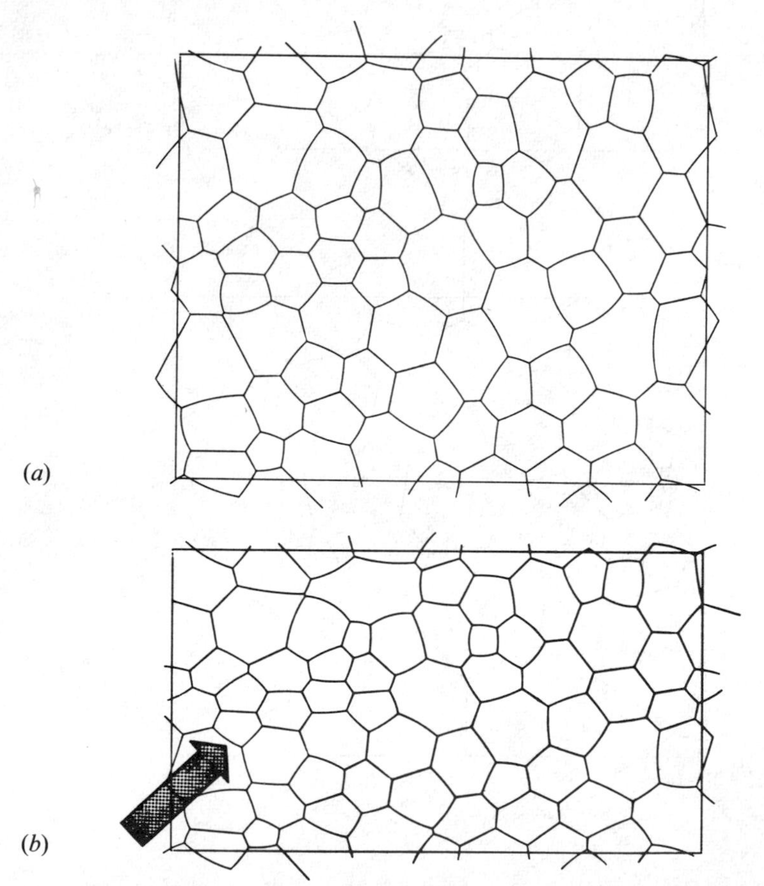

Figure 9 *Example of the variation of structure with shear stress; note the rearrangement of cells in the vicinity of the arrow. (a) Original structure; (b) same structure with an imposed shear strain (rescaled to constant width).*

and their statistics, we refer the reader to papers by Weaire and Kermode,[15,16] and Glazier *et al.*[12]

The idealized model and its simulation may be used for a second purpose, to investigate mechanical properties, if only the *static* relations of stress and strain. This is achieved simply by changing the boundary conditions to represent given strains, recalculating energies and hence stresses.

The results of a large number of such calculations are summarized very schematically in Figure 8. We must distinguish carefully between the *elastic* (recoverable) response to low shear stress and the *plastic* deformation which takes place at higher values of stress. The latter is due to cell rearrangements (Figure 4) which are provoked by shear strain, whenever it happens that the length of a cell wall approaches zero. An example is indicated in Figure 9. Strictly, a disordered froth is likely to have *no* truly elastic response, as some such rearrangements should occur at arbitrarily small strain values for sufficiently large samples. However, they are rare at low values of strain and the concept of an elastic region of a fairly well defined extent is an acceptable one for practical purposes.

Our simulations, although partly obscured by the 'noise' inherent in finite calculations, suggest a *monotonic* stress–strain relation as in Figure 8, so that the maximum (or yield) stress occurs at infinite strain. Such a curve constrasts sharply with the results of calculations based on the hexagonal structure,[13] although the elastic constant (the slope at the origin) is roughly the same.[20]

The imposition of a stress greater than the yield stress entails continuous deformation or *flow*, and viscosity comes into play.[13]

For further details of these calculations and magnitudes of various quantities, see reference 18.

6 Conclusion

The complex structure of a froth makes it difficult to obtain estimates of its properties by analytical methods or 'back-of-an-envelope' calculations. Certain estimates can be based on the hexagonal structure, but the validity for disordered systems must always be checked, for example by simulation.

When stripped to its bare essentials, such a simulation presents little difficulty to a computational treatment in the case of two dimensions. Whether three-dimensional structures can be similarly investigated is doubtful, but other methods[18] may be adaptable to that purpose. Despite a long history, the whole subject is only now beginning to develop an adequate scientific base in well founded concepts and experimental techniques, to which the present author claims only a small contribution. Others may be seen in the recent critical review of foam rheology by Kraynik.[13] Together, as he says, 'the current theories provide a rational basis for developing our intuition and reinforce the need for careful characterization of foam structure and systematic rheological measurements.' Further progress calls for an interdisciplinary approach, if due importance is to be ascribed to all aspects of the problem.

Acknowledgement. Research support from Eolas is acknowledged.

References

1. D. Weaire and N. Rivier, *Contemp. Phys.*, 1984, **25**, 59.
2. C. V. Boys, 'Soap-bubbles,' SPCK, London, 1896. (There are various later editions.)
3. A. S. C. Lawrence, 'Soap Films,' G. Bell, London, 1929.
4. K. Mysels, K. Shinoda, and S. Frankel, 'Soap Films,' Pergamon Press, New York, 1959.
5. J. J. Bikerman, 'Foams,' Springer, New York, 1973.
6. S. Berkman and G. Egloff, 'Emulsions and Foams,' Reinhold, New York, 1941.
7. F. J. Almgren, Jr., *Math. Intellir.*, 1982, **4**, 164.
8. D'A. W. Thompson, 'On Growth and Form,' Cambridge University Press, Cambridge, 1917.
9. D. Weaire, in 'Physical Properties of Amorphorous Materials,' ed. D. Adler, B. B. Schwartz, and M. C. Steele, Plenum Press, New York, 1985, p. 157.
10. C. S. Smith, 'A Search for Structure,' MIT Press, Cambridge, MA, 1981.
11. C. S. Smith, in 'Metal Interfaces,' American Society for Metals, Cleveland, OH, 1952, p. 65.
12. J. A. Glazier, S. P. Gross, and J. Stavans, *Phys. Rev. A*, 1987, **36**, 306.
13. A. M. Kraynik, *Annu. Rev. Fluid Mech.*, 1988, **20**, 325.
14. H. Princen, this volume.
15. D. Weaire and J. P. Kermode, *Philos. Mag.*, 1983, **48**, 245.
16. D. Weaire and J. P. Kermode, *Philos. Mag.*, 1984, **50**, 379.
17. J. Von Neumann, in 'Metal Interfaces,' American Society for Metals, Cleveland, OH, 1952, p. 108.
18. D. Weaire and T. L. Fu, *J. Rheol.*, 1988, **32**, 271.
19. D. Weaire, J. P. Kermode, and J. Wejchert, *Philos. Mag.*, 1986, **53**, L101.

Contribution of Drainage, Coalescence, and Disproportionation to the Stability of Aerated Foodstuffs and the Consequences for the Bubble Size Distribution as Measured by a Newly Developed Optical Glass-fibre Technique

By A. D. Ronteltap and A. Prins

DEPARTMENT OF FOOD SCIENCE, AGRICULTURAL UNIVERSITY, DE DREIJEN 12, 6703 BC WAGENINGEN, THE NETHERLANDS

1 Introduction

Aqueous foams, once formed, are in thermodynamic terms unstable, which means that foams will deteriorate with time. The rate of deterioration is of great interest in many fields of industry and everyday life. Beer foam is one of the most studied examples in the food industry.[1] The rate of deterioration of foam depends mostly on the velocity of three physical phenomena, *i.e.* drainage, coalescence, and disproportionation.

Drainage is the liquid flow from the foam, or the rise of foam bubbles as a consequence of gravity. The main physical parameter influencing drainage is viscosity; a high viscosity gives slow drainage.[2] Foam stability testers are often based on the measurement of the rate of drainage.[3] Drainage leads to a decrease in the total foam volume, because the liquid–foam interface rises, but it does not affect the upper level of the system. Also, the bubble size distribution in the foam does not change as a result of drainage, although the shape of the bubble might be affected.[4] Here for the sake of simplicity the assumption is made that all bubbles have a spherical shape. Another most important but indirect effect of drainage is that the film thickness in the foam decreases with time. As explained later, disproportionation accelerates as a result of drainage and coalescence becomes more likely.

Coalescence of bubbles is the breakage of the thin film between two bubbles. Many mechanisms for coalescence have been proposed,[5] and they all have in common that coalescence occurs preferably when the film thickness is low. Important for the evolution of the bubble size distribution is that coalescence leads to coarsening of the foam. Coalescence gives only large bubbles in the foam, which means that the bubble size distribution will shift to larger bubbles.[6]

Disproportionation is the result of inter-bubble gas diffusion caused by a difference in gas pressure between bubbles. When only one gas is present, this pressure difference corresponds to a difference in Laplace pressure. Gas diffuses from a smaller bubble to larger ones; hence small bubbles shrink, while larger bubbles grow. The rate of disproportionation depends considerably on several parameters, especially the gas solubility and the film thickness, which may have very different values in different foams.[7]

2 Theory

Nowadays the methods for measuring foam stability do not distinguish between the three physical processes mentioned. In some cases total drainage is measured, in others foam height or foam volume. Other aspects of foam stability can also be measured, but always the result is some total effect of drainage, coalescence, and disproportionation. The measurement of bubble-size distributions has not yet been used to distinguish between drainage, coalescence, and disproportionation, although substantial research work has been done on the measurement of foam stability and the determination of bubble-size distributions in aqueous foams.[8,9] Strictly, there is only one objective method to measure foam stability and that is to measure the bubble-size distribution as a function of time.

There is a large difference in the effects of coalescence and disproportionation on the evolution of the bubble size distribution. Only larger bubbles appear as a result of coalescence, whereas disproportionation can result in both larger and smaller bubbles hence a bimodal distribution can be obtained.[10,11]

A distinction can be made between coalescence and disproportionation using gases of different solubility. The rate of disproportionation depends strongly on gas solubility. By using gases with very different solubilities such as nitrous oxide or carbon dioxide on the one hand and nitrogen or oxygen on the other, the rate of disproportionation can be influenced. Coalescence will mostly be independent of the gas used. Using gases of very different solubilities, an estimate can therefore be made of the contribution of coalescence and disproportionation to the evolution of the bubble-size distribution.

3 Materials and Methods

A new method will be presented here which permits the measurement as a function of time of, (i) the upper level of the foam, (ii) the level of the foam–liquid interface, and (iii) the evolution of the bubble-size distributions in foam. The rate of drainage, the changes in gas content, the rate of foam collapse, and the changes in foam volume are thus obtained.

The apparatus consists of three major parts, as shown in Figure 1: firstly a mechanism for moving the fibre up and down the foam at known speed, secondly the fibre itself combined with the opto-electronic unit, and thirdly equipment for data acquisition and for calculation of the bubble-size distributions and other foam properties.

From the opto-electronic unit, light is emitted into the fibre. The end of the fibre consists of a very small rounded tip of diameter *ca.* 20 μm, the diameter of the fibre itself being 200 μm. The essence of the method is that the amount of light

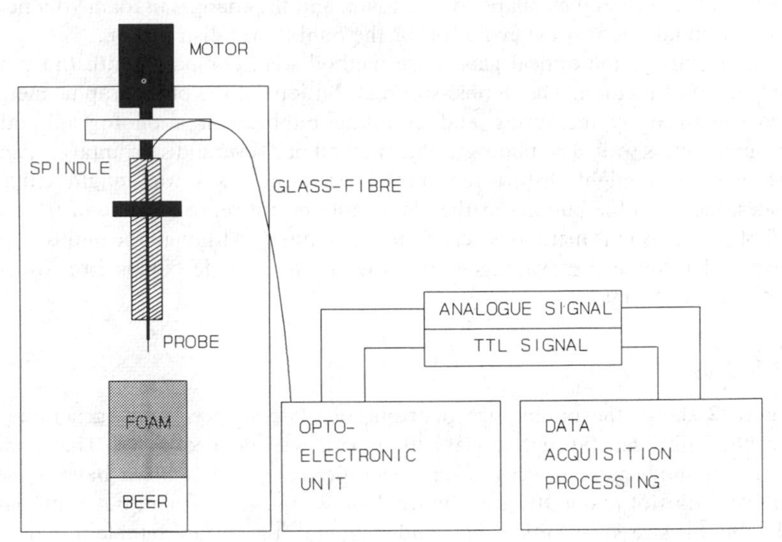

Figure 1 *The optical glass-fibre probe method with the mechanical part, the opto-electronical part and the part for data acquisition and processing*

reflected at the tip of the fibre depends on the refractive index of the medium surrounding the tip. When the refractive index of the medium is approximately the same as that of the glass, almost no light is reflected. However, when the refractive index of the medium is much lower than that of the glass, part of the light is reflected. Therefore, when the tip is surrounded with gas more light is reflected than when the tip is in liquid. A beam splitter separates the returning beam; half is returned to the source and lost and the other half is received by a light-sensitive cell and converted into an electronic signal. The opto-electronic unit also contains an analogue–digital converter to make data acquisition easier.[12,13]

On moving the fibre through a foam, an alternating signal corresponding to gas and liquid is obtained. When the speed of the probe is known, the time travelled in the gas and in the liquid are a measure of the bubble-size distribution in the foam. Either the analogue or the digital signal is used to calculate the bubble-size distribution.

With the calculation of the bubble-size distribution a problem similar to the 'tomato salad' problem is to be solved. The observed one-dimensional gas lengths are not equal to the actual three-dimensional bubble radii because a cross-section of a bubble is hardly ever made through two polar ends. Furthermore, bubbles of large diameter have a greater chance than bubbles of smaller diameter of being pierced by the optical probe. A statistical method has been used to calculate the three-dimensional size distribution from the one-dimensional distribution as described by Weibel.[14] The method makes use of the gas fraction of the foam which follows from the measurement of the upper level of the foam and the level of the liquid–foam interface. When the experiment is done as a function of time, the rate of drainage, the collapse of the foam, and the changes in foam volume are detected in addition to the evolution of the bubble-size distribution.

The results of the optical glass-fibre method were compared with those of a photographic method. The bubble-size distributions in the photographic method were obtained by measuring and counting bubbles on a photograph taken through a glass wall. The photographic method has several disadvantages, *e.g.* (i) the glass wall might distort the bubbles, (ii) the glass wall might enhance coalescence, (iii) the bubbles at the glass wall are not representative of the foam bubbles, and (iv) the method is very time consuming.[7] Although the photographic method has these disadvantages, an order of magnitude comparison of both methods can be made.

4 Results

Figure 2 shows the bubble-size distribution of fresh beer foam generated by sparkling nitrogen through a glass filter with well defined pores. The foam is almost homodisperse directly after generation and the bubbles have a mean bubble radius of about 100 µm. Figure 3 shows the same foam, but 3 min later. The bubble-size distribution has widened and the mean bubble radius has increased. Most important is the observation that no bubbles have become smaller, meaning that disproportionation did not occur.

Using carbon dioxide instead of nitrogen, almost the same bubble-size distribution is obtained directly after generation of the foam (Figure 4). After 3 min however, a completely different picture is obtained. From Figure 5 it is clear that

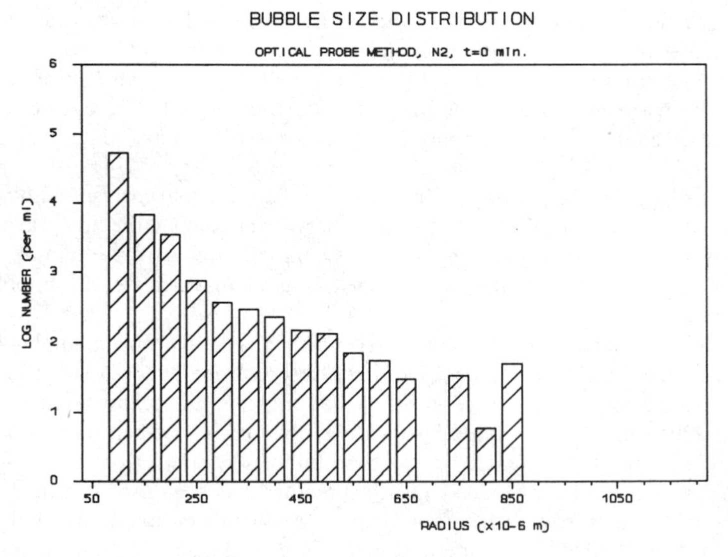

Figure 2 *Bubble-size distribution of nitrogen-filled beer foam at* t = 0 min *as determined with the optical glass-fibre method*

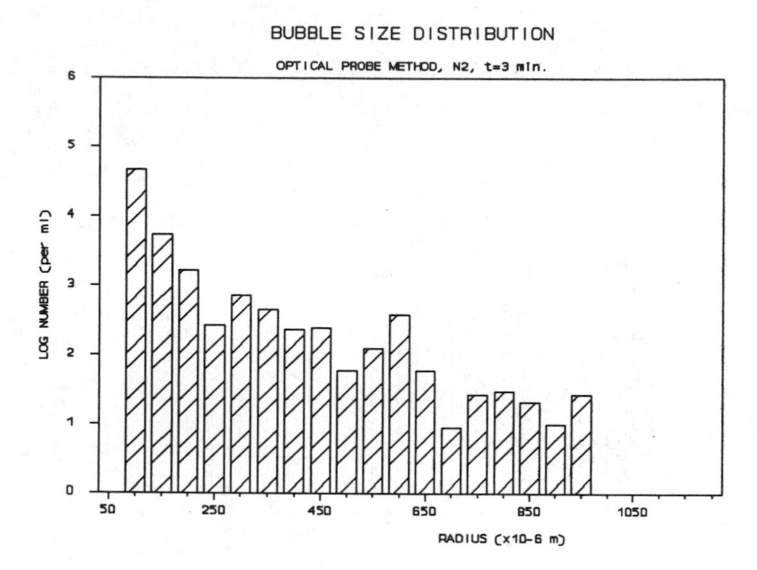

Figure 3 *Bubble-size distribution of nitrogen-filled beer foam at* t = 3 min *as determined with the optical glass-fibre method*

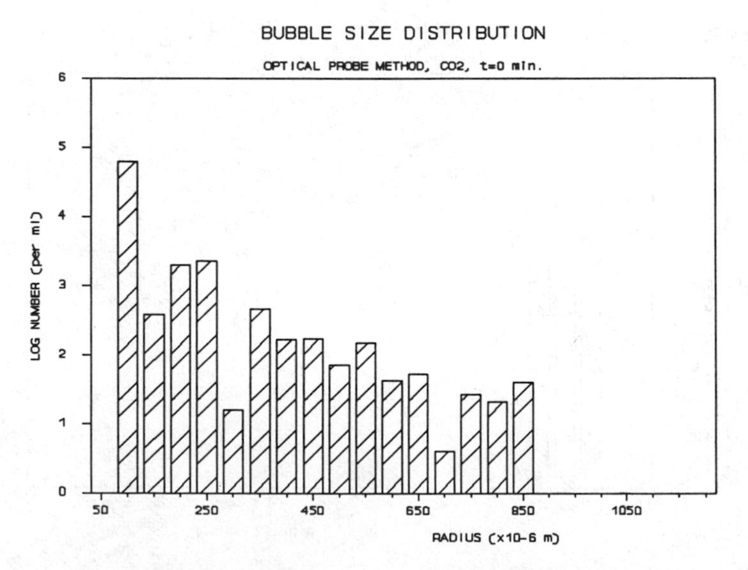

Figure 4 *Bubble-size distribution of carbon dioxide-filled beer foam at* t = 0 min *as determined with the optical glass-fibre method*

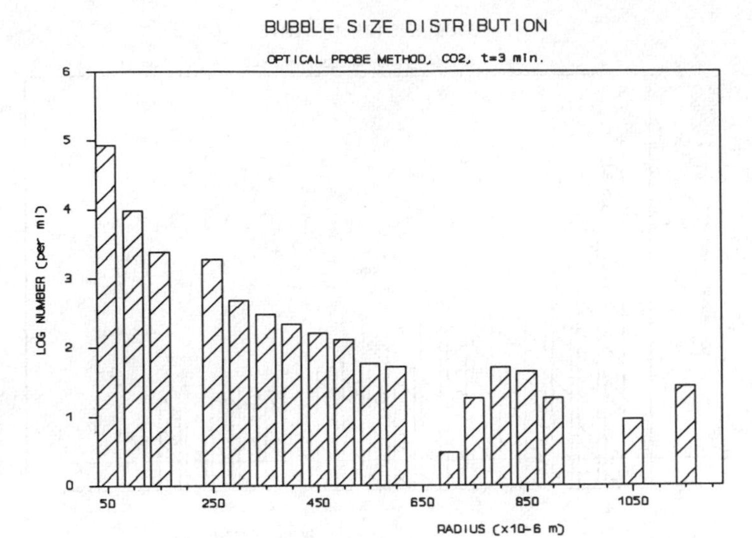

Figure 5 *Bubble-size distribution of carbon dioxide-filled beer foam at* t = 3 min *as determined with the optical glass-fibre method*

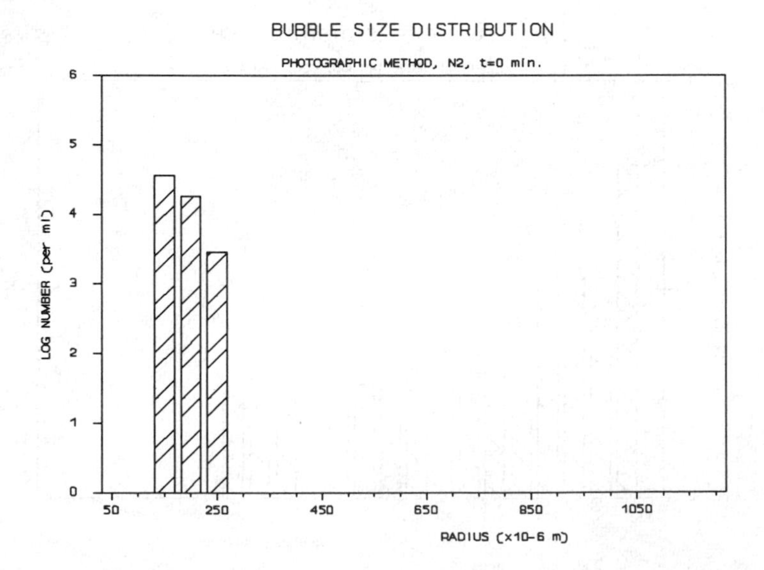

Figure 6 *Bubble-size distribution of nitrogen-filled beer foam at* t = 0 min *as determined with the photographic method*

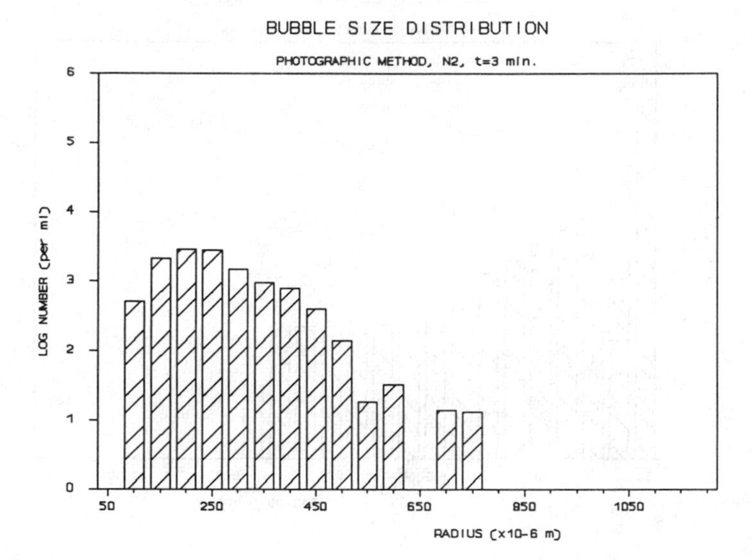

Figure 7 *Bubble-size distribution of nitrogen-filled beer foam at* t = 3 min *as determined with the photographic method*

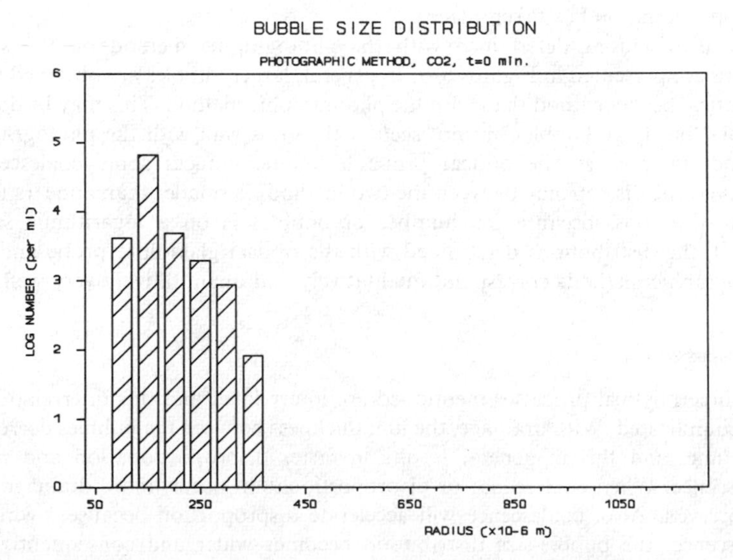

Figure 8 *Bubble-size distribution of carbon dioxide-filled beer foam at* t = 0 min *as determined with the photographic method*

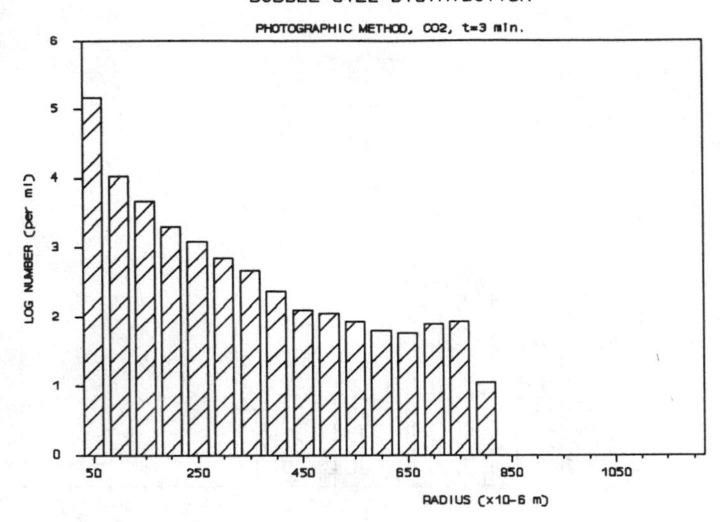

Figure 9 *Bubble-size distribution of carbon dioxide-filled beer foam at* t = 3 min *as determined with the photographic method*

bubbles have shrunk as a consequence of gas diffusion and the bubble size-distribution has widened much more than with nitrogen. It is evident that disproportionation has taken place.

The distributions determined with the photographic method on the same systems are presented in Figures 6–9. In general, larger bubbles are observed with the optical probe method than with the photographic method. This may be due to the fact that large bubbles are not seen at the glass wall with the photographic method, or because the optical probe technique induces some coalescence. However, the discrepancy between the two methods is smaller than appears from the distributions, because the number of bubbles is on a logarithmic scale. Overall, the distributions determined with the optical glass-fibre probe and the photographic methods correspond qualitatively and quantitatively very well.

5 Discussion

The three physical processes mentioned are interrelated and this interrelation is very complicated. With drainage, the film thickness between the bubbles decreases with time, and this in general results in faster disproportionation and more coalescence. When coalescence or disproportionation occur, more drainage will be observed. Also, coalescence will accelerate disproportion because, owing to coalescence, the bubble-size distribution becomes wider and consequently the Laplace pressure differences increase. Furthermore, disproportionation enhances coalescence. As a consequence of disproportionation, rearrangements in the foam might lead to an increase in coalescence. These are only a few of the possible interactions between drainage, coalescence, and disproportionation.

These considerations demonstrate that the measurement of foam stability is complicated. It is difficult to make a quantitative distinction between the different physical processes occurring in the foam. Nevertheless, semi-quantitative results are obtained on measuring the stability of beer foam and the evolution of bubble-size distributions in the foam. With the introduction of this new application of the optical glass-fibre probe technique, a method has been introduced that might increase our knowledge of aqueous foams.

Acknowledgements. The research on the optical glass-fibre technique and on the measurement of bubble-size distributions was performed in cooperation with the department of Physical Technology of the Technical University, Delft, The Netherlands. The work was supported by Heineken Research, Zoeterwoude, The Netherlands.

References

1. C. W. Bamforth, *J. Inst. Brew.*, 1985, **91**, 370.
2. K. B. Kann and V. N. Feklistov, *Colloid J.*, 1985, **46**, 1052.
3. A. D. Rudin, *J. Inst. Brew.*, 1957, **63**, 506.
4. J. A. Kitchener and C. F. Cooper, *Q. Rev. Chem. Soc.*, 1959, **13**, 71.
5. A. Mar and S. G. Mason, *Kolloid Z.*, 1968, **225**, 55.
6. G. M. Nishioka, *Langmuir*, 1986, **2**, 649.
7. A. J. de Vries, in 'Adsorptive Bubble Separation Techniques,' ed. R. Lemlich, Academic Press, New York, London, 1972, p. 7.
8. H. Sasaki, H. Matsukawa, S. Usui, and E. Matijevic, *J. Colloid Interface Sci.*, 1986, **113**, 500.
9. A. Selecki and R. Wasiak, *J. Colloid Interface Sci.*, 1984, **102**, 557.
10. R. Lemlich, *Ind. Eng. Chem. Fundam.*, 1978, **17**, 89.
11. A. Monsalve, R. S. Schechter, *J. Colloid Interface Sci.*, 1984, **97**, 327.
12. J. J. Frijlink, P. A. van Halderen, J. Hofmeester, and M. M. C. G. Warmoeskerken, *I^2-procestechnologie*, 1986, **3**, 19.
13. J. J. Frijlink, 'Physical Aspects of Gassed Suspension Reactors,' *PhD Thesis*, Technical University, Delft, 1987, p. 119.
14. E. R. Weibel, in 'Stereological Methods,' Vol. 2, Academic Press, New York, London, 1989, p. 215.

Beer Foams

By C. W. Bamforth*

BASS PLC, 137 HIGH STREET, BURTON-ON-TRENT, STAFFS., UK

1 Introduction

Although there are regional differences across the UK, generally the presence of a white, stable foam (head) is one of the major quality attributes of beer. Beer foam characteristics can be considered in terms of three major parameters: (a) head formation, (b) head retention and (c) foam lacing (cling, adhesion).

Head formation ('foamability') is largely dependent on the gas content of beer and on that introduced at the point of dispensing, together with the tendency of this gas to leave solution. There should seldom be any difficulty in generating sufficient foam during beer dispensing. Rather, the brewing scientist has concentrated on the factors which determine stability (head retention of a preformed foam). However, those materials which afford the most stable foams are probably also the ones which foam up the most readily.

The sequence of events transpiring during the lifetime of a beer foam have been documented earlier.[1] It is especially important, however, that the structure of beer foam changes during ageing, from being essentially a wet liquid matrix of spherical bubbles into one wherein the bubbles have adopted a polyhedral shape, with thin lamellae between them. On ageing, the foam changes to being more 'solid' in nature, owing to an increase in the surface viscosity. The result is an increased tendency of the foam to be deposited on surfaces, that is, to cling (lace, adhere).

Lacing is arguably the most important aspect of foam quality. Few drinkers assess foam quality on beer, especially that of other drinkers, by judging the depth of foam remaining on top of the liquid beer. Rather they note the foam clinging to the walls of the glass, which testifies to good foam quality throughout the drinking experience—and also probably convinces them that the glass is of pristine cleanliness! The drinker is probably justified in this assumption, for excellent head on beer depends very much on (a) the dispensing conditions—the more head generated, the more opportunity there will be within the lifetime of the foam for 'solidification' to occur; (b) the glass—greasy glasses will not support foam; and (c) the beer—and its relative content of foam-positive and foam-negative components.

* Present address: Bass Brewing (Preston Brook) Ltd., Runcorn, Cheshire WA7 3BN

2 Foam Stability

When considering foam stability, two fundamental difficulties must be taken into consideration: (a) is the method of foam measurement relevant?; and (b) do all foam active substances present *really* have a role to play?

The method of measuring foam presents particular difficulties. Many investigations of the materials responsible for foam stability have relied upon measuring head retention from foams generated in quantities which are vastly in excess of those normally produced during dispensing in the trade. When a beer is totally foamed, materials may gain access which would normally remain in the beer when it is dispensed conventionally. Furthermore, substantial volumes of liquid can drain from foams without noticeable foam collapse. It is the foam *itself* whose survival should really be quantified.

Because a material extractable from beer can, in isolation, afford good foaming properties does not necessarily mean that it contributes significantly to the foam properties of that beer. Moreover, increasing the content of a given material may not necessarily improve the foam, even if that substance does have a major role in foam stabilization, for it may already be present in excess.

3 Foam Measurement

Notwithstanding the difficulties referred to above, most studies of beer foam stability have been made using head retention measured by liquid drainage techniques. Principal amongst these procedures are the methods of Rudin (see[2]) and Ross and Clark.[3] Their advantages are that they do permit ready quantification of trends in gross foam quality and the Rudin procedure, in particular, is a valuable means for quantifying the foaming tendencies of isolated foam materials (see later).

Rudin's procedure uses an attemperated glass column (Figure 1) into which is introduced a defined volume of degassed beer (to the 10-cm mark). The beer is gassed to generate foam to the 32.5-cm mark by bubbling carbon dioxide through the sinter at a defined rate. (Clearly consistency within and between apparatus is ultimately dependent on sinter quality). After ceasing the gas flow, foam collapse is assessed by timing the reformation of liquid beer between the 5- and 7.5-cm marks. The value (in seconds) obtained is the head retention value (HRV).

The procedure of Ross and Clark[3] is recommended by the American Society of Brewing Chemists and is said to reflect more closely the performance of beer foam in the trade. It is, however, of less value than the Rudin procedure[2] as a means for the assessment of isolated foam fractions. In this procedure foam is generated by pouring beer into a funnel. Beer which recollects is drained away at a rate such that only foam is present 90 s after the end of pouring. At a stage which is at least 225–230 s after the first draining period, measurement is made of the volume of beer remaining in the foam. The values are entered into an algebraic expression which affords a Σ value that relates directly to foam stability.

Recently, attempts have been made to design procedures for assessing foam stability which relate more closely to longevity of the head *per se*. Prominent amongst such procedures is the Truefoam method,[4] in which a freshly dispensed

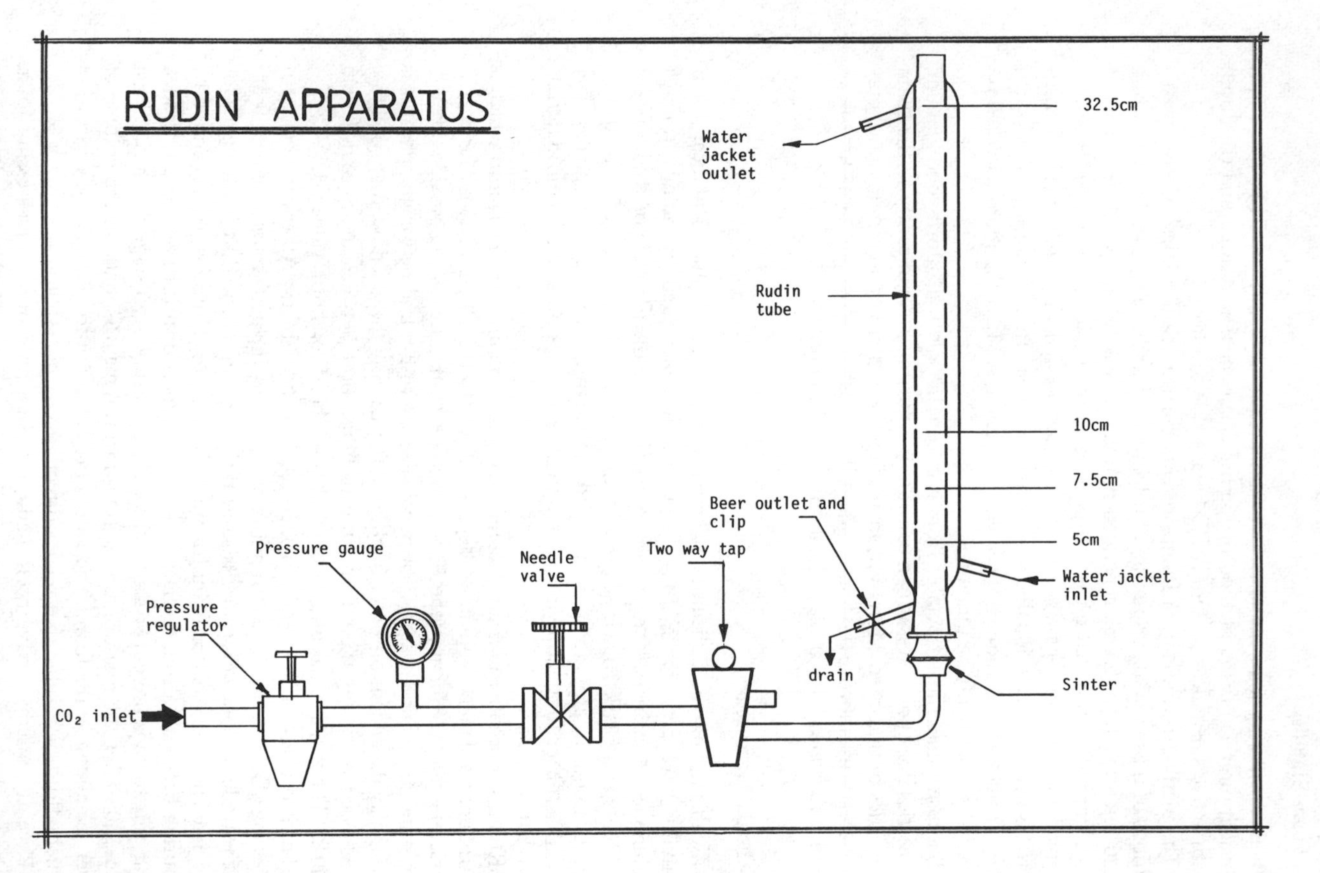

Figure 1 *Measurement of foam stability*

beer is placed between a light source and photocell. Once foam has collapsed to reveal a portion of beer surface not covered by foam, the light reaches the photocell and a timing mechanism is halted. The period expired relates to the time taken for a 'dark spot' to appear on the surface of the beer and is thus a direct measure of foam stability.

One of the few quantitative procedures available for the measurement of foam-cling is the Lacing Index procedure of Jackson and Bamforth.[5] Foam is generated in a standard fashion and then drinking is simulated by periodic drainage of liquid. Once all the liquid has been removed, the foam adhering to the glass is removed as a solution in water, the absorbance (230 nm) of which relative to that of the original beer affords a direct measure of total lacing (Lacing Index). This procedure is one of the few which provides a direct measure of foam itself.

4 Materials Contributing to Beer Foam

Polypeptides—Two principal structural features have been studied with regard to the foaming properties of beer polypeptides: molecular size and hydrophobicity.

For some time it has been accepted that beer polypeptides providing the most stable foams are those of relatively higher molecular weight. More recently, it was concluded that it is the proportion of larger polypeptides relative to the low molecular weight, foam-destabilizing polypeptides which is most relevant, and not the absolute level of the larger polypeptides.[6] However, Bamforth[7] demonstrated that dialysable barley proteins afford foams equally as stable as those from their non-dialysable counterparts at the equivalent protein concentration. Whether the same applies to the beer polypeptides, which are decomposition products of the barley proteins, remains to be seen.

One parameter which incontrovertibly relates to foam stability of both barley and beer proteins is hydrophobicity: proteins which bind strongly to hydrophobic resins such as octyl- or phenyl-Sepharose display increased Rudin HRV values.[8] The importance of surface hydrophobicity in the foaming properties of proteins generally had already been recognized.[9]

Clearly, amphipathic features are of particular relevance to the foaming properties of beer polypeptides. It may be that higher molecular weight polypeptides improve foam stability once they have been denatured to expose previously hidden internal hydrophobic residues. Kato *et al.*[10] have demonstrated that a much better relationship exists between hydrophobicity and foam stability if proteins are first partially denatured. Whether there is a direct relationship between HRV and the level of amphipathic (and 'potentially amphipathic') polypeptides in beer or whether foam stability is determined by the ratio of amphipathic to less amphipathic polypeptides, the latter interfering with foam stability, remains to be ascertained.

Because of their amphipathic character, the foaming polypeptides are those which tend most readily to leave solution and are therefore lost to a greater or lesser extent during beer production and processing. Further, some beers are produced by replacing relatively large quantities of malt with sugars which provide no polypeptides. Accordingly, beers can be relatively deficient in such materials. It is possible, however, to supplement beers with foaming polypeptides.

Notable amongst these materials is hydrolysed albumen.[11] This material is more efficacious in beers which are relatively deficient in endogenous foam polypeptide.

It is particularly important that any material introduced into beer should cause no lessening of beer clarity or stability. Albumen must be hydrolysed to an extent that will permit its use for head improvement but which will preclude the development of haze on pasteurization and storage.

A potentially more desirable source of such exogenous foaming polypeptide would be the raw materials of beer production themselves. This concept of circumventing process losses by extracting foaming proteins from barley and adding them back to the finished product was initially pursued by Bamforth and Cope.[12] However, the yields of suitable proteins were low. Latterly, workers at the Brewing Research Foundation have developed the concept further, via the isolation of wort polypeptides.[13]

Hop Bitter Substances—The enhanced bitterness of beer heads is testimony to the concentration of bitter substances within them. Principal amongst these are the hop iso-α-acids (Figure 2). In particular, it may be noted that unhopped beers do not cling to glassware, owing to the role of the iso-α-acids in promoting the thioxotropic changes referred to earlier. For most beers, 10–20 p.p.m. of iso-α-acid appears to be sufficient to afford satisfactory lacing. However, the brewer cannot simply adjust bitterness to facilitate improved lacing, as there would be an unacceptable shift from the specified bitterness for the product in question.

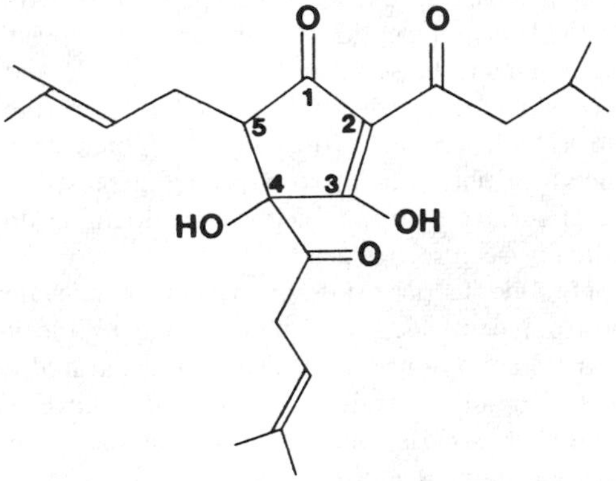

O
O
1
5
2
4
3
HO
OH
O

Isohumulone

Figure 2 *Iso-α-acid. Beer also contains isocohumulone and isoadhumulone, which have isobutyryl and 2-methylbutyryl residues, respectively, substituted on the ring at C-2*

There is no consensus of opinion regarding the nature of the interaction between foam polypeptides and the iso-α-acids. Most probably it involves ionic bonding between the negatively charged acids and positively charged groups on the polypeptide. Acetylation of the lysine groups in beer polypeptides and the use of high salt concentrations and high pH all suppress foaming.[14] This would be consistent with the overcoming of ionic interactions. Such interactions only seem to occur when the polypeptides and iso-α-acids are brought into contact at high concentrations in the bubble lamellae. Studies by Wenn[15] confirmed that foaming polypeptides from beer have high isoelectric points. Certain metal ions, such as copper and iron, promote the iso-α-acid–polypeptide interaction. Regrettably, they also potentiate the oxidative deterioration of beers and must therefore be avoided.

Ethanol—Alcohol-free beers have a reduced tendency to foam, as evidenced by their customary inability to fill a Rudin apparatus completely with foam. The addition of merely 1% (by volume) of ethanol is sufficient to rectify this situation. At alcohol concentrations of up to 3–5%, head retention and lacing will be maximal for the system in question (for example, see Table 1). Further increases in ethanol concentration are to the detriment of foam quality.

Table 1 *Influence of ethanol on Lacing Index*

Sample	*Lacing Index*	*Sample*	*Lacing Index*
Original beer	3.7	DB + 3% ethanol	3.6
Dealcoholized beer (DB)	1.8	DB + 5% ethanol	3.2
DB + 1% ethanol	3.6	DB + 10% ethanol	1.5

Theories advanced for the role of ethanol invoke the reduction of surface tension, decrease in the solubility of carbon dioxide and therefore promotion of bubble release and through a direct influence on the conformation of polypeptides and on their interaction with other surface-active substances in beer. Like bitterness, however, alcohol content is not a parameter which can be freely adjusted by the brewer, as the level is standard for any given product.

Additives to Improve Beer Foam—Two materials are extensively used as a means for supporting the endogenous foam constituents, namely propylene glycol alginate (PGA) and gaseous nitrogen. The former is especially valuable as an agent for protecting beer foams once formed, for example from the deleterious effects of lipids (see later). Curiously, however, it does appear to lessen the ability of beer foams to cling. It may be inferred that this is a result of competition of PGA with iso-α-acids for the amphipathic foam-stabilizing peptides. Jackson *et al.*[16] furnished evidence that the ionic interactions involved in the PGA–peptide association are closely similar to those responsible for iso-α-acid–peptide links.[16]

Nitrogen present at 10–20 p.p.m. greatly increases the stability of beer foams. Beers containing nitrogen give heads with much smaller bubbles, which not only enhances their visual appeal (whiteness) but also significantly increases their longevity. Smaller bubbles rise more slowly, affording an increased opportunity for surface-active molecules to accumulate within their walls. Perhaps more important, liquid drainage from the greatly increased surface area of small bubbles will be much slower. The reasons why nitrogen affords smaller bubbles are not firmly established.

Inhibitors of Foaming—It is actually the case that many beers already contain ample foam-promoting materials to support excellent heads, provided that a sufficiency of foam is generated in the first place. That many beers do not foam sufficiently well is due to the presence of inhibitors ('foam negatives'). These may originate from several sources: (a) brewing raw materials; (b) yeast metabolism during fermentaion; (c) antifoams surviving into the finished product; (d) adventitious process contaminants; and (e) detergents and fat encountered at the point of dispensing.

Perhaps the most common of these is the last. The vigorous efforts of the brewer to provide a beer of maximum foam potential will be to no avail if that beer is subsequently dispensed under adverse conditions—for example, through equipment which may introduce oil into glasses, which in turn are washed in foam-negative detergents and not subsequently rinsed thoroughly in clean water, or if customers themselves introduce lipid from foodstuffs consumed simultaneously with the beer. PGA can afford some degree of protection, but only through diligent 'outside' quality control can the brewer hope to introduce satisfactory conditions of foam dispensing in the trade. (Even under conditions of relative cleanliness, pouring a beer with a full foam is time consuming; less than meticulous barmen will take all manner of shortcuts to dispense beer as rapidly as possible, irrespective of the visual appeal of the resulting pint.)

Beer is a unique case in that foaming is minimized during fermentation to ensure efficient fermenter usage, but a stable head is demanded from the final product on dispensing. Process foaming is curtailed through the use of antifoams, which may include silicone and the esters of fatty acids. These antifoams are then removed—on yeast, by the use of isinglass finings, and also by filtration. If removal is inadequate, however, antifoam will survive and cause severe diminution of beer foam quality. Filter aid selection may be crucial—materials such as kieselguhr are more effective adsorbents than are the perlites, for instance.

It is generally held that there is relatively little chance of lipids from malt surviving the brewing process. Indeed, it has been demonstrated that, whereas many lipids are foam-negative when first introduced into beer, their prolonged contact with other materials (probably protein) in beer renders them less harmful to the head.[17] However, lipids can probably exert their negative influences to a greater extent in beers which are relatively deficient in foam-promoting materials. Such beers will also be less resilient to the introduction of lipophilic materials, for example on dispensing.

The deleterious effect on foam of shorter chain fatty acids is not negated by prolonged contact with beer. These lipids, for example caprylic and caproic acid, are products of yeast metabolism and are particularly inhibitory to cling.[18]

Acknowledgements. The Directors of Bass PLC are thanked for permission to publish this paper.

References

1. C. W. Bamforth, *J. Inst. Brew.*, 1985, **91**, 370.
2. L. R. Bishop, A. L. Whitear, and W. R. Inman, *J. Inst. Brew.*, 1975, **81**, 131.
3. S. Ross and G. L. Clark, *Wallerstein Lab. Commun.*, 1939, **6**, 46.
4. P. J. Wilson and A. P. Mundy, *J. Inst. Brew.*, 1984, **90**, 385.
5. G. Jackson and C. W. Bamforth, *J. Inst. Brew.*, 1982, **88**, 378.
6. F. R. Sharpe, D. Jacques, A. F. Rowsell, and A. L. Whitear, *Proc. Eur. Brew. Conv. Cong., Copenhagen*, 1981, 607.
7. C. W. Bamforth, *J. Inst. Brew.*, 1985, **91**, 154.
8. P. T. Slack and C. W. Bamforth, *J. Inst. Brew.*, 1983, **89**, 397.
9. A. A. Townsend and S. Nakai, *J. Food Sci.*, 1983, **48**, 588.
10. A. Kato, A. Takehashi, N. Matsudomi, K. Kobayashi, and S. Nakai, *Agric. Biol. Chem.*, 1981, **45**, 2755.
11. C. W. Bamforth and R. Cope, *J. Am. Soc. Brew. Chem.*, 1987, **45**, 27.
12. C. W. Bamforth and R. Cope, *Proc. Eur. Brew. Conv. Cong., Helsinki*, 1985, 515.
13. T. M. Morris and I. Slaiding, *Proc. Eur. Brew. Conv. Cong., Madrid*, 1987, 561.
14. K. Asano and N. Hashimoto, *Rep. Res. Lab. Kirin Brew. Co.*, 1976, No. 19,9.
15. R. V. Wenn, *J. Inst. Brew.*, 1972, **78**, 404.
16. G. Jackson, R. T. Roberts, and T. Wainwright, *J. Inst. Brew.*, 1980, **86**, 34.
17. R. T. Roberts, P. J. Keeney, and T. Wainwright, *J. Inst. Brew.*, 1978, **84**, 9.
18. C. W. Bamforth and G. Jackson, *Proc. Eur. Brew. Conv. Cong., London*, 1983, 331.

Solid Foams

By A. C. Smith

AFRC INSTITUTE OF FOOD RESEARCH, NORWICH LABORATORY, COLNEY LANE, NORWICH NR4 7UA, UK

1 Ashby's Model

The understanding of the mechanical properties of solid foams has been enhanced and consolidated in recent years by the work of Ashby and co-workers.[1-3] Their treatment considers the mechanical properties of closed- and open-cell foams which deform elastically, plastically, and by fracture. Briefly, the approach considers three-dimensional square-cell foams (Figure 1). The ratio of the foam density, ρ, to the density of the foam walls, ρ_w, is given in terms of the strut or cell wall thickness, t, and the foam cell length, l:

Closed cells: $\rho/\rho_w = t/l$
Open cells: $\rho/\rho_w = (t/l)^2$

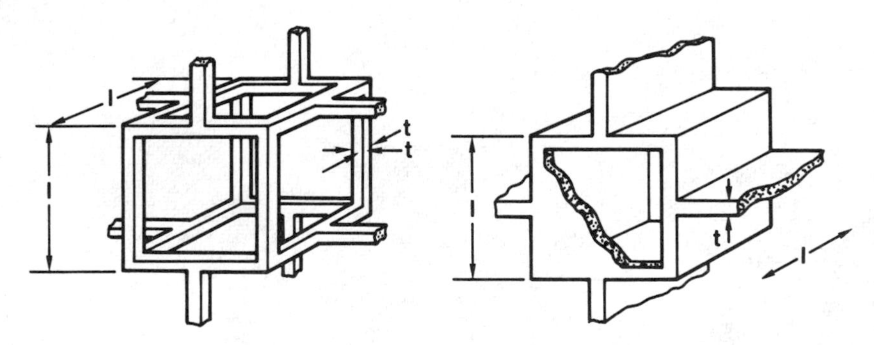

Figure 1 *Ideal closed- and open-foam cells*

The deformation of the foams is understood from the cell structures in two dimensions. Linear elastic deformation of foams is treated by considering the cell walls to be bending beams (Figure 2). The deflection, δ, of a single beam is given by

$$\delta \propto \frac{Fl^3}{12E_wI}$$

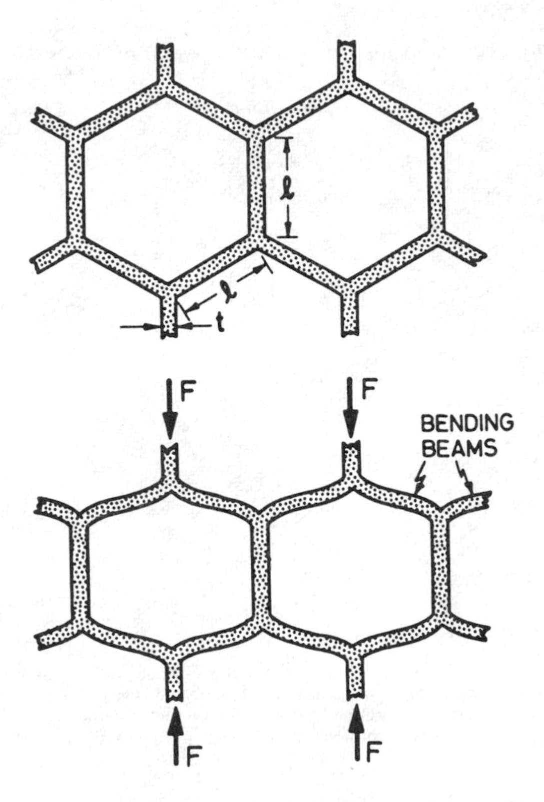

Figure 2 *Linear elastic deformation of a foam (from reference 3)*
(Reproduced by permission of ASM International)

where E_w is the Young's modulus of the foam wall, F is the force applied to the foam, and I is the second moment of the area, which equals $t^4/12$ for an open-cell strut and $t^3l/12$ for a closed-cell side.

The foam modulus $E = s/\varepsilon$, where s is the stress on the foam and ε is the strain; $s = F/l^2$ and $\varepsilon = \delta/l$. Hence $E = E_w t^4/l^4$ for an open-cell foam and $E = E_w t^3/l^3$ for a closed-cell foam. In terms of the relative density, $E/E_w \propto (\rho/\rho_w)^2$ for open cell foams and $E/E_w \propto (\rho/\rho_w)^3$ for closed cell foams.

Non-linear elastic behaviour is considered by describing the cell walls as Euler struts. The same powers of the relative density are predicted, although the foam mechanical property is now the buckling stress (Table 1).

Ashby treated the plastic yielding of foams in terms of the creation of plastic hinges at the cell-wall intersections. A power law is again predicted for the plastic yield stress of the foam (Table 1).

The compressive strength of brittle foams is derived from the moment on a cell wall (Figure 3). A wall of an open cell fails when the moment, M, acting on it is $M = \sigma_{fw}t^3/6$, where σ_{fw} is the modulus of rupture which is identified with the flexural strength. The moment M is equated to Fl, since the force F acts normal to

Table 1 *Ashby's equations for stiffness and strength of foams*

Deformation	*Open-cell foams*[a]	*Closed-cell foams*[a]
Linear elasticity	$\frac{E}{E_w} \propto \left(\frac{\rho}{\rho_w}\right)^2$	$\frac{E}{E_w} \propto \left(\frac{\rho}{\rho_w}\right)^3$
Elastic collapse	$\frac{\sigma_e}{E_w} \propto \left(\frac{\rho}{\rho_w}\right)^2$	$\frac{\sigma_e}{E_w} \propto \left(\frac{\rho}{\rho_w}\right)^3$
Plastic collapse	$\frac{\sigma_p}{\sigma_{pw}} \propto \left(\frac{\rho}{\rho_w}\right)^{3/2}$	$\frac{\sigma}{\sigma_w} \propto \left(\frac{\rho}{\rho_w}\right)^2$
Brittle crushing	$\frac{\sigma_f}{\sigma_{fw}} \propto \left(\frac{\rho}{\rho_w}\right)^{3/2}$	$\frac{\sigma_f}{\sigma_{fw}} \propto \left(\frac{\rho}{\rho_w}\right)^2$
Fracture toughness	$\frac{K_{1c}}{\sigma_{fw}} \propto \left(\frac{\rho}{\rho_w}\right)^{3/2} \sqrt{l}$	$\frac{K_{1c}}{\sigma_{fw}} \propto \left(\frac{\rho}{\rho_w}\right)^2 \sqrt{l}$

[a] E = Young's modulus of foam; E_w = Young's modulus of foam wall material; σ_e = elastic collapse stress of foam; σ_p = plastic collapse stress of foam; σ_{pw} = yield strength of foam wall material; σ_f = crushing strength of foam; σ_{fw} = modulus of rupture of foam wall material; K_{1c} = fracture toughness of foam; l = foam cell size; ρ = density of foam; ρ_w = density of foam wall material.

the wall of length l. The foam strength, σ_f, is given by

$$\sigma_f = \frac{F}{l^2} \propto \frac{M}{l^3} \propto \frac{\sigma_{fw} t^3}{l^3}$$

For open-cell foams,

$$\frac{\sigma_f}{\sigma_{fw}} \propto \left(\frac{\rho}{\rho_w}\right)^{1.5}$$

For closed-cell foams, a similar analysis yields

$$\frac{\sigma_f}{\sigma_{fw}} \propto \left(\frac{\rho}{\rho_w}\right)^2$$

This analysis was extended to describe crack propagation. For a foam, the plane strain toughness, K_{1c}, is given by

$$\frac{K_{1c}}{\sigma_{fw}} \propto \sqrt{\pi l} \left(\frac{\rho}{\rho_w}\right)^{3/2}$$

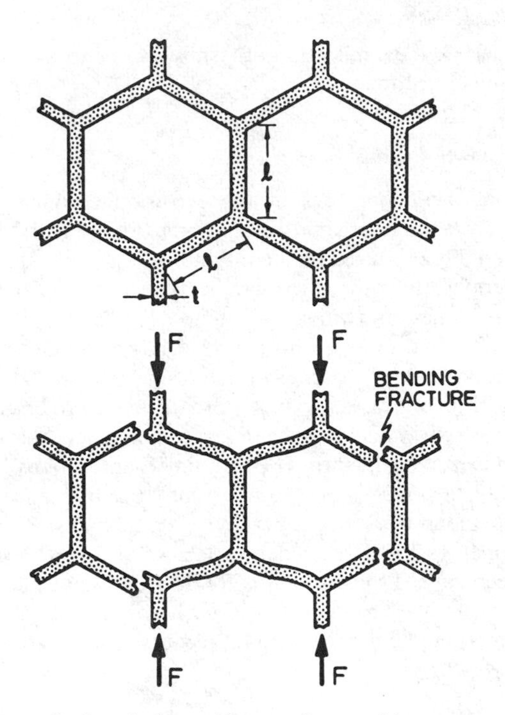

Figure 3 *Crushing of a brittle foam (from reference 3)*
(Reproduced by permission of ASM International)

for an open-cell foam and

$$\frac{K_{1c}}{\sigma_{fw}} \propto \sqrt{\pi l} \left(\frac{\rho}{\rho_w}\right)^2$$

for a closed-cell foam, where σ_{fw} is the modulus of rupture of the cell wall.

In all cases, the mechanical property, σ, of the foam normalized by the wall value scales with a power of the relative density (Table 1). In general,

$$\frac{\sigma}{\sigma_w} \propto \left(\frac{\rho}{\rho_w}\right)^n \tag{1}$$

where σ is a general mechanical property and σ_w is the appropriate wall mechanical property.

Although this treatment has a number of simplifying assumptions, it provides an elegant, unifying approach to the mechanical properties of cellular solids. Ashby provided evidence that the response of foamed ceramics, metals and polymers could be assembled in terms of equation (1). At the time of his review, no evidence was given of the applicability of these equations to food systems. Solid food foams are ubiquitous, however, and the Ashby treatment offers a framework from which to investigate their mechanical properties. The remainder of this chapter will illustrate the application of Ashby's work to the mechanical properties of extrusion cooked foams.

2 Extrusion Cooked Foams

Process—One of the major uses of the extrusion cooking process is in the production of snack foods, crispbreads and breakfast cereals.[4] Low-density solid foams may be readily produced on a continuous basis from starch-based ingredients. In these applications the raw ingredients in the form of solid powder are conveyed by a single- or twin-screw pump. The material passes through a temperature, shear, and pressure history as it moves along the extruder barrel. The solid material becomes a viscous dough or melt, which is then formed by the dies (Figure 4). In many cases expanded products are produced by allowing the included superheated water to escape as steam when the pressure falls to atmospheric on exit from the dies. The resulting foams differ widely in expansion and bulk density, depending on the extrusion variables and the raw material composition. Previous work has generally related expansion, bulk density, and mechanical properties to the extruder variables.[5–8] Most mechanical property studies have been carried out using texture devices[6,7] and the data presented in technique-specific units. The Ashby approach suggests a relationship between mechanical properties and density without recourse to the extrusion conditions in both cases.

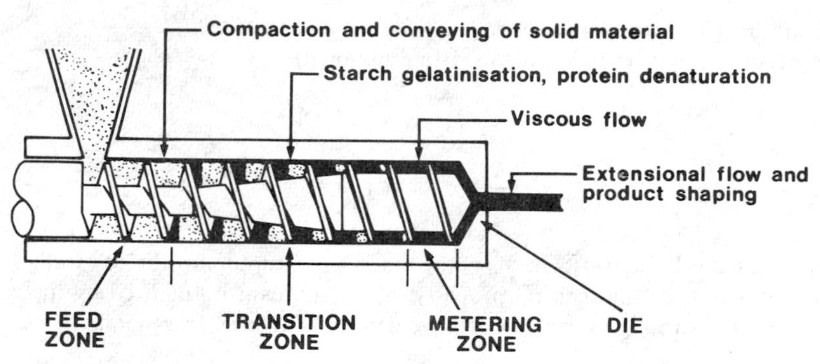

Figure 4 *Schematic extrusion cooking process*

In this work the foams were produced using a Baker Perkins MPF 50D twin-screw, co-rotating extruder with circular dies. A rectangular section die was also used to make some bar-shaped samples. The materials used were maize (Smith Flour Mills, Worksop; YG600 grits) and wheat starch (Tenstar Products, Ashford, Kent; Trustar brand). A range of extruder screw speeds and feed moistures was used to produce foams of different density, although variations in die size and extrusion temperature were used to produce some samples.

Structure—An immediate deviation from the Ashby model is apparent from a consideration of the extruded foam structures in Figure 5. The foam cells are not uniform in size. The width of the pore size distribution varies with extrusion conditions, although the least porous foams tend to have the widest distributions owing to the greater contribution from the surface-affected pores.

Density—The bulk density of the foams may be measured by sand displacement or from their dimensions in the case of simple geometries. The wall density may be measured by pycnometry using a suitable solvent.

Mechanical Properties—The only property in equation (1) that defies straightforward measurement is the wall mechanical property, σ_w. In the case of foamed metals and polymers, σ_w was identified with the bulk value.[3] The analogous use of the mechanical property of an unfoamed food extrudate would be unrepresentative of the foam wall. The different processing conditions in the two cases could lead to marked changes in the material formed. In addition, the use of the bulk value is open to question owing to orientation effects in the formation of the foam wall. The mechanical properties of synthetic polymers are known to vary significantly with orientation.[9]

The mechanical properties of foamed plastics had earlier been related to their bulk density using a power law:[10]

$$\sigma \propto \rho^n \tag{2}$$

This may be seen as an approximation to equation (1) for the case of constant wall properties. In the absence of the foam wall mechanical properties, equation (2) was used to assess the experimental data. Rather than using texture-measuring cells to measure mechanical properties, compression, flexure, and tension tests were carried out according to British Standards for plastics components.[11] An indentation technique was also employed together with high strain rate impact of specimens.

3 Results

The modulus and strength of cylindrical foamed maize extrudates were obtained in compression, tension, and flexure using an Instron universal testing machine.[12–14] The details of the sample dimensions and the equations used are given in Table 2. Representative logarithmic plots of mechanical property against density are given in Figures 6–8. The values of the power law index *n* are recorded in Table 3.

The data described above were obtained at a low strain rate ($5 \times 10^{-3}\ s^{-1}$). Impact tests allow much higher strain rates to be used. An instrumented pendulum has been developed at the Institute of Food Research, Norwich Laboratory, to test extruded products at strain rates of the order of 10–100 s^{-1}, depending on impact energy and sample size.[15] Cylindrical foam specimens were compressed radially by a cylindrical hammer striking them orthogonally. The strength may be calculated from the impact energy and the deformation (Table 2) and was observed to vary with the bulk density according to equation (2) (Figure 9).[16] The value of *n* is recorded in Table 3.

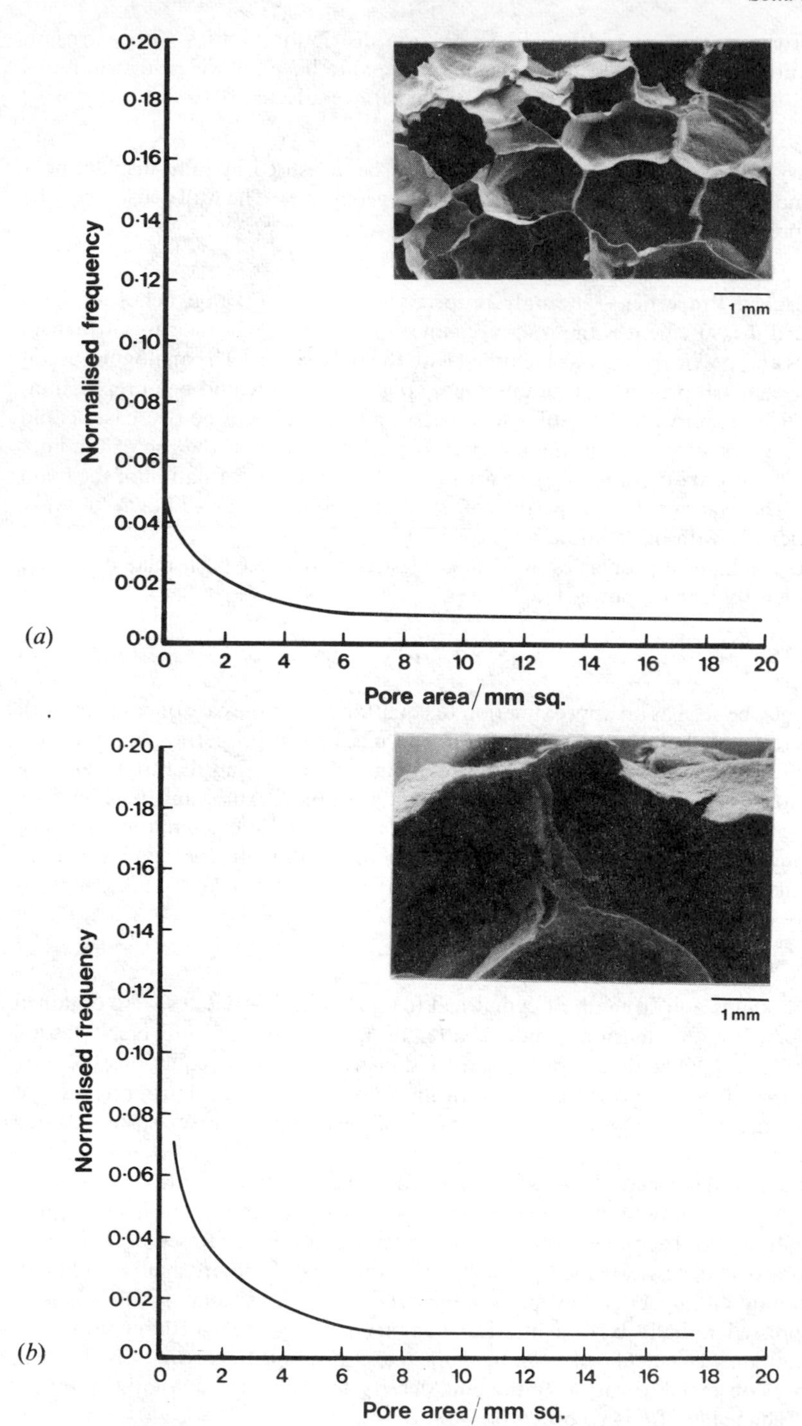
0·20
0·18
0·16
0·14
0·12
0·10
0·08
0·06
0·04
0·02
0·0
Normalised frequency
1 mm
(a)
0 2 4 6 8 10 12 14 16 18 20
Pore area/mm sq.
0·20
0·18
0·16
0·14
0·12
0·10
0·08
0·06
0·04
0·02
0·0
Normalised frequency
1 mm
(b)
0 2 4 6 8 10 12 14 16 18 20
Pore area/mm sq.

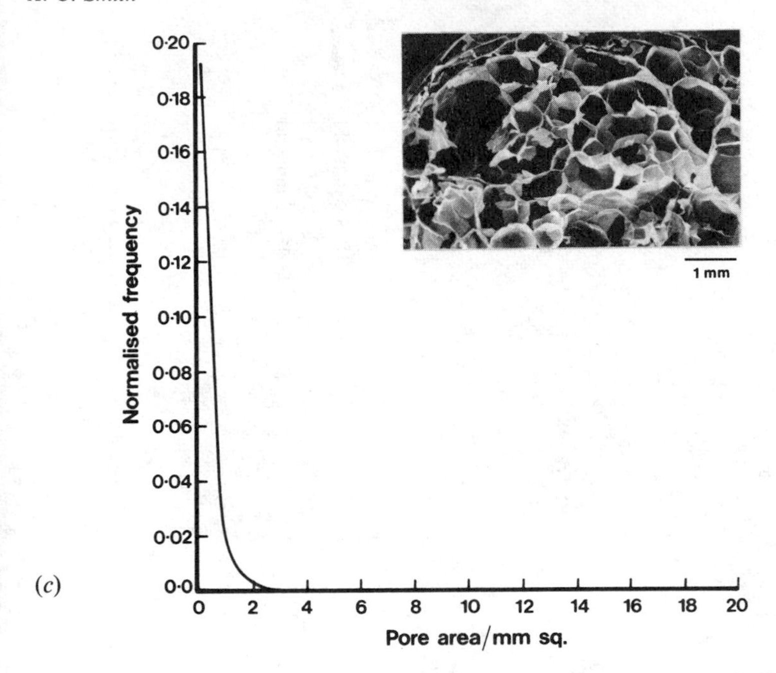

Figure 5 *Pore size distribution for extrusion cooked maize foams. Bulk density:* (a) 210, (b) 60 *and* (c) 40 kg m^{-3}

An intermediate strain rate (10^{-2}–10^{-1} s^{-1}) is obtained by a pin indentation technique based on an Instron testing machine.[12–14] A 0.58 mm diameter pin was driven axially and radially into cylindrical maize extrudates (Figure 10a). Ideally, the force-deformation response takes the form shown in Figure 10b. The modulus and strength may be calculated from the force–deformation response corresponding to the penetration of each pore wall (Table 2).[13,14] These local foam mechanical properties also obey equation (2) (Figure 11) and the values of n are given in Table 3.

The pin indentation approach also gives one-dimensional structural information on the foams. The force peaks (Figure 10) correspond to the breakage of successive foam walls. In some cases the pin breaks the foam walls cleanly, whereas in others the collapse of adjacent walls causes the accumulation of debris. In the latter case the force peaks are superimposed on a ramped base line (Figure 12). Withdrawal of the pin isolates the base line, which may be subtracted to yield the true force–displacement response. The separation of the force peaks was compared with the distance between foam walls from scanning electron micrographs. Although there is some overcounting of force peaks in the absence of any signal filtering, the technique may be used to obtain both structural information and mechanical properties.[13]

The fracture mechanics parameter, G_c, may be obtained from a Charpy test,[17] which involves the measurement of the energy to break specimens notched to different extents. This technique is restrictive as it specifies rectangular bar

Table 2 *Equations and conditions for mechanical testing of cylindrical samples*

Test	*Strength*[a]	*Modulus*[a]	*Strain rate*[a]	
Tension	$\frac{4F}{\pi d^2}$	$\frac{\Delta F}{\Delta Y} \times \frac{4L}{\pi d^2}$	$\frac{V}{L} = 5.5 \times 10^{-3}\,s^{-1}$	$L = 100$ mm Gauge length = 50 mm $V = 5$ mm min^{-1}
Flexure	$\frac{8FL}{\pi d^3}$	$\frac{\Delta F}{\Delta Y} \times \frac{4L^3}{3\pi d^4}$	$\frac{V}{L} = 5.5 \times 10^{-3}\,s^{-1}$	$L = 150$ mm Span = 100 mm $V = 5$ mm min^{-1}
Axial compression	$\frac{4F}{\pi d^2}$	$\frac{\Delta F}{\Delta Y} \times \frac{4L}{\pi d^2}$	$\frac{V}{L} = 5.5 \times 10^{-3}\,s^{-1}$	$L = 15$ mm $V = 5$ mm min^{-1}
Radial compression	$\frac{6F}{\pi dl}$	$\frac{\Delta F}{\Delta Y} \times \frac{6}{\pi l}$	$\frac{V}{D} = 2.5\text{–}5.0 \times 10^{-3}\,s^{-1}$	$L = 57$ mm $V = 5$ mm min^{-1}
Pin indentation	$\frac{4F}{\pi D^2}$[b]	$\frac{\Delta F}{\Delta Y} \times \frac{4L}{\pi d^2}$	$\frac{V}{L} = 10\text{–}92 \times 10^{-3}\,s^{-1}$	$V = 2$ mm min^{-1}
Impact compression	$\frac{E}{xA}$	—	$\frac{x}{dt} = 20\text{–}200\,s^{-1}$	$L = 80$ mm

[a] V = Crosshead speed; E = impact energy; F = force to break; A = hammer projected area; t = penetration time; l = average pore diameter; d = specimen diameter; x = penetration; D = pin diameter; L = specimen length; Y = crosshead movement.
[b] Alternative equations depending on deformation mode are given in reference 14.

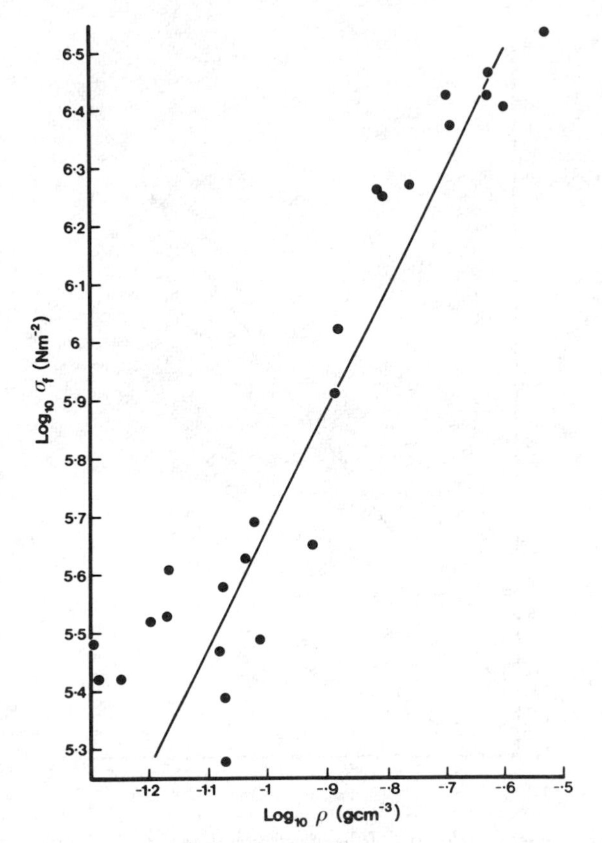

Figure 6 *Flexural strength, σ_F, of maize extrudates as a function of the bulk density, ρ*

samples. Some maize foam specimens were cut from a rectangular section extrudate prior to final cooling and values of G_c were estimated. The method was also modified to allow the testing of cylindrical foams.[18] The value of G_c was obtained from a plot of energy to break, U, against the expression $BD\varphi$, where B and D are the test piece cross-section dimensions and φ is a calibration factor which depends on the notch depth and the span:[19]

$$U = G_c BD\varphi$$

The values of G_c obtained for different density foams are given in Table 4 and compared with other data for synthetic polymers.

Some extruded foams were also equilibrated in a humid environment for different periods prior to testing.[20] Flexural tests using an Instron testing machine were carried out to investigate the dependence of the modulus and strength on density when the density changes were caused by dehydration and hydration.

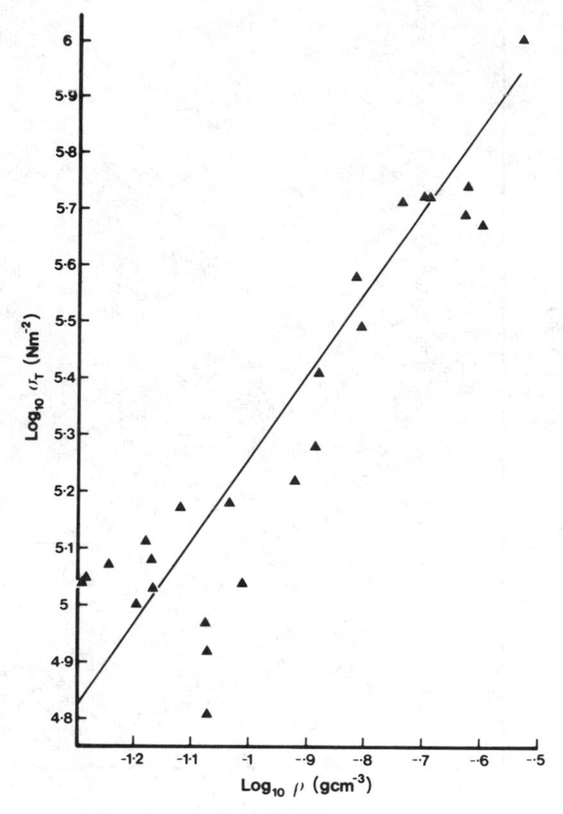

Figure 7 *Tensile strength, σ_T, of maize extrudates as a function of the bulk density, ρ*

Figure 13 compares the effect of density changes due to moisture uptake and extrusion conditions on the modulus.

4 Discussion

The data obtained for a number of test geometries and strain rates conform to the approximate relationship of equation (2). The index for elastic deformation is higher than that for failure, which is consistent with the predictions of the model.[3] The tensile and compressive strengths have been shown to be of a similar magnitude, as predicted by the model, in contrast to solid specimens, for which the compressive strength may be ten times greater than that in tension.[3] The results tend to be closer to the open-cell prediction despite the closed-cell structure of the extrudates. This is in keeping with Ashby's argument that closed-cell foams behave as open-cell foams owing to concentration of material in the edges. For a wider range of foams extruded from maize and wheat starch, the value of n in

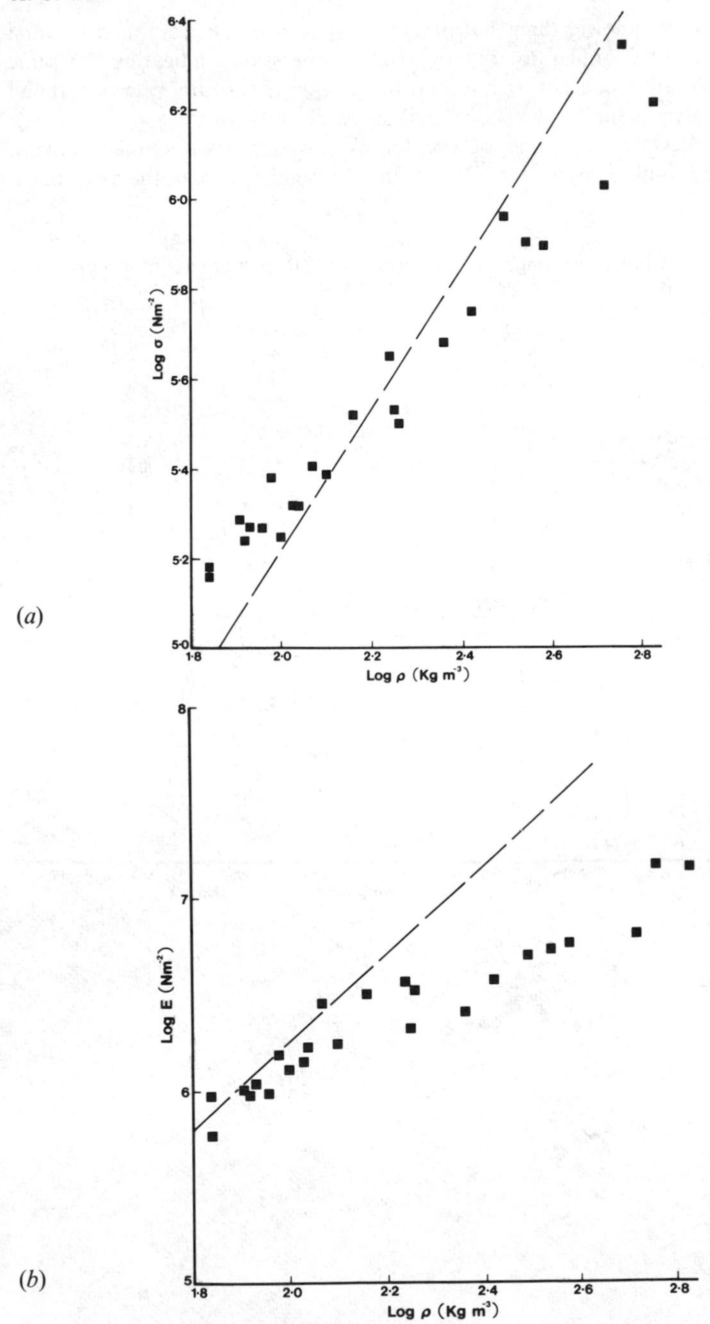

Figure 8 *(a) Radial and axial compressive strength, σ, and (b) modulus,* E, *as a function of the bulk density,* ρ. ———, Axial; ■, radial.

axial compression is lower than that predicted by Ashby. The response in radial compression is very similar to that in axial compression, indicating the same deformation in the bulk of the foam. Increasing the strain rate for radial compression gives a similar dependence of strength on density.

The local mechanical testing of the foams shows a greater anisotropy in properties than bulk compression. Within the centre of the foam the mechanical

Table 3 *Observed power law indices* [n *in equation* (2)] *for maize and wheat starch foams*

Test	*Modulus*	*Strength*
Tension[a]	1.7	1.5
Flexure[a]	2.0	2.0
Axial compression[a]	2.3	1.6
Axial compression	1.4	1.1
Radial compression[a]	1.1	1.1
Radial impact compression	—	1.1
Axial pin indentation	2.9	2.3
Radial pin indentation[a]	—	0.7

[a] Maize foams only.

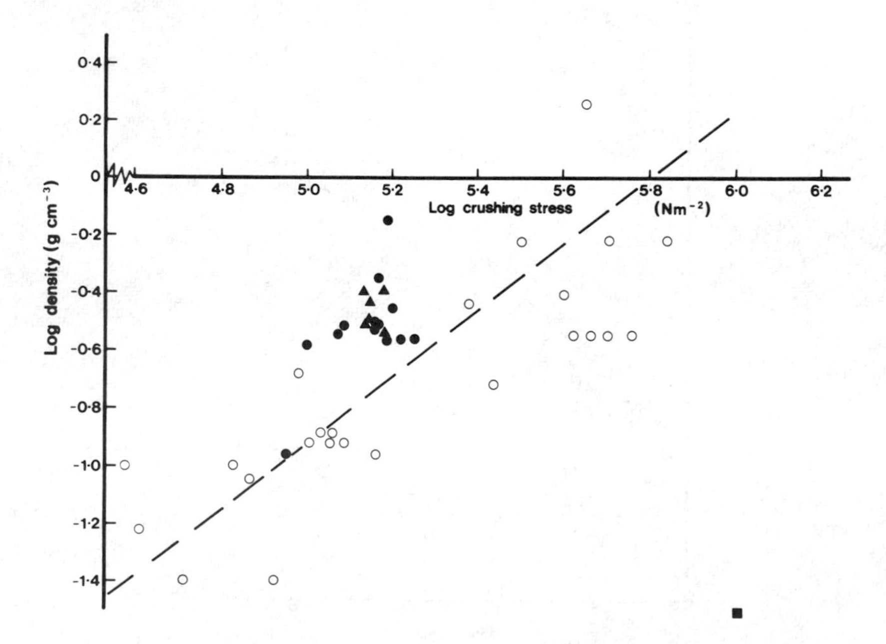

Figure 9 *Compressive impact strength of extrusion cooked foams.* ▲, *Maize starch;* ○, *maize;* ●, *wheat starch;* ■, *foamed polystyrene*

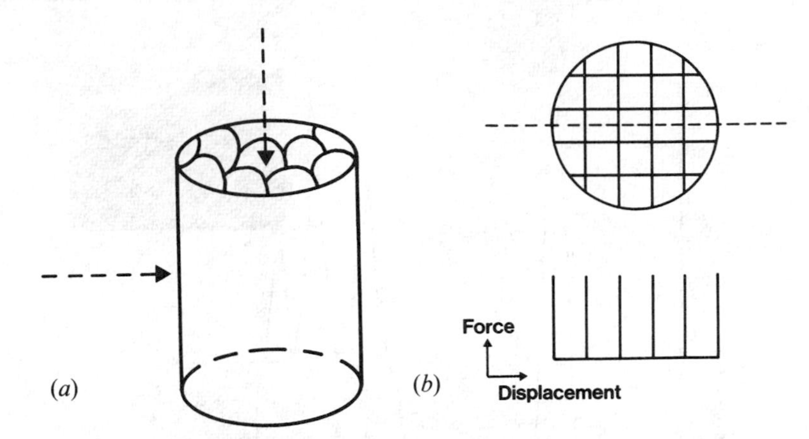

Figure 10 *(a) Radial and axial pin indentation of cylindrical foams. (b) Idealized force–displacement response for penetration of a foam*

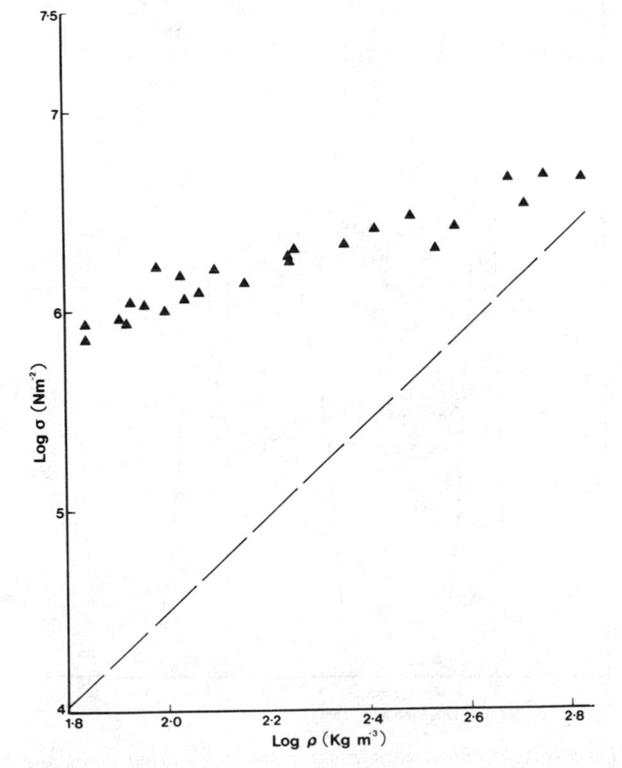

Figure 11 *Strength, σ, of extrusion cooked maize foams obtained by pin indentation of the first pore wall as a function of the bulk density, ρ.*——, Axial; ▲, radial

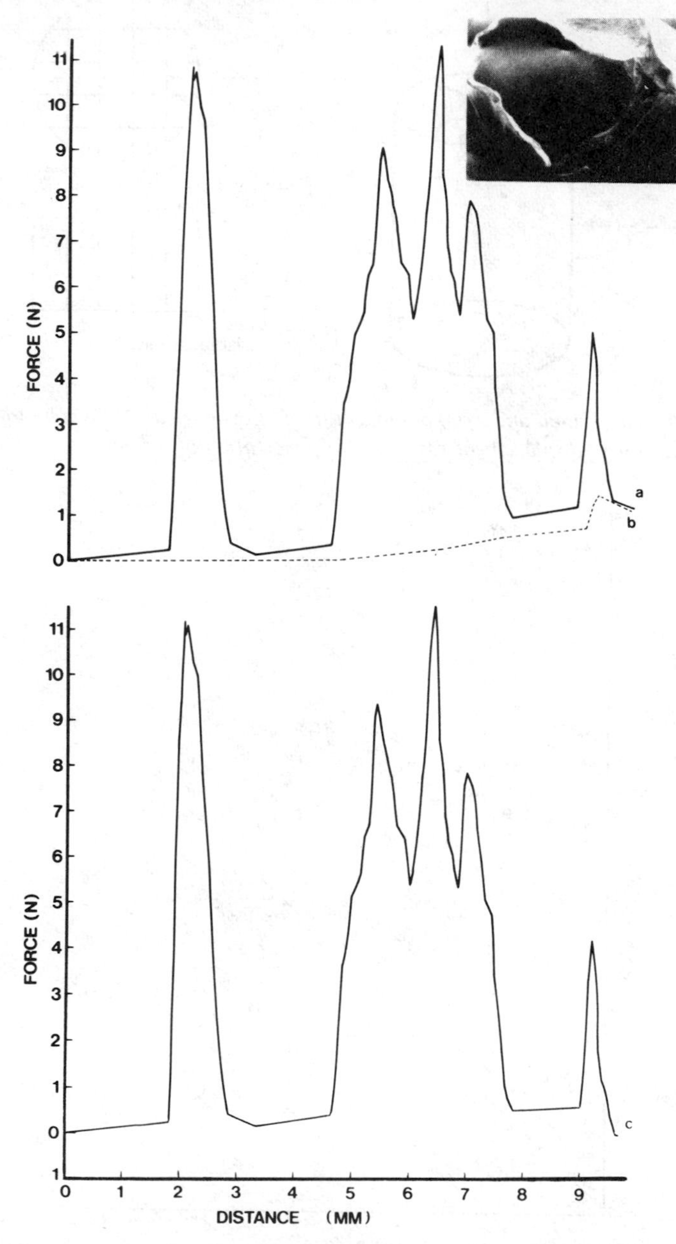

Figure 12 *Force–displacement response for pin indentation along the axis of an extrusion cooked maize foam. (a) Pin driven into specimen; (b) pin withdrawn from specimen; (c) difference of (a) and (b) to give corrected force–displacement curve.*

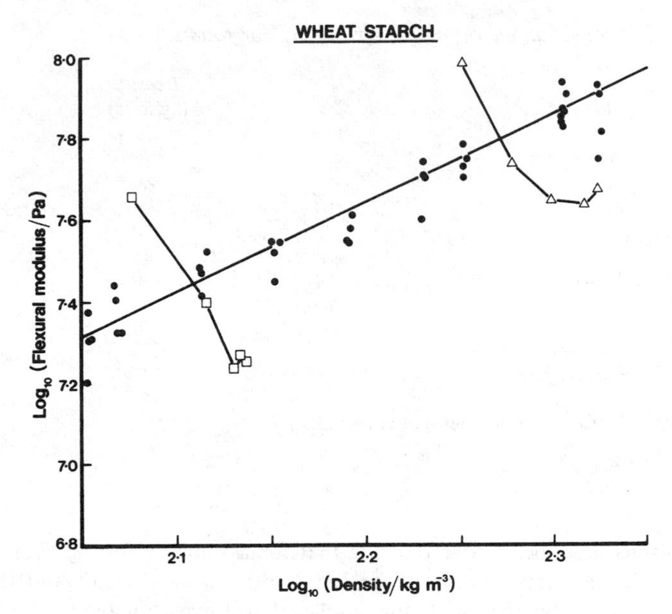

Figure 13 *Flexural modulus of extrusion cooked wheat starch foams as a function of the bulk density.* ●, *Different extrusion conditions;* △,□, *effect of hydration.*

properties increase more rapidly with density, close to Ashby's prediction for closed cells. In contrast, the strength of the surface is almost independent of bulk density (Figure 11). The surface is up to 100 times stronger than the interior of the foam. The surface is composed of a continuous 'shell' of foam walls and their mechanical properties are influenced far less by the surrounding pores than those in the bulk. The local surface strength and stiffness would be expected to represent better the foam wall properties. The actual interpretation of the indentation force depends on the mode of deformation of the pore wall. It may be converted to a stress through considering a 'punch' action of an element of wall or the three-point bend of the wall which is pinned by the adjacent foam walls.[14]

A combination of the expressions for the stress intensity factor, K_{1c}, and the elastic modulus, E, for a foam (Table 1) gives an expression for the strain energy release rate, G_c:

$$G_c \propto \frac{(K_{1c})^2}{E} \propto \frac{\sigma_w{}^2}{E_w} \cdot l \cdot \frac{\rho}{\rho_w}$$

The analogue of equation (2) for constant wall properties is $G_c \propto \rho l$, where l is the pore size. In the absence of a unique pore size, G_c has been found to increase in proportion to the foam density (Table 4). Data for polyurethane foams show that the value of G_c increases with increasing density, although a minimum was found at a bulk density of $100\ \mathrm{kg\,m^{-3}}$.[21]

Table 4 *Critical strain energy release rate,* G_c*, for foams*

G_c ($J\,m^{-2}$)	*Material*	*Density* ($kg\,m^{-3}$)	*Reference*
71	Wheat starch	70	18
206	Maize	280	
335	Maize	640	
347	Maize	660	
359	Maize	720	
392	Maize	790	
88	Polystyrene	30	
295	Polyurethane	37	21
540	Polyurethane	420	
1040	Unfoamed poly(methylmethacrylate)	—	17

The results described above indicate that for a range of testing geometries a simple power law exists between the foam stiffness or strength and the bulk density. In terms of equation (1) this implies that it is reasonable to assume that the ratio $\sigma_w/\rho_w{}^n$ is constant. It is interesting, however, to examine foams for which the wall properties are deliberately changed, in order to check the predictions of equation (1). One particular example is the dehydration and hydration of the foams. Equation (1) may be rewritten as

$$\frac{\sigma}{\sigma_w} \propto \left(\frac{V_w}{V}\right)^n$$

where V is the foam volume and V_w is the volume of the foam wall material.

Hydration will have two principal effects. First, it will cause swelling of the foam wall material. The increase in V_w will be greater than that of V, causing the foam mechanical property to increase. Secondly, it will affect the wall mechanical property, σ_w. The variaton of σ_w with water content may be inferred from the response of unfoamed food materials. For example, measurements of the Young's modulus of wheat pasta have shown a rapid decrease as the water content is increased.[22] This phenomenon may be interpreted by reference to the plasticization of synthetic polymers by solvents, which is accompanied by a marked decrease in mechanical properties.[23] The application of the same concepts to biopolymers[24] indicates the importance of molecular mobility in determining their mechanical properties. The mechanical properties of solid foams would therefore be expected to reflect the foam wall properties when the density is increased by hydration. Data for wheat starch foams show this type of response. If the foam density is increased by changing the extrusion variables the strength and stiffness increase, whereas the same increase in density by hydration causes a decrease in these mechanical properties.

5 Conclusion

In many cases of testing geometry and strain rate, the mechanical properties of extruded foams obey a simple power-law relationship with the bulk density. This is analogous to the description of synthetic polymer foams. Data obtained for different extrusion conditions for starch-based extrudates may be reduced to a single plot, which obviates the need to link mechanical properties to the process variables.

The power-law relationship represents a special case of the Ashby treatment of foam deformation for constant wall mechanical properties and density. Although the extruded foams are non-uniform, the main features of the model are observed. The Ashby treatment indicates the importance of the wall properties in determining the performance of the foams. A particular example is post-extrusion moisture conditioning, which may have as large an effect on the product as varying the extrusion conditions themselves.

Acknowledgement. I thank Professor M. F. Ashby, FRS, for permission to reproduce Figures 2 and 3 from his work.

References

1. L. J. Gibson, M. F. Ashby, G. S. Schajer, and C. I. Robertson, *Proc. R. Soc. London, Ser. A*, 1982, **382**, 25.
2. L. J. Gibson and M. F. Ashby, *Proc. R. Soc. London, Ser. A*, 1982, **382**, 43.
3. M. F. Ashby, *Met. Trans. A*, 1983, **14**, 1755.
4. A. Senouci, A. C. Smith, and P. Richmond, *Chemical Eng. (London)*, 1985, **417**, 30.
5. D. V. Harmann and J. M. Harper, *Trans. ASAE*, 1973, **16**, 1175.
6. J. M. Faubion and R. C. Hoseney, *Cereal Chem.*, 1982, **59**, 529.
7. J. Owusu-Ansah, F. R. de Voort, and D. W. Stanley, *Can. Inst. Food Sci. Technol. J.*, 1984, **17**, 65.
8. B. Launay and J. M. Lisch, *J. Food Eng.*, 1983, **2**, 259.
9. I. M. Ward, 'The Mechanical Properties of Solid Polymers,' Wiley, New York, 1982, p. 273.
10. E. Baer, 'Engineering Design for Plastics,' Chapman and Hall, New York, 1964.
11. B. S. 4370, 'Methods of Test for Rigid Cellular Materials,' Part I, 1968; Part II, 1973, British Standards Institution, London.
12. R. J. Hutchinson, G. D. E. Siodlak, and A. C. Smith, *J. Mater. Sci.*, 1987, **22**, 3956.
13. A.-L. Hayter and A. C. Smith, *J. Mater. Sci.*, 1988, **23**, 736.
14. R. J. Hutchinson, I. Simms, and A. C. Smith, *J. Mater. Sci. Lett.*, 1988, **7**, 666.
15. A.-L. Hayter, E. H. A. Prescott, and A. C. Smith. *Polym. Testing*, 1987, **7**, 27.
16. A.-L. Hayter, A. C. Smith, and P. Richmond, *J. Mater. Sci.*, 1986, **21**, 3729.
17. G. P. Marshall, J. G. Williams, and C. E. Turner, *J. Mater. Sci.*, 1973, **8**, 949.
18. A. R. Kirby and A. C. Smith, *J. Mater. Sci.*, 1988, **23**, 2251.
19. E. Plati and J. G. Williams. *Polym. Eng. Sci.*, 1975, **15**, 470.
20. R. J. Hutchinson, S. Mantle, and A. C. Smith, *J. Mater Sci.*, 1989, In press.
21. A. McIntyre and G. E. Anderton, *Eur. J. Cell Plast.*, 1978, July, 153.
22. J. Andrieu and A. Stamatopoulos, *Lebensm.-Wiss. Technol.*, 1986, **19**, 448.
23. P. B. Bowden, in 'The Physics of Glassy Polymers,' ed. R. N. Haward, Applied Science, Barking, 1973, Ch. 5.
24. K. J. Zeleznak and R. C. Hoseney, *Cereal Chem.*, 1987, **63**, 121.

Competitive Adsorption Between Proteins and Small-molecule Surfactants in Food Emulsions

By Eric Dickinson and Christine M. Woskett

PROCTER DEPARTMENT OF FOOD SCIENCE, UNIVERSITY OF LEEDS, LEEDS LS2 9JT, UK

1 Introduction

Food colloids contain two main classes of surface-active materials: proteinaceous emulsifiers (especially milk or egg proteins) and low molecular weight emulsifiers (lecithin, Spans, Tweens, etc.). How proteins and small-molecule surfactants are distributed between the droplet surface and the bulk oil or aqueous phases is an important factor affecting the stability and texture of edible oil-in-water emulsions such as salad dressing and ice-cream.[1] In order to optimize systematically the formulation of a food product with respect to ease of emulsification and shelf-life on storage, it is useful to have some understanding of whether or not the added ingredient is likely to be at the surface of the dispersed particles, as such knowledge will undoubtedly give an insight into the molecular origin of the structure and rheology of the food colloid. The distribution of proteins and low molecular weight emulsifiers in food emulsions is determined by the competitive adsorption between the two types of molecules at the oil–water interface, and by the nature of protein–surfactant interactions, both at the interface and in bulk aqueous solution.

A large amount of information is available in the literature on the subject of polymer–surfactant interactions and polymer–surfactant complexes.[2] The chief driving force for the association of surfactant molecules into micelles in water is the reduction in hydrocarbon–water contact area of the alkyl chains. This same driving force will favour association of surfactant molecules with the hydrophobic regions of dissolved polymer molecules, whilst at the same time the head groups of ionic surfactants may be involved in attractive interactions with oppositely charged groups along the polymer chain. Proteins contain a mixture of hydrophobic groups and electrically charged groups, and so it is not altogether surprising that many small amphiphilic molecules bind strongly to proteins. An early investigation by Goddard and Pethica[3] demonstrated that when a dissolved protein, such as bovine serum albumin, is on the acid side of its isoelectric point, the addition of an 'equivalent' amount of anionic surfactant leads to the formation of a stoicheiometric precipitate, with an equal number of positive charges from the protein and negative charges from the surfactant. It was found[3] that such a precipitate could be resolubilized by adding excess of surfactant, giving rise to the concept of a second layer of bound surfactant, with charged groups pointing outwards, attached to the primary bound layer through micelle-like hydrocarbon chain association.

This paper selectively reviews the underlying principles and experimental information relevant to an understanding of protein–surfactant interactions in food emulsions. There is a direct experimental link between studies of surfactant–protein binding and the competitive adsorption of proteins and small-molecule surfactants; many of the investigators, starting with the early work of Cockbain[4] on sodium dodecyl sulphate (SDS) + bovine serum albumin and Knox and Parshall[5] on SDS + gelatin, have used interfacial tension measurements (and much more commonly surface tension measurements) to determine the stoicheiometry of protein–surfactant complexes in bulk solution. Most of the experimental studies involve mixtures of a single protein with a single surfactant. The situation in food emulsions is, however, much more complicated than this, as the typical proteinaceous emulsifier consists of a mixture of several different protein components, each of which will be involved in competitive adsorption with the others at the oil–water interface.[6] In addition, the food emulsion will contain a mixture of small-molecule surfactants, both water- and oil-soluble, and these will compete with one another for binding sites on proteins and for adsorption sites at the interface.

We shall first outline the thermodynamic and statistical mechanical principles relevant to the competitive adsorption of small and large molecules at a fluid interface. Of particular practical importance here is the set of conditions under which small molecules may completely displace polymers from the interface. The remainder of the paper will be concerned with problems of trying to relate the simple statistical theory to the complexities of real protein + surfactant systems.

2 Theory of Competitive Adsorption of Polymers and Surfactants

The simultaneous advances in liquid state theory and computer simulation techniques during the past 15–20 years have led to a much increased understanding of the statistical mechanics of adsorption from condensed phases.[7] Realistic continuum models have been developed for systems containing simple spherical solute molecules, but for polymeric systems most current statistical treatments are based on lattice models in which each lattice site, in bulk solution or at the interface, is deemed to be occupied either by a solvent molecule or a segment of the polymer chain. These lattice-based theories are, for the most part, direct descendants of the statistical theory devised by Flory and Huggins to describe the thermodynamics of polymer solutions.[8] An essential feature of the Flory–Huggins theory is that it includes pair interactions only with the first coordination sphere of neighbours about a given solvent molecule or polymer segment; it is therefore inapplicable to systems in which long-range interactions (*e.g.* electrostatic) are important. In this way, the Flory–Huggins theory expresses the thermodynamics of a binary polymer + solvent system in terms of the relative strengths of solvent–solvent, solvent–polymer, and polymer–polymer interactions. In considering the thermodynamics of adsorption, one must also include the strengths of the solvent–surface and polymer–surface interactions. The problem of competitive adsorption between polymer molecules and small surface-active solute molecules requires the additional consideration of surfactant–solvent, surfactant–polymer, and surfactant–surface interactions. It is traditional to describe the strengths or relative strengths of these different types of interactions in terms of the Flory–Huggins parameter χ. So, χ_{po} is proportional to the change in energy when two polymer–solvent pair interactions are formed from solvent–solvent and polymer–polymer interactions, χ_{ps} is proportional to the change in energy when two polymer–surfactant pair interactions are formed from polymer–polymer and surfactant–surfactant interactions, and so on.

In polymer adsorption theory,[9] it is convenient also to define a Flory–Huggins exchange parameter χ_S corresponding to the free energy change associated with the formation of polymer–surface contacts at the expense of solvent–surface contacts. The use of a lattice model means that polymer segment attachment to the surface is accompanied by concomitant detachment of solvent molecules from sites at the interface; $-\chi_S kT$ is the free energy difference between segment–surface and solvent–surface contacts. Theory predicts[10] that long flexible chains do not adsorb unless χ_S exceeds some critical value χ_S^c, the magnitude of which is typically of the order of a fraction of kT. This is because the decrease in free energy arising from the formation of polymer–surface contacts must outweigh the loss in conformational entropy experienced by the polymer molecule upon adsorption.

The parameter χ_S in relation to χ_S^c tells us whether the polymer is on or off the surface. In a system containing adsorbed polymer, any change in χ_S which causes its value to approach or become below χ_S^c will cause polymer displacement from the interface. Such a change may be induced by addition of a low molecular weight surfactant to the system, as illustrated in Figure 1. At a critical composition of the mixed solvent, the excess adsorbed amount of polymer has been reduced to zero, and the polymer has been competitively displaced from the interface.

A detailed statistical treatment of polymer displacement by small-molecule surfactants has been set out[11] and tested[12] by Cohen Stuart and co-workers. By assuming that the interfacial region can be approximated by a single plane of lattice sites, the theory gives simple analytical expressions for the adsorption isotherm and the critical conditions for polymer displacement. To represent properly the train–loop–tail configurations of adsorbed polymer molecules, it is generally necessary to invoke a multilayer model,[10] *i.e.* a lattice consisting of many layers of sites parallel to the planar interface. However, close to the critical point χ_S^c, Cohen Stuart *et al.*[11] postulate that a one-layer approximation is valid, because the polymer surface excess is small for $\chi_S \approx \chi_S^c$, and so the polymer segment density can be equated to its bulk value in all except the first (or surface) layer. In this one-layer approximation, then, the excess adsorbed amount of polymer is taken to be a simple monotonic function of the polymer volume fraction, θ_p, in the surface layer.

In considering adsorption from a ternary system of polymer (p) + solvent (o) + surfactant (s), it is useful to define three binary adsorption energy exchange parameters: $\chi_{S,po}$ ($= -\chi_{S,op}$) for adsorption of polymer from solvent, $\chi_{S,ps}$ ($= -\chi_{S,sp}$) for adsorption of polymer from surfactant (in the absence of solvent), and

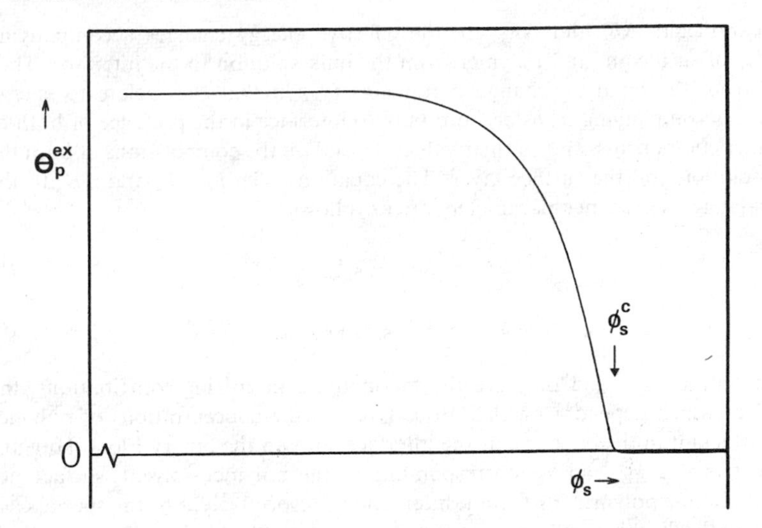

Figure 1 *Displacement of polymer from an interface by surfactant. The polymer surface excess concentration θ_p^{ex} is plotted against the bulk surfactant volume fraction φ_s. Displacement is complete at $\varphi_s = \varphi_s^c$*
(Redrawn from *J. Colloid Interface Sci.* 1984, **97**, 515)

$\chi_{S,so}$ ($= -\chi_{S,os}$) for adsorption of surfactant from solvent (in the absence of polymer). The three binary exchange parameters are related by

$$\chi_{S,po} + \chi_{S,os} + \chi_{S,sp} = 0 \tag{1}$$

Strong displacement of polymer by surfactant occurs if $\chi_{S,ps} \ll 1$. At equilibrium, a fraction θ_p of surface sites is occupied by polymer segments, a fraction θ_s by surfactant molecules (one per site), and a fraction $\theta_o = 1 - \theta_p - \theta_s$ by solvent molecules. Each polymer molecule consists of m identical segments, of which a fraction f are in trains, *i.e.* adsorbed in the surface layer. In bulk solution, the volume (site) fractions of polymer, surfactant, and solvent are defined as φ_p, φ_s, and $\varphi_o = 1 - \varphi_p - \varphi_s$, respectively.

In the one-layer approximation, the adsorption isotherm for a mixed solution of polymer + surfactant has a particularly simple form.[11,13] The total solute surface fraction is given by

$$\theta = \theta_p + \theta_s = [(1 - \theta)^{fm}\varphi_p K_p/(1 - \varphi)^{fm}] + [(1 - \theta)\varphi_s K_s/(1 - \varphi)] \tag{2}$$

where $\varphi = \varphi_p + \varphi_s$ is the total bulk volume fraction of solute, and K_s and K_p are factors which depend on the adsorption energies of the two solute components:

$$K_s = \exp(-\Delta E_s/kT) \tag{3}$$

$$K_p = \exp(-\Delta E_p/kT) \tag{4}$$

The parameters ΔE_s and ΔE_p are the effective energy changes accompanying transfer of surfactant and polymer from the bulk solution to the interface. They differ from the binary exchange parameters (χ_S) in that they relate to energy changes accompanying transfer from bulk to interface in the presence of both of the other components, and so their values depend on the compositions of both the bulk solution and the surface layer. The equations relating ΔE_s and ΔE_p to the Flory–Huggins exchange parameters are as follows:

$$\Delta E_s = \Delta h_{so} - kT\chi_{S,so} \tag{5}$$

$$\Delta E_p = fm(\Delta h_{po} + kT\chi_S^c - kT\chi_{S,po}) - kT\chi_S^c \tag{6}$$

The quantities Δh_{so} and Δh_{po} are the thermodynamic mixing contributions, the values of which depend[11] on the lattice type, on the concentrations of polymer and surfactant in the bulk and at the interface, and on the binary Flory–Huggins parameters χ_{po}, χ_{so}, and χ_{ps} corresponding to the polymer–solvent, surfactant–solvent and the polymer–surfactant interactions, respectively. For the special case of a completely athermal system ($\chi_{po} = \chi_{so} = \chi_{ps} = 0$), Δh_{so} and Δh_{po} are both identically zero; under these conditions, the quantities ΔE_s and ΔE_p are just constants, *i.e.* independent of composition.

Figure 2 shows a theoretical displacement curve calculated by de Feijter *et al.*[13] using equation (2) with $fm = 50$, $\varphi_p = 0.002$, $\Delta E_s = -12\ kT$ and $\Delta E_p = -60kT$ ($-1.2\ kT$ per adsorbed segment). Athermal solvent conditions are assumed for both surfactant and polymer ($\Delta h_{so} = \Delta h_{po} = 0$). The theory predicts that the total surface coverage θ remains essentially constant (*ca.* 0.63) until the polymer is in effect completely displaced from the interface ($\theta_p \rightarrow 0$). Despite the highly idealized assumptions made in this calculation, it is noteworthy that the general form of the displacement curve in Figure 2 is similar to that found for protein displacement from the interface in oil-in-water emulsions by anionic surfactants (see Figure 7).

The critical surfactant volume fraction φ_S^c at which the polymer surface excess concentration just vanishes is given by[11]

$$\varphi_s^c = \frac{\exp[\chi_{S,po} - \chi_s^c - (\Delta h_{po}/kT)] - 1}{\exp[\chi_{S,so} - (\Delta h_{so}/kT)] - 1} \tag{7}$$

Under athermal conditions, and for values of $\chi_{S,po}$ and $\chi_{S,so}$ which are not too small, equations (1) and (7) lead to the simple equation

$$\chi_{S,ps} = \ln\varphi_s^c + \chi_S^c \tag{8}$$

We see from equation (8) that polymer displacement will still occur even if the surfactant and the polymer segments interact equally strongly with the surface. The complete desorption of polymer at $\varphi_s^c = \exp(-\chi_S^c)$ for $\chi_{S,ps} = 0$ reflects the unfavourable effect of conformational change on the adsorption of the polymer chain.

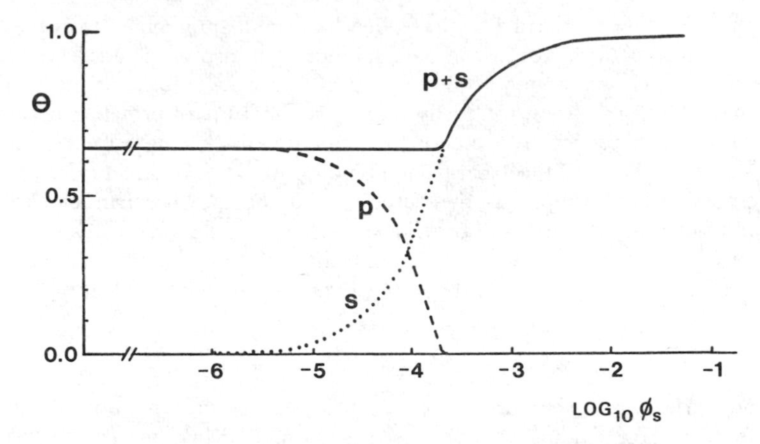

Figure 2 *Theoretical displacement curve from simple statistical theory. The total solute surface fraction* θ ($= \theta_p + \theta_s$) *is plotted against the logarithm of the bulk surfactant volume fraction* φ_s. *Contributions to* θ *from polymer* (p) *and surfactant* (s) *are also shown*
(Redrawn from *Colloids Surf.*, 1987, **27**, 243)

As a treatment of competitive adsorption in protein + surfactant systems, the theory outlined above is extremely idealized. It does not take account of either the detailed macromolecular complexities of protein structure or the micellar aggregation structures formed in solutions of small-molecule surfactants. In a more general sense, however, it can take reasonable account, through the various Flory–Huggins parameters, of the differing affinities of proteins and surfactants for the interface, and of effects of protein–surfactant interactions and solvent quality on the adsorption isotherms. In real systems, these latter effects depend considerably on the nature of the chemical interactions involved in the binding of proteins and surfactants to surfaces and to each other. Before discussing the nature of interactions between proteins and surfactants at an interface, it is first appropriate to consider the strength and nature of interactions which may occur between proteins and small-molecule surfactants in bulk aqueous solution.

3 Protein–Surfactant Interactions in Aqueous Solution

Small-molecule surfactants have a strong binding affinity, not only for fluid interfaces, but also for other dissolved species—small ions, polymers, and other surfactant molecules. The same combination of hydrophilic and hydrophobic segments which gives a protein its high surface activity also makes it an ideal substrate for binding of small amphiphiles, both ionic and nonionic. Surfactant–protein binding affects adsorption isotherms in two main ways: it changes the adsorption energy of the protein for the interface by affecting the net charge or the overall hydrophobicity, and it reduces the amount of free surfactant in the bulk phase which is available for displacing protein from the interface. The actual position is further complicated by the tendency of small amphiphiles to aggregate into micellar structures, not just in bulk solution, but also in the adsorbed state and in complexes with bound protein.

In the following sections, we briefly survey the literature of protein–surfactant interactions with particular reference to those proteins having particular food colloid interest. Some of the experimental binding studies give direct information on competitive adsorption, as the authors used the measurement of surface tension to quantify the extent of protein–surfactant binding in aqueous solution. It is convenient to divide the surfactants into three classes: anionic, cationic, and nonionic. The abbreviations used for the various surfactants are given in the Appendix.

Anionic Surfactants—The surfactant that has been studied in most detail in protein–surfactant binding experiments is sodium dodecyl sulphate (SDS), a reagent widely used by biochemists in the purification and characterization of proteins (*e.g.* in SDS polyacrylamide gel electrophoresis). SDS is a valuable general-purpose solubilizing agent in the characterization of a wide range of hydrophobic macromolecules, although one has to recognize that its presence may influence the analytical results. In molecular weight determination by light scattering, for example, measurements must be restricted to surfactant levels below the critical micelle concentration (cmc).[14] SDS protein binding has been exploited by Kato *et al.*[15] in the determination of protein hydrophobicity. At low surfactant concentrations (a few per cent. of cmc), the SDS binding capacity of proteins was found to be proportional to the surface hydrophobicity, as determined by a fluorescence probe method,[16] over a considerable range of pH, ionic strength, and temperature. The results indicate the importance of surfactant–protein hydrophobic interactions in the binding of SDS to proteins at low surfactant concentrations.

One of the most carefully studied systems has been SDS + lysozyme.[17–22] Figure 3 shows a typical binding isotherm of SDS–lysozyme as determined by Fukushima *et al.*[22] Two distinct binding regions can be identified: the low SDS concentration region, where the binding number, N_b, is less than 10 and the interactions are primarily electrostatic, and the high SDS concentration region ($N_b \geqslant 10$), where the driving force for binding is mainly hydrophobic. The plateau in the isotherm is considered to correspond to the effective saturation of all the high-affinity cationic binding sites (arginyl residues) on the positively charged protein molecule.[17] When the free surfactant concentration approaches the cmc, the binding isotherm shows a dramatic increase in N_b over a relatively narrow range of surfactant concentration. In this non-specific cooperative binding region, the driving force for complex formation is similar to that for the co-operative association of surfactant molecules into micelles in the absence of protein. Comparison of binding isotherms and precipitation curves shows[21] that the number of bound SDS molecules which gives maximum precipitation is consistent with the number of negative ions required to neutralize the net charge on the lysozyme molecule. The precipitation curve becomes narrower with increasing pH as less surfactant is required for charge neutralization. The electrostatic binding of

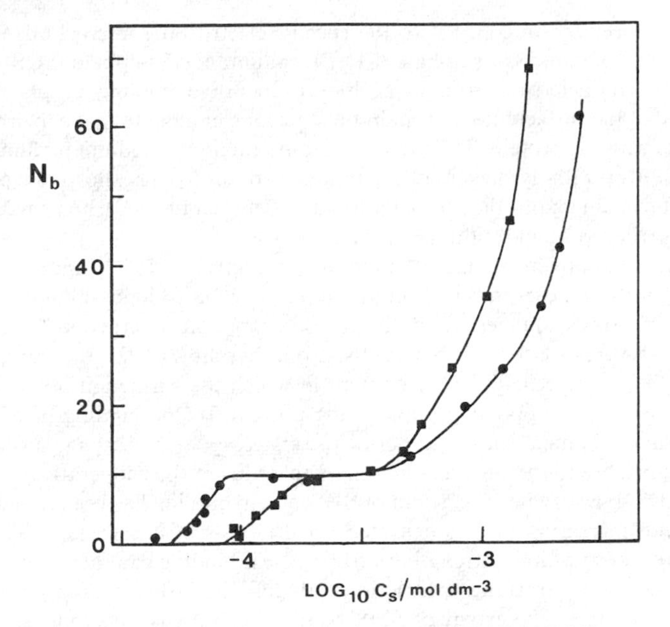

Figure 3 *Binding isotherm for SDS + lysozyme* (25 °C, *pH* 5.8) *at two electrolyte concentrations. The binding number* N_b *is plotted against the logarithm of the surfactant concentration* C_s : ●, 0.01 M *NaCl*; ■, 0.03 M *NaCl*
(Redrawn from *Bull. Chem. Soc. Jpn.*, 1982, **55**, 1376)

SDS to lysozyme increases the surface hydrophobicity of the protein, and this leads to the aggregation of protein–surfactant complexes. At pH 6 and an SDS concentration of 0.03M, an average aggregation number of 3.0 has been inferred.[17]

Evidence for the view that electrostatic interaction with ionic groups is largely responsible for the initial SDS–lysozyme binding region comes from measurements of binding isotherms and enthalpies of interaction of SDS with derivatives of lysozyme in which the cationic residues have been modified.[18] Calorimetric measurements by Jones and Manley[18] have also shown that, at high binding levels, SDS initiates a conformation change in lysozyme; this process is endothermic, in contrast to the initial exothermic interaction at low binding levels. Effects of pH and ionic strength have also been investigated,[19,22] Hydrogen-ion titration curves[19] show that proton binding to SDS–lysozyme complexes is closely linked to surfactant binding. The effect of sodium chloride on the binding isotherm[22] is to produce a narrowing of the plateau region at $N_b \approx 10$ as salt concentration increases. This is attributed[22] to displacement of chloride ions by dodecyl sulphate ions. Subsequent binding of ions to the electrically neutral complex is apparently restricted to hydrophobic regions, and so it has been suggested[20] that the cooperative region of the binding isotherm may involve interactions of surfactant micelles with the protein–surfactant complex. Indeed, under these conditions, the binding number of sodium ions per SDS molecule approaches that for an SDS micelle.[22] In contrast to the specific electrostatic interactions found with SDS and sodium decyl sulphate (SDeS), sodium octyl sulphate (SOS) gives an isotherm characteristic of non-specific co-operative binding at pH 3.2,[17] indicating that the surfactant alkyl chain must be long enough to make hydrophobic contacts with the protein. The even weaker interaction of sodium perfluorooctanoate (SPFO) with lysozyme is attributed[23] both to the shortness of the fluorocarbon chain and to the mutual antipathy between fluorocarbon surfactant and hydrocarbon protein residues.

SDS binding to bovine plasma albumin has been investigated by Inoue *et al.*[24] Like lysozyme, it appears that the initial binding of SDS to high-affinity sites on the protein involves both electrostatic and hydrophobic interactions. At high binding levels ($N_b > 80$), ^{13}C NMR spectroscopy shows that the surfactant molecules are in a micelle-like environment in which the surfactant alkyl chains are associated with non-polar groups of the protein. In the binding of SDS to bovine serum albumin,[25] the isotherm rises steeply at a certain surfactant concentration close to the cmc, in a manner analogous to the cooperative binding region of SDS + lysozyme. The extent of SDS–casein binding has been found[26] to decrease with increasing pH, as expected on the basis of electrostatic binding. However, the effect of ionic strength on SDS–casein binding cannot be explained on the basis of electrostatic interactions alone. Similarly, with gelatin, the observed variation in the extent of SDS binding[27] with pH and ionic strength indicates that the role of non-electrostatic binding forces is not the same under all conditions. The SDS–gelatin binding is greatest at 30 °C; it was suggested by Sen *et al.*[27] that a proportion of binding sites is hidden in the gel matrix at 15 °C, and that the weaker bonds between SDS and gelatin are broken at temperatures substantially above 30 °C. Chattoraj and co-workers[27,28] have also studied the binding of SDS to the binary protein mixtures, bovine serum albumin + casein and bovine serum albumin + gelatin, and to the ternary protein mixture, bovine serum albumin + casein + gelatin. Most of the binding of SDS to myelin basic protein is highly co-operative;[29] precipitation of protein–surfactant complexes is probably caused by cross-linking through hydrophobic interaction between surfactants electrostatically bound to different myelin molecules. Cooperative binding of deoxycholate (DOC) to myelin basic protein at and above the cmc has also been observed.[30]

The binding of anionic surfactants to proteins may produce extensive conformational change. With SDS–lysozyme complexes at low binding number, thermodynamic data[18] are consistent with electrostatic interaction without significant conformational change. In the cooperative binding region, however, it has been inferred from circular dichroism[21] that the amount of α-helical structure is increased at the expense of the β-structure, both eventually reaching constant values. The enhanced helical formation is attributable to the development around the protein of a more strongly hydrophobic environment.

The stepwise nature of the unfolding of bovine serum albumin on addition of SDS leads to two distinct breaks in the binding isotherm at $N_b \approx 40$ and $N_b \approx 120$.[31] Sigmoidal isotherms obtained by equilibrium dialysis[25] exhibit multi-step processes where apparent saturation may occur in two or more conformational states with massive co-operative binding close to the cmc. Recent small-angle neutron scattering experiments[32] on complexes of bovine serum albumin with lithium dodecyl sulphate (LDS) have suggested that, as LDS is continuously added, the protein molecules are gradually transformed into random-coil conformations with a string of constant-size LDS micelles randomly decorating the polypeptide backbone (a 'pearl necklace' structure). Results from stopped-flow kinetic experiments by Takeda[33] have been interpreted in terms of a helix-to-coil transition as SDS binds to bovine serum albumin, with the rate constant for conformational change depending strongly on SDS concentration. Studies on δ-chymotrypsin using a similar technique[34] indicated that a protein cannot adopt various different conformations as reaction intermediates, because the surfactant binds rapidly, causing the transition only to a particular conformation. Recent measurements[35] have shown that SDS and SDeS reduce the α-helical content of bovine serum albumin to a similar degree, although the corresponding increase in β-structure is less for SDeS. Circular dichroism studies of SDS binding on bovine plasma albumin by Inoue *et al.*[36] have revealed that initially exposed phenylalanine residues become shielded from the solvent through hydrophobic interaction with the alkyl chain of the surfactant. The unchanged helical content of the protein is explained[36] by considering that the basic amino acid residues have an enhanced ability to propagate an ordered structure in the presence of anionic surfactant, whereas the non-polar region of the protein changes its conformation by hydrophobic interaction with the alkyl chain of the surfactant. Other studies of the change in protein conformation on binding of anionic surfactant have involved the systems SDS + ovalbumin, sodium dodecylbenzene sulphonate (LAS) + ovalbumin, sodium trioxyethylene dodecyl sulphate (LES) + ovalbumin, sodium lauroylmethyl taurid (LMT) + ovalbumin,[37] DOC + tubulin,[38,39] SDS + porin (membrane protein),[40] sodium *N*-methyl-*N*-oleoyltaurine (SMOT) + gelatin,[41] SDS + soybean glycinin,[42] and SDS + whey proteins.[43]

Cationic Surfactants—Compared with anionic surfactants, there is much less information on the binding of cationic surfactants to proteins. Where data are available, however, the general trends are similar to those found with anionics: high-affinity electrostatic binding at low surfactant concentrations and non-specific cooperative binding at high concentrations. In detail, however, many of the data are difficult to interpret mechanistically. For instance, the binding of n-alkyltrimethylammonium bromides to bovine serum albumin,[25] casein,[26] and gelatin[27] as a function of pH and ionic strength does not show a regular behaviour that is consistent with a purely electrostatic interaction. Intrinsic viscosity measurements suggest[44] that the binding of cetylpyridinium bromide (CPB) or cetyltrimethylammonium bromide (CTAB) to transfusion gelatin* is mainly electrostatic at low surfactant concentrations but non-electrostatic at high concentrations. The increased binding of CPB or CTAB with increased pH is in agreement with deprotonation of side chains.[45] Surfactant cations are bound specifically to protein carboxylate residues, and the additional massive binding is explained[45] by the formation of micellar clusters on the transfusion gelatin via

* A form of gelatin developed as a blood substitute.

hydrophobic interactions with other surfactant molecules and with hydrophobic residues on the protein.

The binding of cationic surfactants to bovine serum albumin produces, as with anionic surfactants, a stepwise unfolding of the protein. Circular dichroism measurements indicate that the surfactant concentration required to disrupt the α-helical structure increases as the length of the surfactant alkyl chain decreases. Other effects of the hydrophobic tail length are, however, not so straightforward to interpret: at 15, 30 and 45 °C, the measured amount of myristyltrimethylammonium bromide (MTAB) bound to bovine serum albumin is considerably greater than that for either CTAB or dodecyltrimethylammonium bromide (DTAB), in contrast to results at 65 °C, where binding of CTAB is greatest;[46] the relative amounts bound to gelatin at 30 °C lie in the order CTAB > DTAB > MTAB;[27] with casein the corresponding order is MTAB > CTAB > DTAB.[26] The deviation from a systematic dependence on surfactant tail length could possibly be explained by irregular folding of bound surfactants.[26] The structure of the cationic head group also appears to be important. For example, greater binding of CPB than of CTAB to transfusion gelatin leads to more extensive conformational change in the protein.[44,45]

In the binding of cationic surfactants to casein, Sadhukhan and Chattoraj[26] have suggested that the steric hindrance of peptide groups by pyrrolidine rings is an important factor. The phenomenon is even more important for gelatin, where almost a third of the residues are proline. The steric effect, which may inhibit surfactant binding to neighbouring residues also, has been confirmed in a study of CTAB binding to polyvinylpyrrolidine in which no interaction at all was observed.[26] Binding of MTAB to casein containing fat is considerably higher than that for defatted casein. This may be explained in terms of lipid–protein interactions replacing protein–protein interactions, and thereby exposing more sites for surfactant binding,[26] or, alternatively, the surfactant may simply bind directly to the fat, perhaps displacing some protein in the process. The binding of CTAB to binary and ternary protein mixtures has also been studied by Chattoraj and co-workers.[27,28]

Nonionic Surfactants—The biochemical study of membrane proteins commonly involves the use of nonionic surfactants. In contrast to ionic surfactants, nonionic surfactants can be used to extract proteins from membrane without loss of biological activity or disruption of the native conformation.[40] The amount of binding of nonionic amphiphiles varies considerably, however, with the protein structure: lipophilic proteins bind co-operatively large quantities of nonionic surfactants near the cmc, but a number of hydrophilic proteins show little interaction.[47] It appears that nonionic surfactants can bind to hydrophilic proteins only if they possess hydrophobic patches. For example, Sukow *et al.*[48] have suggested that, in the binding of a series of Triton X surfactants [$(CH)_3CCH_2(CH_3)_2CC_6H_4O(CH_2CH_2O)_nCH_2OH$] to bovine serum albumin, the site which can accommodate two (or three) surfactant molecules is a hydrophobic pocket in the protein. That the interaction is predominantly hydrophobic is indicated by the small endothermic enthalpy change.[48] With the system lysozyme + hexa(oxyethylene) dodecyl ether ($C_{12}E_6$), the decrease in surfactant binding with increasing protein concentration has been explained[47] in terms of the loss in binding sites as the protein aggregates *via* hydrophobic interactions. The presence of a large hydrophobic surface area on band 3 human erythrocyte membrane protein is indicated[49] by the extensive coverage of the protein surface (up to 64%) by β-octyl glucoside.

Analysis of the state of aggregation of membrane proteins in solutions of nonionic surfactants is now a primary source of information on protein association in biological membranes, although care must be taken to avoid artefacts due to impurities in the surfactants.[50] The 10-fold increase in the amount of band 3 protein extracted by nona(oxyethylene) dodecyl ether ($C_{12}E_9$) as compared with Triton X-100 (n = 9–10) suggests[51] that there are specific interactions involved; while the two surfactants have similar physical properties (cmc, HLB number, *etc.*), the chemical structures of their hydrophobic regions are different. Also significant is the structure of the surfactant hydrophilic region. With a series of Triton X compounds, for instance, the binding strength to bovine serum albumin was found[48] to decrease with the number of oxyethylene units. It is noteworthy that reversibility in protein structural changes[39,40] and sedimentation behaviour[38] can be achieved after saturation with, and subsequent removal of, β-octyl glucoside. Similarly, the enzymic activity of rat brain microsomes was almost completely restored[52] by dialytic removal of disaggregating concentrations of β-octyl glucoside. Other uses of β-octyl glucoside in the extraction of membrane proteins have been described.[53–56] Its main advantage over other nonionic surfactants is its high cmc, which facilitates relatively rapid removal of the surfactant by dialysis. Another application of β-octyl glucoside is in the growth of protein crystals for analysis by X-ray diffraction.[57]

4 Protein Displacement from Interfaces by Surfactants

As a general rule, small amphiphilic molecules are more surface-active than proteins, insofar as they lead to a lower interfacial tension at the oil–water interface at the same concentration by weight. All other things being equal, then, one expects small-molecule surfactants to displace proteins from the surface of emulsion droplets above a certain critical concentration. The question arises[58] as to whether the interfacial composition in emulsions containing low molecular weight emulsifiers + proteins can be established simply on the basis of a 'league table' of interfacial pressures at equivalent concentrations, or whether the ubiquitous protein–surfactant binding just described makes the position in practice very much more complicated. The answer, as we shall see below, is that the interfacial tension is in fact a good general guide to protein displacement by emulsifiers; on the other hand, it appears that the details of competitive adsorption in particular systems are rather sensitive to the nature of protein-surfactant interactions occurring in aqueous solution and at the interface.

Figure 4 shows interfacial tension data[59] for the system SDS + gelatin at an n-tetradecane–water interface (pH 7, ionic strength 0.005 M, 25 °C). The tension is plotted against the logarithm of surfactant concentration at constant bulk protein concentration (0.1 or 0.4 wt.-%). At the highest SDS concentrations (well above the cmc), the tension for the SDS + gelatin system is the same as that for pure

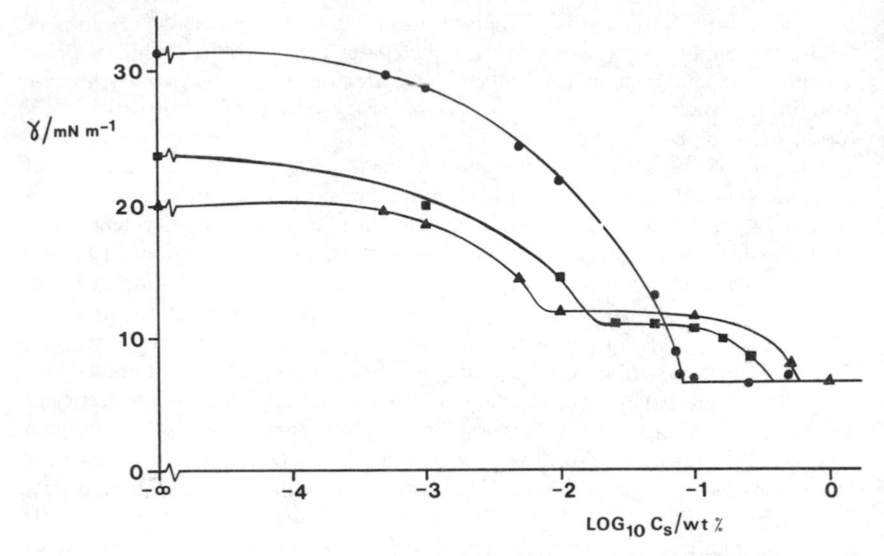

Figure 4 *Interfacial tension for SDS + gelatin at the n-tetradecane–water interface* (25 °C, *pH* 7.2, 0.005 M *phosphate buffer). The tension* γ *is plotted against the logarithm of the surfactant concentration* C_s : ●, *pure SDS*; ■, *SDS* + 0.1 wt.-% *gelatin;* ▲, *SDS* + 0.4 wt.-% *gelatin*

SDS, suggesting that the protein is completely displaced from the interface. At all lower concentrations, however, the situation is more complex: it appears that the interfacial composition is strongly influenced by a balance between surfactant adsorbed at the oil–water interface and surfactant bound to the protein. The form of the curves in Figure 4 is probably typical of many systems of the type protein + anionic surfactant. Curves of similar shape have recently been reported by de Feijter *et al.*[13] with SDS or sodium dodecyl sulphonate (SDSO) as surfactant and β-casein or β-lactoglobulin as protein, at both the air–water and oil–water interfaces.

Assuming that the general form of the plot of tension against logarithm of surfactant concentration for protein + ionic surfactant is as illustrated schematically in Figure 5, it is convenient to divide the plot into five regions (I–V). At very low surfactant concentrations (region I), the tension is essentially the same as that of the protein alone; the protein necessarily predominates at the interface, and what surfactant is present is bound strongly—primarily by electrostatic interactions but reinforced by hydrophobic interactions—to protein molecules in bulk solution and at the interface. In region II, the tension falls more rapidly with increasing surfactant concentration. What we envisage happening in this region is the increasing occupation of surface sites by the adsorption of surfactant molecules in the small gaps between protein train segments caused by the conformational constraints of the adsorbed protein. At the same time, there is probably an additional contribution to the lowering of the tension in region II because the protein–surfactant complex is more hydrophobic than the native protein. Region

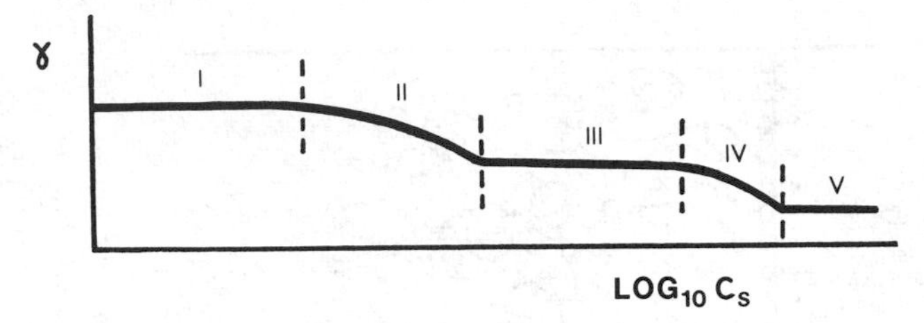

Figure 5 *Sketch of tension γ against logarithm of surfactant concentration* C_s *for system of ionic surfactant + protein. The plot is divided into five distinct regions (I–V)*

III is the intermediate plateau region, extending over a considerable range of surfactant concentration in the vicinity of the cmc of pure surfactant. At some point along this section of the curve, the tension of the mixed protein + surfactant system becomes higher than that of the pure surfactant, despite the fact that the former contains more surface-active material. Presumably, in this plateau region, it is more favourable energetically for surfactant molecules to bind cooperatively on to the protein than it is for them to displace hydrophobic protein segments from the interface. Eventually, however, this free energy balance is reversed, and in region IV the adsorbed protein–surfactant complexes are gradually displaced from the interface in a process which may also be, at least in part, cooperative. Region V corresponds to the situation in which the interface is covered with a surfactant monolayer, and the bulk solution contains a mixture of surfactant monomers, protein–surfactant complexes, and surfactant micelles, some of which are in intimate association with protein molecules. An idealized pictorial representation of these five regions in molecular terms is given in Figure 6.

Although surfactant interacts with protein in bulk solution and at the interface, the extent of binding need not be the same in both cases, and the difference may vary from one protein to another. On a solid alumina surface, it appears that protein–surfactant interaction produces a lateral compression of adsorbed gelatin,[60] but a significant lateral expansion of bovine serum albumin.[61] The effect of protein–surfactant binding is to produce an apparent shift in the cmc of the surfactant, and the higher the protein concentration the greater is the shift, as illustrated in Figure 4 for SDS + gelatin.

In the plateau region, the concentration of free surfactant is increasing much more slowly than the total surfactant concentration. The large increase in protein–surfactant binding in this region coincides also with the surfactant concentration range in which protein displacement in emulsions occurs, as determined by the depletion method for SDS + β-casein, SDS + β-lactoglobulin, or SDSO + β-lactoglobulin.[13] This prompted de Feijter *et al.*[13] to speculate whether protein–surfactant binding in solution might be the main driving force for protein desorption, with increasing surfactant concentration shifting the equilibrium from protein–surface to protein–surfactant interactions.

In interpreting tension data for SDS + gelatin at the air–water interface,

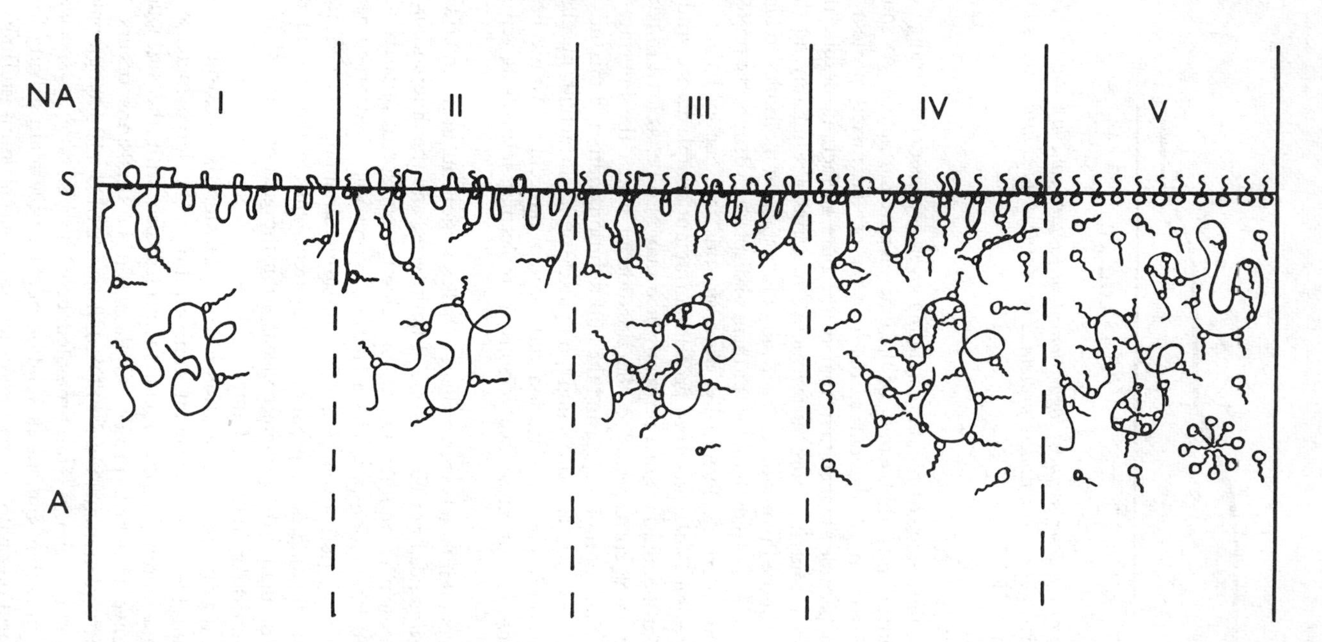

Figure 6 *Pictorial representation of protein–surfactant binding and competitive adsorption as a function of increasing surfactant concentration. Illustrations refer to a mixture of idealized random-chain protein molecules and simple ionic amphiphiles at the surface (S) between aqueous (A) and non-aqueous (NA) phases. Situations I–V correspond to regions I–V in Figure 5*

Wüstneck *et al.*[62] have suggested that the continued binding of surfactant to protein *via* electrostatic interactions leads to complexes that are more surface-active than either species alone. According to this explanation, the beginning of the plateau region (region III in Figure 5) corresponds to the formation of completely hydrophobicized complexes, and the end of the plateau region roughly corresponds to the point at which free SDS molecules begin to displace complexes from the interface.

The detailed shape of the plot of tension against logarithm of surfactant concentration depends on the nature of both the surfactant and the protein. With the cationic surfactant CTAB, no plateau is observed at pH 1.5, but the curves are of the typical form (see Figure 5) at pH 4.9, 7, and 10.[63] The CTAB concentrations at the beginning of the plateau and on the formation of the saturated layer were found[63] to decrease with increasing pH. This contrasts with SDS + gelatin, for which the surfactant concentration at the beginning of the plateau was found[62] to increase with pH. Despite its greater number of negatively than positively charged groups at pH 7, the gelatin was found[62,63] to bind less CTAB than SDS. This difference was attributed to the different chemical structures of the surfactants. Wüstneck *et al.*[64] also observed that the form of the surface tension isotherms with SDS or CTAB is sensitive to both the ash content and the molecular weight distribution of the gelatin sample used. Ericsson and Hegg[65] interpreted the surface tension isotherms for SDS + ovalbumin and SDS + bovine serum albumin in terms of a partial unfolding of the protein in the plateau region. Zourab *et al.*[66] reported that with β-lactoglobulin or insulin the order of efficiency for reducing the interfacial tension is cetylpyridinium chloride (CPC) ≈ CPB ≈ CTAB > laurylpyridinium chloride (LPC), whereas with myoglobin + anionic surfactants the order shows a different dependence on alkyl chain length. Interfacial tension data for SDS + β-lactoglobulin at the corn oil–water interface have also been reported.[67]

With nonionic surfactants, the plot of tension against surfactant concentration has a qualitatively different form from that just described for ionic surfactants. It is generally assumed that the complexes are rather surface-*in*active, since hydrophobic areas of protein and nonionic surfactant come together in the binding. For the system $C_{12}E_6$ + bovine serum albumin, Nishikido *et al.*[47] observed a shift in the cmc which depends on protein concentration. The surface tension isotherm is curved at high protein concentrations, but it approaches the linear form of that for pure surfactant at low concentrations. With lysozyme, on the other hand, the isotherms overlap and each is of the same linear form, the shift in cmc being small and not systematically dependent on the protein concentration. This anomalous behaviour may be explained in terms of protein aggregation being favoured by surfactant binding *via* the flexible hydrocarbon chains.[47] There is also no shift in the cmc for 1-monocaproin + ovalbumin,[65] but simply two distinct regions in the surface tension isotherm: protein dominates the interface at low surfactant concentrations, only to be replaced by surfactant at higher concentrations. Similarly, for glycerol monostearate (GMS) + casein, the surface pressure *versus* area curves for different ratios of GMS and casein have been found[68] to present two distinct regions: the area increases with increasing casein content at low surface pressures, but the isotherms in the high-pressure region are essentially the same as that for pure GMS. In experiments with mixtures of GMS and glycerol distearate (GDS), it is suggested[69] that there exists a greater degree of interfacial association between the mixed surfactants and the disordered casein molecules (mainly α_{s1}- and β-caseins) than with the globular whey proteins, α-lactalbumin or β-lactoglobulin.

Surface rheological measurements are a valuable source of information about interactions between proteins and surfactants at fluid interfaces. Addition of anionic or cationic surfactant to gelatin, for instance, leads to an increase in all the surface rheological parameters.[62,70,71] Increased surfactant binding to gelatin at the air–water interface is accompanied by a reduction in the thickness of foam films stabilized by the protein–surfactant complex,[72] *viz*. 8–12 nm compared with >80 nm for gelatin alone. At concentrations of SDS or CTAB exceeding the binding capacity of the protein for surfactant, the commom black films turn into Newton black films with a thickness comparable to that for the pure surfactant system. Maximum values of the surface rheological parameters in SDS + gelatin and CTAB + gelatin occur[62,70] at surfactant concentrations corresponding to the beginning of the plateau region in the plot of tension against logarithm of surfactant concentration. Above the cmc, as surfactant displaces gelatin from the interface, the surface viscosity drops by several orders of magnitude, and the surface rheology eventually becomes Newtonian at the highest surfactant concentrations (region V in Figure 5). A similar effect has also been observed[73] with the globular protein β-lactoglobulin. In systems containing GMS + GDS + casein, the ratio of GMS to GDS giving the largest surface viscoelastic parameters has been found[74] to correspond also to that giving the maximum emulsion stability.[75]

The most systematic study so far of the relationship between tension measurements and the interfacial composition in emulsions is the work of de Feijter *et al.*[13] using a depletion method to monitor both the protein and the surfactant concentrations. Figure 7 shows experimental results for the separate displacement of β-casein or β-lactoglobulin by SDS from the droplet surface in sunflowerseed oil-in-water emulsions. The results were found to be virtually independent of protein type, the nature of the oil phase (paraffin oil or sunflowerseed oil), or whether the surfactant was added before or after homogenization. Displacement curves for the two dodecyl anionic surfactants, SDS and sodium dodecyl sulphonate (SDSO), were found to be almost coincident, but a much higher concentration (× 100) of sodium heptyl sulphonate (SHSO) was required to produce the same displacement. The shape of the displacement curve with the nonionic surfactant Tween 20 was found to be slightly different from that obtained with the anionic surfactants. With Tween 20, as with the oil-soluble surfactants Span 80 and glycerol monooleate (GMO), displacement of protein was not complete at the highest surfactant concentration studied (0.2 wt.-%).

Comparison of Figures 2 and 7 shows good agreement between the simple statistical theory and the experimental displacement data. Despite the fact that proteins are much more complicated than the model polymers considered in the theory and the fact that the one-layer lattice-based representation of the interface is highly idealized, it is clear that the essential physical features of protein displacement are adequately reproduced. The agreement is, however, no more than qualitative, owing to the arbitrary choice of athermal solvent conditions and

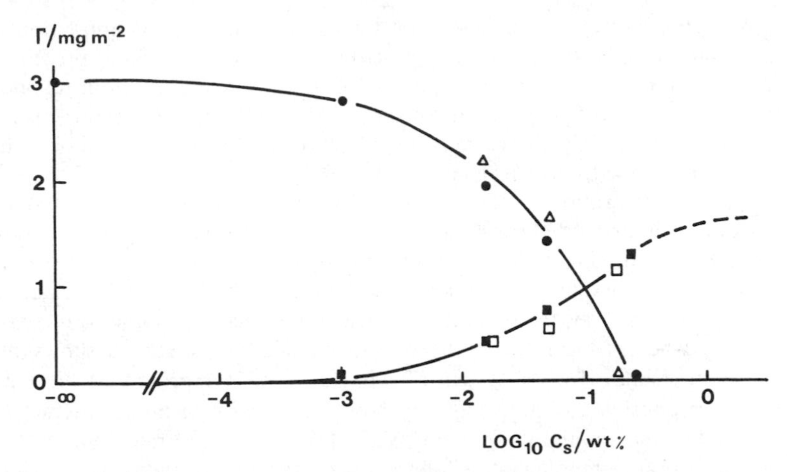

Figure 7 *Displacement of protein from the droplet surface by surfactant in sunflowerseed oil-in-water emulsions (pH* 6.5, 50 wt.-% *oil). The surface concentration* Γ *is plotted against the logarithm of the (final) surfactant concentration* C_s : □, △, *SDS +* β-*casein;* ■, ●, *SDS +* β-*lacto-globulin*
(Redrawn from *Colloids Surf.* 1987, **27**, 243)

the essentially arbitrary choice of parameters ΔE_s and ΔE_p in equation (2). At present quantitative analysis is limited by the lack of detailed information on adsorption energies of surfactants and protein residues at the oil–water interface. The critical displacement concentration φ_s^c for a particular protein–surfactant combination depends on the nature of the protein–surfactant interactions and on the adsorption energies of protein and surfactant. According to de Feijter *et al.*,[13] adsorption energies per molecule for SDS and Tween 20 are similar (*ca.* 12kT), even though the nonionic surfactant is much less effective at displacing protein from the interface. This suggests a difference in displacement behaviour arising from a difference in binding behaviour to protein of the two types of surfactant. To some extent this can be handled in the theory through the Flory–Huggins parameter χ_{ps}, insofar as a more negative value of χ_{ps} implies stronger protein–surfactant attraction and therefore less surfactant available for protein displacement. However, the simple statistical theory described here is in no way adequate to deal with the co-operative aspects of protein–surfactant binding, or the competition between surfactant and protein–surfactant complexes which is undoubtedly the real situation in many surfactant + protein systems. What can be said, however, is that protein displacement by surfactants has the same general form irrespective of the nature and strength of these interactions.

Several displacement studies of proteins by small-molecule surfactants have involved adsorption at solid rather than liquid surfaces. Recently for instance, Brock and Enser[76] assessed the binding strength of proteins to the hydrophobic surface of chemically modified porous glass particles by washing with a buffer solution containing 1 wt.-% SDS. Although all the bound bovine serum albumin could be eluted by washing, this was not the case for the various non-globular proteins considered; with myosin and κ-casein, about half the bound protein is retained on the particles, but the proportions of α_{s1}-casein and β-casein retained are less. It is suggested[76] that the adsorbed layer of the disordered proteins consist of a mixture of reversibly and irreversibly adsorbed molecules, of which only the former are amenable to displacement by SDS.

Competitive adsorption at the alumina–water interface was investigated by Samanta and Chattoraj[60,61] for various surfactants with gelatin or bovine serum albumin. At low surfactant concentrations, the estimated binding ratio of surfactant to protein at the interface was similar to that in the bulk. At high surfactant concentrations, however, the binding ratio at the interface was found to increase enormously whereas the ratio in the bulk remained low. Figure 8 shows the experimental results[60] for the system CTAB + gelatin. It appears that the solid surface is responsible for inducing a massive co-operative binding interaction between the surfactant and the protein. With SDS + bovine serum albumin, this massive co-operative binding at the interface occurs at a surfactant concentration two orders of magnitude lower than the cmc.[61]

Kronberg *et al.*[77] have studied the competitive and co-operative adsorption of polymers and surfactants on kaolinite surfaces. It was observed that small amounts of polyacrylic acid strongly impede the adsorption of the nonionic surfactant decaoxyethylene nonylphenyl ether (NP-EO_{10}), whereas large amounts enhance it. The presence of specific interactions between the polymer and the surfactant was found to be consistent with the flow behaviour of concentrated kaolinite suspensions containing NP-EO_{10} and polyacrylic acid.

Experimental information on competitive adsorption between proteins and low molecular weight emulsifiers in food emulsions is still relatively scarce. In a study of the composition of recombined milk fat globules, Oortwijn and Walstra[78] found that displacement of milk protein from the surface of emulsion droplets is greater with water-soluble Tween 20 than with oil-soluble GMS or GMO. From studies of the effect of emulsifier addition on emulsion stability, it is known[79] that there is considerable variation in the effectiveness of different emulsifiers. This probably arises from differences in the tendency of various emulsifiers to complex with proteins. For instance, the better stability with respect to creaming of emulsions containing sodium stearoyl-2-lactylate (SSL), compared with those containing Tween 20 or Tween 60, has been attributed[79] to interfacial interaction between SSL and caseinate. In a study of simulated cream liqueurs, it has been found[80] that GMS or SSL displaces a significant proportion of caseinate from the surface of the emulsion droplets. Improved stability with respect to creaming on addition of GMS or SSL is attributed to complex formation between low molecular weight emulsifier and protein, both in bulk solution and at the interface.

Finally, we note that competition between proteins and small-molecule surfactants can have a considerable effect on the rheology of food emulsions by changing the state of aggregation of the dispersed droplets. Bridging flocculation between droplets, for instance, may be disrupted by using a surfactant to displace protein from the interface,[58] and thereby reduce the emulsion viscosity. In whipped toppings, on the other hand, displacement of protein from the oil–water

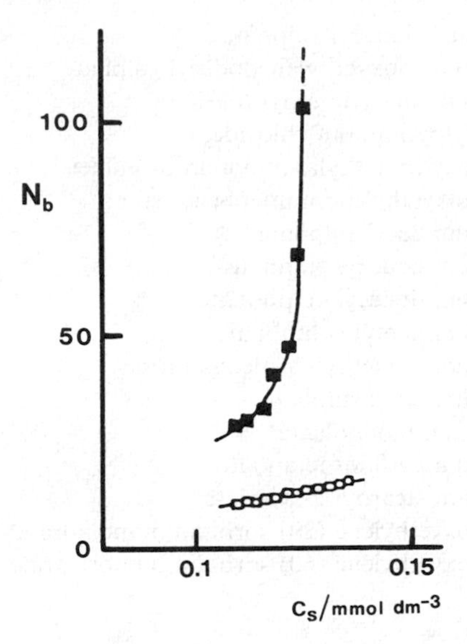

Figure 8 *Comparison of CTAB–gelatin binding at the alumina–water interface with that in bulk solution* (28 °C, *pH* 4.0, *ionic strength* = 0.1 M, *total CTAB concentration* = 10^{-3} M). *The binding number* N_b *is plotted against the equilibrium surfactant concentration* C_s : □, *bulk*; ■, *interface*

(Redrawn from *J. Colloid Interface Sci.*, 1987, **116**, 168)

interface by addition of low molecular weight emulsifiers leads to an increase in bulk rheological properties caused by partial droplet coalescence at the air–water interface.[81]

Acknowledgement. C. M. W. acknowledges receipt of an AFRC Studentship in conjunction with Unilever Research (Colworth House).

Appendix: List of Surfactant Abbreviations Used in Text (Only Those Marked with Asterisks are Permitted Emulsifiers in the UK)

CPB	Cetylpyridinium bromide
CPC	Cetylpyridinium chloride
CTAB	Cetyltrimethylammonium bromide
$C_{12}E_6$	Hexaoxyethylene dodecyl ether
$C_{12}E_9$	Nonaoxyethylene dodecyl ether
DOC	Deoxycholate
DTAB	Dodecyltrimethylammonium bromide
GDS	Glycerol distearate*
GMO	Glycerol monooleate*
GMS	Glycerol monostearate*
LAS	Sodium dodecylbenzene sulphonate

LDS	Lithium dodecyl sulphate
LES	Sodium trioxyethylenedodecyl sulphate
LMT	Sodium lauroylmethyl taurid
LPC	Laurylpyridinium chloride
MTAB	Myristyltrimethylammonium bromide
$NP\text{-}EO_{10}$	Decaoxyethylene nonylphenyl ether
SDeS	Sodium decyl sulphate
SDS	Sodium dodecyl sulphate
SDSO	Sodium dodecyl sulphonate
SHSO	Sodium heptyl sulphonate
SMOT	Sodium *N*-methyl-*N*-oleoyltaurine
SOS	Sodium octyl sulphate
Span 80	Sorbitan monooleate*
SPFO	Sodium perfluoroctanoate
SSL	Sodium stearoyl-2-lactylate*
Tween 20	Polyoxyethylene (20) sorbitan monolaurate*
Tween 60	Polyoxyethylene (60) sorbitan monostearate

References

1. E. Dickinson and G. Stainsby, 'Colloids in Food,' Applied Science, Barking, 1982.
2. E. D. Goddard, *Colloids Surf.*, 1986, **19**, 255, 301.
3. E. D. Goddard and B. A. Pethica, *J. Chem. Soc.*, 1951, 2659.
4. E. G. Cockbain, *Trans. Faraday Soc.*, 1953, **49**, 104.
5. W. J. Knox and T. O. Parshall, *J. Colloid Interface Sci.*, 1970, **33**, 16.
6. E. Dickinson, *Food Hydrocolloids*, 1986, **1**, 3.
7. E. Dickinson and M. Lal, *Adv. Mol. Relaxation Int. Processes*, 1980, **17**, 1.
8. P. J. Flory, 'Principles of Polymer Chemistry,' Cornell University Press, Ithaca, NY, 1953.
9. A. Silberberg, *J. Chem. Phys.*, 1968, **48**, 2835.
10. J. M. H. M. Scheutjens and G. J. Fleer, *J. Phys. Chem.*, 1979, **83**, 1619.
11. M. A. Cohen Stuart, G. J. Fleer, and J. M. H. M. Scheutjens, *J. Colloid Interface Sci.*, 1984, **97**, 515.
12. M. A. Cohen Stuart, G. J. Fleer, and J. M. H. M. Scheutjens, *J. Colloid Interface Sci.*, 1984, **97**, 526.
13. J. A. de Feijter, J. Benjamins, and M. Tamboer, *Colloids Surf.*, 1987, **27**, 243.
14. M. N. Jones and P. J. W. Midgley, *Biochem. Soc. Trans.*, 1984, **12**, 625.
15. A. Kato, T. Matsuda, N. Matsudomi, and K. Kobayashi, *J. Agric. Food Chem.*, 1984, **32**, 284.
16. A. Kato and S. Nakai, *Biochim. Biophys. Acta*, 1980, **624**, 13.
17. M. N. Jones and P. Manley, *J. Chem. Soc., Faraday Trans. 1*, 1979, **75**, 1736.
18. M. N. Jones and P. Manley, *J. Chem. Soc., Faraday Trans. 1*, 1980, **76**, 654.
19. M. N. Jones and P. Manley, *J. Chem. Soc., Faraday Trans. 1*, 1981, **77**, 827.
20. M. N. Jones, P. Manley and P. J. W. Midgley, *J. Colloid Interface Sci.*, 1981, **82**, 257.
21. K. Fukushima, Y. Murata, N. Nishikido, G. Sugihara, and M. Tanaka, *Bull. Chem. Soc. Jpn.*, 1981, **54**, 3122.
22. K. Fukushima, Y. Murata, G. Sugihara, and M. Tanaka, *Bull. Chem. Soc. Jpn.*, 1982, **55**, 1376.
23. K. Fukushima, G. Sugihara, Y. Murata, and M. Tanaka, *Bull. Chem. Soc. Jpn.*, 1982, **55**, 3113.
24. Y. Inoue, S. Sase, R. Chûjô, S. Nagaoka, and M. Sogami, *Biopolymers*, 1979, **18**, 373.
25. M. Sen, S. P. Mitra, and D. K. Chattoraj, *Indian J. Biochem. Biophys.*, 1980, **17**, 370.

26. B. K. Sadhukhan and D. K. Chattoraj, *Indian J. Biochem. Biophys.*, 1983, **20**, 66.
27. M. Sen, S. P. Mitra, and D. K. Chattoraj, *Indian J. Biochem. Biophys.*, 1980, **17**, 405.
28. B. K. Sadhukhan and D. K. Chattoraj, in 'Surfactants in Solution', ed. K. L. Mittal and B. Lindman, Plenum Press, New York, 1984, p. 1249.
29. P. F. Burns, C. W. Campagnoni, I. M. Chaiken, and A. T. Campagnoni, *Biochemistry*, 1981, **20**, 2463.
30. P. F. Burns and A. T. Campagnoni, *Biochim. Biophys. Acta*, 1983, **743**, 379.
31. K. Takeda, M. Miura, and T. Takagi, *J. Colloid Interface Sci.*, 1981, **82**, 38.
32. S.-H. Chen and J. Teixeira, *Phys. Rev. Lett.*, 1986, **57**, 2583.
33. K. Takeda, *Bull. Chem. Soc. Jpn.*, 1983, **56**, 1037.
34. K. Takeda, *Bull. Chem. Soc. Jpn.*, 1982, **55**, 1335.
35. K. Takeda, M. Shigeta, and K. Aoki, *J. Colloid Interface Sci.*, 1987, **117**, 120.
36. Y. Inoue, S. Sase, R. Chûjô, S. Nagaoka, and M. Sogami, *Polym. J.*, 1980, **12**, 139.
37. K. Miyazawa, M. Ogawa, and T. Mitsui, *Int. J. Cosmet. Sci.*, 1984, **6**, 33.
38. J. M. Andreu and J. A. Mañoz, *Biochemistry*, 1986, **25**, 5220.
39. J. M. Andreu, J. de la Torre, and J. L. Carrascosa, *Biochemistry*, 1986, **25**, 5230.
40. Z. Markovic-Housley and R. M. Garavito, *Biochim. Biophys. Acta*, 1986, **869**, 158.
41. B. H. Tavernier, *J. Colloid Interface Sci.*, 1983, **93**, 419.
42. B. R. Sureshchandra, A. G. Appu Rao, and M. S. Narasinga Rao, *J. Agric. Food Chem.*, 1987, **35**, 244.
43. M. Donovan and D. M. Mulvihill, *Ir. J. Food Sci. Technol.*, 1987, **11**, 77.
44. J. P. S. Arora, R. P. Singh, D. Soam, and S. P. Singh, *Bull. Soc. Chim. Fr.*, 1984, **1–2**, 19.
45. J. P. S. Arora, S. P. Singh, and V. K. Singhal, *Tenside Detergents*, 1984, **21**, 197.
46. M. Sen, S. P. Mitra, and D. K. Chattoraj, *Colloids Surf.*, 1981, **2**, 259.
47. N. Nishikido, T. Takahara, H. Kobayashi, and M. Tanaka, *Bull. Chem. Soc. Jpn.*, 1982, **55**, 3085.
48. W. W. Sukow, H. E. Sandberg, E. A. Lewis, D. J. Eatough, and L. D. Hansen, *Biochemistry*, 1980, **19**, 912.
49. P. K. Werner and R. A. F. Reithmeier, *Biochemistry*, 1985, **24**, 6375.
50. G. Pappert and D. Schubert, *Biochim. Biophys. Acta*, 1983, **730**, 32.
51. R. Moriyama and S. Makino, *Biochim. Biophys. Acta*, 1985, **832**, 135.
52. L. Corazzi and G. Arienti, *Biochim. Biophys. Acta*, 1986, **875**, 362.
53. C. Baron and T. E. Thompson, *Biochim. Biophys. Acta*, 1975, **382**, 276.
54. P. Rosevear, T. van Aken, J. Baxter, and S. Ferguson-Miller, *Biochemistry*, 1980, **19**, 4108.
55. R. J. Gould, B. H. Ginsberg, and A. A. Spector, *Biochemistry*, 1981, **20**, 6776.
56. W. L. Dean and R. D. Gray, *J. Biol. Chem.*, 1982, **257**, 14679.
57. A. McPherson, S. Koszelak, H. Axelrod, J. Day, R. Williams, L. Robinson, M. McGrath, and D. Cascio, *J. Biol. Chem.*, 1986, **261**, 1969.
58. D. F. Darling and R. J. Birkett, in 'Food Emulsions and Foams,' ed. E. Dickinson, Special Publication No. 58, Royal Society of Chemistry, London, 1987, p. 1.
59. C. M. Woskett, unpublished results.
60. A. Samanta and D. K. Chattoraj, *J. Colloid Interface Sci.*, 1987, **116**, 168.
61. A. Samanta and D. K. Chattoraj, *Prog. Colloid Polym. Sci.*, 1983, **68**, 144.
62. R. Wüstneck, L. Zastrow, and G. Kretzschmar, *Colloid J. USSR*, 1985, **47**, 387.
63. R. Wüstneck, L. Zastrow, and G. Kretzschmar, *Colloid J. USSR*, 1987, **49**, 6.
64. R. Wüstneck, H. Hermel, and G. Kretzschmar, *Colloid Polym. Sci.*, 1984, **262**, 827.
65. B. Ericsson and P. O. Hegg, *Prog. Colloid Polym. Sci.*, 1985, **70**, 92.
66. S. M. Zourab, S. N. Srivastava, A. Halim, and V. M. Sabet, *Egypt. J. Chem., 1982,* **25**, 141.
67. E. N. Jaynes and M. A. Flood, *J. Disp. Sci. Technol.*, 1985, **6**, 55.
68. P. Paquin, M. Britten, M.-F. Laliberté, and M. Boulet, *ACS Symp. Ser.*, 1987, No. **343**, 677.
69. A. Rahman and P. Sherman, *Colloid Polym. Sci.*, 1982, **260**, 1035.
70. R. Wüstneck, G. Kretzschmar, and L. Zastrow, *Colloid J. USSR*. 1987, **49**, 207.
71. R. Wüstneck, V. V. Krotov, and M. Ziller, *Colloid Polym. Sci.*, 1984, **262**, 67.

72. R. Wüstneck and H.-J. Müller, *Colloid Polym. Sci.*, 1986, **264**, 97.
73. S. M. Zourab, S. N. Srivastava, A. Halim, and V. M. Sabet, *Egypt. J. Chem.*, 1982, **25**, 131.
74. G. Doxastakis and P. Sherman, *Colloid Polym. Sci.*, 1986, **264**, 254.
75. G. Doxastakis and P. Sherman, *Colloid Polym. Sci.*, 1984, **262**, 902.
76. C. J. Brock and M. Enser, *J. Sci. Food Agric.*, 1987, **40**, 263.
77. B. Kronberg, J. Kuortti, and P. Stenius, *Colloids Surf.*, 1986, **18**, 411.
78. H. Oortwijn and P. Walstra, *Neth. Milk Dairy J.*, 1979, **33**, 134.
79. L. M. Smith and T. Dairiki, *J. Dairy Sci.*, 1975, **58**, 1254.
80. E. Dickinson, S. K. Narhan, and G. Stainsby, this volume, p. 372.
81. N. Krog, N. M. Barfod and W. Buchheim, in 'Food Emulsions and Foams,' ed. E. Dickinson, Special Publication No. 58, Royal Society of Chemistry, London, 1987, p. 144.

Electrostatic Interactions Between Proteins and their Effect on Foam Composition and Stability

By David C. Clark, Alan R. Mackie, Linda J. Smith, and David R. Wilson

CHEMICAL PHYSICS DIVISION, AFRC INSTITUTE OF FOOD RESEARCH, NORWICH LABORATORY, COLNEY LANE, NORWICH NR4 7UA, UK

1 Introduction

Processed foods are complicated mixtures of components including proteins, lipids, and carbohydrates. The manner in which these molecules interact with their own type and with the different classes can ultimately determine the structure and stability of the product formed.[1]

The complexity of food presents formidable problems to the direct study of the functional properties of the constituents *in situ* and therefore considerable effort has been directed at examining the isolated components. Progress has been made which has allowed simple mixtures that are more representative of real systems to be investigated.

It is the adsorption of amphipathic proteins at interfaces that has a major influence on the functional properties of foams and emulsions.[2] The adsorption characteristics of isolated proteins of different structural types were systematically studied by Graham and Phillips.[3] More recently, protein mixtures[2] have been examined and it has been shown that adsorption is not simply related to bulk phase concentration. The adsorption of blood plasma proteins is not only surface dependent[4] but also competitive, as components that are present in comparatively low concentrations, *e.g.* fibrinogen and haemoglobin, are preferentially adsorbed.[5] Milk proteins show similar behaviour with β-casein preferentially adsorbed at fluid interfaces.[6]

Co-operative protein–protein interactions also influence adsorption in mixed systems. In whole egg albumen the relative concentrations of the components do have a significant effect on foaming properties.[7] Enhanced foam stabilization occurs between oppositely charged proteins.[8] Increased levels of lysozyme, which is positively charged at neutral pH, only improve foam stability in the absence of ovomucin, as the complex formed between these proteins has poor solubility.

The foaming properties of mixtures of bovine serum albumin (BSA) and lysozyme have been used to model systems of interacting species.[8,9] Using whipping methods, it was established that the stable and highly expanded foams obtained with BSA near its isoelectric point were also observed over a much wider range of pH in the presence of equimolar concentrations of lysozyme. This latter protein is a poor foamer in isolation[10] but is incorporated in the foams obtained with the protein mixtures.

In a recent paper we described a spectroscopic assay which allows the analysis of the composition of mixtures of BSA and lysozyme.[9] In this study we examined the influence of solution conditions on the electrostatic interaction between lysozyme and BSA in solution and extended our investigation of the composition of foams generated from mixtures under different conditions. We also studied the formation, drainage, and thickness of suspended microscopic films of BSA and lysozyme under selected conditions and these observations are discussed in terms of foam stability.

2 Materials and Methods

Fatty acid- and globulin-free BSA and chicken egg lysozyme were obtained from Sigma. The purity of the proteins was checked by sodium dodecyl sulphate (SDS) polyacrylamide gel electrophoresis. Both BSA and lysozyme were found to contain low levels ($<5\%$) of high molecular weight contaminants. Electrophoresis of BSA under non-denaturing conditions[11] revealed that approximately 20% of the total BSA was present as a dimer. Surface tension isotherms were recorded for both proteins using the Wilhelmy plate technique and showed no evidence of contaminating surface-active species.[12] All other chemicals were of AnalaR grade from BDH. All experiments were carried out at room temperature unless specified otherwise.

Protein stock solutions were prepared in high-quality water (surface tension $>72.8\ \mathrm{mN\,m^{-1}}$ at 20 ± 0.1 °C) and adjusted to pH 8 over a calibrated pH electrode by addition of small volumes of 0.1 M NaOH with gentle stirring. Protein mixtures were prepared by taking an appropriate volume of 5 M NaCl solution to provide the desired ionic strength followed by addition of the required volumes of BSA and then lysozyme stock solutions. The mixtures were used immediately unless specified otherwise. The concentration of stock protein solutions was determined spectrophotometrically using absorbance coefficients at 280 nm of $0.66\ \mathrm{ml\,mg^{-1}\,cm^{-1}}$ for BSA[13] and $2.58\ \mathrm{ml\,mg^{-1}\,cm^{-1}}$ for lysozyme.[14]

Protein-stabilized foams were prepared by a modification[11] of the method of Weil.[15] Briefly, approximately 50 ml of protein solution were loaded above a sintered-glass frit (17–40 μm porosity) in a 2.5-cm diameter chromatographic column. The sample was sparged with a regulated supply of oxygen-free (white-spot) nitrogen at a flow-rate of $22–23\ \mathrm{cm^3\,min^{-1}}$. The protein-stabilized foam generated was slowly forced up a 100-cm column by the gas pressure. The foam was then driven around a U-bend and was finally collapsed by bubbling through a known volume of salt solution in a collection vessel. The foams generated by this method were characterized and average bubble diameters of 3.69 ± 1.92 mm were observed.[12]

The compositions of protein mixtures and solubilized foams were measured using a spectroscopic assay described previously.[9] The assay is based on the marked differences in the absorption spectra of tyrosine-rich BSA and tryptophan-rich lysozyme. The spectrum of the mixture is fitted by minimizing the sum of the squares of the differences between it and a computer-generated fit obtained using the following expression:

$$\mathrm{Fit(I)} = X_1 Y_1(\mathrm{I}) + X_2 Y_2(\mathrm{I}) + X_3 S(\mathrm{I})$$

where Y_1 and Y_2 are the spectra of BSA and lysozyme, respectively, X_i is the fraction of each component, and $S(\mathrm{I})$ is a correction factor for the contribution of scatter from the complex which is based on the λ^4 dependence of spherical scattering species.[9,16]

Spectra were recorded using a Perkin-Elmer Lambda 9 spectrophotometer in the absorbance range 0.5–2.0 using cuvettes of appropriate path length and were normalized to a 1-cm path length prior to analysis. Spectrophotometer control and data acquisition were by a BBC microcomputer via an RS232-C interface. Data resolution was one point per nanometre.

Suspended Thin Films—Suspended thin films were investigated in a specially constructed cell consisting of a brass housing into which a film ring could be inserted. In a typical experiment a droplet of protein solution was introduced into the film ring [a ground glass annulus (3 mm diameter) with a capillary side arm]. The film ring was then inserted into the brass holder, which contained water channels to allow thermostating. An observation chamber in the centre of the holder, sealed top and bottom by 1-in diameter glass windows, allowed the film to be viewed using a microscope. The chamber also contained a horseshoe-shaped trough with a filter-paper wick which could be filled with solution to promote vapour saturation. After equilibration, the volume of the protein droplet could be reduced to allow thin film formation (<0.5 mm diameter) by withdrawal of solution down the capillary by manipulation of an adjustable micropipette.

Observation of film thinning was performed with the thin-film cell set on the stage of a Nikon Diaphot-TMD inverted microscope equipped with an epi-illumination attachment. Long working distance objectives (magnification $\times 20$ and $\times 40$) were routinely used and photographs were taken using a Nikon FE2 camera with Fujichrome DX 100D colour transparency film.

The thickness of films was measured using the apparatus shown in Figure 1. Laser light (He–Ne, 3 mW) at a wavelength of 633 nm was passed through a chopper to a beam splitter where it was reflected vertically down and focused on to the film using a microscope objective (magnification $\times 40$). The resulting spot size was approximately 10 μm. Light reflected up from the two interfaces of the film (see inset, Figure 1) passed back up through the lens and beam splitter. The reflected beam was filtered by a 633-nm narrow band-pass optical filter and detected by a photodiode. The signal from the photodiode was amplified and fed into a phase-sensitive detector (PSD) along with a signal from the chopper. A PSD was used to reduce the noise in the system and increase stability. By applying a constant negative voltage to the signal, the PSD backed off any background reflected light to maximize the change in intensity due to reflectance of the film. The output from the PSD was fed into the analogue-to-digital converter on a computer. This allowed the film reflectance to be monitored with a maximum resolution of 1 in 4096.

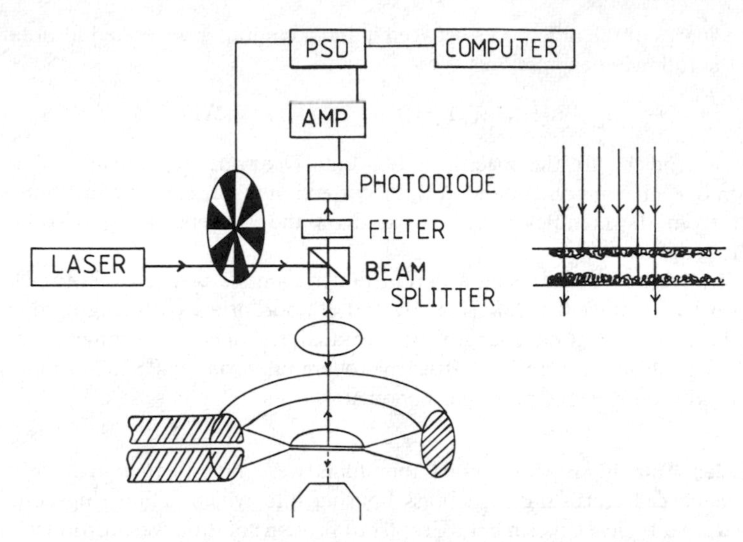

Figure 1 *Schematic diagram of the reflectance apparatus for film thickness measurement. The side-arm capillary and ground-glass annulus containing the film are shown in cross-section above the objective lens of an inverted microscope. The inset on the right shows incident light transmitted and reflected from the two interfaces of the thin film.*

The thickness (h) is calculated using the equation

$$h = \frac{\lambda}{2\pi n} \sin^{-1} \left\{ \frac{I/I_m}{1 + [4R/(1+R)^2]\,[1 - (I/I_m)]} \right\}^{\frac{1}{2}} \tag{1}$$

where

$$R = (n-1)^2/(n+1)^2 \tag{2}$$

λ is the wavelength of the laser, n is the refractive index of the film, I is the intensity at equilibrium and I_m is the intensity at a maximum. A refractive index of 1.335 was used throughout this study.[17] The value of the intensity at a minimum is the baseline subtracted from both I and I_m. In order to measure I and I_m, it was necessary to take the logarithm of the intensity as the film drained and of the reflectance as it went through maxima and minima. Intensity values for maxima, minima, and equilibrium must be established. The accuracy of the apparatus was checked by measuring the thickness of black films of 0.50% SDS in 0.1 M NaCl. Average thicknesses of 11.72 ± 0.71 nm were obtained, which compare favourably with the literature.[17,18]

3 Results

Effect of Solution Conditions on the Interaction of BSA and Lysozyme—The involvement of electrostatic interaction in the aggregation of BSA and lysozyme

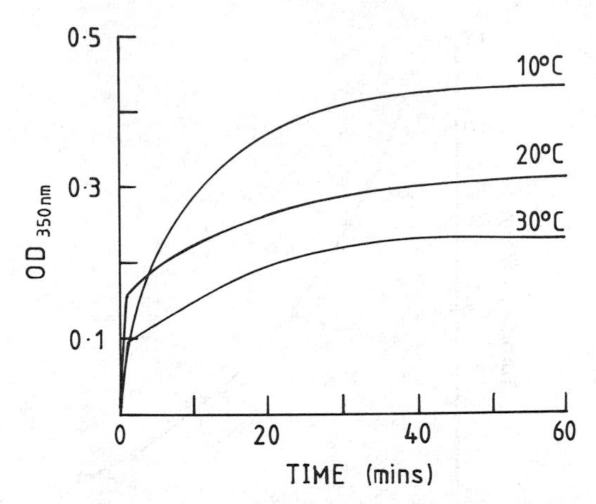

Figure 2 *Plot of turbidity at* 350 nm *as a function of time for mixtures of BSA and lysozyme (final concentrations each* 1 mg ml^{-1}) *after mixing at* 4 °C. *The cuvette holder of the spectrophotometer was pre-equilibrated at the indicated temperature*

has been demonstrated previously.[8,9] Earlier work[9] revealed that the interaction was temperature dependent, and this is confirmed in Figure 2. In this experiment, equal volumes of solutions of BSA and lysozyme (2 mg ml^{-1} in water at pH 8 equilibrated at 4 °C) were rapidly mixed in a 1-cm cell and placed in the thermostated cell holder of the spectrophotometer, which had been equilibrated at the appropriate temperature. Aggregation measured by turbidity at 350 nm was monitored at three different temperatures (Figure 2). The plateau value of the turbidity increased with decreasing temperature but the rate of increase in turbidity (3.78 ± 0.60 h^{-1}) was independent of temperature. The plateau was reached after approximately 1 h and then the signal began to decrease with time. This effect is demonstrated in Figure 3, where the two protein components (2 mg ml^{-1}), equilibrated at room temperature, were mixed and the turbidity at 350 nm was monitored as a function of time. A very large reduction (>6-fold) in turbidity occurred over a period of 7 h.

In a separate experiment, the time dependence of the protein composition of the supernatant obtained by centrifugation (10 min, 13 500 g, MSE Microcentaur) of a mixture of BSA and lysozyme (2.5 mg ml^{-1} of each) was measured using the spectroscopic assay. The results are shown in Figure 4. The total protein in the supernatant remained constant at 3.94 ± 0.04 mg ml^{-1}, but there was a redistribution of the protein components between the supernatant and pelletable fractions. Initially the supernatant contained 2.13 mg ml^{-1} BSA and 1.81 mg ml^{-1} lysozyme, giving a BSA:lysozyme ratio of 1.18. However, this changed slowly with time so that after 3 h the level of BSA in the supernatant had risen to 2.26 mg ml^{-1} where the lysozyme concentration had dropped to 1.66 mg ml^{-1}, giving a BSA:lysozyme ratio of 1.36. From these results, it can be deduced that the composition of the pelletable aggregates also changed. Initially this aggregate fraction contained elevated levels of BSA and lower levels of lysozyme, but this

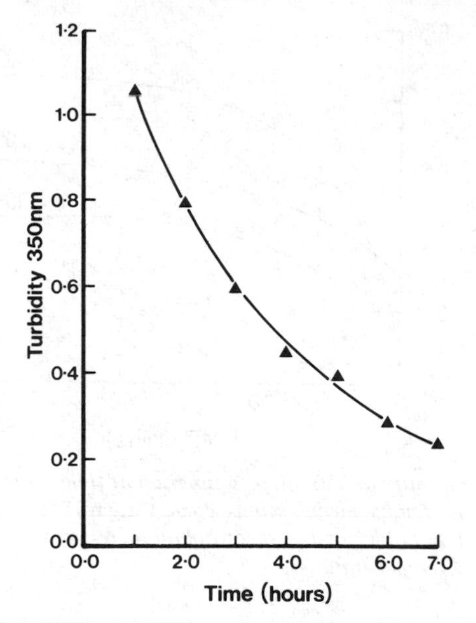

Figure 3 *Plot of turbidity at* 350 nm versus *time for a mixture of BSA and lysozyme each* (1 mg ml^{-1} *final concentration) at pH* 8. *The temperature was* 23 °C.

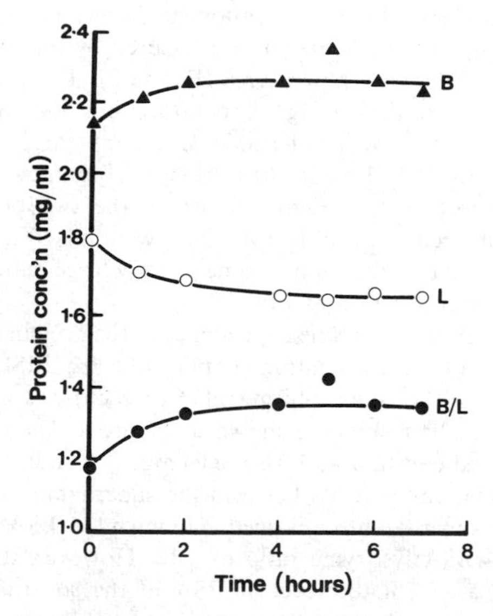

Figure 4 *Plot of the protein composition of the supernatant obtained after centrifugation of a mixture of BSA and lysozyme (starting concentrations each* 2.5 mg ml^{-1}) *as a function of time. B = BSA; L = lysozyme; B/L = BSA to lysozyme ratio*

changed with time so that more BSA was released into solution while more lysozyme was incorporated into the complex. The time dependence of this effect was almost three times faster ($0.89 \pm 0.03\ h^{-1}$) than that observed for the change in turbidity with time (Figure 3), which had a rate of $0.33 \pm 0.02\ h^{-1}$. Therefore, although these observations may be related, they do not appear to be measuring the same process.

A possible explanation for the different rates was a change in particle size in parallel with redistribution of protein between the aggregates and the supernatant. Particle size analysis using a Malvern 2600HSD particle sizer of a mixture of BSA and lysozyme (both $0.5\ mg\ ml^{-1}$) is shown in Figure 5. A time-dependent change in particle size distribution was observed with at least four species of aggregate present at time zero. These have approximate sizes of 40, 19, 5, and 2 μm. The concentration of the three large components reduced with time at rates that decreased with increasing aggregate size. The rate of disappearance of the 19-μm aggregate was very similar to that of the change in protein composition of the supernatant (Figure 4). Ionic strength is known to reduce the extent of interaction of BSA and lysozyme in solution. Particle size data obtained with solutions containing 0.025 M NaCl revealed that at time zero the size distribution was similar to that in the absence of salt but the extent of aggregation was much reduced (a 14-fold concentration increase was required to give an equivalent signal). The large 40-μm aggregate disappeared faster than at zero ionic strength but the 19-μm particle persisted for longer (*ca.* 100 min).

Influence of Solution Properties on Foam Composition—The influence of the time dependence of the supernatant composition (Figure 4) and extent of aggregation (Figure 5) on foam composition was examined using the spectroscopic assay. A mixture containing $1\ mg\ ml^{-1}$ of both lysozyme and BSA was prepared and an aliquot was foamed immediately. Several hours later, when the mixture was clear, a second aliquot was foamed. The results obtained (Table 1) show that the incorporation of lysozyme in the foam increases with time.

Table 1 *Protein composition of foam as a function of time after mixing*

Sample	*Time* (h)	*Temperature* (°C)	*Appearance*	*[BSA]* (mg/ml)	*[Lysozyme]* (mg/ml)	*Lysozyme:BSA*
1	0	Room	Turbid	1.23	1.08	0.88
2	6	Room	Clear	0.50	1.01	2.01
3	0	15	Turbid	1.20	1.14	0.95
4	6	15	Intermediate	1.03	1.27	1.23

The influence of pH was examined in the presence and absence of 0.2 M NaCl. In the presence of salt the interaction between BSA and lysozyme is much reduced, as determined by turbidity.[9] This was reflected in the low levels of lysozyme incorporation in the foam throughout the range of pH investigated (Figure 6). In the absence of salt, the extent of interaction as measured by turbidity was low between pH 6 and 7 but increased dramatically in the pH range 7–7.5 and was maximal in the region of pH 8 (Figure 6). High levels of lysozyme

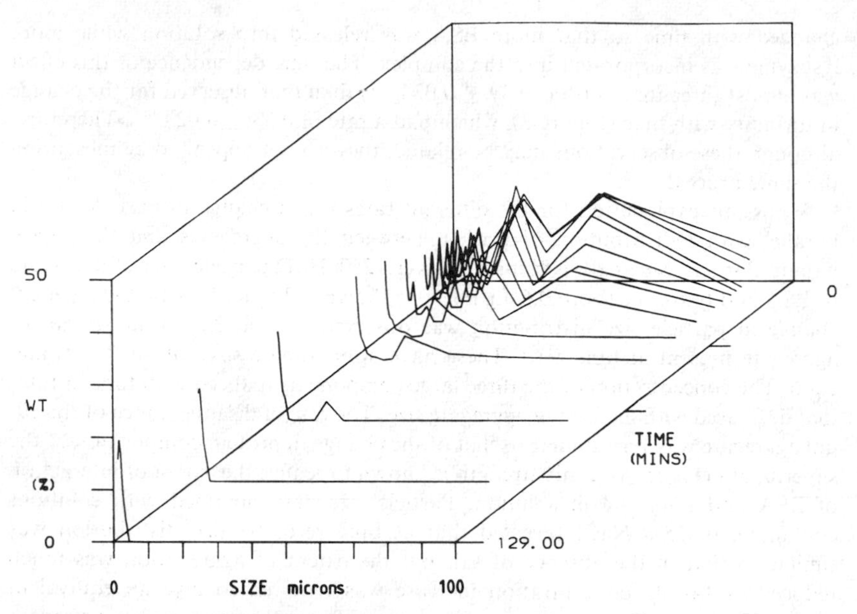

Figure 5 *Time dependence of the particle size distribution for a mixture of BSA and lysozyme (final concentrations* 0.5 mg ml^{-1} *of each) obtained at room temperature using a Malvern* 2600*HSD instrument. The distribution is expressed as weight percentage of the total particulate fraction found in the size bands of average sizes* 86.7, 44.3, 28.7, 20.7, 15.7, 12.1, 9.4, 7.3, 5.7, 4.5, 3.5, 2.7, 2.2, 1.7 *and* 1.3 μm

incorporation into the foam were observed betweeen pH 7 and 9, where the interaction is high.

The ionic strength dependence of the composition of foams formed from BSA–lysozyme mixtures is shown in Figure 7. Incorporation of lysozyme was maximal at an ionic strength of 0.025. Under these conditions, the interaction between the protein is considerably reduced, as judged by turbidity, compared with that at zero ionic strength.[9]

Observations with Suspended Films—The drainage of suspended thin films of BSA (0.5 mg ml^{-1}, pH 8) was found to be dependent on ionic strength. In the absence of salt, BSA formed unstable silver grey films which did not drain to black. Films formed in 0.025 M NaCl were aggregate free and produced common black films after 10–15 min. Drainage followed a regular pattern. First, a uniform grey film formed around the perimeter of the film and then black spots appeared within this region. The black spots remained localized at the film periphery and did not undergo rapid motion within the film, as is often observed with SDS films. The black regions expanded slowly until they fused to form a continuous ring around the film. This resulted in the encirclement and hence trapping of a thick region in

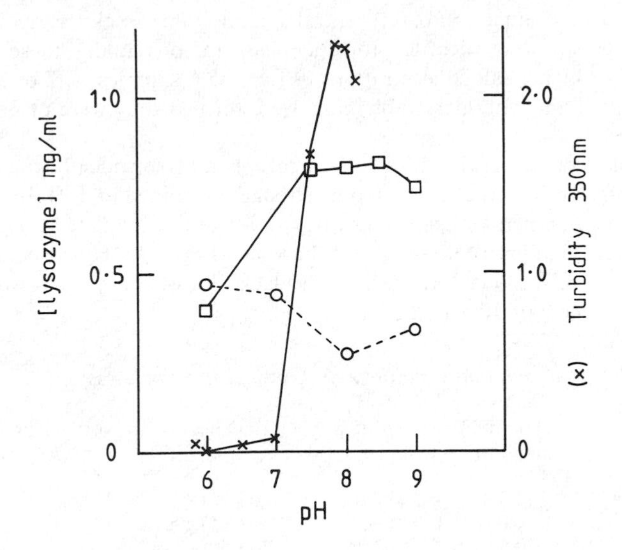

Figure 6 *pH dependence of the composition of the foam formed in the presence (○) and absence (□) of* 0.2 M *NaCl. Results are expressed as the ratio of lysozyme to BSA detected in the foam,* i.e. *the number of milligrams of lysozyme present per milligram of BSA.* X, *Sample turbidity at* 350 nm *in the absence of salt*

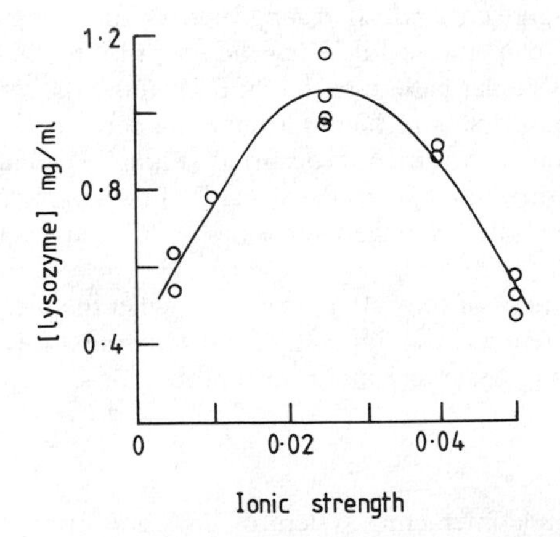

Figure 7 *Ionic strength dependence of the composition of foams formed from mixtures of BSA and lysozyme* (1 mg ml^{-1} *of each) at pH* 8. *The composition is expressed as the ratio of lysozyme to BSA detected in the foam,* i.e. *the number of milligrams of lysozyme present per milligram of BSA.*

the centre of the film which became thicker as the area of black film increased and caused further drainage into the central region. Film thicknesses in the range 15.25 ± 2.66 nm were recorded for the black regions under these conditions (Table 2). At higher salt concentrations (0.1 M), BSA samples were often found to be aggregated and although stable films were formed they were of non-uniform thickness.

Black films of mixtures of BSA (0.5 mg ml^{-1}) and lysozyme (0.5 mg ml^{-1}) were readily obtained over the range of ionic strength examined (0–0.1). In the absence of salt, large numbers of aggregates (mainly 2–5 μm in diameter) were observed, which resulted in films of non-uniform thickness (Figure 8). However, unlike films formed from BSA alone, local regions of black film were rapidly formed with an average thickness of 30.6 nm (Table 2).

Table 2 *Effect of solution conditions on limiting film thickness*

[*BSA*] (mg/ml)	[*Lysozyme*] (mg/ml)	[*NaCl*] (M)	Thickness (nm)	Number of *determinations*
1.0	1.0	—	30.60 (1.72)[a]	7
1.0	1.0	0.025	22.16 (1.06)	4
0.5	0.5	0.025	16.46 (1.50)	4
1.0	1.0	0.100	25.03 (2.59)	4
1.0	—	—	Unstable	
0.5	—	0.025	15.25 (2.66)	3
0.5	—	0.100	15.65 (3.39)	2

[a] The figures in parentheses are standard deviations.

Films formed immediately after mixing from samples containing 0.025 M NaCl contained a few aggregates which disappeared if the sample was allowed to incubate at room temperature for a few hours prior to film formation. Film thinning followed a similar pattern to that of BSA under the same conditions with black film (average thickness 16.46 nm) around the periphery and film thickening in the central region as constriction occurred (Figure 8). Black films were also obtained with mixtures containing 0.1 M NaCl. Few aggregates were observed under these conditions and average thicknesses of 26.07 nm were obtained for the black regions (Table 2).

Re-formation of films of BSA after rupture revealed the presence of aggregates which were not present in fresh films. BSA–lysozyme mixtures showed a reduced tendency towards aggregate formation on rupture.

4 Discussion

The apparently simple interacting system of BSA and lysozyme shows complex ionic strength, pH, temperature, and time dependences in solution. All these factors contribute to the extent and composition of the aggregates present in mixtures of these proteins and they influence the composition of the foam generated. We attempted to standardize our approach and used freshly prepared mixtures immediately in our foaming experiments unless stated otherwise.

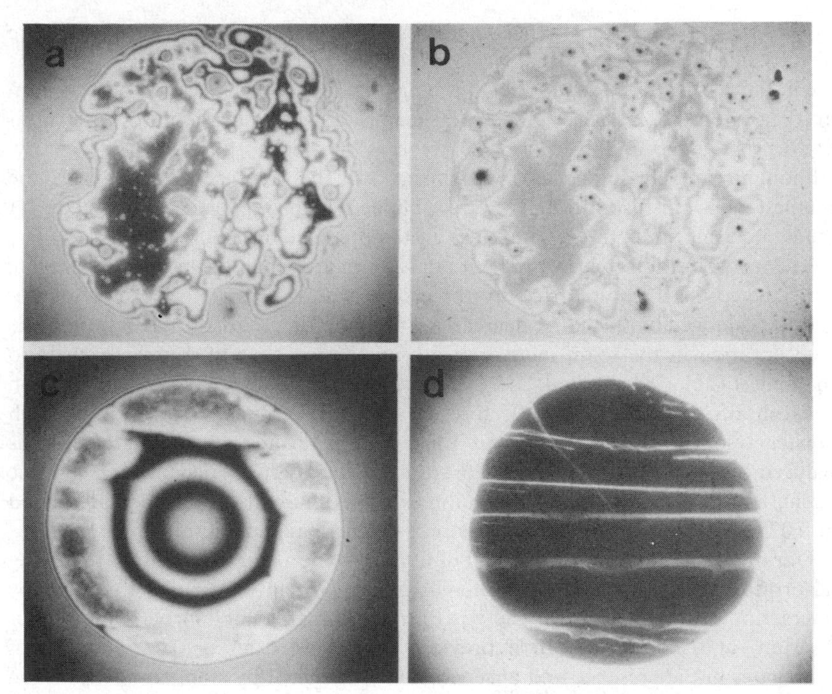

Figure 8 *Photographs of thin films. (a) A thin film formed from a BSA–lysozyme mixture (*0.5 mg ml^{-1} *of each) in the absence of salt immediately after mixing. Note non-uniform thinning and black regions. (b) The same film as (a) photographed under phase contrast illumination to highlight the aggregates present. (c) A film formed under similar conditions to (a) but in the presence of* 0.025 M *salt. Note the reduction in aggregates, black regions forming around the periphery, and thickening of the centre of the film. (d) The same film as (c)* 10 min *later. This film was unusual as it formed a complete black film owing to drainage of the central regions through the striations in the film*

The foam that is collected is well drained, having been slowly forced up a 100 cm column by gas pressure, but inevitably contains some bulk solution in the Plateau borders which is detected in the compositional assay. Thus these results are distorted by a contribution from non-adsorbed protein. This problem also affects other methods of quantification of the protein concentration of foams, *e.g.* polyacrylamide gel electrophoresis. We are currently investigating the preparation of dry foams by forced drainage under reduced pressure.

The limiting thickness of films of BSA has been reported previously but not under the conditions of pH and ionic strength reported here. The thicknesses we obtain for BSA at pH 8 in 0.025 M NaCl of 13.95 nm are considerably greater than 8.9 and 4.35 nm reported for BSA at pH 7 in the absence and presence of 0.05 M NaCl, respectively.[18] The difference in the drainage properties and thickness of films of BSA alone and BSA–lysozyme mixtures implicates the involvement of lysozyme in film formation in the mixed system.

It is unlikely that the large aggregates (2–40 μm) identified in solution are directly involved in film formation. Most of the aggregates observed in the films are in the 2–5-μm size range. The 40-μm aggregates are never seen and the 19-μm aggregates only occasionally. The presence of aggregates in the film causes non-uniform drainage and formation of regions of thicker film in their vicinity. Although relatively stable films were formed in the presence of aggregates, the thinnest and most stable films were obtained with aggregate-free systems at moderate ionic strength (*e.g.* mixtures in the presence of 0.025 M NaCl).

The incorporation of lysozyme in foams is greatest when the levels of aggregate are reduced. This is supported by three observations. First, the level of aggregation, as judged by turbidity, decreases with time (Figures 3 and 5) and lysozyme incorporation in the foam increases with incubation time after mixing prior to foaming (Table 1). Second, the temperature dependence of the interaction between BSA and lysozyme (Figure 2) also affects foam composition,[9] with greater lysozyme incorporation at higher temperatures where aggregation is reduced. Finally, lysozyme incorporation in the foam is dependent on ionic strength but shows unusual behaviour in that maximum incorporation is observed at 0.025 M NaCl and not zero ionic strength where aggregation is maximal.

Our observations support a model of film formation and stabilization involving adsorption of the individual non-aggregated protein molecules followed by interaction at the interface or the adsorption of small soluble protein complexes ($\ll 2\,\mu m$) at the interface. In a previous study[9] we examined the stability of aggregates in the presence and absence of 0.025 M NaCl by stopped-flow dilution experiments. Our results indicated that aggregate dissociation was three times faster in salt and this destabilization of the aggregates enhanced lysozyme incorporation in the foam (Figure 7). However, the rate of dissociation even in salt ($8.84 \times 10\,s^{-1}$) is far too slow to support a model involving aggregate adsorption followed by dissociation. Therefore, it is our view that the presence of large aggregates inhibits foaming and reduces the incorporation of lysozyme into the foam, possibly by physical disruption of the films in the case of the 40-μm species and/or by reducing the concentration of the soluble protein fraction. However, simple depletion of the soluble lysozyme by aggregate formation cannot account solely for reduced lysozyme uptake as we observe increased lysozyme incorporation in foam with time (Table 1) whereas we know that the concentration of free lysozyme decreases with time (Figure 4). It is more likely that the ratio of soluble BSA to lysozyme is a controlling factor. Experiments are in hand to examine this possibility.

5 Conclusions

1. The electrostatic interaction between BSA and lysozyme is dependent on ionic strength, pH, temperature, and time.
2. The composition and size of the aggregates formed change with time.
3. Incorporation of lysozyme in foams is maximal under conditions where interaction occurs but where large aggregates are minimized.
4. The most stable and thinnest films of BSA–lysozyme mixtures are formed under conditions where lysozyme incorporation in foams is maximal.

5. Black films are formed more rapidly in mixtures than by BSA alone but, under conditions where both samples form black films, those formed by BSA are significantly thinner.

Acknowledgements. The authors acknowledge the involvment of Dr. A. C. Pinder and Mr. W. Worts in the design and construction of the thin-film apparatus, Liz Russell for technical assistance and Drs. J. Mingins and M. Coke for helpful discussions during the course of this work.

References

1. P. J. Halling, *CRC Crit. Rev. Food Sci. Nutr.*, 1981, **15**, 155.
2. E. Dickinson, *Food Hydrocolloids*, 1986, **3**, 23.
3. D. E. Graham and M. C. Phillips, *J. Colloid Interface Sci.*, 1979, **70**, 403, 415, 427.
4. J. L. Brash and S. Uniyal, *J. Polym. Sci. Polym. Phys. Ed.*, 1979, **66**, 377
5. B. W. Morrisey, *Ann. N. Y. Acad. Sci.*, 1977, **283**, 50, 80
6. E. Pegson, E. W. Robson, and E. Stainsby, *Colloids Surf.*, 1985, **14**, 135.
7. T. M. Johnston and M. E. Zabit, *J. Food Sci.*, 1981, **46**, 1226
8. S. Poole, S. I. West, and C. L. Walters, *J. Sci. Food Agric.*, 1984, **35**, 701
9. D. C. Clark, A. R. Mackie, L. J. Smith, and D. R. Wilson, *Food Hydrocolloids*, 1988, **2**, 209.
10. M. C. Phillips, *Food Technol.*, 1981, **35**, 50.
11. D. C. Clark, L. J. Smith, and D. R. Wilson, *J. Colloid Interface Sci.*, 1988, **121**, 136.
12. D. C. Clark, J. Mingins, F. E. Sloan, L. J. Smith and D. R. Wilson, in 'Food Emulsions and Foams,' ed. E. Dickinson, Special Publication No. 58, Royal Society of Chemistry, London, 1986, p. 110.
13. E. Daniel and G. Weber, *Biochemistry*, 1966, **5**, 1893.
14. S. E. Halford, *Biochem. J.*, 1975, **149**, 411.
15. I. Weil, *J. Phys. Chem.*, 1966, **70**, 133.
16. R. O. Camerine-Otero and L. A. Day, *Biopolymers*, 1978, **17**, 2241.
17. D. Exerowa, T. Kolarov, and K. Khristov, *Colloids Surf.*, 1987, **22**, 171.
18. P. R. Mussellwhite and J. A. Kitchener, *J. Colloid Interface Sci.*, 1967, **24**, 80.

Steric Stabilization and Flow Properties of Concentrated Casein Micellar Suspensions*

By Mary C. Ambrose Griffin

AFRC INSTITUTE OF FOOD RESEARCH, READING LABORATORY, SHINFIELD, READING BERKS. RG2 9AT, UK

Milk is an important food source from which cheese, butter, yoghourt, and other dairy products are made. Milk (liquid or dried) and milk components (*e.g.* caseinate and whey powders) find uses as ingredients in other food products such as custards, ice cream, meringues, confectionery, baked products, and oven-ready dishes containing sauces or batters. In this paper, studies of the structure of a major component of milk, casein micelles, are described and the effect of alterations of the surface structure of casein micelles in suspension on a macroscopic property, the viscosity, is demonstrated.

Casein micelles form 80% of the protein in bovine milk;[1] they are large (20–600 nm in diameter[2]) colloidal aggregates of the casein polypeptides (α_{s1}-, α_{s2}-, β- and κ-casein) bound together by association with Ca^{2+}, phosphate, and citrate (which comprise 7% of the dry weight of casein micelles).[3,4] The micelles are highly hydrated and deformable.[5] Electron micrographs indicate that they are approximately spherical and that the casein micelles are composed of a large number of small aggregates or submicelles.[3,4] Neutron scattering data are consistent with there being regular repeat units in the micelles of a similar size to the submicelles (15–20 nm).[6,7]

Milk is a highly stable liquid suspension; however, it will clot or gel as a result of various different treatments such as acidification[8] (which provides the basis for yoghourt preparation), incubation with rennet, a mixture of the enzymes chymosin (E.C. 3.4.23.4) and pepsin (E.C. 3.4.23.1),[9] and addition of ethanol.[10] The physical changes which take place in the milk as a result of these treatments arise from coagulation of the casein micelles. The stability of the casein micelles in skimmed milk has been considered by several workers in terms of DLVO (Derjaguin, Landau, Verwey, and Overbeek) theory.[11–13] Green and Crutchfield[11] measured the ζ-potential of native and rennet-treated casein micelles and found that the charge on the casein micelles was reduced by approximately half as a result of renneting. They calculated the potential energy of a pair of casein micelles of radius 53 nm as a function of inter-particle distance using DLVO theory, estimating the double-layer thickness as approximately 1 nm and the Hamaker constant as 1×10^{-21} J. The surface potential was set equal to the

* Royal Society of Chemistry Food Chemistry Group Junior Medal (1987) Lecture.

measured ζ-potential. Their calculations showed that the energy barrier to close approach would be sufficient to prevent aggregation of native micelles, and would be much lower for renneted micelles. However, Payens[12] commented, on the basis of similar calculations, that the energy barrier to approach of two casein micelles would operate only at such short inter-particle separations (*ca.* 0.1 nm) that short-range chemical effects would be dominant. He also suggested that the behaviour of casein micelles in the presence of different electrolytes was inconsistent with purely electrostatic stabilization.[13] However, Derjaguin *et al.*[14] have recently discussed the role of hydration of adsorption layers of sterically stabilized colloids. There is evidence that the solubilities of ions in the water in such adsorption layers may be significantly reduced; one result of this is an increase in the Debye length, which may exceed the thickness of the adsorption layer, in which case there may indeed be an electrostatic barrier to approach of the colloidal particles. As is found for casein micelles,[13] the Schulze–Hardy rule does not hold for such colloidal particles. Derjaguin *et al.*[14] presented calculations of potentials for colloidal particles stabilized by adsorption layers of *non-ionic*, but hydrophilic, polymers; similarly, structuring of water due to solvation of a charged layer might well be expected. In view of these conditions, the role of electrostatic factors in casein micelle stability should be reappraised.[15]

On treatment of casein micelles with chymosin or pepsin the κ-casein polypeptide is cleaved almost specifically at the Phe_{105}—Met_{106} bond.[16] The two products of cleavage, *para*-κ-casein and macropeptide, are very different chemically: the macropeptide appears in the aqueous phase during the clotting of milk. The *N*-terminal *para*-κ-casein is insoluble except in the presence of denaturing agents and has an unusually high concentration of aromatic amino acids. The *C*-terminal macropeptide is glycosylated to varying extents; it has many negatively charged residues, including sialic acid groups. Macropeptide is highly soluble, even in 2% trichloroacetic acid, and is negatively charged over a wide pH range, whereas *para*-κ-casein is positively charged. The amphipathic nature of κ-casein was pointed out by Hill and Wake[17] (this was reflected in some results of ours from Langmuir trough experiments, where differing abilities of the *para*-κ-casein and macropeptide sections of κ-casein to interact with and alter the surface pressure of dimyristoylphosphatidylcholine monolayers were manifested[18]). It has been suggested that κ-casein exists in solution as micellar aggregates[19] (and super-aggregates[20]).

κ-Casein has also been shown to prevent precipitation of the other caseins in the presence of Ca^{2+}; it is said to solubilize the other caseins.[21] The quantities of κ-casein found in different size classes of casein micelles are proportional to micellar surface area;[22] it has therefore been suggested that κ-casein occupies a position on the surface of the micelles.[21,22]

A model was proposed by Walstra[23] for casein micelle structure in which the micelles are sterically stabilized by the protrusion of hydrophilic regions of the casein polypeptides, particularly the macropeptide segments of κ-casein (illustrated in Figure 1 in highly schematic form). (Similar ideas had previously been discussed by Holt[24] and Schmidt *et al.*[25]) This model is consistent with the properties of casein micelle suspensions; *e.g.* treatment with rennet causes removal of the steric stabilization layer, thus resulting in micellar coagulation. It also

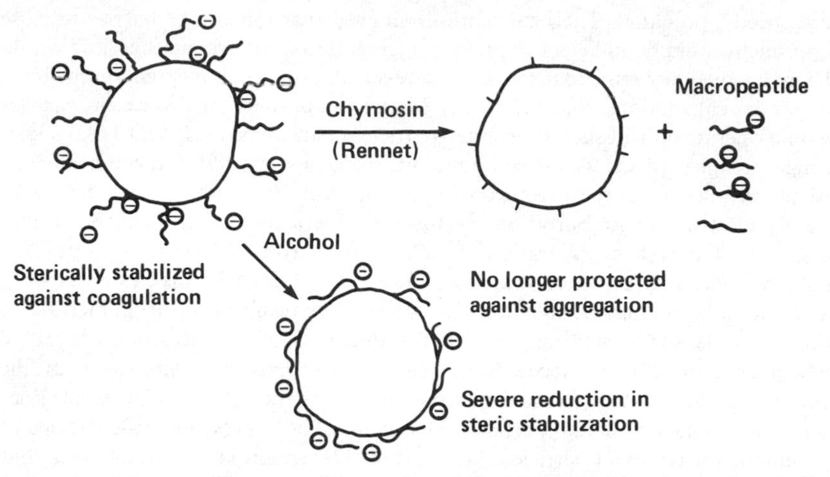

Figure 1 *Schematic diagram of the steric stabilization model for the casein micelle* (After ideas of Walstra[23] and Horne[33])

explains the results of Scott Blair and Oosthuizen,[26] who found that the intrinsic viscosity of a reconstituted fat-free dried milk decreased initially by approximately 17% as a result of rennet treatment, before the onset of coagulation of the milk, after which the viscosity increased. The voluminosity, and hence the volume fraction, of micelles sterically stabilized by a layer of macropeptide segments would decrease as a result of removal of this layer.

Walstra *et al.*[27] provided further experimental evidence of changes in the hydrodynamic size of casein micelles as a result of rennet treatment. Photon correlation spectroscopy (PCS) was used to measure the hydrodynamic diameter of casein micelles in diluted skimmed milk at 30 °C as a function of time after treatment with chymosin. The hydrodynamic diameter (which is an intensity-weighted average), measured at a single scattering vector, was found to decrease initially by approximately 10 nm, followed by an increase in size as coagulation commenced. In view of possible ambiguities in interpretation of PCS data from such a polydisperse sample as casein micelles from skimmed milk, a more detailed study of diluted, native casein micelles, before and after treatment with chymosin, was carried out in our laboratory, data being collected over a wide range of scattering vectors (Figure 2).[28] The sample was maintained at 4 °C throughout the experiment so that aggregation of renneted casein micelles would be negligible[29] (the hydrodynamic diameter of native casein micelles was not affected by temperature between 4 and 35 °C). The increase in the intensity-weighted average diffusion coefficient observed at each scattering vector, K, after treatment with chymosin was consistent with a decrease in hydrodynamic diameter of 8.0 ± 1.3 nm.

Other experimental evidence in support of the steric stabilization model has come from ^{1}H NMR spectroscopy. It has been possible, using ^{1}H NMR spectroscopy, to observe regions of polypeptide of high mobility in proteins, *e.g.*

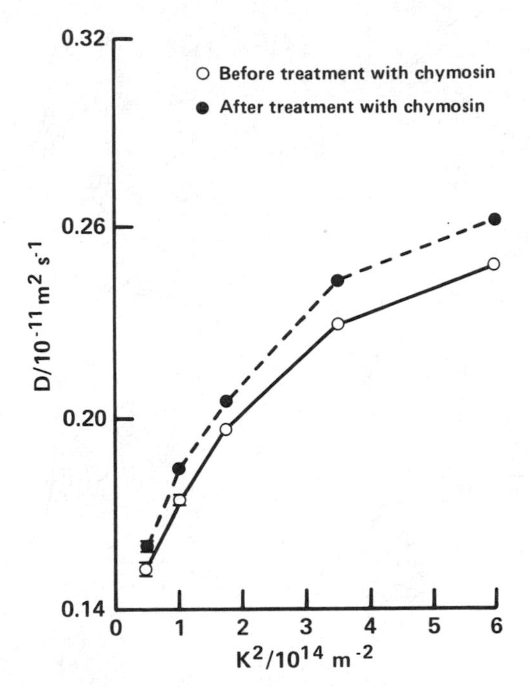

Figure 2 *Angular variation of the measured intensity-weighted average diffusion coefficient,* D_o*, of casein micelles before (○) and after (●) treatment with chymosin.* D_o *was corrected for temperature as described in reference 28. (It should be pointed out that in Figure 3 in reference 28 the error bars represent the standard deviations of the measurements. Here, the standard error has been indicated, where it is larger than the dimensions of the symbols on the graph)*

at the active site of the E2 subunits in the pyruvate dehydrogenase multienzyme complex.[30] If the casein micelles were sterically stabilized by a layer of polypeptide, the protons in the segments forming this layer might be expected to have greater conformational mobility than the core regions of the casein micelles. Figure 3 shows ^1H NMR spectra obtained from casein micelles suspended in a simulated milk salts solution prepared in ^2H_2O. As is discussed in detail previously,[31] the linewidths of the moderately sharp lines visible above the broad envelope of resonances in Figure 3a indicate that there are regions of polypeptide with considerable mobility. Treatment of the casein micelles with chymosin caused the sample to gel in the NMR tube; the corresponding ^1H NMR spectrum is shown in Figure 3c. The moderately sharp lines in Figure 3a have become narrower, but the chemical shifts and relative intensities of these signals are essentially unchanged (*cf.* Figure 3b, a convolution difference spectrum of casein micelles, and Figure 3c). The underlying broad signals have become much broader as a result of the immobilization of the *para*-casein micelle centres due to coagulation.

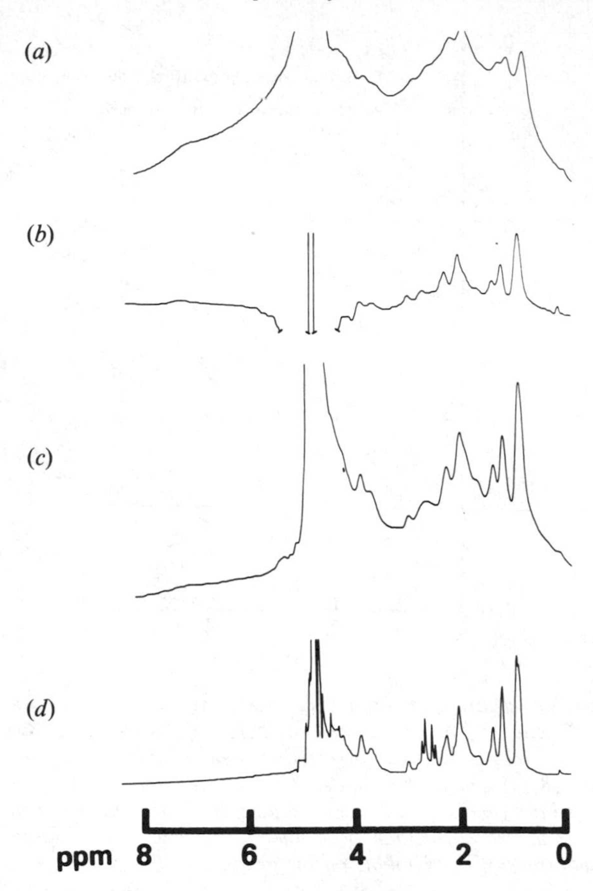

Figure 3 270-MHz ^{1}H *NMR specta of (a) casein micelles, (b) casein micelles (convolution difference spectrum), (c) casein micelles after addition of chymosin and (d) macropeptide isolated from chymosin-treated casein micelles*[31]

The spectrum of the isolated macropeptide is shown in Figure 3d. The peaks are much sharper than those in Figures 3a–c, but the relative intensities of the different resonances, especially in the regions $\delta = 0$–2.3 p.pm., are very similar to those in Figures 3b and c. No sharp peaks are observed in any of the spectra in the region to low field of the water signal ($\delta \approx 4.6$ p.p.m.). The flexible segments of polypeptide must therefore be very poor in aromatic amino acid residues; macropeptide is a 64-residue stretch of polypeptide without any aromatic groups. There is thus a similarity in chemical composition between the macropeptide and the mobile regions of the casein micelles; this is consistent with the identification of the macropeptide segments as forming the casein micelle outer layer. (The quartet centred at 2.63 p.p.m. in the macropeptide spectrum is due to the citrate present in the buffer. The broadening of these peaks in the presence of casein

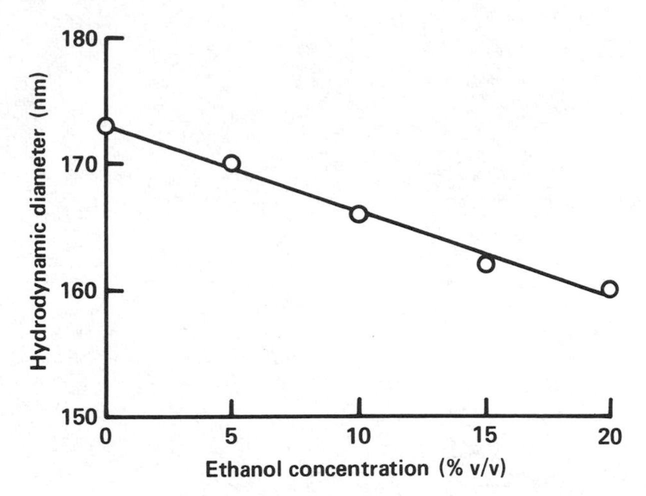

Figure 4 *Effect of ethanol on the intensity-weighted average hydrodynamic diameter of casein micelles after dilution of skimmed milk into* 10 mM *imidazole buffer* (*pH* 7.0) *containing* 5 mM *$CaCl_2$*

micelles may be due to exchange of citrate in free solution with citrate bound to the micelles.)

Ethanol will precipitate milk; a variation of the alcohol stability test has been developed to give an indication of the extent of proteolysis of κ-casein, which may result from bacterial spoilage.[32] Horne[33] showed that the addition of ethanol (up to 20% v/v) to a casein micelle fraction of narrow size distribution caused a progressive decrease in the hydrodynamic radius of the micelles; Figure 4 shows results from our experiments[34] in which skimmed milk was diluted into the same buffer system containing ethanol that Horne used. According to Horne,[33] the effect of alcohol on the casein micelle is to cause collapse of the steric stabilization layer, thus initiating coagulation (Figure 1). This idea could also be examined using ^{1}H NMR spectroscopy, and Figure 5 shows spectra of casein micelles with and without ethanol; Figure 5a is as Figure 3a, and Figures 5b and 5c show the effects on the spectrum of adding [$^{2}H_6$]ethanol. There is progressive broadening on addition of [$^{2}H_6$]ethanol, and at 18% v/v [$^{2}H_6$]ethanol only a broad envelope of resonance due to slowly moving protons, is observed (apart from the appearance of the sharp peaks due to traces of ^{1}H in the [$^{2}H_6$]ethanol in the mixture). These results are entirely consistent with the model drawn schematically in Figure 1. [It should be noted from Figure 4 that the reduction in hydrodynamic diameter of casein micelles in the presence of 20% ethanol is larger than that caused by chymosin treatment; Horne made a similar observation. The difference may be due to effects of ethanol on the internal structure of the casein micelle or on the surface structure, through alterations of the conformation of the κ-casein (see below)].

Alcohols are known, from studies of optical rotatory dispersion,[35,36] to affect the secondary structure of proteins, resulting in an increase in the proportion of

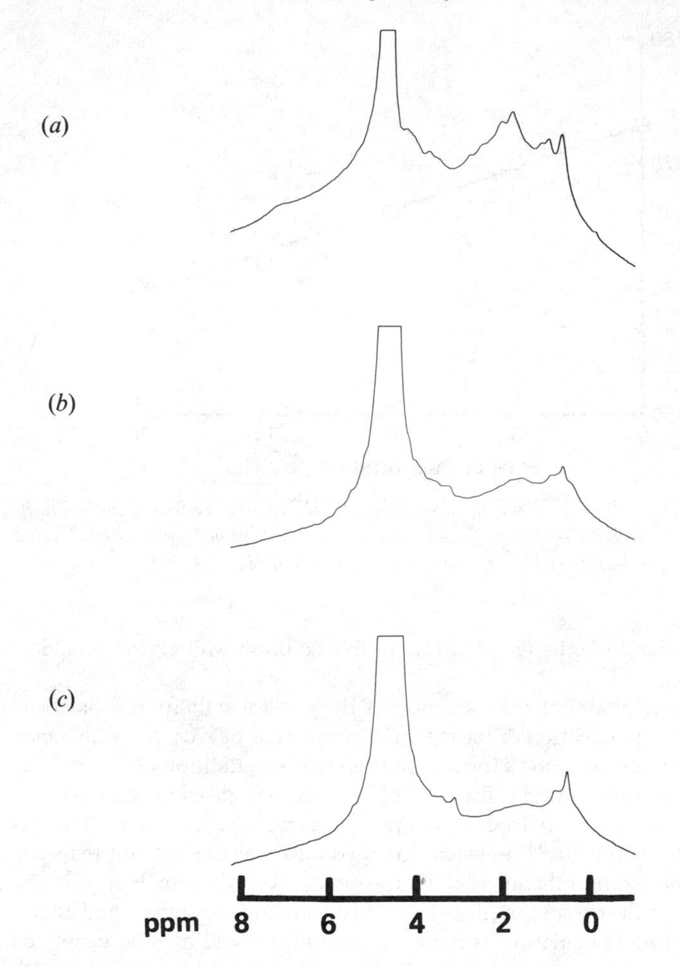

Figure 5 270 MHz ^1H *NMR spectra of casein micelles to which (a)* 0%, *(b)* 9%, *and (c)* 18% v/v [$^2H_6$]*ethanol had been added*

α-helix structure, the higher homologues having a greater effect at the same molar concentration. In order to investigate whether the reduction in thickness of the macropeptide surface layer caused by the presence of ethanol was related to this effect, studies of κ-casein and macropeptide conformation were carried out by means of circular dichroism (CD) spectroscopy.[37] The CD spectrum of κ-casein in phosphate buffer at pH 7.2 (Figure 6) corresponds to a protein with low helix content (6.3%) and significant amounts of β-sheet (37.3%). CD spectra were also obtained in the presence of increasing amounts of four different alcohols, methanol, ethanol, propan-1-ol, and 2,2,2-trifluoroethanol (TFE) (which is known to be a very effective inducer of α-helix conformation[38]). The effects of the four alcohols were qualitatively similar; Figure 6 also shows the CD spectra

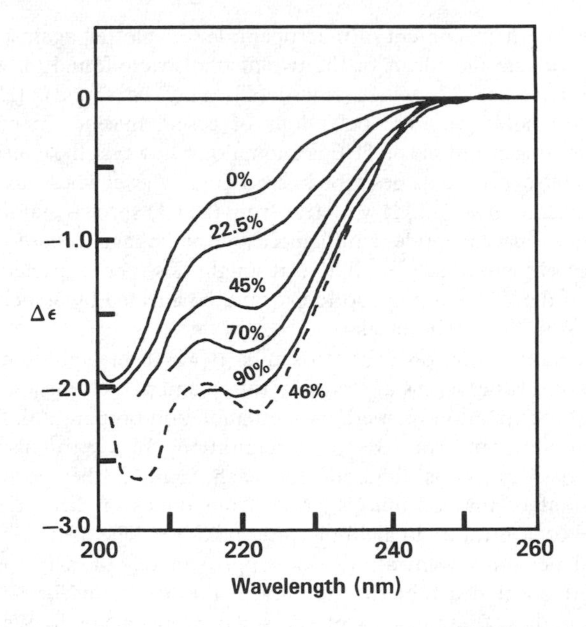

Figure 6 *Effect of ethanol (——) and TFE (- - -) on the far-UV CD spectrum of κ-casein in* 10 mM *phosphate buffer (pH* 7.2*) Alcohol concentrations (*% v/v*) are indicated*

obtained as a result of addition of ethanol and, for comparison, of a saturating amount of TFE. As the concentration of alcohol increased, the intensity of the spectrum increased, the changes being consistent with increasing amounts of α-helix. The molar circular dichroism at 222 nm, $\Delta\varepsilon_{222\ \mathrm{nm}}$, became increasingly negative, the final saturating value obtained for $\Delta\varepsilon_{222\ \mathrm{nm}}$ being the same for each alcohol, suggesting that the same structural change was being induced (below this wavelength there is strong absorption of light by ethanol, resulting in noise in the CD spectra; this is the reason why the spectra in Figure 6, of samples containing high concentrations of ethanol, are truncated at low wavelengths). The efficiency in inducing the decrease in $\Delta\varepsilon_{222\ \mathrm{nm}}$ was in the order TFE > propan-1-ol > ethanol > methanol. The hydrodynamic diameter of casein micelles from skimmed milk diluted into aqueous buffer–alcohol mixtures was measured by PCS (*cf.* Figure 4); in Figure 7 the *change* in hydrodynamic diameter, relative to the diameter at 0% alcohol, is plotted against the change in α-helix content, for each alcohol concentration used for the PCS measurements. The change in α-helix content is given as the percentage of the total change in α-helix content obtained for saturating amounts of alcohol, and was obtained by interpolation of the measurements of $\Delta\varepsilon_{222\ \mathrm{nm}}$ as a function of alcohol concentration (*cf.* Figure 2 in reference 37). The data for the different alcohols lie on different lines, indicating that the decrease in casein micelle size and the induction of α-helix structure in κ-casein caused by the presence of alcohols are not related in a simple way. CD spectra were also run of macropeptide in the presence of ethanol and TFE. In a similar way the change in α-helix content of macropeptide was plotted against change in casein micelle size, and the effects of the two alcohols were found to be different (Figure 6a in reference 37, where full details may be found). [Horne and Davidson[39] have noted that the behaviour of casein micelles in milk in the presence of high concentrations of TFE is anomalous; however the concentrations of TFE used in the experiments described here were at a level where no anomalies were reported (below 16% v/v).] It was clear from the CD spectra that the solvent-induced folding of macropeptide is influenced by the linkage to *para*-κ-casein in the whole κ-casein molecule. Similarly, it might also be expected that the conformation of the κ-casein polypeptides would be affected by association with the other caseins in the casein micelle.

The lowest concentration of alcohol in milk at which precipitation of casein micelles is observed varies with Ca^{2+} concentration and pH. Horne and Parker,[40] from a study of precipitation by methanol, ethanol, and propan-2-ol, found that the lowest concentration required for precipitation, at a particular pH and concentration of Ca^{2+}, was different for each alcohol; they estimated the dielectric constant of the continuous phase from tables of dielectric constant against alcohol concentration in alcohol–water mixtures. The dielectric constants of these critical alcohol concentrations were approximately the same; Horne and Parker therefore concluded that the stability of the casein micelles is governed primarily by the dielectric constant of the surrounding medium. We therefore plotted (not shown) our values of $\Delta\varepsilon_{222\ \mathrm{nm}}$ for κ-casein against dielectric constant of the medium. Dielectric constants of TFE–water mixtures were measured by Clark.[41] The graph showed that the data for methanol and ethanol lay, within experimental error, on the same curve, whereas the two sets of data for propan-1-ol and TFE[41] formed distinct and different curves. We do not have any data for propan-2-ol to know whether they would lie on the same curve as methanol and ethanol. Other workers have commented that the effects of alcohols on protein conformation cannot be correlated with changes in dielectric constant (see discussion in reference 40). If indeed the precipitation of casein micelles by alcohols is caused primarily by the alteration of the dielectric constant of the medium, then it is unlikely that the precipitation process is related in a simple way to the induction of α-helix structure in the casein polypeptides.

As CD spectroscopy showed that the solvent-induced folding of macropeptide was influenced by the proximity to *para*-κ-casein in the whole κ-casein molecule, it might also be expected that the conformation of the κ-casein polypeptides would be affected by association with the other caseins in the casein micelle. Although the decrease in casein micelle hydrodynamic diameter caused by the presence of alcohols cannot be explained simply in terms of the induction of α-helix structure measured with purified κ-casein (Figure 7), such conformational changes in the micelle-bound κ-casein may still be important in determining casein micelle hydrodynamic diameter and stability.

Whatever the exact conformational changes which take place in the macropeptide segments in the presence of alcohols may be, casein micelles are reduced in size in the presence of subcritical concentrations of alcohol[33] (Figures 4 and 7). It was therefore expected that the relative viscosities, η_r ($\eta_r = \eta/\eta_0$, where η is the viscosity of the suspension and η_0 is the viscosity of the continuous phase), of

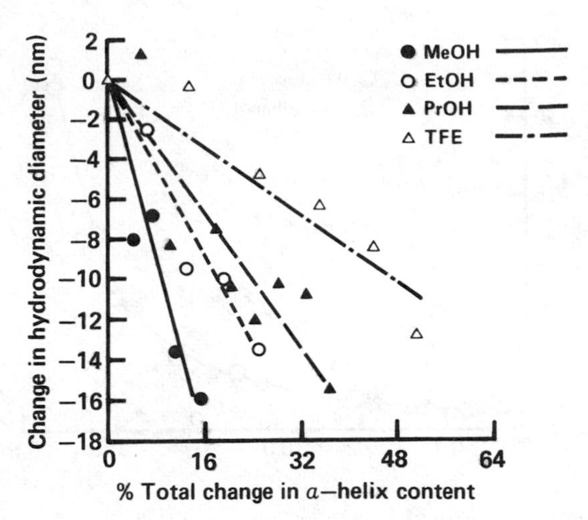

Figure 7 *Change in intensity-weighted average hydrodynamic diameter* (nm) *of casein micelles against the percentage of the total change in α-helix content for purified κ-casein*

casein micelle suspensions, such as skimmed milks of varying concentrations, would be reduced by the inclusion of subcritical concentrations of ethanol. Skimmed milk was therefore concentrated approximately 4.2-fold by ultrafiltration; two series of dilutions were prepared. To one of these series of dilutions ethanol was added to give a concentration of 9.8% v/v in the continuous phase. Viscosities were measured with Ostwald U-tube viscometers. From earlier experience with similar samples of skimmed milks in which we measured viscosities with a Contraves low-shear viscometer, we expected the viscosities of the more dilute samples to be Newtonian, within experimental error. For comparison, the viscosities of the more concentrated samples were also measured over a range of applied stress on a Carrimed controlled-stress rheometer; the U-tube measurements were in good agreement with the low shear plateau value of the viscosity (full details are given in reference 42). It was found, by means of size-exclusion chromatography,[43] that incubation with ethanol did not cause dissociation or aggregation of the casein micelles. It therefore seemed reasonable to attempt to explain the changes in viscosity that were observed as a result of the inclusion of ethanol in terms of a reduction in volume fraction of the casein micelles.

The viscometric data were fitted to a form of the equation of Krieger and Dougherty,[44] modified by expressing the volume fraction of the casein micelles as cV, where c is the concentration of casein and V the voluminosity of casein,[42] to give

$$\eta_r = (1 - cV/\varphi_\infty)^{-2.5\varphi_\infty} \qquad (1)$$

where φ_∞ is the volume fraction at which the relative viscosity would become infinite.

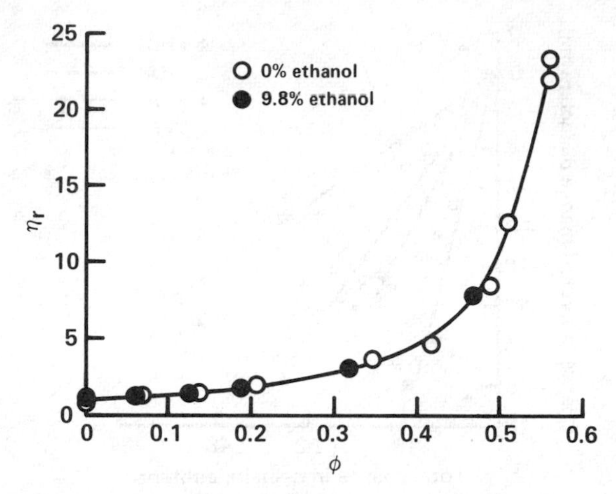

Figure 8 *Relative viscosities of skimmed milks, η_r, plotted against volume fraction, φ obtained as described in the text and in reference* 42. *The line is the least-squares fit to equation* (1)

The voluminosities of the casein in natural serum and in the presence of 9.8% ethanol were found to be 4.60 and 4.23 ml g^{-1}, respectively, while the value of φ_∞ was 0.644. In Figure 8, η_r is plotted against casein volume fraction for both sets of samples; the data points lie on the same curve, which justifies this simple approach to the interpretation of the data. As expected for a polydisperse material, because smaller particles can fit into the interstices between the larger particles, the value for φ_∞ obtained is larger than the value of 0.60 quoted for monodisperse hard sphere colloids.[45]

The reduction in voluminosity of casein micelles in the presence of 9.8% ethanol corresponds to a decrease in micellar hydrodynamic diameter of approximately 5 nm,[42] which should be compared with the decrease in hydrodynamic diameter of 7 nm observed by PCS (Figure 4) in dilute solution. It is possible that the small difference between these two estimates for the decrease in hydrodynamic diameter reflects a slight difference in micelle conformation between the dilute and concentrated suspensions, *e.g.* differing flexibility of the surface coating in the two solvent environments. However, both estimates are approximate, and the two techniques give different averages; we concluded that the measurements made in dilute and concentrated suspension by the two different methods are in fairly good agreement.

There is still disagreement in the literature concerning both the quaternary structure of the casein polypeptides in the casein micelle[3,4,46] and the crystal form of the colloidal calcium phosphate.[47,48] However, as has been described in this paper, there is considerable experimental evidence from light scattering and NMR spectroscopy to show that the κ-casein polypeptides are a major structural component at the surface of the casein micelles, where the macropeptide segments probably form a steric stabilization layer. The effects of alcohols in altering this outer layer have been examined by the same techniques and by CD spectroscopy. The details concerning the distribution of κ-casein at the micelle surface and the internal structure of the micelle remain as fascinating unsolved problems.

The main motivation for food chemists to study the structure of food components at a macromolecular level is the potential to understand and predict macroscopic properties. Here, we have been able to predict and confirm experimentally the nature of the changes in the rheological behaviour of casein micellar suspensions as a result of structural changes caused by altering the solvent environment. The technological benefits of such understanding are self-evident.

Acknowledgements. I thank all my collaborators, especially William Griffin, for criticisms and comments. The data in Figures 5, 6, and 7 were obtained in collaboration with Professor G. C. K. Roberts and Dr S. R. Martin. I thank Jaci Geens and John Price for assistance with the preparation of the figures.

References

1. R. Jenness, in 'Milk Proteins, Chemistry and Molecular Biology. I' ed. H. A. McKenzie, Academic Press, New York, 1970, p. 17.
2. T. C. A. McGann, W. J. Donnelly, R. D. Kearney, and W. Buchheim, *Biochim. Biophys. Acta*, 1980, **630**, 261.
3. D. G. Schmidt, *Neth. Milk Dairy J.*, 1980, **34**, 42.
4. D. G. Schmidt, in 'Developments in Dairy Chemistry. I. Proteins, ed. P. F. Fox, Applied Science, Barking, 1982, p. 61.
5. V. A. Bloomfield and C. V. Morr, *Neth. Milk Dairy J.*, 1973, **27**, 103.
6. P. H. Stothart and D. J. Cebula, *J. Mol. Biol.*, 1982, **160**, 391.
7. P. H. Stothart, submitted for publication.
8. B. E. Brooker, *Dairy Ind. Int.*, 1987, August, 17.
9. D. G. Dalgleish, in 'Developments in Dairy Chemistry. I. Proteins, ed. P. F. Fox, Applied Science, Barking, 1982, p. 157.
10. R. T. Eddison, O. Kempthorne, Z. Hosking, H. Barkworth, C. P. Cox, and A. Rowlands, *J. Dairy Res.*, 1951, **18**, 43.
11. M. L. Green and G. Crutchfield, *J. Dairy Res.*, 1971, **38**, 151.
12. T. A. J. Payens, *Biophys. Chem.*, 1977, **6**, 263
13. T. A. J. Payens, *J. Dairy Res.*, 1979, **46**, 291.
14. B. V. Derjaguin, S. S. Dukhin, and A. E. Yaroshchuk, *J. Colloid Interface Sci.*, 1987, **115**, 234.
15. W. G. Griffin and M. C. A. Griffin, in preparation.
16. J. Jollès, C. Alais and P. Jollès, *Biochim. Biophys. Acta*, 1968, **168**, 591.
17. R. J. Hill and R. G. Wake, *Nature (London)*, 1969, **221**, 635.
18. M. C. A. Griffin, R. B. Infante, and R. A. Klein *Chem. Phys. Lipids*, 1984, **36**, 91.
19. H. J. Vreeman, *J. Dairy Res.*, 1979, **46**, 271.
20. A. Thurn, W. Burchard, and R. Niki, *Colloid Polym. Sci.*, 1987, **265**, 653.
21. D. F. Waugh, In 'Milk Proteins, Chemistry and Molecular Biology, II,' ed. H. A. McKenzie, Academic Press, New York, 1971, p. 3.
22. W. J. Donnelly, G. P. McNeill, W. Buchheim, and T. C. A. McGann, *Biochim. Biophys. Acta*, 1984, **789**, 136.
23. P. Walstra, *J. Dairy Res.*, 1979, **46**, 317.
24. C. Holt, 'Proceedings, International Conference on Colloid and Surface Science, Budapest', ed. E. Wolfram, Akadémai Kiadó, Budapest, 1975, p. 641.
25. D. G. Schmidt, P. Both, B. W. Van Markwijk, and W. Buchheim, *Biochim. Biophys. Acta*, 1974, **365**, 72.
26. G. W. Scott Blair and J. C. Oosthuizen, *J. Dairy Res.*, 1961, **28**, 165.

27. P. Walstra, V. A. Bloomfield, G. J. Wei, and R. Jenness, *Biochim. Biophys. Acta*, 1981, **669**, 258.
28. M. C. A. Griffin, *J. Colloid Interface Sci.*, 1987, **115**, 499.
29. N. J. Berridge, *Nature (London)*, 1942, **149**, 194.
30. R. N. Perham, H. W. Duckworth, and G. C. K. Roberts, *Nature (London)*, 1981, **292**, 474.
31. M. C. A. Griffin and G. C. K. Robaerts, *Biochem. J.*, 1985, **228**, 273.
32. E. Valles, R. Mourgues, J. Vandeweghe, and L. Thiebaud, *Tech. Lait. Marketing*, 1986, December, 33.
33. D. S. Horne, *Biopolymers*, 1984, **23**, 939.
34. J. C. Price and M. C. A. Griffin, unpublished results.
35. B. J. Jirgensons, *J. Biol. Chem.*, 1967, **262**, 912.
36. T. T. Herskovits, B. Gadegbeku, and H. Jaillet, *J. Biol. Chem.*, 1970, **245**, 2588.
37. M. C. A. Griffin, J. C. Price, and S. R. Martin, *Int. J. Biol. Macromol.*, 1986, **8**, 367.
38. P. M. Bayley, D. C. Clark, and S. R. Martin, *Biopolymers*, 1983, **22**, 87.
39. D. S. Horne and C. M. Davidson, *Milchwissenschaft*, 1987, **42**, 509.
40. D. S. Horne and T. G. Parker, *Int. J. Biol. Macromol.*, 1981, **3**, 399.
41. D. C. Clark, personal communication.
42. M. C. A. Griffin, J. C. Price, and W. G. Griffin, *J. Colloid Interface Sci.*, 1989, **128**, 223.
43. M. C. A. Griffin and M. Anderson, *Biochim. Biophys. Acta*, 1983, **748**, 453.
44. I. M. Krieger and T. J. Dougherty, *Trans. Soc. Rheol.*, 1959, **3**, 137.
45. R. C. Ball and P. Richmond, *Phys. Chem. Liq.*, 1980, **9**, 99.
46. I. Heertje, J. Visser, and P. Smits, *Food Microstruct.*, 1985, **4**, 267.
47. C. Holt, S. S. Hasnain, and D. W. L. Hukins, *Biochim. Biophys. Acta*, 1982, **719**, 299.
48. R. L. J. Lyster, S. Mann, S. B. Parker, and R. J. P. Williams, *Biochim. Biophys. Acta*, 1984, **801**, 315.

Effect of Adsorbed Proteins on Interactions Between Emulsion Droplets

By L. R. Fisher and E. E. Mitchell

CSIRO DIVISON OF FOOD PROCESSING, P.O. BOX 52, NORTH RYDE, N.S.W. 2113, AUSTRALIA

1 Introduction

For emulsion droplets to coalesce, the film of continuous-phase liquid between them must drain to a thickness where it will collapse. Emulsion stabilizers adsorb at the droplet–continuous phase interface, reducing the chance of collapse of the draining film and/or subsequent coalescence. The exact mechanism of stabilization appears to depend on the type of stabilizer used.[1]

For proteins, a popular model is that 'the high viscoelasticity of the protein film opposes the surface deformations (either in shear or dilation) that are required for the latter stages of drainage and for rupture of the lamellae.[2] There is, however, very little hard evidence for this model, or indeed for any other model of emulsion stabilization by proteins.

We have now used optical reflectance to measure the thickness of the draining film between approaching triglyceride droplets in aqueous protein solutions as a function of time. We studied two proteins (lysozyme and β-casein) in some detail, and here compare the effects of the adsorbed proteins on the drainage and rupture behaviour of the aqueous lamellae between the droplets.

2 Materials

The triglyceride chosen was MCT810 ('medium chain triglyceride,' mean M.W. = 480; Nihon Oils and Fats), composed of glycerol esterified with a mixture of octanoic and decanoic acids. It is a pure synthetic triglyceride whose interfacial tension and other physical properties are well characterized, in contrast to the situation with the vegetable oils.[3]

Water was taken from a Milli-Q water purification system (Millipore, Bedford, MA, USA), with the final purification step being filtration through a 0.22-μm filter. The conductance of the water was <0.9 μS.

Hydrochloric acid (Ajax Chemicals, Auburn, Australia) and sodium chloride (BDH Chemicals, Australia) were of AnalaR grade. 2-Mercaptoethanol was obtained from Sigma Chemical (St. Louis, MO, USA). Sodium dodecyl sulphate was of 'specially pure' grade (BDH Chemicals, Poole, UK). Lysozyme was obtained from Sigma Chemical or, for some experiments, as a purified preparation from hen's eggs (a gift of Ms J. Back, CSIRO Division of Food Research). Purified β-casein was a gift from Dr. R. J. Pearce, CSIRO Division of Food Processing.

3 Methods

Interfacial tension measurements were performed using the pendant drop technique.[3] Optical reflectance measurements of the thickness of the draining film were performed using the apparatus illustrated in Figure 1. The apparatus consists of two hollow Perspex septa filled with the triglyceride and connected via Teflon tubing to syringes containing the triglyceride. Hemispherical triglyceride droplets, each of radius 1.23 mm, are extruded from the septa by depressing the syringes, a fresh pair of droplets being created for each experiment. The septa are mounted on stainless-steel support plates which can be aligned micrometrically and moved towards each other at controlled speeds by a motor drive. After extrusion and equilibration with the aqueous protein solution for a defined period, the droplets are driven towards each other until they become mutually

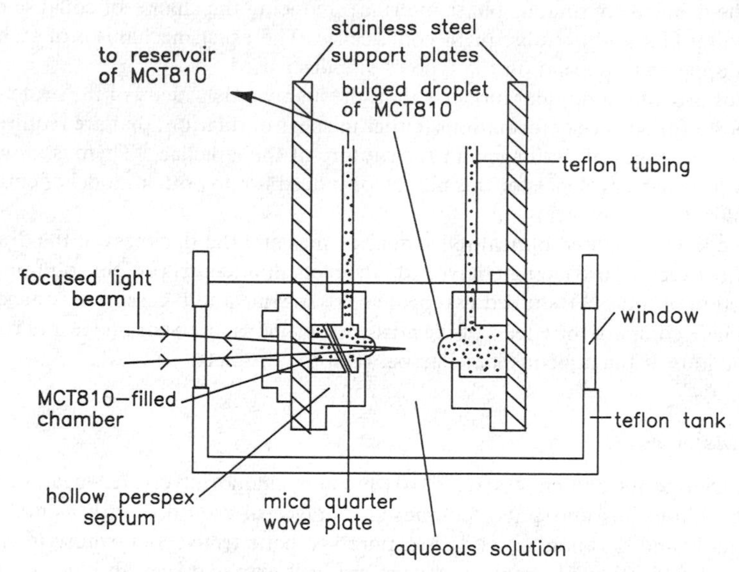

Figure 1 *Median vertical section of apparatus. Transparent Perspex septa, fed by a syringe, contain the oil phase. The transparent septa make the observation of air or water bubbles easy. Stainless-steel face plates are fitted to the septa to provide a lyophobic surface for the efficient capture of the triglyceride droplets. The right-hand septum can be driven towards the left-hand septum at a constant rate by a geared-down stepping motor. The incident laser beam is coincident with the symmetry axis of the two droplets. The detailed optical arrangement, including the function of the illustrated mica quarter-wave plate, is described in reference* 4

flattened at the apex. Once a sufficiently large flattened area is established (diameter 0.15–0.3 mm), producing a draining film with plan-parallel surfaces, the motor drive is switched off. The process of approach and deformation of the droplets can be viewed side-on with a calibrated microscope. All experiments were performed at laboratory ambient temperature (23–27 °C).

The film thickness is calculated from the reflectance of a 2 mW He–Ne laser beam (wavelength 632.8 nm), focused in the plane of the draining film to a spot of *ca.* 30 μm diameter. The lateral resolution is thus limited to this order of magnitude. The resolution in film *thickness*, though, varies from ± 1 to ± 5 nm, depending on the actual value of the film thickness. A particular advantage of the actual experimental arrangement (Figure 1) is that the focused spot can be traversed across the film, so that lateral variations in film thickness can be measured.

Detailed descriptions of the apparatus and of the methods for calculating film thickness from the measured reflectance are given elsewhere.[4-6]

4 Results and Discussion

To select appropriate conditions for protein concentration, solution age, pH, *etc.*, studies were first performed on the time dependence of the interfacial tension between the various protein solutions and MCT810. The results of these studies are summarized in Appendix 1. The conditions chosen for most experiments were protein concentration 10^{-3}% w/w and pH = 5. Salt concentration and the ages of the protein solution and the adsorbed protein film were treated as variables.

Effect of Adsorbed Lysozyme on Thin Film Drainage and Droplet Coalescence— Figure 2a (left hand side) shows a typical record of reflectance as a function of time when two triglyceride droplets approach in aqueous lysozyme solution. There is initially a rapid series of oscillations in the intensity of the reflected beam (Fabry–Perot interference fringes). A complete oscillation corresponds to a change in the distance between the droplets of $\lambda/2n = 632.8/(2 \times 1.333) = 237.4$ nm, where λ is the wavelength of the incident light *in vacuo* and n is the refractive index of the continuous phase liquid.

When the droplets are sufficiently close, they begin to flatten at the apex and the rate of approach becomes limited by the rate at which the continuous phase liquid can drain from between them. The profile of the draining film is necessarily that of a dimple,[7] with the film being thickest at the centre. This shape is a consequence of the requirement for a decreasing pressure from the centre outwards if drainage is to proceed. Nevertheless, for a sufficiently slow rate of approach the shape of the dimple is closely approximated by a plane-parallel film.[8] The rate of approach used in the present experiments (0.14 μm s^{-1}) was slow enough to ensure that this criterion was met [reference 5 and Figure 2*a* (right hand side)].

The drainage of films stabilized by small-molecule surfactants typically follows a pattern where the film thins continuously with time until it reaches a 'critical' thickness, at which stage the film ruptures.[5,9,10] Rupture is thought to be a consequence of constructive interference between the ubiquitous thermal capillary

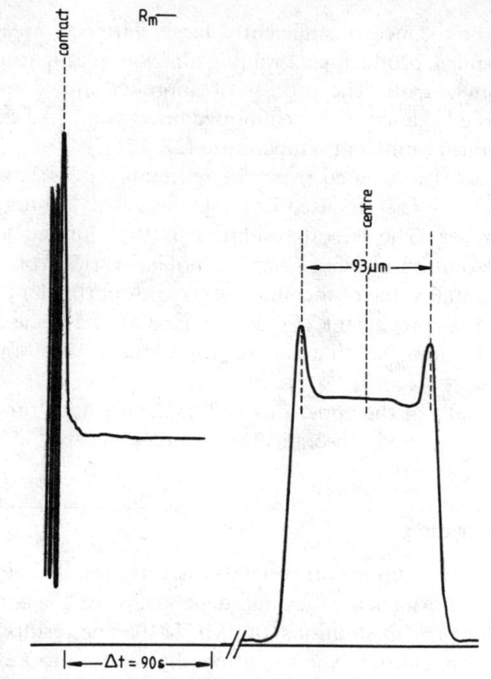
contact
R_m
centre
93 μm
(*a*)
Δt = 90s

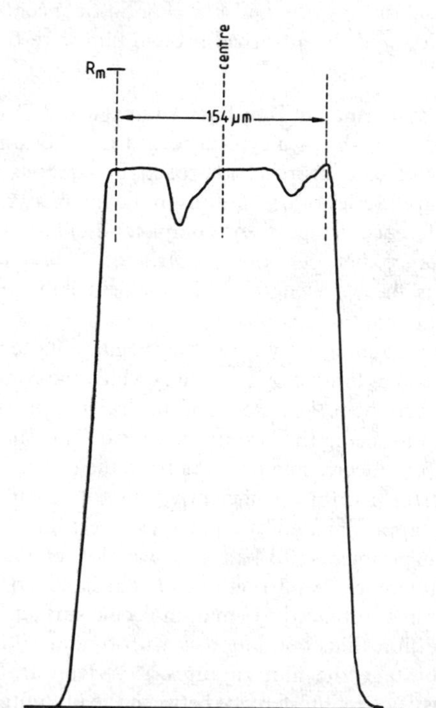
R_m
centre
154 μm
(*b*)

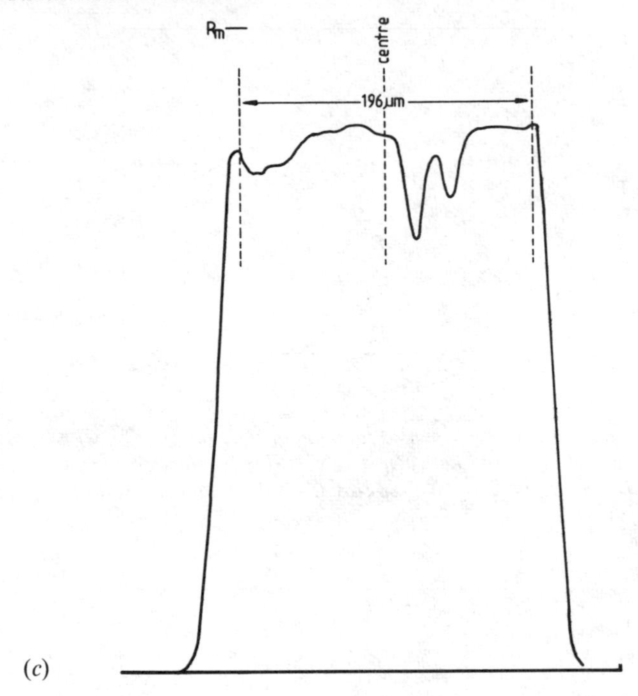

Figure 2 *(a) Left: typical chart record showing the change in intensity of the reflected laser beam (ordinate, arbitrary units) with time when two MCT*810 *droplets in* 10^{-3}% w/w *lysozyme solution are brought together at* 0.14 μm s^{-1}. *Time is increasing from left to right.* R_m *is the maximum possible intensity of the reflected beam from the draining film. The experimental maxima do not reach* R_m *because of the curvature of the droplets and the finite diameter* (ca. 30 μm) *of the focused laser spot. The final plateau in the reflected intensity represents a constant for the draining film with time, the thickness in this case being* 48.9 nm. *The driving motor was still running at this stage. Right: reflected intensity as a function of lateral position for the constant thickness draining film described above. The variation in thickness with lateral position over the flat region of the film is* ±0.7 nm. *(b) Reflected intensity as a function of lateral positon for the film shown in (a) after* 11 min. *[the driving motor was stopped* 2 min *after the scan in (a) was taken]. If variations in film thickness alone are responsible for the variations in intensity, the thickness of the film varies between* 79.7 *and* 89.2 nm. *(c) The same film as in (b),* 134 min *later. At this stage the film thickness varies between* 66.0 *and* 80.5 nm

waves present at both film interfaces, the wave amplitude growing catastrophically when the distance between the interfaces reaches some critical value. As shown in Figure 3, draining films of aqueous lysozyme solution between triglyceride droplets do not follow this pattern. Rather, the films drain to some constant metastable surface separation *J* and then remain at that thickness for a time which

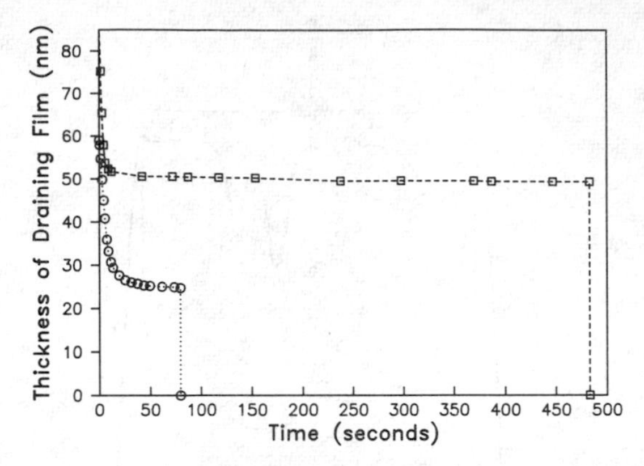

Figure 3 *Thickness of draining film as a function of time for* 10^{-3}% w/w *lysozyme in aqueous solution between triglyceride droplets. Squares, aqueous solution contains* 2×10^{-3} M *NaCl; circles, aqueous solution contains* 10^{-2} M *NaCl*

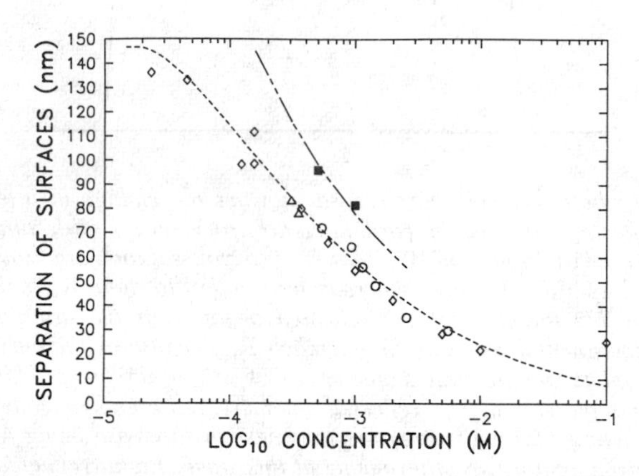

Figure 4 *Surface separation (denoted* J *in text) as a function of ionic concentration in the aqueous phase for aqueous films containing* 10^{-3}% w/w *lysozyme (open symbols) or* 10^{-3}% w/w *sodium dodecyl sulphate (closed symbols) between MCT*810 *droplets. The dashed (theoretical) lines are calculated from a simple force balance as described in Appendix* 2. *Full details are given in reference* 6

can vary from minutes to hours, after which the film collapses abruptly, producng droplet coalescence.

The value of the constant separation J varies with the concentration of added salt in the aqueous phase (Figure 4), and in fact the actual value of J is entirely accounted for on a simple model in which the (attractive) forces deriving from the capillary pressure in the droplets and the van der Waals attraction across the aqueous film are balanced by double layer repulsion (Figure 4 and Appendix 2).

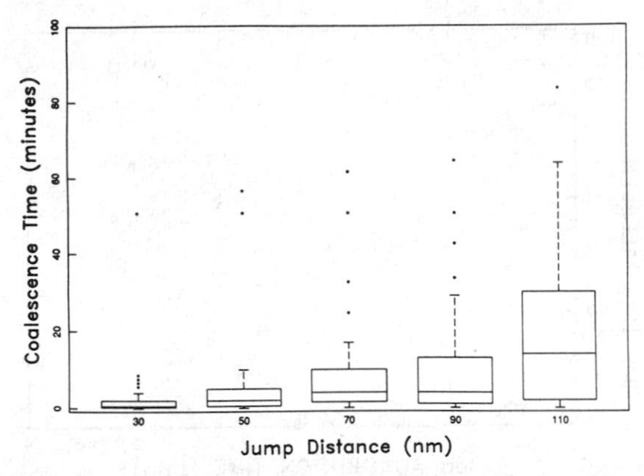

Figure 5 *Box plot of median time to coalescence as a function of draining film thickness at the time of coalescence for aqueous films containing lysozyme* (10^{-3}% w/w) *and different concentrations of sodium chloride to vary the repulsive double layer forces between the interfaces of the film. The centre bar of the boxes gives the median coalescence time. The upper and lower bounds of the boxes represent the third and first quartiles, respectively, that is,* 50% *of the measured points lie between these lines. Vertical whiskers show* 1.5 × *interquartile range. Experimental points outside this range are indicated by asterisks*

The lifetime of the metastable aqueous film depends on a number of factors. One factor is the actual thickness of the film (Figure 5)—the thicker the film, the greater is the lifetime. A second factor is the age of the adsorbed protein film. For adsorbed films less than 5 min old, coalescence occurs within a few seconds. For adsorbed films greater than 2 h old, the lifetime of the draining film can be many hours. Between these limits there is a wide range of lifetimes for the draining film.

The mechanical strength (as measured, for example, by the surface shear viscosity[11]) of adsorbed protein films increases with film age, suggesting that the chance of draining film collapse is indeed related to the mechanical properties of the adsorbed protein film, as claimed by a number of workers recently.[12–16] This conclusion is reinforced by the observation (reference 17 and Figure 6b) that direct physical compression of the adsorbed protein film also increases the lifetime of the draining film, with a dramatic increase being observed beyond a critical compression ratio, thought to correspond to interfacial coagulation of the adsorbed film.

We thus arrive at a picture where collapse of the draining film, leading to coalescence of the triglyceride droplets, is initiated by mechanical failure of the adsorbed protein film rather than by a spontaneous growth in thickness fluctuations at some critical draining film thickness. The chance of mechanical failure depends inversely both on the mechanical strength of the protein film and on the distance between the interfaces of the draining film.

What sort of mechanism could give rise to this picture? We are far from having a complete answer to this question, a hampering factor being the stochastic nature

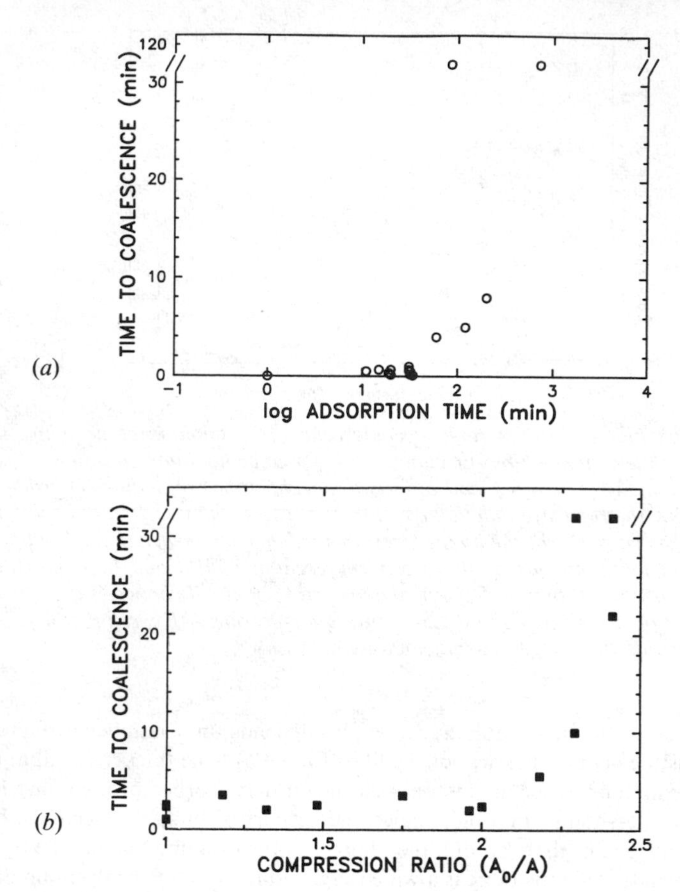

Figure 6 *(a) Time from initial droplet deformation to coalescence for MCT*810 *droplets in* 10^{-3}% w/w *lysozyme solution* +0.1 M *NaCl as a function of the age of the adsorbed film prior to droplet deformation. For the longest adsorption time, the droplets still had not coalesced* ca. 12 h *later. (b) Time from initial droplet deformation to coalescence for the same solution as in (a), but where the age of the adsorbed film was constant* (10 min) *and the droplets were compressed from an initial area* A_0 *to a final area* A *just prior to bringing the droplets together*

of the draining film collapse. One possibility is that collapse is initiated by small dust particles. This is hard to disprove. There are, however, several lines of evidence against it, the strongest being that, in parallel studies with draining films stabilized by sodium dodecyl sulphate, much longer film lifetimes were found (typically many hours). It is difficult to conceive of a particle-induced coalescence mechanism which would produce such a difference between films stabilized by sodium dodecyl sulphate and by lysozyme.

A plausible mechanism, which we are currently investigating, is that some part of the protein film becomes subjected to compressive stresses, eventually producing buckling instabilities which destabilize the film. There are at least two ways in which compressive stresses might arise: (i) thinning of the aqueous film under conditions of constant potential requires an increase in counter ion binding, displacement of (charged) protein molecules from the region of close apposition, or both. Protein displacement could produce a 'pile-up' of protein molecules in the aqueous phase immediately outside the region of close apposition, and will in any case produce compressive stresses in this zone, since adsorbed protein films possess elastic properties.[18] (ii) Once the draining film has thinned to its metastable thickness *J*, it will require little further adsorption for the remaining protein to be depleted from the aqueous film, and approximate calculations show that it will take many hours to replace this protein by diffusion through the aqueous phase. If surface diffusion is also retarded as a consequence of the immobility of the adsorbed protein layer, this will lead to a situation where protein adsorption may continue at the droplet surfaces outside the draining film, but not at the interfaces which bound the draining film. A differential increase in surface pressure will thus arise which will lead to compressive stresses in the zone between the draining film and the rest of the droplet surface.

For both of the mechanisms above, the compressive stresses induced are expected to be greater for thinner films, in accordance with our experimental observation that thinner films have shorter lifetimes.

We have often actually observed corrugations in the adsorbed protein films (Figure 3b and c, with the corrugations appearing some time after the draining film has thinned and increasing in amplitude with time. It is not clear, though, that the corrugations which we observe are related to either of the mechanisms discussed above, as their appearance and amplitude do not correspond with an increased chance of collapse of the draining film.

Effect of Adsorbed β-Casein on Thin Film Draining and Droplet Coalescence—In contrast to the situation with lysozyme, the thickness of draining films stabilized by adsorbed β-casein decreases monotonically with time over several minutes until they reach a thickness at which they suddenly collapse. Examples of such draining patterns are given in Figure 7. This behaviour follows the classical pattern first discussed by de Vries,[10] and presumably arises from the same cause, that is, at some critical film thickness there is sufficient reinforcement of thermally excited capillary waves in the two interfaces which bound the draining film to produce catastrophic film collapse.

For young β-casein films (less than 1 h old), collapse of the draining film generally produces immediate coalescence of the droplets. However, for β-casein films adsorbed from freshly prepared aqueous solutions (less than 3 h old) and then allowed to age for 1–2 h before the droplets are brought together, there is a significant variation of the above pattern. The draining film still collapses at a critical thickness, the value of which is highly reproducible and equal to 39 ± 4 nm, but droplet coalescence does not immediately ensue. Instead, the draining film thickness decreases abruptly to 4 ± 2 nm (Figure 7, open squares)

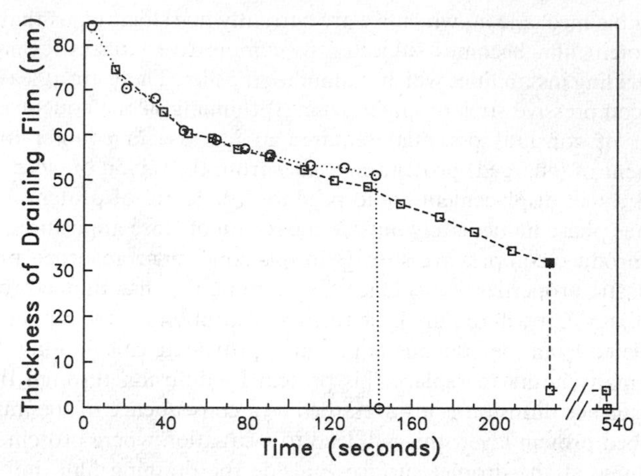

Figure 7 *Thickness of the draining film as a function of time for an aqueous solution of* β*-casein* (10^{-3}% w/w). *Circles, collapse of draining film leads to coalescence; squares, collapse of draining film leads to adhesion, with coalescence occurring* ca. 320 s *later. The filled square is an extrapolated point. Note the monotonic decrease in thickness prior to draining film collapse, in contrast to the situation for lysozyme* (*Figure* 3)

while, concomitantly, its area increases (Figure 8a and b). The thickness of a β-casein monolayer at the air–water interface is approximately 5 nm[19] (measurements at other interfaces are not available), and if this figure is applicable to the triglyceride–water interface, the draining film thickness after collapse is compatible with the thickness of a single β-casein monolayer or two interpenetrating β-casein monolayers. This fact, together with the increase in film area, strongly suggests that, after draining film collapse, the two interfaces of the draining film are in adhesive contact. They remain in this condition for times ranging from 0.25 to 10 min before coalescence abruptly occurs (Figure 8c).

It therefore appears that adsorbed β-casein films which have been allowed to age for 1–2 h do not (unlike adsorbed lysozyme films) possess the requisite mechanical properties to damp capillary waves sufficiently to prevent critical collapse of the draining film. They do, though, develop sufficient mechanical strength to resist disruption during the collapse of the draining film.

The mechanism by which the adsorbed β-casein films develop adhesive contact is of some interest. β-Casein is a protein which contains a relatively large number of hydrophobic amino acid residues,[20] and not all of these residues may reach the plane of the triglyceride–water interface during the adsorption process. There may, for example, be competition for interfacial space between new protein molecules arriving at the interface and the unfolding of previously adsorbed molecules. Adhesion may then arise as a result of interactions between those hydrophobic residues which have remained in the aqueous phase on the apposed interfaces.

An alternative, and more attractive, hypothesis is that adhesion arises as a

(*a*)

(*b*)

(*c*)

Figure 8 *MCT*810 *droplets* (*radius* 1.23 nm) *in* 10^{-3}% w/w β-*casein solution. (a) Droplets brought together with a small deformation at the apex. (b) Appearance of droplets after draining film has collapsed, leading adhesion (note increase in contact area). The contact diameter measured by scanning the laser beam across the film, the boundaries of the adherent zone being defined as those points where the reflectance increased abruptly. (c) Appearance of droplets after coalescence*

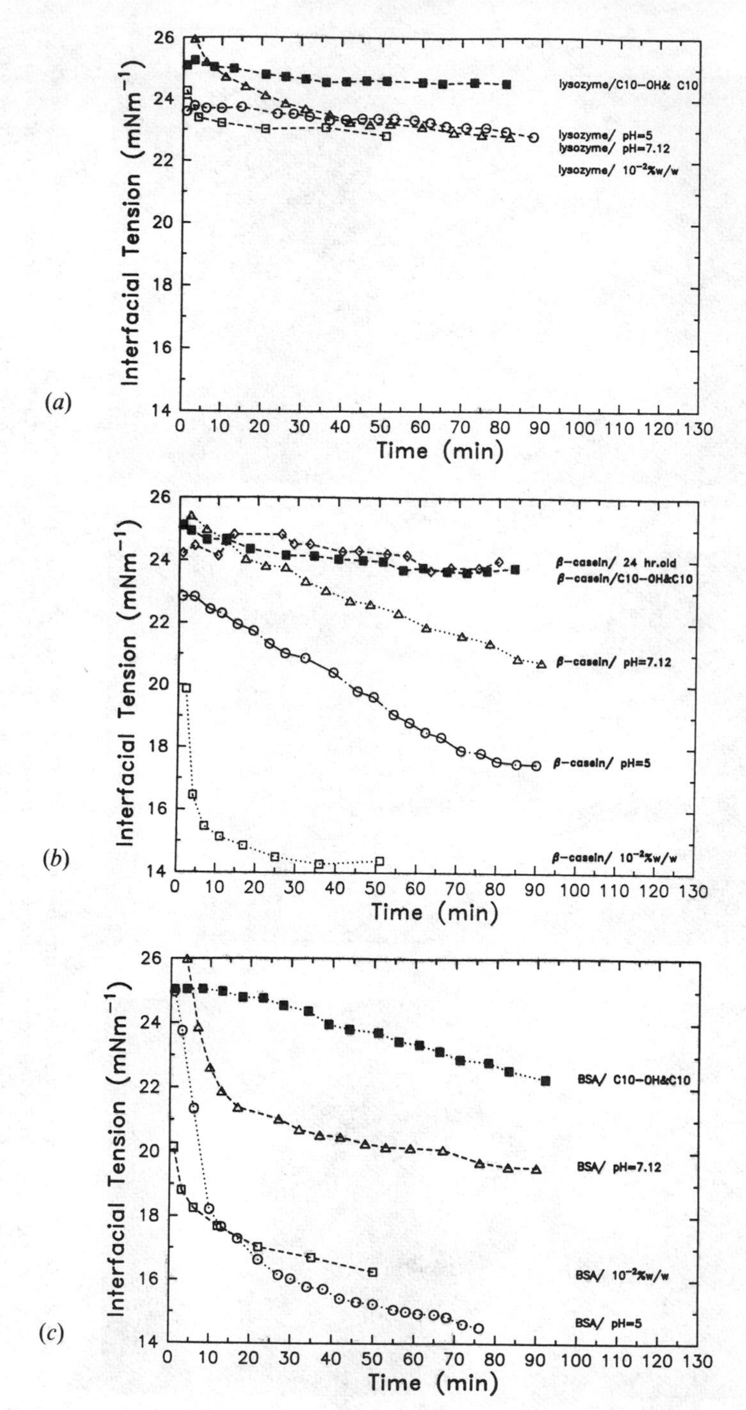
Interfacial Tension (mNm⁻¹)
Time (min)
lysozyme/C10-OH& C10
lysozyme/ pH=5
lysozyme/ pH=7.12
lysozyme/ 10⁻²%w/w
(a)
β-casein/ 24 hr.old
β-casein/C10-OH&C10
β-casein/ pH=7.12
β-casein/ pH=5
β-casein/ 10⁻²%w/w
(b)
BSA/ C10-OH&C10
BSA/ pH=7.12
BSA/ 10⁻²%w/w
BSA/ pH=5
(c)

consequence of protein bridging between the apposed interfaces. Rapid bridging by polymers does indeed occur between sparsely populated interfaces, but would appear to be counter-indicated in the present experiments by the time allowed for β-casein adsorption before bringing the surfaces together. However, if draining film collapse proceeds by the de Vries mechanism,[10] then growth of the capillary waves may produce a sufficient transient, surface expansion to allow space for protein molecules already adsorbed on one surface also to adsorb to the other.

A final variation in the behaviour of draining films stabilized by β-casein arises if the aqueous β-casein solution has been allowed to remain at room temperature ($23 \pm 2\,^{\circ}C$) for more than 3 h. The draining film collapses soon after initial droplet deformation, and droplet coalescence is immediate, with no intermediate adhesion step. It is well known that β-casein aggregates rapidly in solution at these temperatures[21] and the clear conclusion is that the aggregates either do not adsorb (supported by the lack of effect on the interfacial tension (Figure 9b) or, if they do, that they do not exert a stabilizing influence on the draining film.

5 Conclusions

Draining films stabilized by lysozyme or β-casein show few similarities in their behaviour. β-Casein-stabilized films reveal a classic drainage pattern where they thin to a critical thickness and then collapse. The actual value of the critical thickness and whether collapse leads to adhesion or coalescence depend on the age of the adsorbed film and the aqueous protein solution.

Lysozyme-stabilized draining films do not show the above pattern. Rather, the films drain to a thickness given by a balance of attractive capillary and van der Waals forces with repulsive double layer forces and then remain at this thickness for times ranging from minutes to hours, after which they collapse to produce droplet coalescence.

The above results can be rationalized on the basis of the presumed mechanical properties of the adsorbed protein films, although direct evidence is still needed to support these rationalizations.

6 Appendix 1: Time Dependence of Interfacial Tension

During the course of this work, we have had occasion to measure the time dependence of the interfacial tension between MCT810 and aqueous protein solutions under various conditions. Representative data for lysozyme, β-casein, and bovine serum albumin are given in Figure 9. Such data do not give directly

---

Figure 9 *Interfacial tension as a function of time for pendant droplets of aqueous protein solutions in contact with oil phase. (a) Lysozyme; (b) β-casein; (c) bovine serum albumin. All protein concentrations are* $10^{-3}\%$ w/w *unless stated otherwise. The oil phase is MCT*810 *except for curves with the legend* C10-OH + C10, *where the oil phase is a mixture of n-decane and n-decanol in proportions chosen to give an interfacial tension of* $25\ mN\ m^{-1}$ *against water*

the amount of adsorbed protein or the state of that protein, and hence the data were not introduced into the main discussion. They are brought together here since several points arise which are germane not only to the present studies but also to other studies involving the interfacial properties of these proteins.

i. The temporal rate of change and long-term value of the interfacial tension between MCT810 and aqueous β-casein solutions depend very much on the age of the protein solution, with the reduction in interfacial tension becoming negligible for solutions more than 24 h old. This is probably a reflection of the aggregation of the β-casein in solution with time.[21] It does not necessarily imply that the aggregates are not adsorbed, although if they are then they are not effective in preventing coalescence.

ii. The proteins (β-casein and bovine serum albumin) which adopt random-coil configurations in aqueous solution have a more rapid and more pronounced effect on the interfacial tension than does the compactly folded lysozyme. This parallels their behaviour at the n-decane/water interface.[20]

iii. Interfaces were formed between the aqueous protein solutions and a mixture of n-decane and n-decanol. The composition of the mixture was so adjusted that the interfacial tension between pure water and the mixture was 25 mN m^{-1}. The change in surface pressure with time at this interface for all three proteins was much less than that at the MCT810–water interface, despite the similarity of interfacial tensions for the clean interfaces. It therefore appears necessary to take into consideration the detailed chemical nature of an interface when considering the mechanism of protein adsorption and unfolding at that interface.

7 Appendix 2: The Balance of Forces in Films Stabilized by Adsorbed Lysozyme

The dominant attractive force is provided by the capillary pressure P_c, given by $P_c = 2\sigma/r$, where σ is the interfacial tension between the triglyceride and the aqueous solution and r is the measured radius of the triglyceride droplet. The contribution P_{vdw} of van der Waals forces to the attractive pressure is approximated by $P_{vdw} = A/(6\pi D^3)$, where A is the Hamaker constant (taken as 5×10^{-21} J)[22] and D is the thickness of the aqueous film. The balancing repulsive pressure P_e due to double layer overlap is given to within 1% under the conditions of our experiments by[22] $P_e = 1.59 \times 10^8[\mathrm{NaCl}]\gamma^2 \exp(-\kappa D)$, where P_e is the pressure in N m^{-2} $1/\kappa$ is the Debye length, $\gamma = \tanh(e\psi/4kT) \approx \psi/103$, e being the charge on the electron and ψ the surface potential in mV, and [NaCl] is the molar concentration of sodium chloride. The surface potential was taken to be equal to the zeta potential, obtained experimentally.[23]

References

1. L. R. Fisher and N. S. Parker, in 'Advances in Food Emulsions and Foams,' eds. E. Dickinson and G. Stainsby, Elsevier Applied Science, Barking, London New York, 1988, p. 45.

2. P. J. Halling, *CRC Crit. Rev. Food Sci. Nutr.*, 1981, **15**, 155.
3. L. R. Fisher, E. E. Mitchell, and N. S. Parker, *J. Food Sci.*, 1985, **50**, 1201.
4. L. R. Fisher, N. S. Parker, and F. Sharples, *Opt. Eng.*, 1980, **19**, 798.
5. L. R. Fisher, N. S. Parker, and D. A. Haydon, *Faraday Discuss. Roy. Soc. Chem.*, 1986, **81**, 249.
6. L. R. Fisher, E. E. Mitchell, and N. S. Parker, *J. Colloid Interface Sci.*, 1989, **128**, 35.
7. J. D. Chen, *J. Colloid Interface Sci.*, 1984, **98**, 329.
8. B. P. Radoev, A. D. Scheludko, and E. D. Manev, *J. Colloid Interface Sci.*, 1983, **95**, 254.
9. A. Scheludko, *Proc. K. Ned. Akad. Wet.*, 1962, **B65**, 87.
10. A. de Vries, in 'Third International Congress on Detergency, Cologne, 1960,' Vol. 2, p. 566.
11. E. Dickinson, B. S. Murray, and G. Stainsby, *J. Chem. Soc., Faraday Trans. I*, 1988, **84**, 871.
12. V. D. Kiosseoglu and P. Sherman, *Colloid Polym. Sci.*, 1983, **261**, 520.
13. H. J. Rivas and P. Sherman, *Colloids Surf.*, 1984, **11**, 155.
14. M. Nakamura, *Yukagaku*, 1986, **35**, 554.
15. V. M. Sabet and S. M. Zourab, *Indian J. Chem.*, 1982, **21A**, 677.
16. S. M. Zourab, S. N. Srivastava, F. Abdel-Halim, and V. M. Sabet, *Egypt. J. Chem.*, 1982, **25**, 131.
17. L. R. Fisher, E. E. Mitchell, and N. S. Parker, *J. Colloid Interface Sci.*, 1987, **119**, 592.
18. F. Macritchie, *Adv. Protein Chem.*, 1978, **32**, 283.
19. M. C. Phillips, *Food Technol.*, 1981, January, 50.
20. D. E. Graham and M. C. Phillips, *J. Colloid Interface Sci.*, 1979, **70**, 403.
21. M. Noelken and M. Reibstein, *Arch. Biochem. Biophys.*, 1968, **123**, 397.
22. J. N. Israelachvili, 'Intermolecular and Surface Forces,' Academic Press, London, 1985.
23. R. W. O'Brien and R. J. Hunter, *Can. J. Chem.*, 1981, **59**, 1878.

Theoretical Studies of the Solid–Fluid Interface

By G. Rickayzen

PHYSICS LABORATORY, UNIVERSITY OF KENT AT CANTERBURY, CANTERBURY, KENT CT2 7NR, UK

1 Introduction

The aim of this paper is to survey a variety of the theoretical and computational techniques which have been applied to the study of the structure of inhomogeneous fluids, in particular fluids near solid surfaces, and to relate the structure to the properties of the fluids. The work to be described concentrates on the physical forces between molecules, whether in the fluid or the wall, and leaves out of account chemical bonding; molecules bound rigidly to the wall by chemical bonds are regarded as part of the wall. The survey is not comprehensive and reflects the author's interests. The systems and properties studied so far include those listed in Table 1.

Table 1 *Applications*

System	*Properties*
Simple fluid	Adsorption Wetting Phase equilibrium in pores Phase transitions in pores
Molecular fluid	Orientational structure Dielectric properties
Mixed neutral fluid	Relative adsorption
Charged fluid	Electrical double layer Colloidal particles

For the most part, the molecules are modelled very simply by hard spheres (the hard-sphere fluid) or by molecules interacting through the Lennard–Jones potential (L–J fluid);

$$V_{\mathrm{LJ}}(\mathbf{r}) = 4\varepsilon[(\sigma/r)^{12} - (\sigma/r)^{6}] \quad (1)$$

Additionally, the molecules may have charges or dipoles at their centres. Non-spherical molecules may be modelled by overlapping hard spheres or cigar shapes. A major difference between the hard-sphere and the L–J fluids is that the latter includes a long-range attraction. As a consequence, the bulk L–J fluid suffers a liquid–gas-phase transition at a temperature T_c, whereas the hard-sphere fluid does not. Although this makes the hard-sphere fluid unrealistic as a model of a real fluid, it is often useful to separate the molecular forces into a hard-sphere part and a long-range part and, by the use of perturbation theory to treat the latter, to relate the properties of the real fluid to those of the hard-sphere fluid. This idea goes back to van der Waals and is the basis of his equation of state.

2 Methods

Computer Simulation—There are two main methods of simulation, molecular dynamics (MD) and Monte Carlo (MC). In the former case one replaces the equations of motion by finite-difference equations and solves them, whereas in the latter one evaluates the partition function of the system by appropriate sampling of the integrand. Thus MD simulation can provide both dynamic and equilibrium properties. MC simulation leads only to static properties. In principle, both methods are exact, but in practice the accuracy is limited by the computing power and time available. An accuracy of better than 1% in the equilibrium properties can be achieved. Thus computer simulation is usually regarded as providing results against which theories can be tested. This is particularly important in a subject where present theories relate to simple model systems, still a long way from the observed real systems.

A major disadvantage of simulation is the cost and available computing power; these limit the systems studied. At present this means that systems of more than *ca.* 1000–2000 molecules cannot be handled. Hence simulation cannot be applied to the electrolytes of practical colloids because the ratio of ions to solvent molecules is usually less than 1 : 1000. Further, the long-range van der Waals forces increase the time it takes for the system to come to equilibrium. It is therefore usual to truncate these forces and to treat the difference by perturbation theory.

Theories—The theoretical approaches fall into two groups, with some overlap. One approach starts from an approximate grand potential for the system, Ω, which depends on the number-density of the molecules, $\rho(\mathbf{r})$. In equilibrium, $\rho(\mathbf{r})$ minimizes Ω. In the other approach, one starts from an exact set of equations such as the Bogoliubov–Born–Green–Yvon (BBGY) equations and approximates and solves them. The major problem for the theory is that in a simple system there is no small parameter. This means that approximations are usually not controlled and, at the outset, the accuracy of the theory is unknown. Hence the testing of the theory by reference to simulation is very important. However, theory has the advantage that it can be applied to complex systems which may include long-range forces. It also needs less computing power although, as the theories become more sophisticated and the system more complex, the required computing power increases substantially.

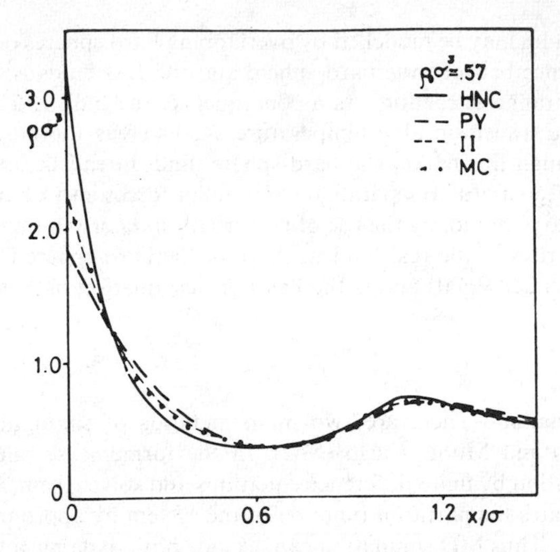

Figure 1 *Density profile of a fluid of hard spheres of diameter* σ, *bulk density* $\rho_o = 0.57\sigma^{-3}$ *near a hard wall. The distance* $\times$ *is measured into the fluid from the position of closest approach. The points come from the Monte Carlo simulation of Snook and Henderson.*[4] *The curves denote different theories as follows:* ——, *HNC;* - - , *PY;* - - - , *reference* 3

This paper will be confined to results from density-functional theories because these cover the area in which the author has worked. In general,

$$\Omega = \Omega_E + \Omega_I \tag{2}$$

where Ω_E is the entropic contribution given by

$$\Omega_E = kT\int d^3r\rho(\mathbf{r})\,[\ln \Lambda\,\rho(\mathbf{r}) - 1] \tag{3}$$

The oldest and most frequently used approximations are the HNC and PY. The HNC approximation is obtained if Ω_I is expanded about the bulk density ρ_B to second order in $(\rho - \rho_B)$. If the entropic term is also expanded in this way, one obtains the PY approximation. If the wall-fluid potential were weak, these theories would be accurate, but unfortunately this is not the case and the theories fail near the wall (*cf.* Figure 1). In fact, it has been emerging that to obtain a good theory of the structure it is necessary for the geometrical properties and the thermodynamics to be correct. The HNC and PY approximations do seem to contain satisfactory geometrical features in Ω_I but they are inconsistent thermodynamically. For example, for a single planar, structureless, hard wall there are no forces of finite range between the molecules of the fluid and those of the wall. Hence, the pressure on the wall, P_ω, is given by kinetic theory as

$$P_\omega = \rho_\omega\, kT, \tag{4}$$

where ρ_ω is the particle density at the wall.

However, the pressure is the same throughout the fluid in a one-dimensional geometry. Hence the number-density at the wall is related to the bulk pressure, P, by

$$\rho_\omega = P/kT \tag{5}$$

This is a general relationship that must be satisfied, but it is not satisfied by either the HNC or PY approximations. It is particularly unfortunate that the PY theory does not satisfy equation (5) because it leads to linear equations which are comparatively easy to solve by computer and which can be solved analytically in some important cases.

We return to two approximations which do ensure that equation (4) is satisfied. The first is due to Tarazona and co-workers[1,2] and is now called the smoothed density approximation (SDA). This divides Ω_I into a hard-sphere part Ω_{hs} and a long-range contribution Ω_{lr} in the spirit of van der Waals. Then,

$$\Omega_{hs} = \int d^3 r \omega(\mathbf{r})$$

where $\omega(\mathbf{r})$ is a function of a local coarse-grained density obtained as a weighted average of $\rho(\mathbf{r}^1)$ within a molecular radius of $\mathbf{r}$. The function $\omega(\mathbf{r})$ is the grand potential per unit volume with bulk density $\bar{\rho}(\mathbf{r})$. This ensures that the local thermodynamics of the hard-sphere fluid are correctly included. A deficiency of the approximation is that it assumes that the bulk equation of state is given satisfactorily by a van der Waals theory.

The other approximation proposed by Rickayzen and Augousti[3] (RA) adds a third-order fluctuation to the HNC grand potential. This is chosen to ensure that equation (4) is satisfied. The method has the advantage that it uses the exact bulk equation of state when this is known but it has the disadvantage that it does not ensure that the local thermodynamics are satisfied everywhere.

3 Applications to Neutral Fluids

This survey is confined to fluids near or bounded by structureless walls which influence the fluid through potentials which depend only on the distance from the wall. This covers the majority of the systems studied.

Hard Sphere Fluid–Hard Wall Interface—In Figure 1 we compare the results of computer simulation with those derived from PY, HNC, and RA theories for the density of a fluid with bulk reduced number-density, $\rho\sigma^3$ (σ is a molecular diameter), of 0.57. All the results show the characteristic oscillatory behaviour which reflects the short-range order and layering of the molecules in the fluid. As expected, the PY and HNC results are poor close to the wall, but are otherwise very good. Agreement between the RA theory and simulation is good throughout the range. Similar good agreement is obtained with the SDA.

L–J Fluid–Hard Wall Interface—Figures 2 and 3 show results obtained by Powles *et al.*[5] for the structure of an L–J fluid near a hard wall at temperatures above and

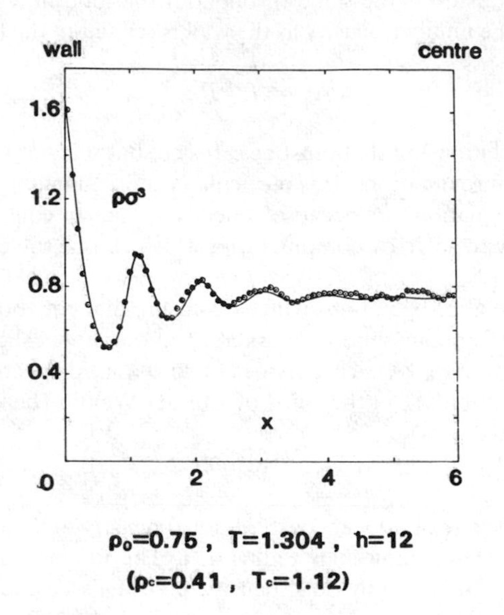

Figure 2 *Density profile of a fluid of L–J molecules with* $\rho_o\sigma^3 = 0.75$ *confined between two hard walls separated by* 12σ *at a reduced temperature of* 1.3. *The curve is from the theory of reference* 3; *the circles are the result of MC simulation*

(Reproduced with permission from *Mol. Phys.*, 1988, **64**, 33)

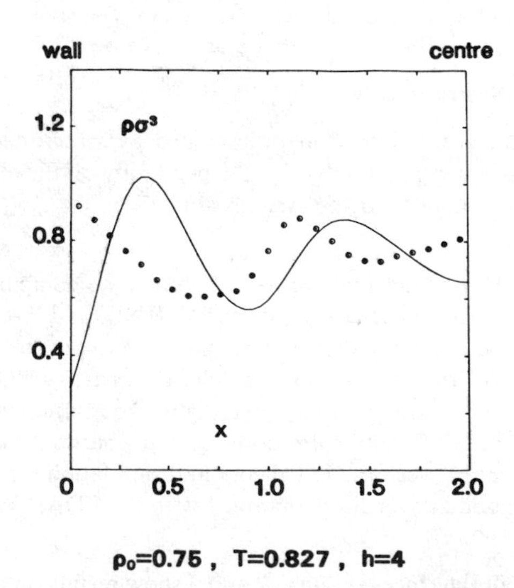

Figure 3 *As Figure 2, but with separation* 4σ *and reduced temperature* 0.827

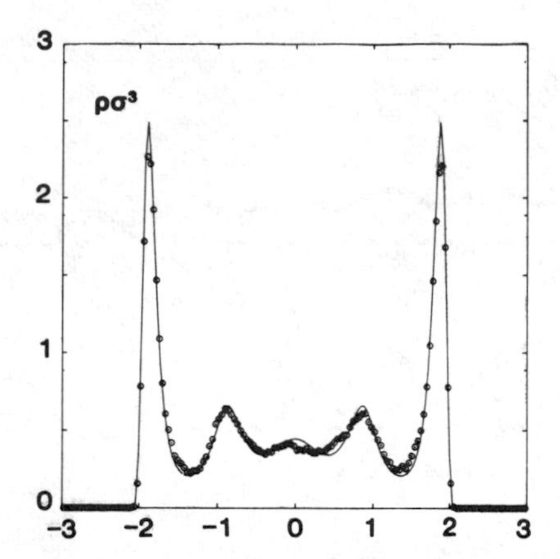

Figure 4 *Density profile of a fluid of L–J molecules with* $\rho_0\sigma^3 = 0.2797$ *confined between two L–J walls at a reduced temperature of* 1.35. *The solid curve was obtained by Powles* et al.[5] *using the theory of reference* 3. *The circles are the simulation results of Walton and Quirke*[6]

below T_c for the bulk fluid. The results derived from RA theory are compared with Monte Carlo simulation. It is seen that whereas the agreement is good for $T > T_c$, it is poor for $T < T_c$. At the lower temperature the liquid tends to peel away from the wall, leaving a gas layer, but, because the theory does not include the correct local thermodynamics at this temperature, it does not show this feature. From results obtained so far, it appears that RA theory provides reliable results for $T > T_c$ and possibly for $T \approx T_c$, but not for lower temperatures, at least when two phases are present.

L–J Fluid–L–J Wall—*Single phase.* Figure 4 shows the theoretical results of Powles *et al.*[5] compared with the results of a Monte Carlo simulation by Walton and Quirke[6] for the structure of an L–J fluid adjacent to an L–J wall with parameters chosen to present ethylene in a graphic slit. There are no free parameters in the theory and the agreement is remarkably good.

Phase equilibria. The groups at Cornell[7] and at Bristol[8] have used the SDA to study the phase transition and structure of a confined L–J fluid. Ball and Evans[8] have studied the structure of the two phases of an L–J fluid confined by a cylinder interacting with the fluid through an L–J interaction. They used parameters to model liquid argon confined by solid carbon dioxide. Their simulation followed a technique due to Panagiotopoulos[9] which ensures that the two phases are in equilibrium with each other without relating them to the bulk. The results are shown in Figure 5, where again the agreement is very good. So far the main deficiency to emerge is in the bulk parameters. Ball and Evans[8] separately carried out a simulation to equilibrate the bulk and pore phases at $R_c = 3.0$ and found that capillary co-existence occurred at $p/p_{sat} \approx 0.14$. This discrepancy is no doubt

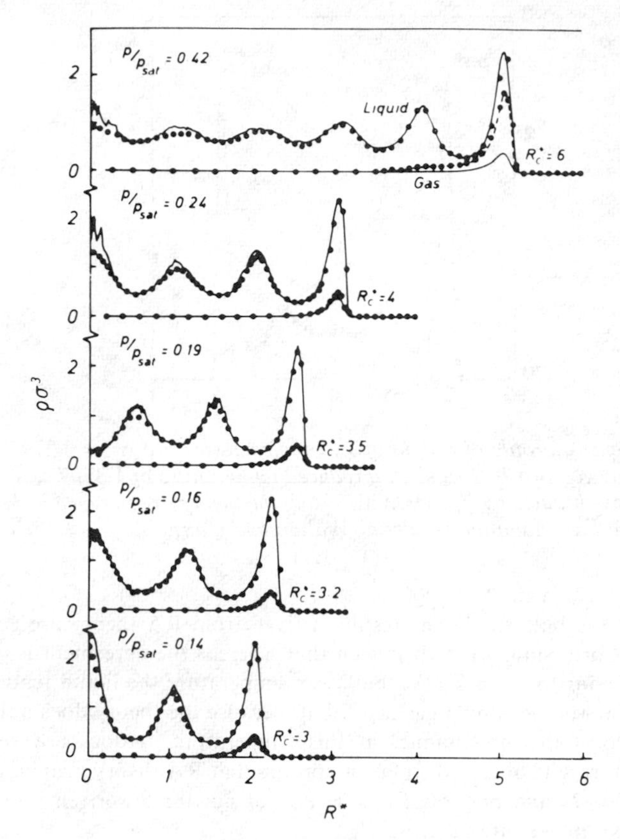

Figure 5 *Density profiles of a fluid of L–J molecules confined to cylindrical pores of radii* R_c *in gaseous and liquid states in equilibrium. The solid curves were obtained using the SDA and the dots are from computer simulation. Radial distances are measured from the axes of the pores*
(Reproduced with permission from *Mol. Phys.*, 1988, **63**, 159)

connected with the fact that the bulk equation of state they use does not accurately describe the bulk fluid.

Mixtures—Very few studies have been made of the structure of confined fluid mixtures. However, in recent work Rickayzen and Moradi[10] compared the results of density functional theory with those of computer simulation for a hard-sphere mixture of fluids with different radii confined by parallel hard walls. They found agreement as good as for the pure fluid.

4 Applications to Charged and Dipolar Fluids

In the case of fluids containing charged ions and dipolar molecules, the forces are long range and cannot be cut off and corrected by perturbation theory. This causes difficulty for both computer simulation and theory. Nevertheless, progress has been made. In the case of charged ions, the force due to one ion is screened by all the others at distances of the order of the Debye length, λ_D. The effective force between ions is therefore short range, although in dilute electrolytes λ_D is many molecular diameters in length. In the case of dipoles the screening is by the dielectric constant which, although it may be large, means that the effective interaction between molecules is still long range. This makes the dipolar fluid more difficult to treat than the charged fluid. For colloids, one needs to include both charged and dipolar molecules.

In many colloidal systems, the colloidal particles are much larger than λ_D. In such cases, the forces between the particles can be obtained from those between parallel planar walls. As this is a much simpler geometry to work with, all research so far has been confined to such systems.

The Charged Fluid—Torrie and co-workers[11–13] and van Megen and Snook[14] have made extensive studies by Monte Carlo simulation of the model fluid consisting of a mixture of hard spheres of the same diameter but carrying charges Ze or $Z'e$ at their centres in a structureless medium of dielectric constant ε, and confined between charged parallel planar hard walls.

Some of the results of Torrie and Valleau are presented in Figures 6 and 7, which show the concentration profiles of the two ions and the profile of the mean electrostatic potential. The results are compared with those obtained from what they call the modified Gouy–Chapman (MGC) theory. This is essentially the normal diffuse layer theory with a Stern layer of thickness equal to the radius of an ion. One can see that for the 1–1 'electrolyte,' even at a concentration of 1 M, the agreement between MGC theory and simulation is very good. As other work has shown,[15] this agreement is also reflected in the force between the walls. For the 2–1 fluid on the other hand, (where counter ions are doubly charged), the agreement is poorer even at the lower concentration of 0.05 M. In Figure 8 are shown the results of MGC theory as a dashed line and the result of a more sophisticated theory (MPB theory) due to Buiyan *et al.*,[16] which tries to improve the treatment of the electrostatic interaction between ions. In MGC theory, each ion is affected only by the average electrostatic potential at its position. In MPB theory, some account is taken of the fact that an individual ion attracts oppositely charged ions and repels like ions. This causes the potential acting on an individual ion to deviate from the average potential. Certainly, the work of Torrie and Valleau shows that this is the most important effect neglected in MGC theory.

The results of Torrie and Valleau for 2–1 electrolytes at various concentrations are shown in Figure 9. Here, the electrostatic potential at the edge of the Stern layer is plotted against the charge on an individual wall or plate for various concentrations M. Also shown are the results of MGC theory (dashed line) and MPB theory (solid line). It is clear that especially when the counter ions are doubly charged, MGC theory is not very satisfactory. MPB theory is much better. Note that in the figures the potential and surface charge density are given in the reduced units $\sigma^* = \sigma d^2/e$ and $\psi^* = e\psi/kT$. The calculations were performed with $d = 4.25$ Å, $\varepsilon = 78.5$ and $T = 298$ K. Then, $\sigma = 88.7\sigma^*\mu C_m{}^{-2}$ and $\psi = 25.7\psi^*$ mV.

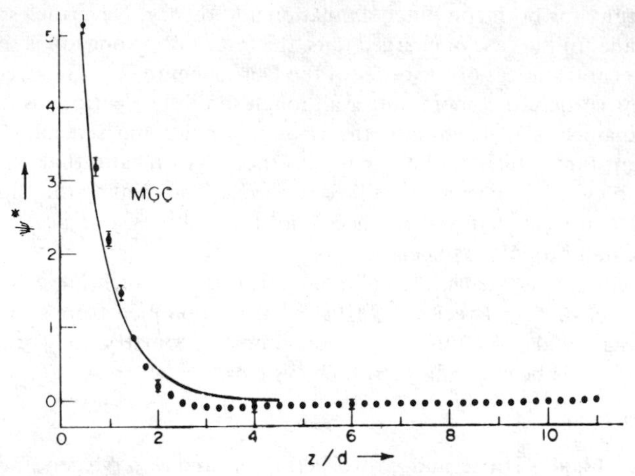

Figure 6 *Profile of the mean electrostatic potential in a* 1–1 *electrolyte of concentration* 1.0 M, $\sigma^* = 0.70$. *The solid curve is MGC theory, the dots are from MC simulation*

(Reproduced with permission from *J. Chem. Phys.*, 1980, **73**, 5807)

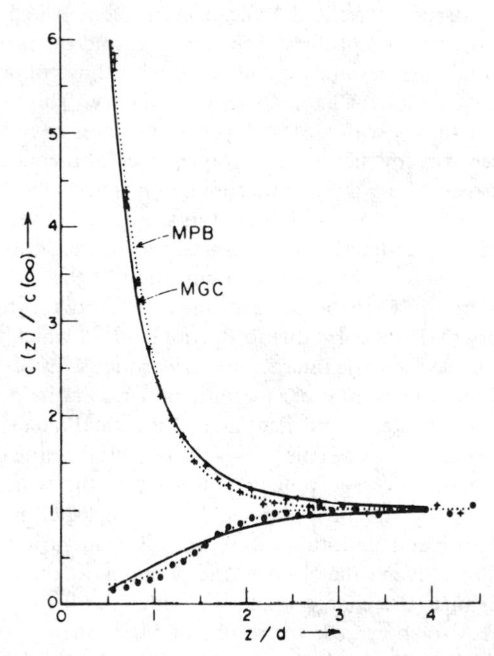

Figure 7 *Ion concentration profiles of the ions of a* 1–1 *electrolyte of concentration* 1·0 M *with* $\sigma^* = 0.141$. *The solid curve is from MGC theory, the dotted curve is from a modified Poisson–Boltzmann (MPB) theory of Buiyan* et al.[16] *The dots are the result of MC simulation*

(Reproduced with permission from *J. Chem. Phys.*, 1980, **73**, 5807)

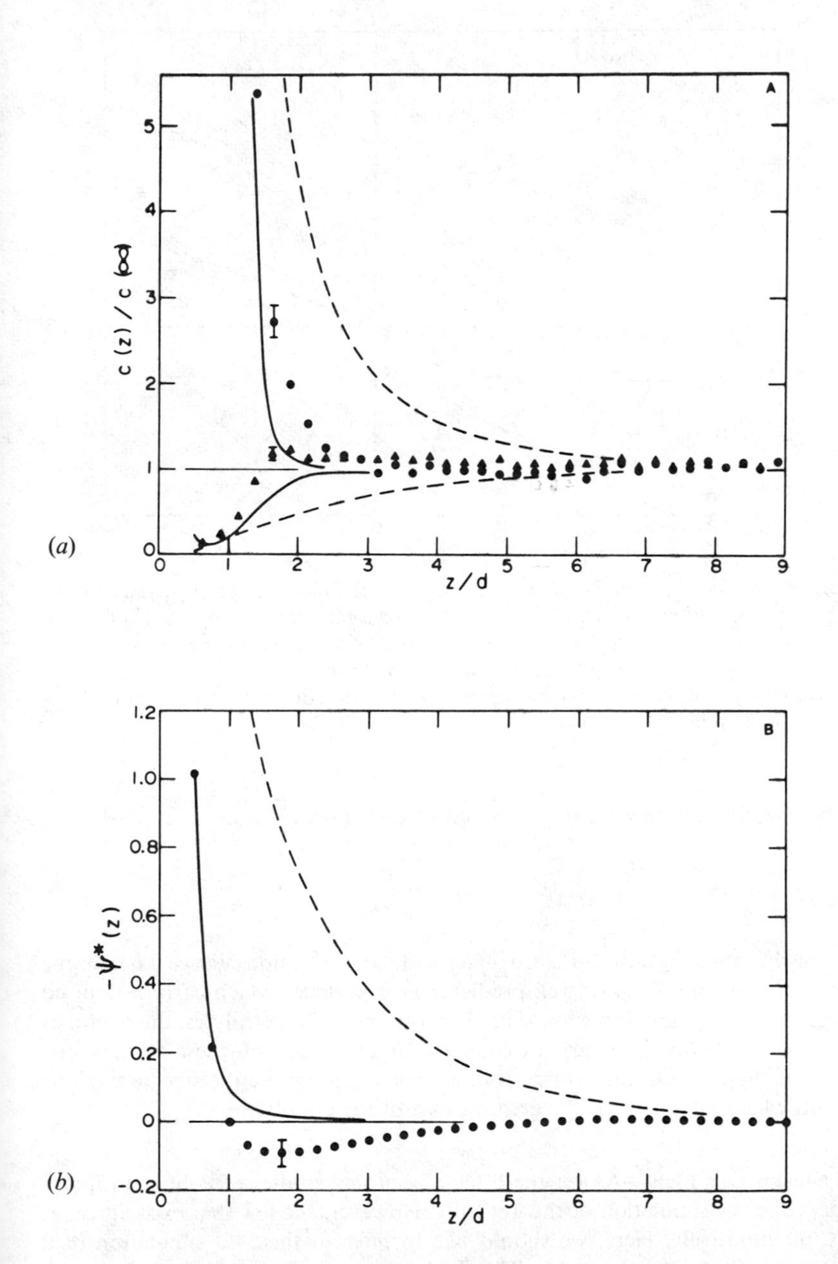

Figure 8 *As Figures 6 and 7, but for a 2–1 fluid with doubly charged counter ions at a concentration of* 0.05 M *with* $\sigma^* = -0.284$
(Reprinted with permission from *J. Phys. Chem.*, 1982, **86**, 3251. Copyright (1982) American Chemical Society)

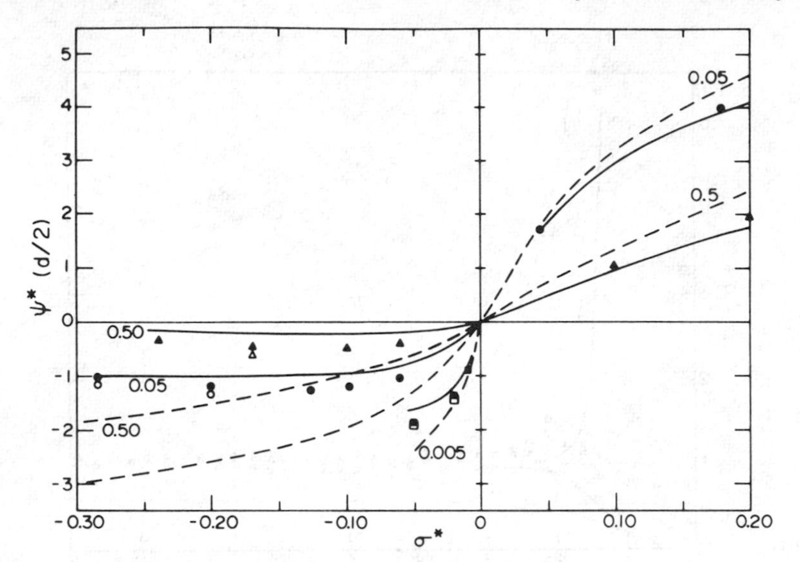

Figure 9 *Reduced potential* ψ^* *at the edge of the diffuse layer as a function of reduced wall charge* σ^* *for various concentrations of a* 2–1 *electrolyte. The points denote the results of simulation, the dashed curves are MGC theory, and the solid curves are MPB theory*
(Reprinted with permission from *J. Phys. Chem.*, 1982, **86**, 3251. Copyright (1982) American Chemical Society)

The potential at the wall, ψ_ω, is related to $\psi(d/2)$ by

$$\psi_\omega = \psi(d/2) + \frac{\sigma d}{2\varepsilon\varepsilon_o} \tag{6}$$

The conclusion is that for 2–1 electrolytes with concentrations above 0.005 M, the electrostatic potential is not well predicted near surfaces which carry a reduced charge density greater than $\sigma^* = 0.05$. For 1–1 and 2–2 electrolytes, the region of applicability of MGC theory extends to larger values of these parameters. However, these conclusions come from models which ignore the structure of the solvent, which, of course, is the greatest part of the electrolyte.

The Stockmayer Fluid—As yet there have been few studies of confined dipolar fluids, either by simulation or theoretically. However, the few that exist agree, at least, quantitatively. Here, we should like to refer to the MD simulation of a Stockmayer fluid by Lee *et al.*[17] The fluid is composed of L–J molecules with permanent dipoles of magnitude μ and was placed in a uniform electric field E perpendicular to the walls. Some of their results for the density and polarization density profiles are shown in Figure 10. These show clearly that the polarization profile mimics the density profile and that oscillations in both are evident

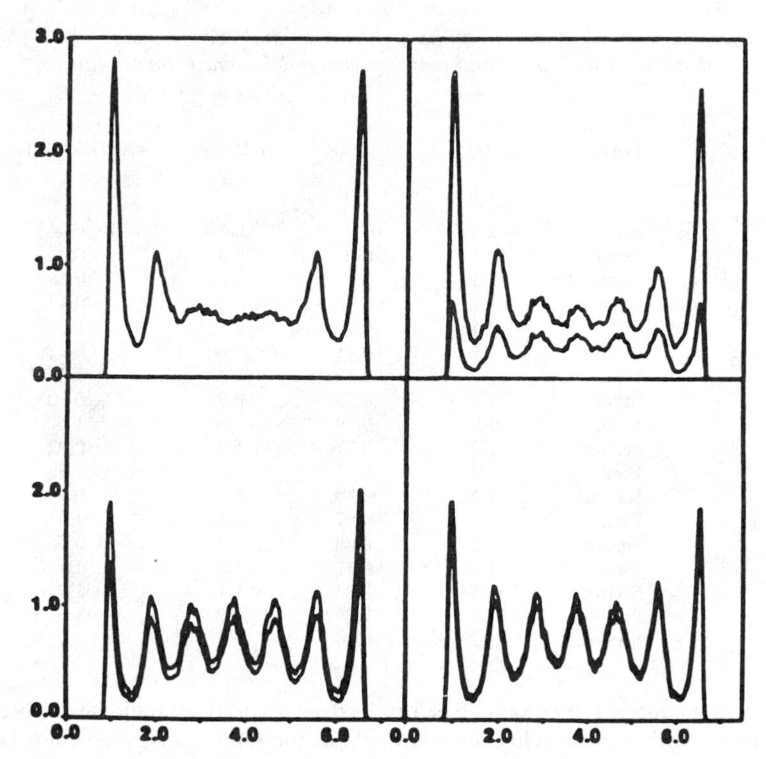

Figure 10 *Density and polarization profiles of a Stockmayer fluid confined between two parallel walls in an electric field* E *perpendicular to the walls. The values of* E *are* 0.0, 0.5×10^9, 1.5×10^9, $2.5 \times 10^9\ \mathrm{V\,m^{-1}}$ *and refer in order to the figures from left to right and top to bottom. The results were obtained using computer simulation*
(Reproduced with permission from *J. Chem. Phys.*, 1986, **85**, 5232)

throughout the fluid which is confined to a slit of width 7.5σ. As Table 2 shows, they find that the equation

$$\langle P_z(z) \rangle = \langle \rho(z) \rangle \ \langle \mu_z(z) \rangle \tag{7}$$

is obeyed fairly well. In this equation, the averages are taken over the first and second peaks as one moves away from a wall.

Figure 10 and Table 2 also show that in the higher electric fields (above $10^9\ \mathrm{V\,m^{-1}}$) the density at the wall tends to decrease. It seems that the alignment of the dipoles in the direction of the field tends to increase the attraction between them; the wall molecules are consequently attracted to the bulk. For their parameters, $\rho^* = 0.5605$, $T^* = 1.18$, and $\mu = 1.36$ D, they found that the polarization is linear in the electric fields for field strengths below $5 \times 10^8\ \mathrm{V\,m^{-1}}$. Above this value the polarization starts to saturate. As for colloidal systems the electric fields are usually less than $5 \times 10^8\ \mathrm{V\,m^{-1}}$, a linear treatment of the

Table 2 *Molecular dyanamics results at the first and second peaks of the local density for a dipolar fluid between charged Lennard–Jones plates*

E ($10^9 V m^{-1}$)	*Layers*	$<\rho(z)>$	$\frac{<\mu_z(z)>}{\mu}$	$\frac{<P_z(z)>}{\mu}$	$\frac{<\rho(z)><\mu_z(z)>}{\mu}$
0.0	First	2.6	0.0	0.0	0.0
	Second	1.1	0.0	0.0	0.0
0.5	First	2.5	0.27	0.64	0.68
	Second	1.1	0.38	0.46	0.42
1.0	First	2.4	0.50	0.1	1.2
	Second	1.1	0.65	0.80	0.72
1.5	First	2.0	0.70	1.4	1.4
	Second	1.1	0.73	0.91	0.80
2.0	First	1.9	0.79	1.6	1.5
	Second	1.2	0.77	1.0	0.92
2.5	First	1.9	0.83	1.6	1.6
	Second	1.2	0.80	1.1	0.93
3.0	First	1.9	0.87	1.6	1.6
	Second	1.2	0.80	1.0	0.96
3.5	First	1.8	0.87	1.6	1.6
	Second	1.2	0.83	1.1	1.0
4.0	First	1.8	0.89	1.7	1.6
	Second	1.2	0.84	1.1	1.0

polarization appears to be valid. The results from theoretical studies are in broad agreement with these conclusions, although, as the same systems have not been studied by both simulation and theory, it is not yet possible to make quantitative comparisons.

The Civilized Model—The simulation of a model consisting of positive and negative ions and also dipolar molecules at realistic densities is not yet a practical proposition. Even at a concentration of 0.5 M only about one molecule in 1000 is an ion. Therefore, for such systems, one has to fall back upon theory. Because of the complexity of the system, the most extensive study has been carried out using the completely linear theory which, for reasons already given, is not entirely satisfactory, especially for the profiles near the walls. Nevertheless, the results show features which one can expect to be retained in improved theories.

The model which has been most thoroughly studied is that consisting of equally sized hard spheres which carry charges $\pm e$ or dipoles μ at their centres, the civilized model. For low ion concentrations, the theory[18,19] leads to a differential capacitance, C_d, for the electrical double layer given by

$$C_d^{-1} = \frac{\lambda_D}{\varepsilon\varepsilon_0}\left[1 + \frac{\sigma}{2\lambda_D}\left(1 + \frac{\varepsilon_r^{-1}}{\lambda}\right)\right] \tag{8}$$

where λ satisfies

$$\varepsilon = \lambda^2(1 + \lambda)^4/16$$

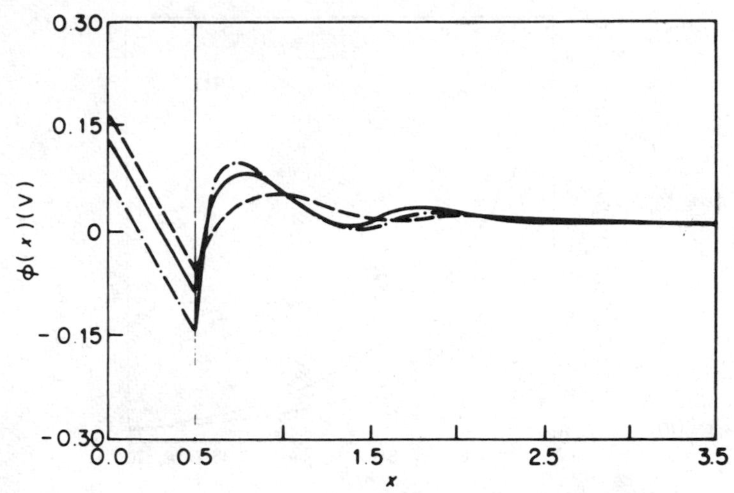

Figure 11 *Profile of the electrostatic potential in the civilized model of a fluid of dipoles and ions near a charged hard wall. The curves correspond to three slightly different approximate solutions*: —, *GMSA–GMSA*; ·—·—·, *GMSA–MSA*; - - -, *MSA–MSA*
(Reproduced from *J. Electroanal. Chem.*, 1983, **150**, 315)

For $\varepsilon = 78$, $\lambda = 2.65$. The result in equation (8) should be compared with that given by Stern layer theory of

$$C_d^{-1} = \frac{\lambda_D}{\varepsilon\varepsilon_0}\left(1 + \frac{\varepsilon_r}{\varepsilon_s} \cdot \frac{\sigma_s}{2\lambda_D}\right) \tag{9}$$

where ε_s is the dielectric constant of the Stern layer and $r_s/2$ its width. From the point of view of the differential capacitance, the theory provides a value for ε_s/σ_s. If $\sigma_s = \sigma$ and $\varepsilon = 78$, then $\varepsilon_s = 2.6$.

However, the profile of the electrostatic potential shown in Figure 11 shows that the effects of including the solvent structure are more profound. These do not conform to those given by MGC theory within three or four molecular diameters of the wall. Therefore, although Stern layer theory gives a reasonable result for the differential capacitance, provided that one makes the correct choice for ε_s, it is unlikely to give a consistent picture of all effects. The fact that one has had to invent several different layers to account for the properties of colloids confirms this view. Further, according to conventional Stern layer theory, the drop in potential across the Stern layer is $\sigma/2\varepsilon_s\varepsilon_o$, whereas according to the microscopic theory it is $\sigma_s/2\varepsilon_o$, a factor of 2.6 greater.

As Figure 12 shows, the calculation of the force between two charged walls deviates from MGC theory in the same way.

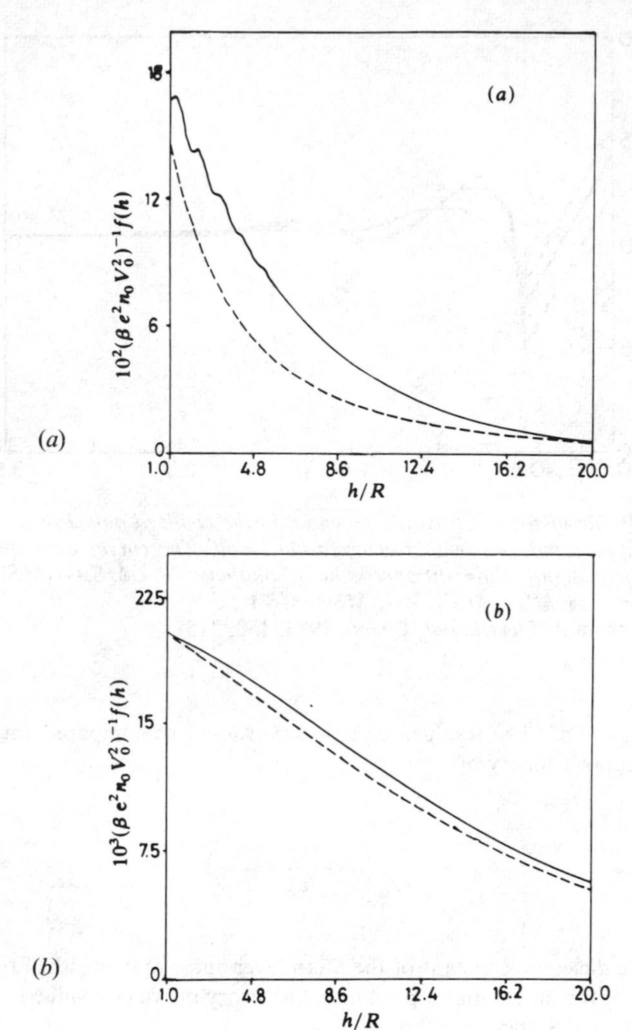

Figure 12 *Force,* f(h), *between two charged walls separated by the civilized model fluid as a function of* h, *the separation of the walls. (a)* $\varepsilon_r = 78$; *(b)* $\varepsilon_r = 7.8$, *from the linear theory; - - -, from MGC theory*
(From reference 21)

5 Conclusions

Significant advances in both the simulation and the theory of confined fluids has been made in recent years. At least for simple neutral fluids, both can be applied. As long as only one phase is present, theory is now able to provide results in agreement with simulation. When two phases are present, such as in wetting phenomena, theory is not entirely satisfactory quantitatively, although it can provide useful qualitative results.

As yet it is not possible to simulate good models of the electrical double layer and theory is still the only approach. For the usual values of the relevant parameters which occur in colloids, it appears that the greatest weakness of classical theory is in its treatment of the solvent as a structureless medium. The structure of the solvent does appear to be important not only for the existence of the Stern layer but also for several layers beyond. Significant departures from the classical DLVO theory, for example, have emerged, although as the comparison with experiment is usually made in terms of arbitrary parameters the discrepancy has not in practice been noted. It seems that the greatest discrepancies are to be expected in non-symmetric rather than in symmetric electrolytes, especially where the counter ions carry the higher charge, for in these cases the counter ions accumulate most strongly near the charged surface in the region where classical theory fails.

Current theories do not yet include the dynamics but, as the structure of the fluid is affected by the solid surface to a distance of several molecular diameters, one must expect that the dynamics will also be affected. It may be that theory in the end will supply only values for phenomenological parameters such as the thickness of the Helmholtz layer, the electrokinetic potential, and the viscosity near the surface, but it is also possible that there are qualitative effects waiting to be discovered.

Note added in proof: the SDA has now been applied[22] to hard sphere mixtures with promising results for mixtures with size ratio, σ_2/σ_1, up to about 3.

References

1. P. Tarazona, *Phys. Rev. A*, 1985, **31**, 2672.
2. P. Tarazona, U. Marini Bettolo Marconi, and R. Evans, *Mol. Phys.*, 1987, **60**, 573.
3. G. Rickayzen and A. Augousti, *Mol. Phys.*, 1984, **52**, 1355.
4. I. K. Snook and D. Henderson, *J. Chem. Phys.*, 1978, **68**, 2134.
5. J. G. Powles, G. Rickayzen, and M. Williams, *Mol. Phys.*, 1988, **64**, 33.
6. J. P. R. B. Walton and N. Quirke, *Chem. Phys. Lett.*, 1986, **129**, 382.
7. B. K. Peterson, K. E. Gibbins, G. S. Heffelfinger, U. Marini Bettolo Marconi, and F. van Swol, *J. Chem. Phys.*, 1988, **88**, 6487.
8. P. C. Ball and R. Evans, *Mol. Phys.*, 1988, **63**, 159.
9. A. Z. Panagiotopoulos, *Mol. Phys.*, 1987, **61**, 813.
10. G. Rickayzen and M. Moradi, *Mol Phys.*, 1989, **66**, 143.
11. G. M. Torrie and J. P. Valleau, *J. Chem. Phys.*, 1980, **73**, 5807.
12. G. M. Torrie, J. P. Valleau, and G. N. Patey, *J. Chem. Phys.*, 1982, **76**, 4615.
13. G. M. Torrie and J. P. Valleau, *J. Phys. Chem.*, 1982, **86**, 3251.
14. W. van Megan and I. K. Snook, *J. Chem. Phys.*, 1980, **73**, 3151.
15. M. J. Grimson and G. Rickayzen, *Mol. Phys.*, 1982, **45**, 221.
16. L. B. Buiyan, C. W. Outhwaite, and S. Levine, *Mol. Phys.*, 1981, **42**, 1271.
17. S. H. Lee, J. C. Rasaiah and J. B. Hubbard, *J. Chem. Phys.*, 1986, **85**, 5232.
18. S. L. Carnie and D. Y. C. Chan, *J. Chem. Phys.*, 1980, **73**, 2949.
19. D. Henderson and L. Blum, *Faraday Sym. Chem. Soc.*, 1981, **16**, 151.
20. F. Vericat, L. Blum and D. Henderson, *J. Electroanal. Chem.*, 1983, **150**, 315.
21. A. Augousti and G. Rickayzen, *J. Chem. Soc., Faraday Trans. 2*, 1984, **80**, 141.
22. Z. Tan, U. Marini Bettolo Marcoui, F. van Swol, and K. E. Gubbins, *J. Chem. Phys.*, 1989, **90**, 3704.

Electrochemical Approach to Studies of Binding and Electrostatic Interaction in Concentrated Food Dispersions

By A. H. Clark, A. Lips, and P. M. Hart

UNILEVER RESEARCH, COLWORTH LABORATORY, COLWORTH HOUSE, SHARNBROOK, BEDFORD MK44 1LQ, UK

1 Introduction

The characterization of concentrated food colloids by non-invasive methods is a challenging task. Such methods as can be applied (*e.g.* NMR, scattering techniques, mechanical spectroscopy) are often difficult to interpret in view of the geometrical complexities of the constituent colloidal units and the diverse nature of the interactions between them. Recently, we introduced a new approach based on potentiometric studies.[1,2] This electrochemical method can be applied directly to concentrated systems providing estimates of ion binding and of the magnitude of electrostatic interactions between the colloidal units.

Ion binding is a major determinant of aggregation effects, be they between biopolymer molecules or colloidal particles. Although such interactions may be monitored by a variety of physical techniques (rheological, spectroscopic, *etc.*), we have shown that an electrochemical approach can provide powerful, complementary information. In addition, it offers advantages in ease of use, in conceptual simplicity, and, most important, in direct (indeed preferred) application to concentrated systems and systems involving mixtures of simple ions.

In essence the method is a 'back-to-back' multi-electrode measurement with data interpretation on the basis of a Donnan calculation by a procedure proposed by Hall.[3-6] The chemical potentials of the common ions (*viz.* Na^+, K^+, Ca^{2+}, H^+, Cl^-) typically encountered in food products are probed directly by an appropriate selection of ion-specific electrodes; the need for 'buffering' the activity of any of the ions is thereby circumvented. Typically, a study would concern the interaction of calcium ions with hydrocolloids (*e.g.* casein at 15%) and the experiment would involve the progressive addition of calcium chloride.

Theoretical considerations have shown that our approach is valid in situations where the electrostatic potentials are small everywhere in the colloidal system or, failing this, situations where electric fields are weak. The first condition is generally implied for biopolymers in the use of Manning's theory.[7] The second condition is likely in concentrated systems and so in general the application of our method to concentrated systems is preferable.

The response of ion-selective electrodes, in concentrated systems in particular, is governed not solely by ion binding but also by the nature of the double layer between the colloidal particles and hence their electrostatic interactions. The Donnan potential can be considered as an average measure for this effect. Our electrochemical method purports to provide estimates of both ion binding and Donnan potentials. If the objective is to measure ion binding, and provided at least one of the above conditions is met, there is no *a priori* difficulty in interpreting electrochemical measurements on concentrated systems. In particular, the analysis for ion binding is not dependent on the geometrical details of the colloidal subunits. The interpretation of Donnan potentials, on the other hand, is dependent on a knowledge of geometrical factors such as particle size, interparticle separations, or preferably the complete radial distribution of particle centres in the system. If such information is available, it is then possible to estimate from the 'average' Donnan potential the surface potential of the colloidal units and their average electrostatic interaction stress, equivalent to the contribution of the diffusible ions to the Donnan osmotic pressure of the concentrated system.[1,2]

Although structural information is desirable in the interpretation of Donnan potentials, it is important to note that this average potential serves as a powerful probe in its own right for stability aspects of concentrated colloidal systems. In common with electrophoretic measurements, it provides an indication of the sign and approximate magnitude of the electrostatic charge on the colloidal particles and it can be used for locating isoelectric points directly in concentrated systems. Moreover, where it can be assumed that the colloidal units are randomly distributed, the Donnan potential can yield a reasonable, albeit approximate, estimate for the contribution of diffusible ions to the Donnan osmotic pressure.

It is an aim of this paper to illustrate the success to date in applying our combined ion binding–Donnan potential measurement to concentrated food colloids. It will be shown that the method holds promise for structurally irregular systems and, as such, directly for many food systems. A second aim is to present measurements on model concentrated systems which are already well characterized by other methods and in which the structural arrangement between the colloidal units is highly regular. Two reasons have prompted the selection of systems with well defined geometry. First, the measured Donnan potentials can then be interpreted more fully and internal consistency can be ascertained, *e.g.* between inferred surface potentials and those determined by other methods (ion binding, electrophoresis), so providing justification for the method. Second, direct measurement of the electrostatic interaction stress is then possible, which can assist in characterizing other interactive forces in colloidal systems, *e.g.* in phospholipid bilayers the magnitude of the van der Waals force balancing the electrostatic stress might be inferred. The model systems selected were bovine serum albumin and egg lecithin, both at 25%, the former representing a lattice cell arrangement of colloidal macro-ions and the latter a flat-plate geometry with constant inter-lamellar spacing. Sodium caseinate systems (8% and 15%) served as examples of less regular structures more typical of food systems.

2 Principle of Method

The Poisson–Boltzmann (P–B) equation is the obvious reference point for theoretical descriptions of electrostatic interactions between charged species, be they colloidal particles, polyelectrolytes, surfactant micelles, or small ions. It is employed, for example, in Manning's theory for polyelectrolyte solutions[7] where the feature of counter-ion condensation is introduced to reduce a high primary surface charge to an effective critical charge per unit length of polyion; the distribution of uncondensed counter ions and co-ions is then describable by a linearized (and mathematically more tractable) form of the P–B equation. In general, mathematical problems limit the application of the P–B equation to systems with relatively simple, well defined geometries. Typical systems, however, are of complex geometry in which the description of electrostatics at best is limited to suitable averages rather than the complete spatial distribution of the various charged constituents. Hall[3–6] has recently suggested such an approach is based on the concept of the Donnan equilibrium between a solution of a charged colloid plus electrolyte and a solution of electrolyte alone. The treatment has been very successful in the description of micellar solutions of ionic surfactants and of the characteristic behaviour of dilute polyelectrolyte solutions.

The requirement of equal chemical potentials in the system under study (superscript s) and the Donnan solution in osmotic equilibrium (superscript D), in respect of any electrically neutral combination of diffusible co-ions (subscript $-$) and counter ions (subscript $+$), is expressed in the theory just described by

$$(C^{D}_{z_+})^{\nu_+} (C^{D}_{z_-})^{\nu_-} (\gamma_{\pm}{}^{D})^{\nu_+ + \nu_-} = \langle C^{s}_{z_+} \rangle^{\nu_+} \langle C^{s}_{z_-} \rangle^{\nu_-} (\gamma_{\pm}{}^{s})^{\nu_+ + \nu_-} \quad (1)$$

where C denotes ionic concentration, $\gamma_{\pm}$ mean activity coefficient and ν the stoicheiometric coefficients for the neutral electrolyte dissociating according to $A_{\nu_+}B_{\nu_-} \rightarrow \nu_+A^{z+} + \nu_-B^{z-}$, with z representing the number of charges of the ion and $\nu_+z_+ + \nu_-z_- = 0$. By diffusible ions are meant all ionic species other than the charged colloid particles or macro-ions in the case of hydrocolloids. The average concentration of a particular species of diffusible ion in the system $\langle C^{s} \rangle$ is taken to be its total concentration minus those fractions which are either specifically bound or electrostatically condensed on the colloid macro-ions. The Debye–Hückel equation is prescribed for estimating the mean activity coefficient, the calculation extending to all diffusible ions but not the macro-ion. By introducing the concept of an effective degree of dissociation for the macro-ion, a predictive capability is provided for relations of the above type. In addition, excluded volume effects can be incorporated in the theory.

From an experimental viewpoint, the chemical potentials of electrically neutral combinations of diffusible co-ions and counter ions are readily measured by electromotive-force (e.m.f.) methods. The preferred approach is to use cells with electrodes reversible to counter ion and co-ion in which liquid junction effects cancel and which therefore permit reliable measurements in concentrated colloids, circumventing the problems (*viz.* suspension effect[8]) and interpretation of the more commonly employed single ion activity measurements. If such back-to-back measurements are performed for all the neutral combinations of diffusible ions (*i.e.* neutral salts) in the system, the average concentration of the ions in both system and Donnan solution can be inferred by reference to the known activity products for standard solutions of the salts. In such calculations, it is necessary to assume that one of the ionic species, usually a co-ion species, does not bind specifically to the colloidal macro-ions. If the total concentrations of the ions in the system are known, it is then straightforward to determine their site-specific (excluding double-layer effects) binding to the macro-ion.

Theoretical arguments[9,10] and experimental evidence[11,12] indicate that, in general, the details of the spatial distribution of ions should be considered in equations for Donnan equilibria. However, our simple formulation in terms of average concentrations of the mobile ions can be justified as follows. The P–B equation links local concentrations of ions with local electrostatic potentials according to

$$\begin{aligned} C_j^{\mathrm{L}} &= C_j^{\mathrm{D}} \exp(-z_je\psi^{\mathrm{L}}/kT) \\ &= C_j^{\mathrm{D}} \exp(-z_je\psi^0/kT) \exp[-z_je(\psi^{\mathrm{L}} - \psi^0)/kT] \end{aligned} \tag{2}$$

where C_j^{L} is the local concentration of ionic species j with charge number z_j, C_j^{D} is the corresponding concentration in the Donnan solution, ψ^{L} is the local electrical potential, ψ^0 the potential at the surface of the macro-ion, e is the electronic charge, k the Boltzmann constant, and T the absolute temperature. We are interested in obtaining average values for the local concentrations of ions in the solution $<C_j^{\mathrm{L}}>$, the average being taken in the first place over the distance between a pair of macro-ions and then over all macro-ion separations. Thus,

$$\begin{aligned} <C_j^{\mathrm{L}}> &= C_j^{\mathrm{D}} <\exp(-z_je\psi^{\mathrm{L}}/kT> \\ &= C_j^{\mathrm{D}} <\exp(-z_je\psi^0/kT)\exp[-z_je(\psi^{\mathrm{L}} - \psi^0)/kT]> \end{aligned} \tag{3}$$

If $z_je\psi^{\mathrm{L}}/kT \ll 1$, or $z_je(\psi^{\mathrm{L}} - \psi^0)/kT \ll 1$, conditions which are fulfilled either when the macro-ion charge is small (*i.e.* ψ^0 is small) or when the solution is sufficiently concentrated that variation of ψ^{L} in relation to ψ^0 (ψ^{L}, ψ^0 can be large) is small, then the exponentials involving ψ^{L} can be expanded and truncated at the linear term to give

$$<C_j^{\mathrm{L}}> = C_j^{\mathrm{D}} \exp(-z_je<\psi>/kT) \tag{4}$$

i.e. in both situations the average local concentrations can be expressed in terms of the average value of $\psi = <\psi>$. In the above, averaging has been assumed to refer to the variation of ψ between a pair of macro-ions, but there is no reason why equation (4) should not also imply averaging over all possible configurations of macro-ions. As the C_j^{D} in the Donnan solution are fixed, it follows, that within the confines of the linear P–B regime, or in weak electric fields such as occur in concentrated systems $z_je(\psi^{\mathrm{L}\,6}\,\psi^0)/kT \ll 1$, at positions where local potentials ψ^{L} equal the average potential $< \psi >$, the local concentrations of both co-ions and counter ions are equal to their average concentrations. The use of equation (1) is thus justified and can be considered to relate to positions in the system where local potentials have the numerical value of the average potential.

As indicated, for example by electrokinetic methods, many polyelectrolyte systems have sufficiently small potentials (or in concentrated systems show sufficiently small variation of the potential in the space between particles) to warrant the use of a linearized P–B expression. Moreover, to the extent that the Manning theory has found wide application, the assumption of low potentials in the double layer surrounding the macro-ion is not unreasonable.

The average potential $<\psi>$ defined by equation (4) is the so-called Donnan potential. It is readily obtained from the comparison between the concentration of reference (non-adsorbing) ions in the Donnan solution, as indicated by the potentiometric measurements, and the known average concentrations in the system. The average potential can be a useful starting point for estimating the surface potential of the particles and, perhaps more important, the magnitude of the electrostatic interaction stress between the particles. The latter is in effect the contribution of all the diffusible ions to the Donnan osmotic pressure Π_{el} and may formally be related to local ion concentrations at positions x in the system where the potential $\psi^{L} = \psi^{M}$ is at a minimum and the electric field strength $\partial\psi^{L}/\partial x$ is zero. The result is

$$\Pi_{el} = RT\sum_{j} C_j^{D}[\exp(-z_j e\psi^{M}/kT)-1] \tag{5}$$

or, allowing for configurational averaging in the real situation,

$$\Pi_{el} = RT\left\langle\sum_{j} C_j^{D}[\exp(-z_j e\psi^{M}/kT)-1]\right\rangle_{\text{config.}} \tag{6}$$

Strictly, a knowledge of the structural arrangement of the colloidal subunits in the system is required for the evaluation of equation (6). This, of course, is not available for other than model systems. Typical food colloids, however, have irregular structures for which we strictly cannot determine the configurational average in equation (6) even when ψ^{M} is small enough to allow truncation of the exponential, *i.e.* $<\psi^{M}>_{\text{config.}}$ is usually unknown. Despite this limitation, an approximate estimate may still be possible in situations where charges are small and configurations tend to be weighted equally. Then the approximation $<\psi^{M}> \approx <\psi>$ is valid (*i.e.* we may use the average concentration of diffusible ions in the solution to estimate Π_{el}), but we emphasize that this step is not rigorously correct, even in cases where it is supported by a situation of low macro-ion charge. Thus, whilst for irregular structures ion binding can be calculated rigorously by the methods discussed above, use of the equation

$$\Pi_{el} = RT\sum_{j} C_j^{D}[\exp(-z_j e<\psi>/kT)-1] \tag{7}$$

to calculate the osmotic pressure is more approximate.

In the following sections, we present detailed examples to illustrate the application of the method described here.

3 Experimental

A commercial sample of sodium caseinate (DeMelkindustrie Veghel) was used without purification. Chemical analysis of the chosen sample indicated the presence of significant levels of sodium (1.4% w/w), calcium (0.08% w/w), and potassium ions (0.00063% w/w), with chloride ion (0.23% w/w) the only diffusible anion.

A purified form of bovine serum albumin (BSA), essentially fatty acid free, was supplied by Sigma (Code A-6003). The levels of diffusible ions were determined as calcium (0.018% w/w), potassium (0.021% w/w), sodium (0.46% w/w), chloride (0.59% w/w), and magnesium (0.006% w/w).

Egg lecithin, 60% L-α-phosphatidylcholine Type IX-E, was supplied by Sigma (Code P-8640); the levels of ions were calcium (0.031% w/w), potassium (0.0025% w/w), sodium (0.36% w/w), chloride (0.36% w/w), and magnesium (0.009% w/w).

Some small-angle X-ray measurements were performed using a Kratky camera with position-sensitive counting. Egg lecithin systems at 25% w/w were studied in the presence of increasing levels of calcium chloride and interlamellar spacings were measured. This information was used in the interpretation of parallel electrochemical measurements.

The detailed application of our electrochemical approach is first illustrated for sodium caseinate. The natural pH of solutions of the caseinate (as supplied) was in the range of 6.5–7, and to such unbuffered solutions were added known amounts of calcium chloride. E.m.f. measurements were performed at 22 °C involving reversible electrodes for sodium, calcium, potassium, and hydrogen ions each sharing a reversible chloride electrode as the common reference electrode. Calibration graphs were constructed for standard solutions of the various chloride salts. Conformity to near-Nernstian behaviour in the performance of each of the electrodes could be demonstrated and it was ascertained that cross-selectivity to cations was not significant for any of the electrodes under the conditions of study. The potentiometric measurements were carried out with an Orion EA 940 ion analyser with high-impedance detection for both the monitor and reference electrodes. The back-to-back e.m.f. measurements scale with the logarithm of the activity products $(C_{z+})^{\nu_+} (C_{z-})^{\nu_-} (\gamma_\pm)^{\nu_+ + \nu_-}$, which are readily quantified through measurements on standard solutions of the various salts.

For the Donnan solution the concentrations of the ionic species Na^+, Ca^{2+}, K^+, and H^+ and Cl^- are then specified by four such measurements together with the condition of electrical neutrality, $[Na^+]^D + 2[Ca^{2+}]^D + [H^+]^D + [K^+]^D = [Cl^-]^D$. Numerical solution of these equations is straightforward; an initial estimate is obtained by assuming that all the activity coefficients are unity. Next, the Davies extension of the Debye–Hückel equation is used to obtain an improved estimate of activity coefficients:

$$\log\gamma_i = \frac{-0.509 z_i^2 \sqrt{I}}{1 + \sqrt{I}} + 0.13 z_i^2 I \qquad (8)$$

in terms of $I\ (= 1/2\Sigma z_i^2 C_i)$ the ionic strength calculated on the basis of the first estimate of the ionic concentrations. Iteration readily yields convergent solutions for the ionic concentrations and the single ion activity coefficients γ_+^D and γ_-^D $(\gamma_+^{\nu_+}\gamma_-^{\nu_-} = \gamma_\pm^{\nu_+ + \nu_-})$.

The diffusible ions of the systems under study, $<C^s>$, are not themselves subject to electrical neutrality, rather their net charge $[<Na^+>^s + 2<Ca^{2+}>^s + <K^+>^s + <H^+>^s - <Cl^->^s]$ balances the negative charge Z of the polyions. Since for pH > 5 caseinate is substantially negatively charged, it is not unreasonable to assume that the chloride ions are co-ions and as such unbound, and their average concentrations can therefore be taken as known. Average concentrations of free counter ions and their activity coefficients are then estimated simply from measured activity products, again following a simple iterative procedure based on equation (8). As the total concentrations of diffusible ions are known, the levels of bound counter ions can be inferred. The total concentrations of diffusible ions can also be used in conjunction with the corresponding concentrations of ions in the Donnan solution to produce so called 'apparent' values for counter-ion binding. These are usually greater than the true (or 'real') values obtained by the procedures just described and are equivalent to the results which might emerge from other types of experiment, *e.g.* equilibrium dialysis.

In the case of bovine serum albumin, the experiment involved the progressive addition of sodium chloride to a 26.8% w/w solution at 22 °C. In view of the low levels of ions other than sodium and chloride, it was sufficient to limit the selection of electrodes to sodium, chloride, and hydrogen only. Despite being a co-ion, chloride is known to bind strongly to serum albumins[13] to a much greater extent, in fact, than the sodium counter ion. In data analysis [equation (8) and the ion balance as above], we therefore took sodium as our reference, non-adsorbing ion.

Egg lecithin was studied at 25% w/w and 22 °C. In this case increasing amounts of calcium chloride were added and the chemical potentials of sodium chloride, hydrochloric acid, and calcium chloride were probed by suitable 'back-to-back' combinations of ion-selective electrodes. In data analysis the choice of non-adsorbing reference ion was chloride for low levels of addition of calcium chloride and sodium at high free calcium levels when the Donnan potential was positive.

4 Results and Discussion

Dilute Caseinate Solutions—The concentration of caseinate selected was 0.45% w/w, a level representative of previous binding studies.[14–17] Figure 1 shows two binding isotherms obtained by the present method. To facilitate comparison with previous studies, the levels of calcium binding are expressed in numbers of ions per amino acid residue, taking an average molecular weight of 112 for the latter based on the known sequence data for α,β, and κ-casein[18–20] and the established ratio of these components in caseinate.[21] The concentration of calcium ions chosen as the abscissa in Figure 1 is that of the Donnan solution (rather than of the system). This choice provides a common reference and parallels the quantity measured by equilibrium dialysis. The isotherms are in reasonable agreement with previous work[14–17] showing the expected levels of calcium binding and the

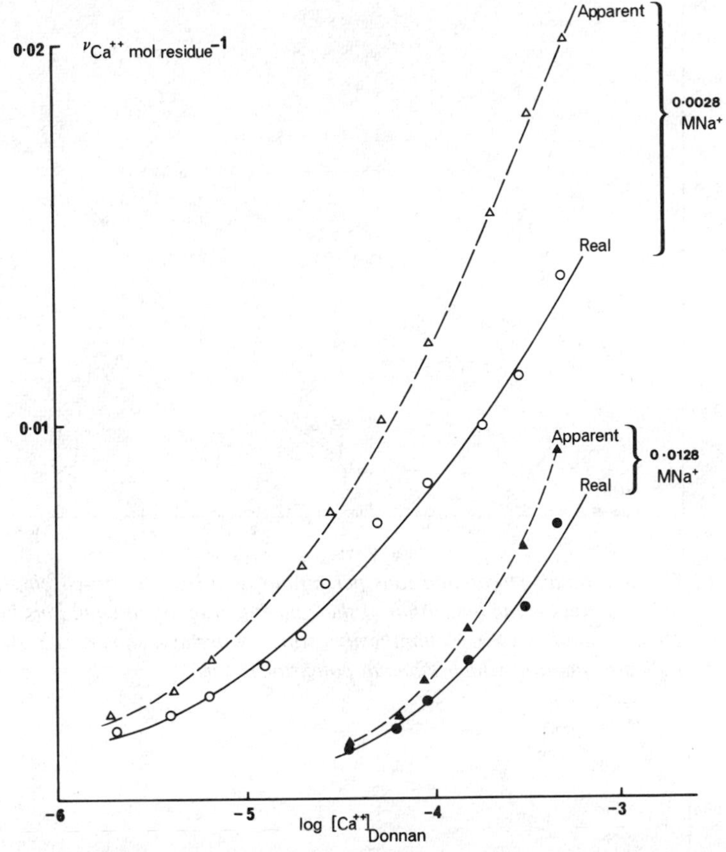

Figure 1 *Real and apparent binding (calcium ions per amino acid residue) to caseinate at* 0.45% w/w versus *the logarithm of the concentration of calcium ions in the Donnan solution. Total sodium concentrations:* △, ○, 2.83×10^{-3} mol dm^{-3}; ▲, ●, 1.28×10^{-2} mol dm^{-3}

sensitivity of binding to ionic strength. For purposes of comparison, Figure 1 also shows corresponding apparent binding isotherms calculated as described previously.

Concentrated Caseinate Solutions—Reproducible e.m.f. measurements could be obtained with systems as concentrated as 15% w/w. Interpretation of the data in terms of the above considerations alone, however, yielded unrealistic (negative) binding of sodium ions. Inclusion of the concept of excluded volume can remove this discrepancy. The effective concentrations of the diffusible ions are taken as greater than the nominal values because a fraction of the water is inaccessible to the ions by a combination of hydration of the caseinate and micellization. We

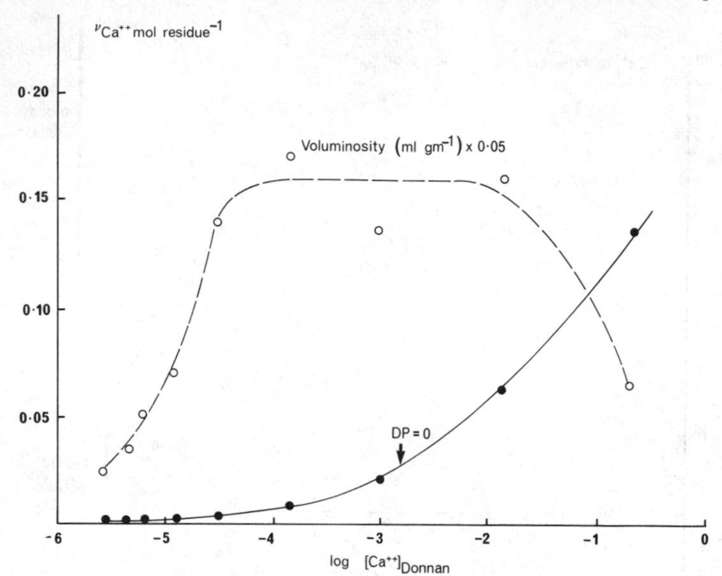

Figure 2 *Calcium binding (calcium ions per amino acid residue) to caseinate at* 8% w/w versus *the logarithm of the concentration of calcium ions in the Donnan solution* (●). *Voluminosity estimates are also shown* (○). *Arrow indicates point at which Donnan potential is zero*

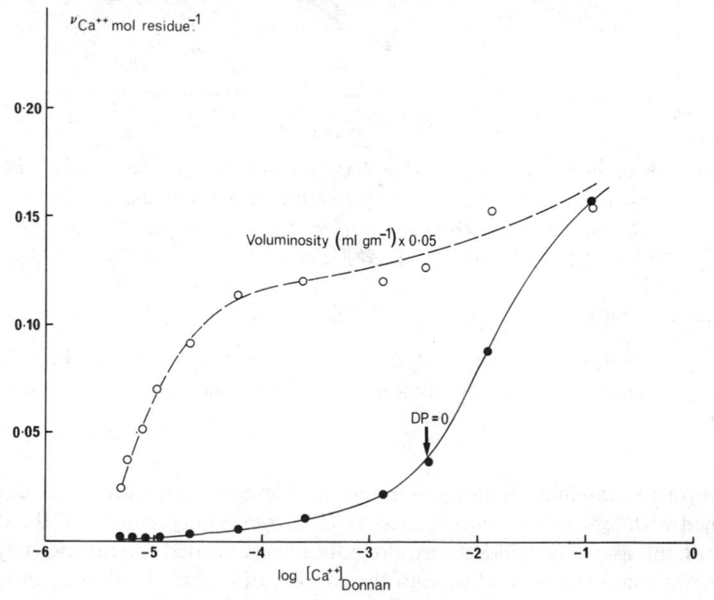

Figure 3 *Calcium binding (calcium ions per amino acid residue) to caseinate at* 15% w/w versus *the logarithm of the concentration of calcium ions in the Donnan solution* (●). *Voluminosity estimates are also shown* (○). *Arrow indicates point at which Donnan potential is zero*

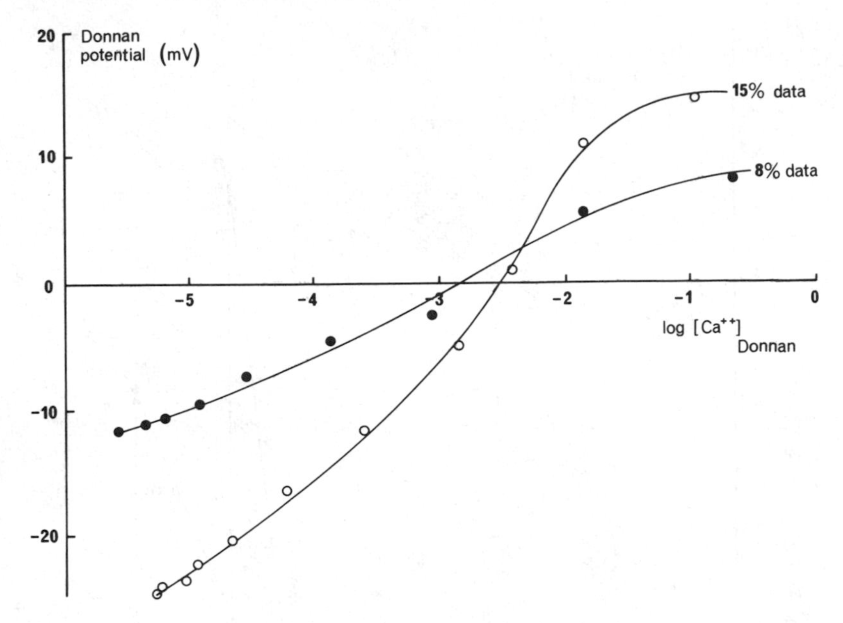

Figure 4 *Inferred Donnan potentials* versus *the logarithm of the concentration of calcium ions in the Donnan solution for concentrated caseinate solutions*

represent the excluded volume effect in terms of V (=volume of water inaccessible to ions per gram of caseinate) and this is expected to be of a similar order to the range of values 2–5 ml g^{-1} observed in studies of casein voluminosity.[22–24] In the analysis V is treated as a variable parameter and binding isotherms are generated for calcium and sodium ions. It is known[15] that sodium ions have *ca.* 1/100th of the affinity of calcium ions for binding to casein. Moreover, the intrinsic binding constants to organic phosphate groups (which are the predominant substrates in ion binding to casein) suggest that the binding of sodium ions is too weak to be easily measured in the dilute caseinate solution. However, at 15% and 8% the local concentrations of sodium ions are high and significant fractions can be expected to be bound. In fact, prior to the addition of calcium chloride mainly sodium ions are bound.

Optimization in the choice of values of V was dictated by the requirements of consistency with the known binding affinities for sodium and calcium ions in dilute systems where excluded volume effects are of no consequence. We find that a model of initial ion exchange (two sodium ions for one calcium ion) followed by more extensive calcium binding past the stage of complete desorption of sodium ions can meet these criteria. In particular, realistic values are found for the amounts of calcium bound (Figures 2 and 3) and corresponding results for V are similar to estimates of voluminosity quoted for casein micelles.

Average Donnan potentials $<\psi>$ obtained using equation (4) are shown in Figure 4 as functions of the amount of added calcium for both the 8% and 15% caseinate systems. The values obtained are within the range expected from

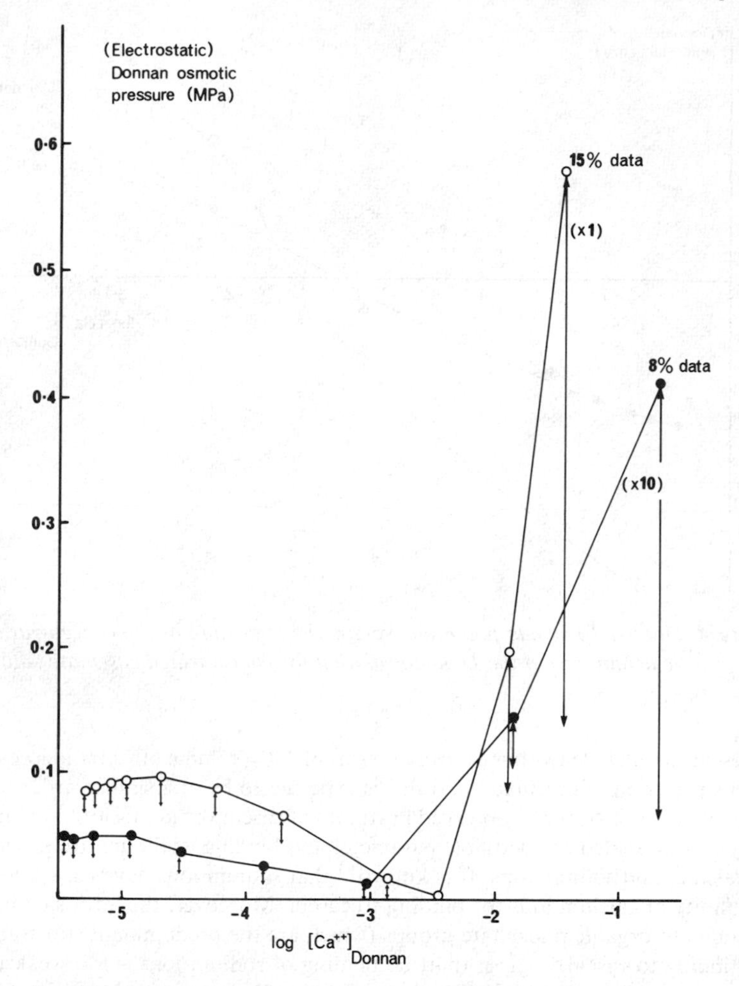

Figure 5 *Electrostatic contribution to the Donnan osmotic pressure* versus *the logarithm of the concentration of calcium ions in the Donnan solution for concentrated caseinate solutions. Error bars indicate upper bounds of uncertainties generated by using average rather than mid point concentrations in calculations*

electrokinetic studies.[22] As can be seen, the Donnan potentials increase with calcium addition, become zero at the precipitation point, then become positive as more calcium is bound. On the basis of equations (7) and (4), the measured Donnan potentials provide approximate estimates for the contribution of the diffusible ions to the Donnan osmotic pressure Π_{el} if it is assumed that electrostatic repulsion between the caseinate molecules is insufficient to create lattice-like order, *i.e.* that all possible configurations between the colloidal macro-ions are weighted equally.

The results in Figure 5 show that the osmotic pressure term Π_{el} decreases with calcium addition and vanishes at the point of precipitation. After this, particularly for the 15% casein system, it rises again sharply, a feature which is unexpected and requires closer examination.

As has been emphasized in Section 2, whereas the Donnan potentials in Figure 4 and the binding data in previous figures are well determined quantities (at least within the limits of our assumptions about stoicheiometry and voluminosity), considerable uncertainty attaches to the results of the osmotic pressure calculations. The very high pressures obtained at the highest levels of bound calcium seem to reflect this, and in consequence the following simple calculation was performed to reveal the level of uncertainty likely to accompany use of equation (7). Remembering that $<\psi^M>$ is the quantity of significance in determining the osmotic pressure, not the Donnan potential $<\psi>$, we may examine how different $<\psi>$ and $<\psi^M>$ are likely to be. To do this we approximate the experimental situation of spherical particles spaced randomly apart, by an average parallel plate geometry (Figure 6), and, following a solution for such a configuration (valid for low potentials), relate $<\psi>$, $<\psi^M>$ to the surface potential ψ^0 by the equations[25]

$$\psi^0 = \kappa d<\psi>/\tanh(\kappa d) \tag{8}$$

and

$$\psi^M = \kappa d<\psi>/\sinh(\kappa d) \tag{9}$$

where d is half the parallel plate spacing and κ equals $8\pi e^2 I/\varepsilon kT$, with ε the dielectric constant of the solution. For various assumed values of d, we may calculate values for ψ^0 and ψ^M from the experimental Donnan potentials $<\psi>$ and hence corresponding osmotic pressures from the ψ^M values. In this way, estimates may be arrived at for the uncertainties introduced by using $<\psi>$ values in equation (7) rather than corresponding ψ^M.

The error bars in Figure 5 indicate the outcome of this calculation and confirm that osmotic pressures at high calcium levels are particularly prone to uncertainty and are likely to have been overestimated by equation (7). However, in the real situation of spherical particles, randomly distributed, errors may be less and the uncertainty limits in Figure 5 should be regarded as upper bounds. Perhaps also relevant to this issue is recent theoretical work by Jonsson and co-workers[26,27] suggesting that Poisson–Boltzmann theory can be expected to overestimate the electrostatic repulsion in the presence of divalent ions.

Concentrated Egg Lecithin Solutions—The binding isotherms for calcium and chloride ions to lecithin at 25% w/w are shown in Figure 7. As with the concentrated caseinate systems, it was necessary in data analysis to allow for the excluded volume of the colloidal macro-ions. This was best accounted for in terms of a partial specific volume of 1 $cm^3\ g^{-1}$ for the non-polar regions of the lipid bilayers with negligible contributions from inaccessible hydration layers. On this

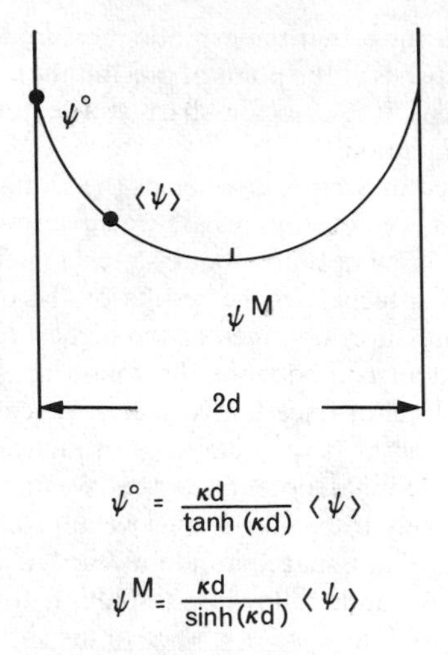

Figure 6 *Relationship between average potential $<\psi>$, surface potential ψ^0, and mid-point potential ψ^M pertaining to the parallel plate model for charge colloid particle–particle interaction*

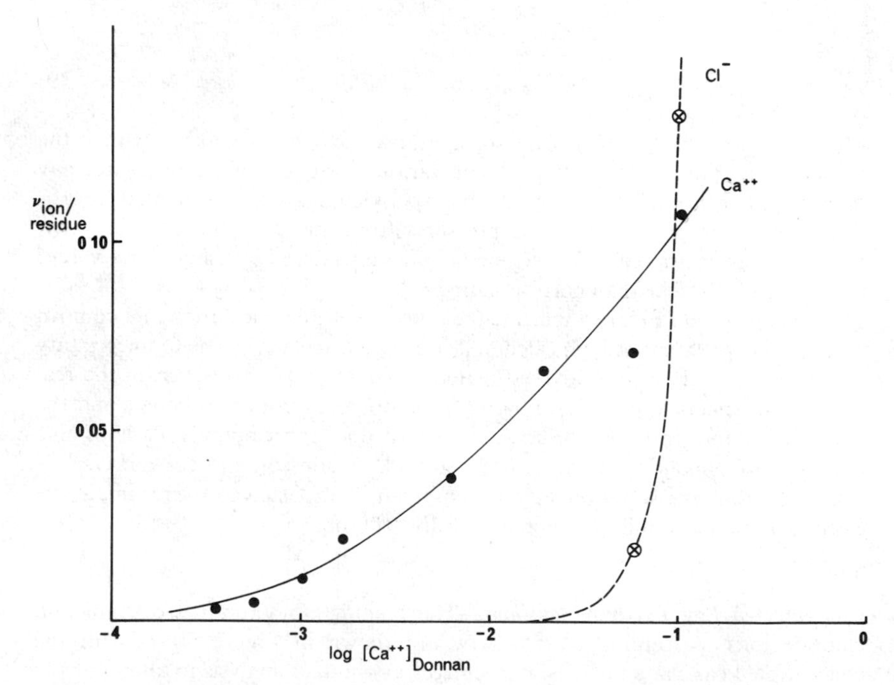

Figure 7 *Binding isotherms (ions per residue) for calcium (●) and chloride (⊗ ions to lecithin at 25%* w/w versus *the logarithm of the concentration of calcium ions in the Donnan solution*

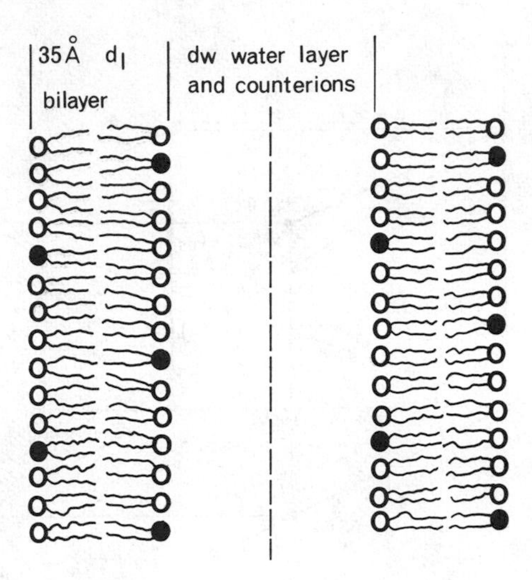

Figure 8 *Schematic representation of phospholipid bilayer*

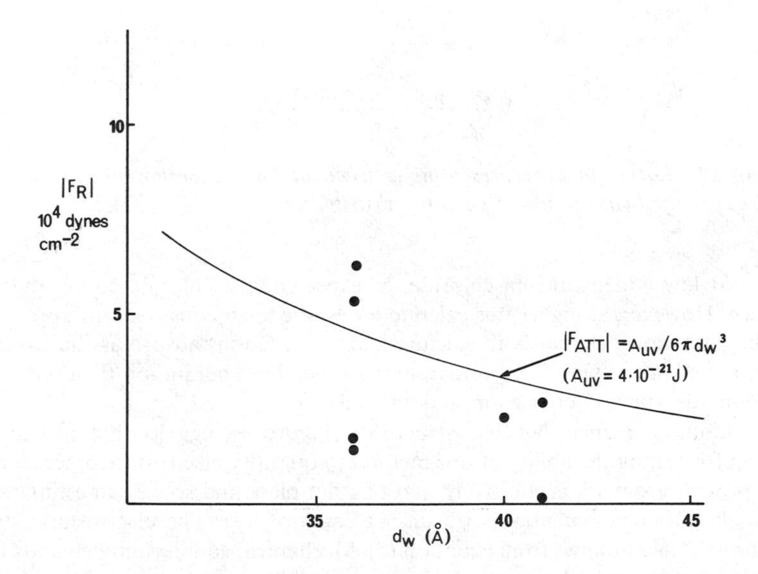

Figure 9 *Electrostatic interaction stress* $|F_R|$ *between lecithin bilayers* versus *surface separation* d_w. *The full curve represents the best estimate of the balancing attractive pressure* $|F_{ATT}|$ *due to van der Waals forces*

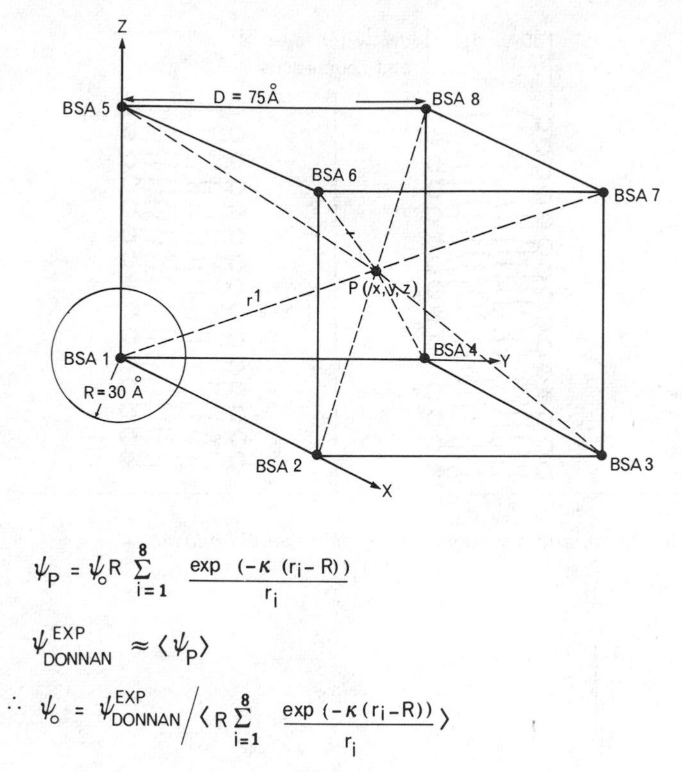

Figure 10 *Lattice model representing concentrated bovine serum albumin solution and calculation of potential distribution*

basis, at low added calcium chloride, as expected, the chloride co-ion did not adsorb. However, at higher free calcium levels, the most consistent interpretation of the data indicates that both calcium and chloride ions adsorb as the Donnan potential changes sign from positive to negative. The sodium ion then served as the non-adsorbing reference ion in data analysis.

In adopting regular lamellar structures (Figure 8), egg lecithin provides a system for testing the ability of our method to quantify electrostatic interactions. The present geometry is effectively that of a flat plate and so we can estimate ψ^M from the Donnan potential $<\psi>$ using equation (9). The electrostatic stress $|F_R|$ ($\equiv \Pi_{el}$) then follows from equation (5). Mechanical equilibrium demands that the electrostatic repulsion stress $|F_R|$ between the bilayers is balanced by a van der Waals attractive pressure which scales as $|F_{ATT}| = Auv/6\pi d_w^3$, where d_w is defined in Figure 8 and A_{uv} is the familiar Hamaker dispersion constant.[28] At the relatively large separations studied here, short-range interactions, *e.g.* hydration forces, are effectively screened and need not be considered.[28] The repulsive stress F_R inferred from the electrode measurements is correlated in Figure 9 with d_w measured by small-angle X-ray scattering. Also shown is a theoretical prediction for a

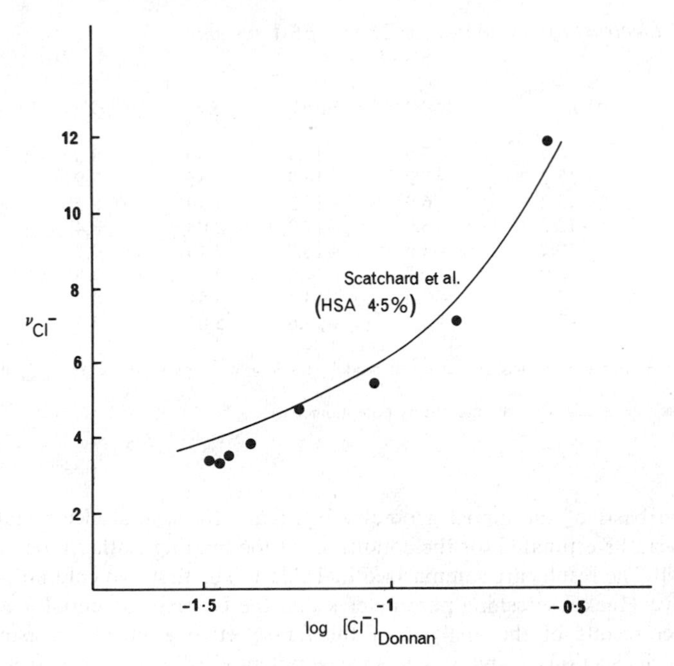

Figure 11 *Binding of chloride (ions per amino acid residue) to bovine serum albumin at* 26.8% w/w versus *the logarithm of the concentration of chloride ions in the Donnan solution. Full curve represents literature data by Scatchard* et al.[13]

Hamaker constant $A_{uv} = 4 \times 10^{-21}$ J. Although there is considerable scatter in the data the mean value of the inferred Hamaker constant of $(2.8 \pm 1.8) \times 10^{-21}$ J is close to the best current estimate for phospholipids. This agreement clearly provides support for our method which may find more general use in the characterization of interaction forces.

Concentrated Bovine Serum Albumin Solution—The globular protein bovine serum albumin has been extensively characterized by a range of techniques. The hydrodynamic size and shape of the BSA molecule are fairly well established and correspond to a mean spherical dimension of 30 Å.[29] At the high concentration (26.8% w/w) studied here, the molecules are expected to be in an ordered arrangement[30] which we will model as a cubic lattice of spheres of diameter 30 Å and mean centre–centre separation of 75 Å (Figure 10).

Our binding data for chloride ions are shown in Figure 11. These are in good agreement with measurements by Scatchard *et al.*[13] for human serum albumin at 4.5% w/w. In data analysis, the sodium ion was constrained to be non-adsorbing and excluded volume was accounted for in terms of the known partial specific volume of BSA of 0.734 $cm^3\ g^{-1}$ and a level of hydration, 0.3 g per gram of dry BSA,[30] taken as inaccessible to the diffusible ions.

Table 1 *Electrostatic estimates for* 26.8% *BSA systems*

κ (10^6 cm^{-1})	$<\psi>_{Donnan}$ (mV)	ψ_m (mV)	ψ^0 (mV)	κR	Q^a	E^b
6.0	−14.1	−7.9	−17.4	1.80	8.0	7.0
6.2	−13.3	−7.2	−16.9	1.85	7.9	6.9
6.3	−13.2	−6.9	−17.2	1.90	8.2	7.1
6.8	−12.2	−5.9	−17.0	2.03	8.4	7.6
7.9	−10.2	−4.0	−16.7	2.40	9.2	8.4
9.9	−6.9	−1.7	−14.6	2.96	9.5	9.1
12.7	−4.8	−0.6	−13.4	3.81	10.5	10.7
16.8	−3.6	−0.1	−13.6	5.03	13.5	14.8

[a] Q = Particle charge estimated on cubic cell model with R = 30 Å for κ and $<\psi>_{\mathrm{Donnan}}$ defined by potentiometry.
[b] E = Particle net charge directly inferred by potentiometry.

On the basis of the model geometry in Figure 10, several electrostatic parameters can be estimated for the conditions of the binding isotherm described in Figure 11. The results are summarized in Table 1. The first two columns, namely the Debye–Hückel screening parameter κ and the Donnan potential $<\psi>$, are the direct results of the analysis of the ion-selective electrode measurements described in Sections 2 and 3. The surface potential ψ^0 is obtained from κ and $<\psi>$ through the equation

$$\psi^0 = <\psi> / <R \sum_{j=1}^{8} \frac{\exp(-\kappa(r_j - R))}{r_j}> \tag{10}$$

where it is implicit (i) that the local potential $\psi^{\mathrm{L}}(r)$ between isolated particles scales[25] as $\psi^{\mathrm{L}}(r) = \psi^0 R\exp(-\kappa(r-R))/r$, (ii) that in a concentrated system of particles the local potential can be represented by a linear superposition of isolated pair potentials, and (iii) that the cell geometry of Figure 10 applies with the summation restricted to the first unit cell. The second approximation is valid for low potentials in the linear P–B regime.[25] The approach of equation (10) is relevant for the calculation of the complete potential distribution and in particular for the determination of ψ^{M} at the position of zero electric field. As already discussed, this potential permits the calculation of the electrostatic interaction stress Π_{el}.

The surface potential ψ^0 provides a basis for estimating the surface charge Q of the BSA particles, since[25]

$$Q = R\varepsilon(1 + \kappa R)\psi^0 \tag{11}$$

where ε is the dielectric constant of the medium (= 78.5 for water). The values of Q (in electronic units per particle) so determined are compared with E, the particle charge indicated directly from our Donnan analysis of the multi electrode measurements. The agreement between Q and E is remarkably close. It confirms internal consistency in our method and greatly strengthens the case for its wider use.

References

1. A. Lips, A. H. Clark, and D. G. Hall *Food Hydrocolloids*, 1988, **2**, 95.
2. P. M. Hart, A. H. Clark and A. Lips, in 'Gums and Stabilisers for the Food Industry', ed. G. O. Phillips, D. J. Wedlock and P. A. Williams, Vol. 4, IRL Press, Oxford, Washington, 1988, p. 135.
3. D. G. Hall, *J. Chem. Soc., Faraday Trans. 1*, 1981, **77**, 1121.
4. D. G. Hall, *Colloids Surf.*, 1982, **4**, 367.
5. D. G. Hall, *J. Chem. Soc., Faraday Trans. 1*, 1984, **80**, 1193.
6. D. G. Hall, *J. Chem. Soc., Faraday Trans. 1*, 1985, **81**, 885.
7. G. S. Manning, *J. Chem. Phys.*, 1969, **51**, 924, 3249.
8. H. Pallman, *Kolloidchem. Beih.* 1930, **30**, 334.
9. J. Th. G. Overbeek, in '*Colloid Science*,' ed. H. R. Kruyt, Vol. 1, Elsevier, Amsterdam, 1952, pp. 191–193.
10. J. Th. G. Overbeek, *Prog. Biophys. Chem.*, 1956, **6**, 58.
11. H. Hammerstein, *Biochem. Z.*, 1924, **147**, 481.
12. F. W. Klaarenbeek, 'Over Donnen evenwichten bij solen van Arabische Gom,' *Thesis*, Utrecht, 1946.
13. G. Scatchard, I. H. Scheinberg and S. H. Armstrong, Jr., *J. Am. Chem. Soc.*, 1950, **72**, 535.
14. D. F. Waugh, C. W. Slattery, and L. K. Creamer, *Biochemistry*, 1971, **10**, 817.
15. C. W. Slattery and D. F. Waugh, *Biophys. Chem.*, 1973, **1**, 104.
16. D. G. Dalgleish and T. G. Parker, *J. Dairy Res.*, 1980, **47**, 113.
17. T. G. Parker and D. G. Dalgleish, *J. Dairy Res.*, 1981, **48**, 71.
18. J. C. Mercier, F. Grosslande, and B. Ribadeau-Dumas, *Eur. J. Biochem.*, 1971, **23**, 42.
19. B. Ribadeau-Dumas, G. Brignon, F. Grosslande, and J. C. Mercier, *Eur. J. Biochem.*, 1972, **25**, 505.
20. J. C. Mercier, G. Brignon, and B. Ribadeau-Dumas, *Eur. J. Biochem.*, 1973, **35**, 222.
21. A. H. Clark and S. B. Ross-Murphy, *Adv. Polym. Sci.*, 1987, **83**, 57
22. D. G. Dalgleish, *J. Dairy Res.*, 1984, **51**, 425.
23. T. A. J. Payens, *J. Dairy Res.*, 1979, **46**, 291.
24. P. Walstra, *J. Dairy Res.*, 1979, **46**, 317.
25. J. Th. G. Overbeek, in '*Colloid Science*,' ed. H. R. Kruyt, Vol. 1, Elsevier, Amsterdam, 1952, pp. 249–256.
26. L. Goldbrand, B. Jonsson, H. Wennerstrom, and P. Linse, *J. Chem. Phys.*, 1984, **80**, 2221.
27. B. Svensson and B. Jonsson, *Chem. Phys. Lett.*, 1984, **108**, 580.
28. D. M. Le Neveu, R. P. Rand, V. A. Parsegian, and D. Gingell, *Biophys. J.*, 1977, **18**, 209.
29. P. G. Squire, P. Moser, and C. T. O'Konski, *Biochemistry*, 1968, **7**, 4261.
30. D. Benedouch and Sow-Hsin Chen, *J. Phys. Chem.*, 1983, **87**, 1473.

Lecithin-stabilized Silica Dispersions

By A. C. Mackie* and M. J. Hey

DEPARTMENT OF CHEMISTRY, UNIVERSITY OF NOTTINGHAM, NOTTINGHAM NG7 2RD, UK

and J. R. Mitchell

DEPARTMENT OF APPLIED BIOCHEMISTRY AND FOOD SCIENCE, SCHOOL OF AGRICULTURE, UNIVERSITY OF NOTTINGHAM, SUTTON BONINGTON, NR. LOUGHBOROUGH, LEICS. LE12 5RD, UK

1 Introduction

One of the most important applications of lecithins in foods is in the control of the rheological properties of molten chocolate. Lecithin is added to lower chocolate viscosity, thus allowing the level of fat in the formulation to be reduced with a cost saving. It is generally stated that the addition of up to about 0.3% of lecithin reduces the level of cocoa butter required by ten times the weight of added lecithin.[1,2]

The aim of this work was to gain some insight into the mechanism of this effect by studying a model system consisting of lecithin-stabilized silica particles in a variety of non-aqueous media. Silica particles were chosen because they have a large, easily determined surface area and owing to chemisorbed water they have a surface consisting of hydroxyl groups which resembles the surface of a sugar particle.

The objectives of the study were (i) to determine adsorption isotherms for lecithins on silica particles; (ii) to measure the effect of adsorbed lecithin on inter-particle association; and (iii) to determine the relationship of the adsorption and aggregation behaviour to the rheological properties of the dispersions.

2 Materials and Methods

Materials—Dipalmitoyl-DL-α-phosphatidylcholine (DPPC), dioleoyl-L-α-phosphatidylcholine (DOPC), dilinoleoyl-L-α-phosphatidylcholine (DLPC), and soya-bean lecithin, all of 99% purity, were obtained from Sigma Chemical (Poole, Dorset).

Methanol and ethanol were spectroscopically pure products (>99.5%) from Burroughs (FAD) (Poole, Dorset) and n-propanol was supplied by BDH (Poole, Dorset) and was of AnalaR grade. n-Dodecane (99% pure) was supplied by Aldrich Chemical (Gillingham, Dorset).

*Present address: Thorn Lighting Ltd., Lincoln Road, Enfield EN1 1SB, UK.

Three types of pyrogenic non-porous silica powder were used. For the lecithin adsorption experiments, a product (TK800) was obtained from the National Physical Laboratory (Teddington, Middlesex). This had a reported specific surface area of 152.4 $m^2 g^{-1}$, obtained from nitrogen BET measurement. For the photon correlation spectroscopic and rheological measurements, the products Aerosil 200 and OX 50 were obtained from Degussa (Frankfurt, FRG). The density of the amorphous silica was assumed to be 2.2 $g cm^{-3}$ when determining volume fractions.

Methods—*Adsorption Isotherms.* The adsorption isotherms were determined from the equilibrium concentration of lecithin in the presence of the silica particles. Lecithin concentration was measured using a spectroscopic method involving the formation of a complex between molybdate and inorganic phosphate liberated from the lecithin. The experimental details have been reported previously.[3]

Preparation of Dispersions. Dispersions in alcohols were prepared in 30-ml glass bottles by adding a known volume of lecithin in alcohol or alcohol alone to a mass of dried silica. The lecithin concentration needed to give the required surface coverage was calculated from the adsorption isotherms. The sample bottles were tightly sealed and ultrasonicated for 2 h by suspension in a Dawes Model 125 ultrasonic bath (125 W) for 2 h. The samples were equilibrated for approximately 16 h at 25.8 °C. Immediately prior to measurement, a further sonication was carried out for 5 min in a smaller bath (Dawes, Model 6441A, 80 W). These severe ultrasonication procedures were required to disperse gel-like aggregates.

To prepare dispersions in the absence of lecithin, the required mass of silica was placed in a round-bottomed flask which allowed samples to be easily dispersed by using a glass pestle. Solvent was added to give a known volume fraction of silica. The resulting gel was dispersed by the pestle and transferred to a tube. The sample was sonicated by inserting an ultrasonic probe into the unsealed tube, which was suspended in an ice water bath to minimize the temperature rise and consequent solvent evaporation caused by the hot probe. A power of 60 W was used for 5 min. Great care was taken at all times to ensure that stray water drops did not contaminate the sample.

Because lecithin was insoluble in dodecane, the silica particles had first to be coated with lecithin from an ethanol dispersion. The ethanol was then removed and the coated particles were dispersed in the dodecane. The procedure was as follows.

Lecithin solution in ethanol was added to a mass of silica in a 35-ml centrifuge tube, which was then sealed. Sonication and equilibration were carried out as described above for the preparation of dispersions in ethanol. The samples were then centrifuged at 3000 g until completely sedimented. The longest centrifugation time required was 3 h for Aerosil 200 dispersions at low surface coverages; other systems took less time. With long centrifugation times the samples heated significantly. This was countered by cooling the tubes for 20 min in a water-bath at 25.8 °C after centrifugation for 1 h. After the final period of centrifugation, the samples were again kept at 25.8 °C for 30 min before the supernatant liquid was removed with a pipette. The samples were then dried under vacuum at room temperature to constant weight, a process that took about 2 h. After drying the samples were thoroughly crushed to a powder using a glass pestle and stored at −4 °C. Dispersions were then prepared in dodecane as described above.

Rheology. Viscosity measurements were made at 25.0 °C with a constant-stress Deer rheometer equipped with concentric cylinder geometry (R_1 and R_2 = 0.6 and 0.75 cm or 2.8 and 2.9 cm). To prevent evaporation of the alcohols, a foam-filled cap was designed. This was soaked with the appropriate alcohol and fitted over the gap between the two cylinders. Flow curves were obtained by increasing, then decreasing, the shear stress stepwise, leaving 30 s between each change of shear stress and recording the rotational speed at the end of this 30 s period. Dynamic rheological measurements were made on dispersions in dodecane using a Weissenberg Model R19 rheogoniometer, also at 25 °C.

Photon Correlation Spectroscopy (PCS). Measurements of the second-order autocorrelation function were performed on a Malvern 4700 light-scattering apparatus. This set-up was provided with an integral 40-mW He–Ne laser of wavelength 632.8 nm. The thermostatted water-bath which surrounded the sample contained water which was passed through a 0.45-μm Millipore filter for 10 min prior to measurements. Light scattered by the dispersion under investigation was detected by a photo-multiplier passing signals to an EMI Malvern Model K7032-OS auto-correlator. The hydrodynamic diameter of the equivalent spherical particle was evaluated from the *z*-average diffusion coefficient. The volume fraction was 0.01.

Sedimentation. Sedimentation studies were carried out at room temperature at volume fractions of 0.02 in tubes of volume 4 or 9 ml on lecithin-coated unsonicated and sonicated silica dispersions in dodecane. For the unsonicated dispersion the movement of the sedimenting boundary under gravity was followed with time and the equilibrium height determined as a fraction of the initial suspension height. In the case of the sonicated dispersion, mild centrifugation (4000 *g*) was necessary in order to obtain uniform sedimentation.

3 Results

Alcohol Dispersions—*Adsorption Behaviour*. These results, which were obtained with the TK 800 silica, have been reported previously.[3] The essential features of the data are shown in Figures 1 and 2 and Table 1.

Figure 1 shows the isotherms obtained for the four different lecithins adsorbed from ethanol and Figure 2 refers to the adsorption of the DPPC from three different alcohols. These isotherms were fitted to the Langmuir equation, which enables the limiting molecular area and the adsorption coefficient to be estimated. The values obtained are shown in Table 1.

There are three features of interest:

(i) Decreasing the polarity of the solvent leads to an increase in the strength of binding, as evidenced by the increase in adsorption coefficient.

(ii) The limiting molecular area on adsorption from ethanol increases with the extent of unsaturation of the synthetic lecithins. This limiting area is always greater than the area expected if only the polar head group adsorbs and

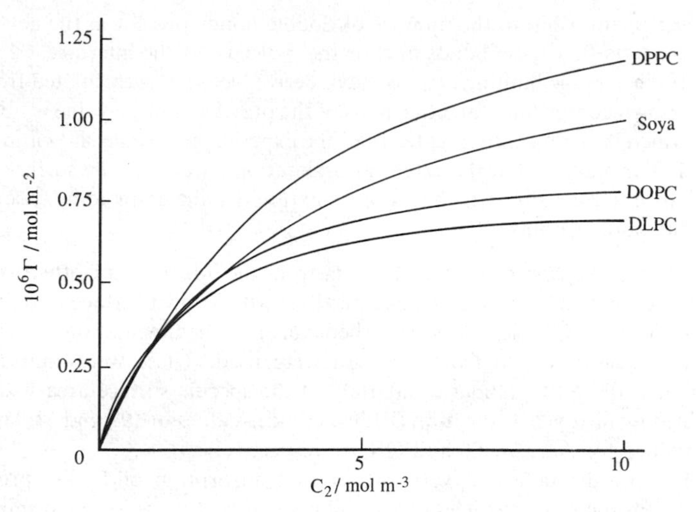

Figure 1 *Isotherms for lecithins adsorbed on silica from ethanol at* 25.8 °C

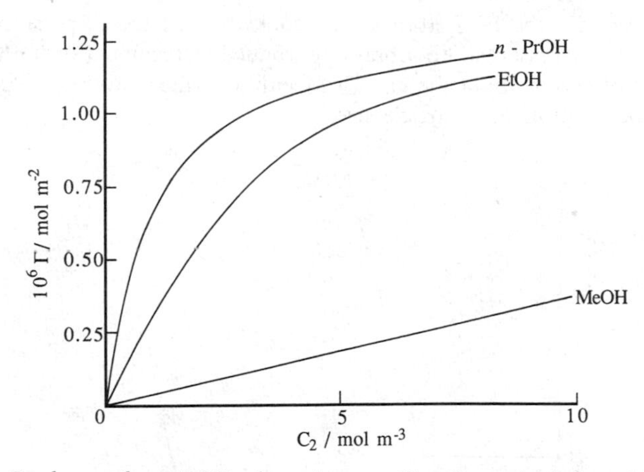

Figure 2 *Isotherms for DPPC adsorption on silica from alcohols at* 25.8 °C

Table 1 *Limiting molecular areas and Langmuir adsorption coefficients*

Lecithin	*Solvent*	*Area* (Å^2 molecule^{-1})	*Adsorption coefficient* (m^3 mol^{-1})
DPPC	Methanol	112	0.029
DPPC	Ethanol	102 ± 6	0.29 ± 0.03
DPPC	Propanol	125 ± 10	1.18 ± 0.29
DOPC	Ethanol	172 ± 17	0.56 ± 0.11
DLPC	Ethanol	210 ± 31	1.00 ± 0.48
Soya	Ethanol	123 ± 11	0.28 ± 0.04

the relationship to the number of double bonds present in the acyl chain suggests that these bonds anchor the molecule at the interface.

(iii) If the average limiting area for soya-derived lecithin is calculated from the known composition and the areas of the pure lecithins, a value is obtained which is significantly greater than the experimental value shown in Table 1. This implies that the extent of orientation parallel to the surface of the adsorbed unsaturated chains in a mixture is reduced in comparison with the pure lecithins.

Photon Correlation Spectroscopy. The data from the adsorption isotherms allow the preparation of dispersions of silica particles with known surface coverages of lecithins. For the PCS and subsequent rheological and sedimentation studies, two pyrogenic silicas obtained from Degussa were used. These were considerably cheaper than the NPL standard material but the specific surface area had to be determined by nitrogen adsorption (BET method). Values of 199 and 31.9 $m^2\ g^{-1}$ were obtained for Aerosil 200 and OX 50, respectively.

The PCS results indicated very clearly that adsorption of DPPC promotes association of the silica particles. This is illustrated by Figure 3, which shows the average hydrodynamic radius of the aggregates as a function of fractional surface coverage of DPPC.

Rheology. As with the PCS studies, ultrasonication of the dispersions prior to measurement was essential to obtain reproducible results. For DPPC, information was obtained about the effect of lecithin surface coverage, alcohol type, silica volume fraction, and particle size.

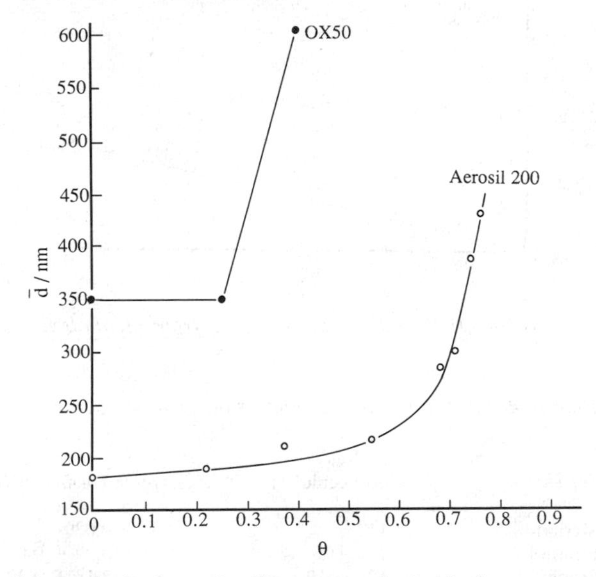

Figure 3 *Mean hydrodynamic diameters obtained from photon correlation spectroscopy for sonicated, ethanolic dispersions of silica with various fractional coverages of DPPC*

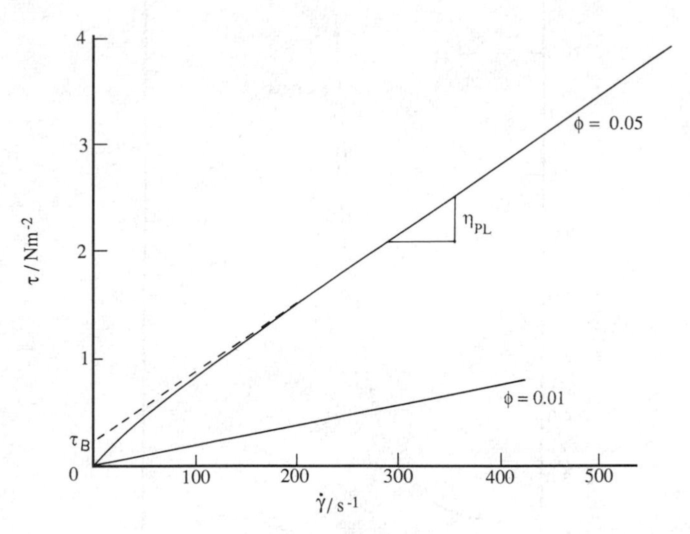

Figure 4 *Flow curves obtained for dispersions of Aerosil 200 in methanol illustrating plastic viscosity (η_{PL}) and Bingham yield stress (τ_B) obtained for high-volume fractions of silica*

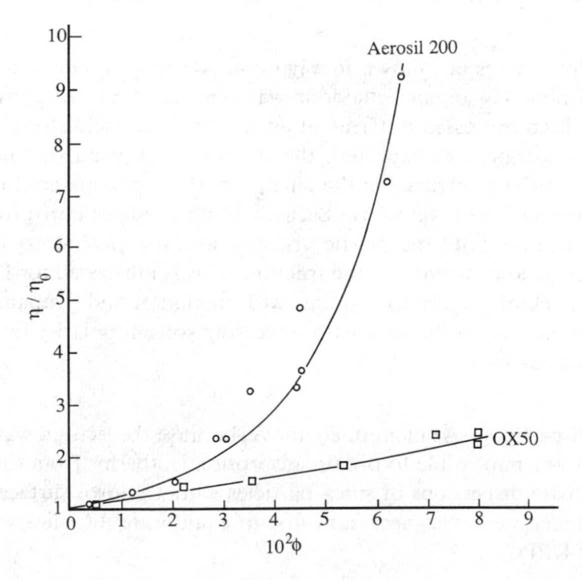

Figure 5 *Relative viscosities of dispersions of Aerosil 200 and OX 50 in methanol as a function of volume fraction*

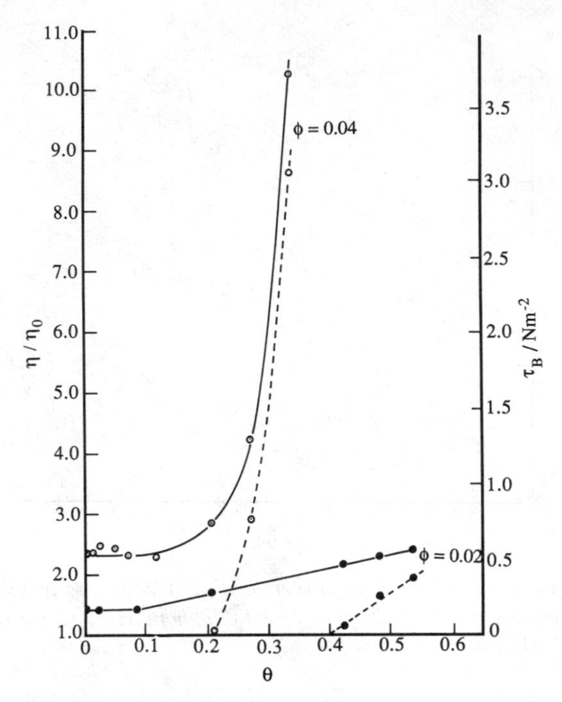

Figure 6 *Relative viscosities and Bingham yield stresses for dispersions of Aerosil* 200 *in ethanol as a function of surface coverage of DPPC*

Typical flow curves are shown in Figure 4. Above a certain volume fraction, pronounced non-Newtonian behaviour was observed and, for convenience, the results have been expressed in terms of an extrapolated yield stress and a plastic viscosity as illustrated. As expected, the viscosity at a given volume fraction of uncovered particles is higher for the smaller particle size material (Aerosil 200) than the large OX 50 (Figure 5). Because lecithin adsorption promotes inter-particle association, both the plastic viscosity and the yield stress increase with surface coverage at constant volume fraction. This is illustrated for DPPC-coated particles in ethanol (Figure 6). Studies with methanol and propanol show that viscosity and yield stress increase with increasing solvent polarity for both coated and uncoated particles.

Dodecane Dispersions—As mentioned above, because the lecithin was insoluble in dodecane, it was impossible to obtain adsorption isotherms from this solvent. In order to prepare dispersions of silica particles with a known surface coverage of lecithin in dodecane, it was necessary first to equilibrate the silica with ethanolic solutions of DPPC.

Photon Correlation Spectroscopy and Rheology. Silica dispersions in dodecane behaved very differently to dispersions in the alcohols. In the absence of lecithin,

Table 2 *Yield stress against DPPC coverage for OX 50–dodecane*

Coverage (monolayers)	*Yield stress* ($N\,m^{-2}$)
0	3.3
0.27	2.6
0.88	1.7
1.26	2.1
1.52	0.7
1.96	0.6

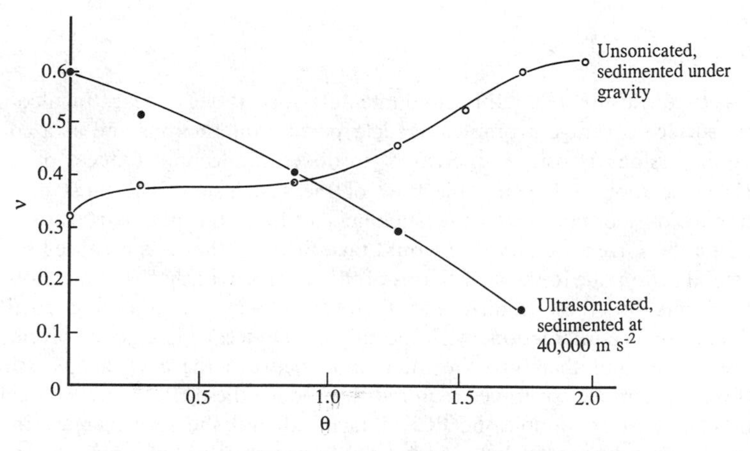

Figure 7 *Sedimented volumes for sonicated and unsonicated dispersions of OX 50 in n-dodecane as a function of monolayer coverage of DPPC*

the Aerosil 200 dispersions were gel-like at volume fractions greater than about 0.01. Dynamic rheological measurements at frequencies around 1 Hz gave values of tan δ of the order of 0.1, confirming the elastic nature of the dispersion. However, in contrast to polymer gels, non-linear behaviour was found down to the lowest accessible strains (around 0.01). At lower volume fractions the dispersions had the appearance of sedimented gel-like aggregates. This contrasted with the ultrasonicated dispersions in alcohol, which were liquid-like and did not sediment even at low volume fractions. For this reason it was considerably more difficult to carry out PCS and rheological measurements on dodecane dispersions. Both visual observation and the measurements that could be made demonstrated very clearly that lecithin coverage inhibited particle–particle association and enhanced flow. For example, Table 2 displays the dependence of the yield stress, measured directly with the Deer rheometer, on lecithin coverage at a volume fraction of 0.02 for OX 50 in dodecane.

Sedimentation Studies. The difference between the two types of solvent system is illustrated most dramatically by measurements of sedimented volume fraction. Figure 7 compares the sedimented volumes of sonicated and unsonicated dispersions of OX 50 as a function of lecithin coverage. In both cases the silica volume fraction was 0.02. The decrease in sedimented volume with lecithin coverage for the sonicated sample indicates a more closely packed aggregate structure which reflects a stabilization of the particles by the lecithin. On the other hand, the unsonicated dispersion shows an increase in sedimented volume with lecithin coverage, indicating destabilization of the silica particles with lecithin coverage. In this case the floc structure is characteristic of that present in the original ethanol dispersion from which the coated particles were prepared. This structure has to be destroyed by ultrasonication before behaviour characteristic of a non-polar solvent can be observed.

4 Discussion

The basic conclusion from this fundamental work is very clear. In alcohols, lecithin surface coverage promotes particle–particle interactions and as a consequence dispersion viscosity is increased; in dodecane, lecithin reduces particle–particle association, enhancing the flow of the suspension. The most obvious question to consider first is why the 'stickiness' of lecithin-coated particles should depend on the solvent quality. We would take the view that this is caused by the steric stabilization due to the acyl chains of the adsorbed lecithin. The alcohols are poor solvents for these chains and therefore chains on adjoining particles associate to reduce interactions with the solvent. Dodecane is a good solvent for the acyl chains and therefore the interaction between the acyl chains will be repulsive. The surface coverages required to produce these effects are well below one monolayer; for example the PCS data in ethanol show an increase in the hydrodynamic radius at lecithin coverages as low as one third of a monolayer and at the higher volume fractions significant changes in rheological properties are observed at this level of coverage. The adsorption isotherm data suggest that at monolayer coverage the saturated acyl chains are orientated perpendicular to the surface. The fact that steric stabilization or association effects are observed at low coverages suggests that below monolayer coverage the acyl chains are orientated similarly. A corollary of this argument is that higher surface coverages of the unsaturated lecithins would be required to produce comparable effects.

The molecular weight of lecithin is low compared with those of the polymeric surfactants for which a steric stabilization mechanism is usually invoked. For this reason we consider that the interaction is predominantly enthalpic rather than entropic in nature. In the case of the uncoated silica particles in alcohol, a significant stabilizing factor will be hydrogen bonding of the alcohol to the silanol groups on the silica surface. When the particles are coated with lecithin, this interaction will no longer occur. Hence the loss of the hydrogen bonded solvent will 'reinforce' the steric association between the lecithin chains.

In developing a quantitative interpretation of the flow behaviour, it is first necessary to appreciate that there are several levels of organization in these dispersions. The dimensions of the primary particles would be of the order of 10 nm. However, electron microscopic studies have shown that these primary particles always form aggregates which cannot be disrupted even by ultrasonication. Primary aggregates can flocculate reversibly to give flocs which can be broken down by ultrasonication to smaller particles and ultimately to the primary aggregates. For the uncoated particles in the alcohols, reassociation will not occur because of hydrogen bonding to the surface of the silica by the solvent. The diameter of these particles following ultrasonication was measured as 190 nm for Aerosil 200, which is consistent with previous electron microscopic studies.[4]

In our experiments in alcohols, the lecithin-coated particles are ultrasonicated, which breaks the flocs down into primary aggregates. When ultrasonication is stopped, the primary particles will reassociate into flocs. The dispersion at this stage does not appear to change its properties with time. On shearing, the flocs will be progressively broken down, the mean size decreasing with increasing shear rate. It has been suggested[5] that the flow size will be proportional to $1/\dot{\gamma}^{0.4}$.

Below a certain critical shear rate the flocs will associate. At this critical shear rate the tensile force exerted by the shear field is just sufficient to break apart a pair of colliding flocs.[6] Above this shear rate there is a linear relationship between shear stress and shear rate and the degree of openness of the flocs remains constant. In the elastic floc model this degree of openness is specified by the parameter C_{fp}, which is defined by φ_f/φ_p, where, φ_p is the volume fraction occupied by the primary particles and φ_f is the volume fraction occupied by the flocs. In our case, since the primary aggregates cannot be broken down, these are the constituent units of the floc.

An upper estimate of the floc volume fraction can be obtained from the sedimented volume measurements. How close this is to the true value will depend on how much additional solvent there is in the sediment compared with the flocs, *i.e.* how the flocs pack within the sediment. The maximum value for C_{fp} is given by the ratio of the space filling volume fraction to the particle volume fraction.

Table 3 *Values of* $C_{fp\ max}$ *against DPPC coverage for 2% OX 50–dodecane*

	$C_{fp\ max}$	
Coverage (monolayers)	*Unsonicated (gravity)*	*Sonicated (centrifugation)*
0	16.5	30
0.27	19	26
0.87	19	20.5
1.26	23	15
1.52	26.5	—
1.73	30	7.5
1.96	31	—

Table 3 displays this parameter for the ethanol-type structure and the dodecane-type structure. In evaluating this we have assumed, as previously discussed, that the non-ultrasonicated structure in dodecane was representative of that present in the ethanol when the lecithin coated particles were prepared.

Another approach to the evaluation of C_{fp} is to use the relationship between volume fraction and viscosity of the type shown in Figure 5. This can be fitted to semi-empirical equations such as the Krieger–Dougherty and Mooney equations.

Table 4 C_{fp} *values derived from viscosities*

	C_{fp}	
Alcohol	*Aerosil* 200	*OX* 50
Methanol	13.7	5.4
Ethanol	9.2	5.3
Propanol	6.8	2.8

At low volume fractions these equations reduce to the expression ln η/η_o = 2.5 × volume fraction. The effective volume fraction for dispersions that contain flocs will be larger than the particle volume fraction by a factor equal to C_{fp} because of the immobilized liquid within the flocs. Values obtained in this way for silica dispersions in alcohols are listed in Table 4.

If the sedimentation data for the unsonicated dispersions in dodecane reflect the structure which is present in ethanol, then we can compare the C_{fp} values obtained from the viscosity data with $C_{fp\ max}$ obtained from sedimentation. This composition suggests that about one third of the sedimented volume is taken up by the flocs.

A comparison of the viscosity C_{fp} values for the three alcohols reveals that particles which are dispersed in propanol form more compact flocs. This is indicative of a weaker particle attraction compared with the other two alcohols. At first sight this result is surprising in view of the relative polarities of the alcohols. However, solvation of the surface hydroxyl groups of the silica must involve disruption of the hydrogen-bonded structures in the bulk liquid and therefore we would predict that solvation in propanol would be more favourable, as the extent of hydrogen bonding in this solvent is less than in the other two alcohols.

The most appropriate choice of rheological model to use in interpreting the extrapolated yield stresses is the elastic floc model of Hunter.[6] This gives an expression for the Bingham yield stress in terms of C_{fp} and particle volume fraction. For uncoated particles, our data show the predicted linear relationship between yield stress and volume fraction squared. Substitution of $C_{fp\ max}$ and the measured yield stress, together with appropriate estimates of the other parameters in the equation, leads to a value for the floc radius which is two to three orders of magnitude greater than that measured by PCS.

5 Conclusions and Implications for Chocolate Processing

The work reported here has thrown some light on how pure lecithins modify the rheology of particulate dispersions. It has shown that lecithins can promote or prevent particle–particle association, depending on the quality of the solvent with respect to the acyl groups present in the molecule. These ideas can be extended to other solvent systems. Essentially, if the surfactant has a low solubility in the solvent it will adsorb readily to the surface but will not be effective in stabilizing the particles, whereas a surfactant with a high solubility will adsorb only weakly but will provide good stabilization. An ideal emulsifier needs to be strongly adsorbed while at the same time having protruding chains for which the continuous phase is a good solvent. Although these requirements appear contradictory, they could be achieved by a suitable choice of head group.

It has been reported that lecithin can influence chocolate viscosity at coverages below a monolayer.[1] Our work provides some support for this, as we have shown in the model system that dispersion viscosity is influenced at lecithin concentrations as low as 0.3 of a monolayer. With commercial mixed emulsifiers it has been found that viscosity and yield stress first decrease and then increase with lecithin concentration. By demonstrating that lecithin can both increase and decrease viscosity, it is possible to understand this effect if the surface composition changes with the concentration of the mixed surfactant in the continuous phase. An alternative interpretation is that the rheology of the continuous phase changes with increasing surfactant concentration and at higher surfactant concentrations this starts to dominate.

Acknowledgement. Support from the Agricultural and Food Research Council is gratefully acknowledged.

References

1. W. Bartusch, in 'Proceedings of First International Congress on Cocoa and Chocolate Research, Munich,' 1974, p. 153.
2. J. Chevalley, *J. Texture Stud.*, 1975, **6**, 177.
3. M. J. Hey, A. C. Mackie, and J. R. Mitchell, *J. Colloid Interface Sci.*, 1986, **114**, 286.
4. F. Ehrburger, V. Guerin, and J. Lahaye, *Colloids Surf.*, 1984, **9**, 371.
5. W. Russel, *J. Rheol.*, 1980, **24**, 287.
6. R. J. Hunter, *Adv. Colloid Interface Sci.*, 1982, **17**, 197.

Suppression of Perceived Flavour and Taste by Food Hydrocolloids

By Z. V. Baines and E. R. Morris

DEPARTMENT OF FOOD RESEARCH AND TECHNOLOGY, CRANFIELD INSTITUTE OF TECHNOLOGY, SILSOE COLLEGE, SILSOE, BEDFORD MK45 4DT, UK

1 Introduction

The texture (or consistency) of food products is an important factor in their acceptability to the consumer, and is often generated or controlled by the use of hydrocolloid thickeners or gelling agents. However, the objective level of flavouring required to give an acceptable subjective intensity of perceived flavour in thickened or structured products is usually much higher than in more fluid systems. The efficiency of 'flavour release' from such products can be of considerable commercial significance, as flavouring is often one of the major cost components in manufactured foods. Previous research in this area has concentrated mainly on the effect of food hydrocolloids on perceived intensity of the four basic tastes (sweet, sour, bitter, and salty) rather than on flavour (aroma) perception, and has usually been restricted to investigation of differences between different gums over a small range of concentrations or viscosities.[1–4] In this work we have attempted to trace the molecular origin of suppression of both flavour and taste (sweetness) in hydrocolloid matrices of three different types: solutions of disordered 'random coil' polysaccharides varying in both chemical composition and molecular weight, a conformationally ordered 'weak gel' system (xanthan), and a 'true gel' system (calcium alginate), in which the degree of cross-linking can be varied by varying the Ca^{2+} concentration.

2 Quantification of Perceived Sensory Attributes

To those unfamiliar with the area, it may seem difficult or impossible to quantify something as inherently subjective as perceived flavour or taste. In practice, however, although individuals can differ greatly in their subjective preferences, they show remarkable agreement in judging attribute intensity. For example, although some people like sweet drinks and some do not, almost everyone will agree on which are sweeter than others, and their estimates of *how much* sweeter show an amazingly good correlation with actual sugar content. In general,[5] the relationship between the objective intensity of an external stimulus (S) and its perceived, subjective intensity (response, R) follows a 'power law' of the form

$$R = \alpha S^{\beta}$$

so that a double-logarithmic plot of R *versus* S yields a straight line of gradient β (which is characteristic of the particular attribute involved, but can vary widely for different attributes) and intercept α (which has no fundamental significance, but simply reflects the scoring procedure adopted in rating R).

In this work, samples were rated for specific attributes such as perceived 'in-mouth' viscosity (or 'thickness'), flavour intensity, and sweetness by the technique of magnitude estimation against a fixed control.[6] The control sample (which was chosen to be near the middle of the range, to reduce any possible bias or scale distortion) was assigned a value of 100 for each attribute under investigation. Volunteers (typically 15–20 in each investigation) were then asked to give appropriate scores to the other samples. For example, a sample judged to be twice as sweet as the control would be given a 'sweetness' score of 200, one judged to have half the intensity of flavour would be given a 'flavour' score of 50, and so on. Each attribute was assessed at a separate panel session to minimize 'halo' effects. Panellists were familiar with the magnitude estimation technique, but were otherwise untrained (*i.e.* it was left to each individual to assess the samples in whatever way seemed most natural). Because it was not possible for every panellist to assess every sample, owing to the large number of samples being assessed, a balanced incomplete block design was used, and scores were adjusted for differences between panellists by analysis of variance.[7]

3 'Random Coil' Solutions

Procedure—The effect of conformationally disordered ('random coil') hydrocolloids on the perceived intensity of flavour and taste was investigated using six different polysaccharide samples: sodium alginate (Kelco Manucol DH), guar gum samples of high, medium, and low molecular weight (designated H, M, and L) which were, respectively, Hercules TH 225, Meyhall Meyprogat 90, and Meyhall Meyprogat 60, and carboxymethylcellulose (CMC) of high (H) and medium (M) molecular weight (Hercules Blanose gums 9HF and 9M31XF, respectively). The intrinsic viscosity ([η]) of each sample (Table 1) was determined at 25°C on low-shear concentric cylinder couette viscometer using solutions with relative viscosities in the range 1.3–2.0. In all instances a linear dependence of shear stress on shear rate (*i.e.* Newtonian behaviour) was verified. Extrapolation to intrinsic viscosity was by a combined Huggins, Kraemer, single-point method.[8]

Table 1 *Intrinsic viscosity* ([η]; 25 °C) *of 'random coil' polysaccharide samples*

Sample	[η] (dl g^{-1})	*Sample*	[η] (dl g^{-1})
Guar gum H	16.7	CMC H[a]	25.0
Guar gum M	8.4	CMC M[a]	12.5
Guar gum L	5.0	Alginate[a]	7.0

[a] Dialysed against 0.01 M NaCl.

Solutions for sensory analysis were prepared using a wide range of polysaccharide concentrations, both above and below the onset of coil overlap and entanglement (the c^* transition). The 'zero shear' viscosity (η_o) of each solution was determined at 25 °C from steady-shear measurements on a Sangamo Viscoelastic Analyser, using cone and plate geometry with a 0.035 rad cone angle and 50 mm diameter, and the coil overlap concentration (c^*) was determined from the associated change in slope of a standard plot of log ($\eta_0 - \eta_s$) *versus* log c (where η_s is the viscosity of the solvent).

All solutions incorporated a fixed concentration of sucrose (10% w/w) and flavouring (500244E; P.F.W. Ltd, 0.2% w/w). With the charged polysaccharides (CMC and alginate), a fixed concentration of salt (10 mM NaCl) was also incorporated to minimize changes in coil dimensions on varying the polymer concentration, and 0.1% trisodium citrate was included in the alginate solutions as a sequestrant for any residual free calcium in the sample. Solutions were prepared using an overhead stirrer for initial dispersion and a Silverson high-shear mixer to obtain true solutions, and were presented to the sensory panel at ambient temperature (*ca.* 25 °C).

Results—Figure 1 shows the variation in sensory panel scores (S) for perceived thickness, sweetness, and flavour intensity of samples incorporating the same objective concentrations of sucrose and flavouring, but different concentrations of the same polysaccharide. The sharp break in the concentration dependence of perceived thickness corresponds to the onset of coil overlap and entanglement

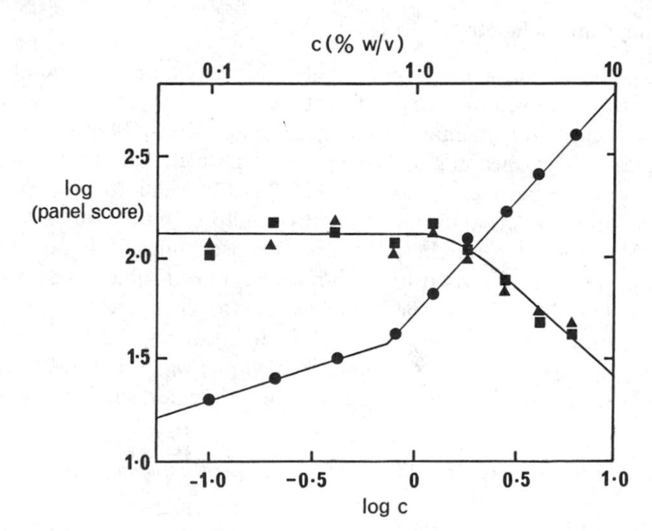

Figure 1 *Perceived 'thickness'* (●), *sweetness* (■) *and flavour intensity* (▲) *of samples incorporating fixed concentrations of sucrose and flavouring, but thickened with different concentrations* (c) *of the same polysaccharide (guar gum L)*

(Reproduced with permission from *Food Hydrocolloids*, 1987, **1**, 197)

(the c^* transition) and is paralleled by a corresponding change in the concentration dependence of the objective viscosity. The flavour and sweetness scores are essentially independent of polymer concentration below c^*, but decrease steeply at higher concentrations, with both attributes showing the same extent of suppression.

The reduction in perceived intensity of flavour and sweetness with increasing polymer concentration (c) can be expressed quantitatively by a simple equation:

$$S = S_0/[1 + (c/c_{\frac{1}{2}})^p] \qquad (1)$$

where S denotes attribute scores from the sensory panel, S_o is the constant value of S at low polymer concentrations, $c_{\frac{1}{2}}$ is the concentration at which $S = S_o/2$, and p is the absolute value of the terminal slope of log S *vs.* log c at high concentration.

For all samples studied, p remained approximately constant at *ca.* 1.66, but fitted values of $c_{\frac{1}{2}}$ varied substantially from sample to sample (*i.e.* the onset, and subsequent concentration dependence, of flavour/taste suppression is different for different polysaccharides and for different molecular weights of the same polysaccharide). In all instances, however, $c_{\frac{1}{2}} \approx 9c^*$, so that when the absolute concentrations of each sample are expressed relative to the coil-overlap concentration (*i.e.* as c/c^*), panel scores for different 'random coil' polysaccharides became closely superimposable, as shown in Figure 2.

4 Xanthan 'Weak Gels'

Under most conditions (*i.e.* except at high temperature and/or very low ionic strength), xanthan exists in solution in a rigid, ordered conformation, which

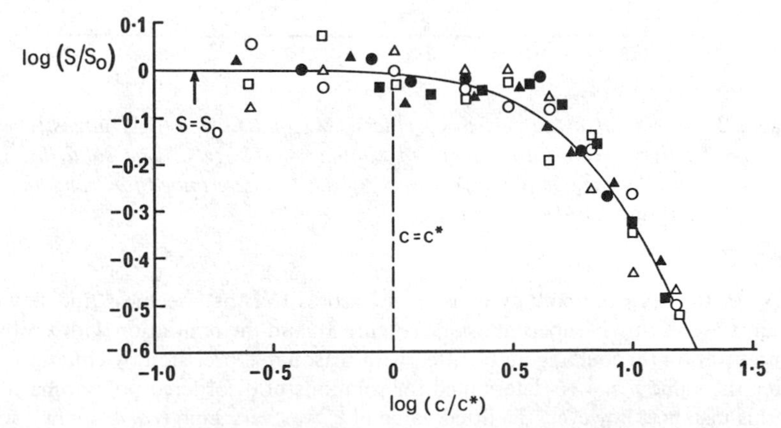

Figure 2 *Flavour/taste suppression by 'random coil' polysaccharides. Results (mean values of panel scores for sweetness and flavour intensity) are shown for alginate (▲), carboxymethylcellulose of high (■) and medium (●) molecular weight, and guar gum of high (□), medium (○), and low (△) molecular weight*

confers unusual solution properties. In particular, under conditions of small deformation (*e.g.* low-amplitude oscillation) xanthan solutions show rheological behaviour characteristic of a gel[8] ($G' > G''$ with little frequency dependence in either modulus), but at higher deformation (*e.g.* under rotation) the network breaks and the solution viscosity shows much greater shear-rate dependence than for entangled 'random coils.'

Taste-panel studies of xanthan 'weak gels' were carried out following a similar experimental procedure to that adopted for random coil solutions. Again the concentrations of sucrose and flavouring were held constant (at 10 and 0.2% w/w, respectively), and the concentration of xanthan (Keltrol F; Kelco Inc.) was varied within the range 0.125–3.0% w/w.

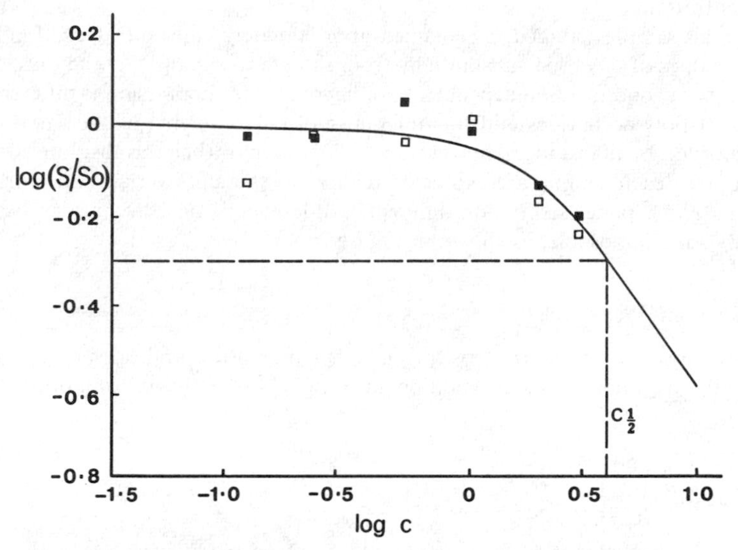

Figure 3 *Variation of perceived sweetness* (■) *and flavour* (□) *intensity with polymer concentration* (c) *in xanthan 'weak gels.' The solid line was fitted using equation* (1) *with* p = 1.66 *(as determined for 'random coil' polysaccharide solutions)*

As in the 'random coil' systems, panel scores (S) for sweetness and flavour intensity were closely superimposable (Figure 3), and their variation with xanthan concentration (c) could again be fitted with reasonable precision by equation (1), using the value $p \approx 1.66$ determined for solutions of disordered polysaccharides. In this instance, however, the fitted value of $c_{\frac{1}{2}}$ was very high (*ca.* 4% w/w), with no significant suppression of flavour/taste intensity at xanthan concentrations below *ca.* 1%. Thus, over the concentration range of practical importance for food applications, xanthan has no detectable effect on the perceived intensity of flavour and taste, consistent with its industrial reputation of giving exceptionally good 'flavour release.'

5 Calcium Alginate Gels

Alginate is a linear copolymer of two negatively charged sugars, β-D-mannuronate and α-L-guluronate, which occur in long homopolymeric sequences of both types ('M-blocks' and 'G-blocks') and in mixed heteropolymeric sequences.[10] The relative proportions of the two constituent sugars can vary widely, and samples are often designated as 'high-M' or 'high-G' to denote a preponderance of mannuronate or guluronate, respectively.

Sodium alginate exists in solution as a conformationally disordered 'random coil' (and was included as one of the samples in our study of flavour/taste release from random coil polysaccharides). However, if calcium ions are present, alginate can also form gels. The poly-L-guluronate chain sequences are the only regions of the polymer involved in the formation of junction zones in the gel matrix. Calcium ions are sandwiched between two polyguluronate segments to form the junctions, giving a structure analogous to eggs in an egg-box. Polyguluronate sequences within these dimeric 'egg-box' structures are locked in a regular, buckled, two-fold conformation, with only the inner faces of each chain participating in Ca^{2+} binding.[11] Hence the calcium-ion concentration required for complete gelation is equivalent to half the stoicheiometric requirement of the polyguluronate regions of the polymer (*e.g.* 25% of the total stoicheiometric equivalent for an alginate containing 50% polyguluronate).

Procedure—As the extent of cross-linking in alginate gels can be controlled by the amount of calcium present, two types of investigation were carried out on 'flavour release' from such systems. First, the effect of progressive formation of an alginate gel matrix on the perceived intensity of flavour and sweetness was investigated by keeping the polymer concentration constant and varying the calcium concentration. Second, calcium-ion concentration (in terms of the stoicheiometric equivalence of calcium to alginate) was held constant and the alginate concentration was varied.

The taste panel studies in which calcium content was varied were carried out using two different alginate samples: a high-M alginate (Kelco Manucol DH) and a high-G alginate (Kelco Manugel DJB) of approximately the same molecular weight. The samples for each series of taste panels contained the same amount of polymer, sucrose (20% w/w), and flavouring (0.02% Blackcurrant D1407 NA; Bush Boake Allen) but different stoicheiometric equivalences of calcium. Two polymer concentrations were studied, 1.0 and 1.5% w/w, at calcium stoicheiometric equivalences ranging from 0 to 30%.

The effect of varying the polymer concentration at a fixed stoicheiometry of Ca^{2+} was investigated for a single alginate sample (Manucol DH, which was also included in the 'random coil' studies), using a Ca^{2+} concentration sufficient for saturation of polyguluronate (30% of the total stoicheiometric requirement of the sample).

Results—The perceived intensity of flavour and sweetness showed a marked decrease with increasing calcium-ion concentration for both alginates (Figure 4).

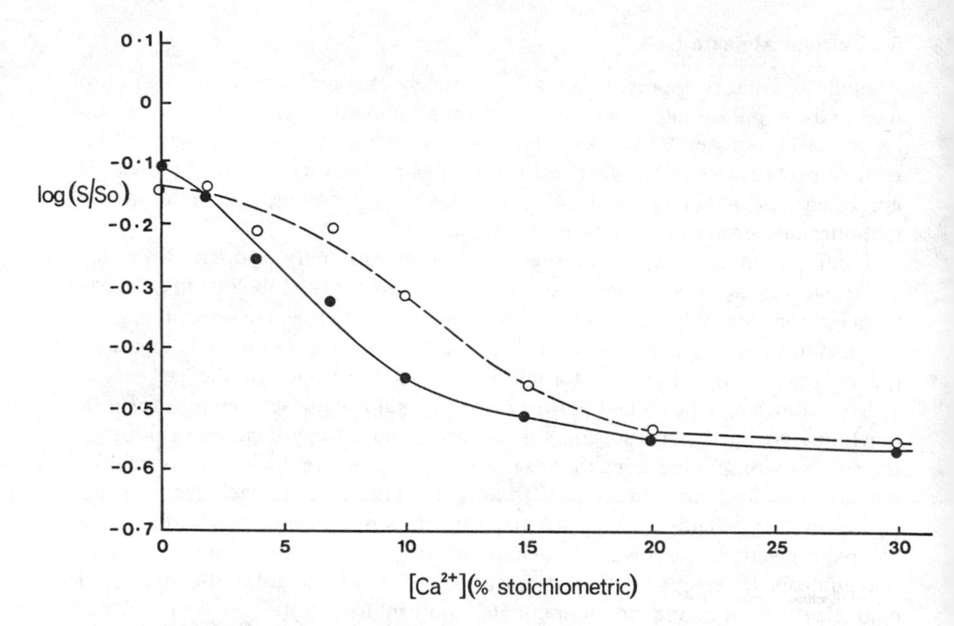

Figure 4 *Calcium-induced gelation of high-G (○) and high-M (●) alginates, showing the associated reduction in perceived intensity of sweetness and flavour (mean score for both attributes)*

For each sample the degree of suppression was independent of polymer concentration at equivalent concentrations of Ca^{2+} although, as shown in Figure 4, there were significant differences in calcium dependence between the high-M and high-G samples. As in the solution studies, panel scores (S) for sweetness and flavour intensity were also identical, within experimental error, and the results shown in Figure 4 are mean values of both, expressed relative to the maximum perceived intensity in dilute polysaccharide solutions (S_o). For both alginates the perceived flavour/taste intensity reaches a final minimum 'plateau' value at high Ca^{2+} concentrations where junction zone formation is complete but, as might be expected, the plateau is reached at lower stoicheiometric equivalence for the high-M sample, which contains a lower proportion of the polyguluronate sequences responsible for both Ca^{2+} binding and gel network formation.

On increasing the concentration of alginate (c) at a fixed Ca^{2+}/alginate ratio within the 'plateau' region, panel scores (S) for perceived intensity of flavour and taste decrease steeply, with both attributes again being suppressed to the same extent. In contrast to the characteristic concentration dependence observed for solutions [see Figure 2 and equation (1)], log S in the calcium alginate gel system decreases linearly with increasing log c (Figure 5) and the perceived suppression of flavour/taste intensity is very much higher than for equivalent concentrations of the same alginate sample in the 'random coil' solution state.

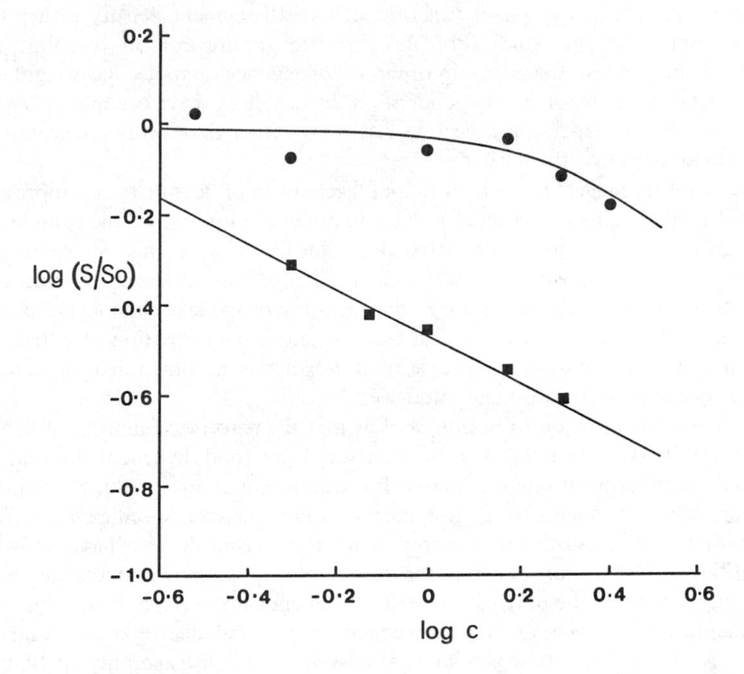

Figure 5 *Suppression of flavour/taste intensity (mean scores for both attributes) with increasing concentration* (c) *of alginate (high-M sample) in the 'random coil' solution state* (●) *and in gels* (■) *formed at a fixed Ca^{2+}/alginate ratio within the 'plateau' region from Figure 4* (30% *of total stoicheiometric equivalence*)

6 Discussion and Conclusions

In all the systems studied ('random coil' solutions, 'weak gels', and 'true' gels), the suppression of perceived intensity was the same for flavour and taste (sweetness). As 'flavour volatiles' and sucrose are very different chemically, this argues against direct binding of these small molecules to the polymer having any significant role in flavour/taste suppression. Similarly, as flavour and taste are perceived through different receptors (in the nasal air spaces and on the surface of the tongue, respectively), direct interference by the polymer in receptor–substrate interactions can also be eliminated.

A more likely mechanism is hindered transport of flavour/taste components from the interior of the sample (where they cannot be perceived) to the surface (where they may come into direct contact with taste receptors on the tongue or, for flavour volatiles, be released into the oral/nasal air-space). Restricted diffusion is unlikely to be a significant factor in reduced mobility, as the diffusion coefficient of small molecules within a polymer matrix[12] decreases sharply with increasing polymer concentration, but is independent of the molecular weight of the polymer (*i.e.* the dominant effect is the probability of a small molecule hitting a chain segment, and this is a function of overall segment density rather than chain length). In our studies of flavour/taste suppression in 'random coil' solutions, in contrast, there was an obvious dependence on molecular weight, and the extent of suppression was dependent on the degree of space occupancy (*i.e.* the extent of coil overlap, as in objective viscosity) rather than on polymer concentration alone (as in diffusion).

We therefore suggest that the principal mechanism of flavour/taste suppression in thickened systems is restricted mixing between the interior of the sample and the surface, due to increased viscosity. Consistent with this interpretation, xanthan 'weak gels,' which show much more pronounced shear thinning than entangled 'random coils' at the same concentration or space occupancy, also show much less suppression of flavour and taste, whereas the formation of a 'true' gel network, with consequent suppression of mixing, is accompanied by a much greater decrease in flavour/taste intensity.

The overall conclusion from this work is that the perceived intensity of flavour and taste in systems thickened or structured by food hydrocolloids can be understood in terms of polymer network properties, without invoking binding of flavour/taste components to the polymer, or any other such specific effects. With conformationally disordered 'random coil' polysaccharides we have shown a quantitative relationship between the extent of suppression and the degree of space occupancy by the polymer. The direct dependence of flavour/taste intensity in calcium alginate gels on polymer concentration and degree of cross-linking indicates that hydrocolloid gels may also show a simple generality of 'flavour release' properties analogous to that already established for entangled coils.

Acknowledgements. We thank the AFRC for the award of a research studentship to Miss Z. V. Baines, Kelco, Meyhall and Hercules for providing the gum samples used, Bush Boake Allen and P. F. W. Ltd. for flavourings, and Miss N. L. Stanley for experimental collaboration.

References

1. R. M. Pangborn, Z. M. Gibbs, and C. Tassan, *J. Texture Stud.* 1978, **9**, 415
2. T. Izutou, S. Taneya, E. Kikuchi, and T. Sone, *J. Texture Stud.*, 1981, **12**, 259
3. H. Stone and S. Oliver, *J. Food Sci.*, 1966, **31**, 129
4. M. Vaisey, R. Brunon, and J. Cooper, *J. Food Sci.*, 1969, **34**, 397
5. S. S. Stevens, 'Psychophysics,' ed. G. Stevens, Wiley, New York, 1975
6. A. N. Cutler, E. R. Morris, and L. J. Taylor, *J. Texture Stud.*, 1983, **14**, 377
7. M. C. Gacula, Jr., and J. Singh, 'Statistical Methods in Food and Consumer Research,' Academic Press, London, 1984, pp. 141–175
8. E. R. Morris, in 'Gums and Stabilisers for the Food Industry. 2—Applications of Hydrocolloids,' eds. G. O. Phillips, D. J. Wedlock, and P. A. Williams, Pergamon Press, Oxford, 1984, pp. 57–78
9. Z. V. Baines and E. R. Morris, *Food Hydrocolloids*, 1987, **1**, 197
10. O. Smidsrød, *Faraday Discuss. Chem. Soc.*, 1974, **57**, 263
11. E. R. Morris, D. A. Rees, D. Thom, and J. Boyd, *Carbohydr. Res.*, 1978, **66**, 145
12. B. N. Preston, T. C. Laurent and W. D. Comper, in 'Molecular Biophysics of the Extracellular Matrix,' eds. S. Arnott, D. A. Rees, and E. R. Morris, Humana Press, Clifton, NJ, 1984, pp. 119–162.

Effect of Surfactants and Non-aqueous Phases on the Kinetics of Reactions of Food Preservatives

By B. L. Wedzicha and A. Zeb

PROCTER DEPARTMENT OF FOOD SCIENCE, UNIVERSITY OF LEEDS, LEEDS LS2 9JT, UK

1 Introduction

The function of food preservatives is primarily to control microbial spoilage in foods; this action depends on preservative molecules possessing some non-polar character to allow passage through biological membranes. Hence we find that although the range of compounds used as preservatives is diverse, including organic acids, esters, and ionic inorganic compounds such as nitrite or sulphite, they all show appreciable solubility in organic solvents. With ionic preservatives this property arises from non-ionic forms (*e.g.* N_2O_3 from nitrite ion and SO_2 from sulphite ion) in equilibrium with the ionic species. The organic solvents may include dispersed non-aqueous phases such as fats or essential oils. It is expected, therefore, that when added to foods, such additives will partition between water and any non-aqueous phases present.

2 Solute Partitioning

A quantitative description of the equilibrium distribution of a solute between two phases, α and β, is described in terms of the distribution or partition coefficient, P^x, defined here in terms of molar fractions x_i^α and x_i^β of solute component i in each phase, as

$$P^x = x_i^\alpha / x_i^\beta \quad (1)$$

The conventional partition coefficient, P, is expressed in terms of solute concentrations c_i^α and c_i^β in each of the phases as

$$P = c_i^\alpha / c_i^\beta \quad (2)$$

the relationship between P^x and P being

$$P = P^x(V^{0,\beta}/V^{0,\alpha}) \quad (3)$$

where $V^{0,\alpha}$ and $V^{0,\beta}$ are the molar volumes of phases α and β. In this paper, phase α will be regarded as the non-aqueous (oil) phase and β the aqueous component.

The behaviour of solutes in bulk two-phase systems (*e.g.* oil–water) or bulk three-phase systems (*e.g.* air–oil–water) represents a relatively simple problem, even if solute–solute interactions occur, such as formation of solute dimers or trimers, or solute–solvent interactions, such as hydration equilibria, need to be taken into consideration. In food systems we are, of course, usually concerned not with bulk aqueous and non-aqueous phases but dispersed phases. It follows that any solute which is soluble in water and the organic phase will also be surface active and there will be an excess solute concentration at the interface. If we consider a dilute emulsion, say with volume fraction 0.05, consisting of 2-μm droplets, the interfacial area is 150 $m^2\ dm^{-3}$. Such an emulsion could accommodate at the interface all the solute when added at concentrations of say 0.015 wt.-%, if the capacity of the interface is taken as 1 $mg\ m^{-2}$. Organic acids and esters may be required for food preservation at concentrations up to 0.1 wt.-% and a significant proportion of that added could reside at the interface in concentrated emulsions. In practice, emulsions will be stabilized by other surfactants which are expected to be more strongly adsorbed at the interface than the solutes in question here and should have little effect on the properties of the two phases at the interface. Provided that no interfacial adsorption contributes to solute partitioning, there is at first sight no reason why the partition coefficient of a solute in an emulsion should differ from that in the bulk two-phase system.

A very large number of measurements have been made of partition coefficients for solutes in macroscopic two-phase systems,[1] but the only available measurements in emulsions are those determined by Texter's group[2,3] for substituted *p*-phenylenediamines in gelatin-stabilized emulsions (0.5% gelatin, 40 °C) containing up to 5 wt.-% of dibutyl phthalate, *N*,*N*-diethyldodecanamide, octanol, undecanol, trihexyl phosphate, and dodecane as the non-aqueous phases dispersed in water. Despite the dissimilarity of these solvents from dispersed phases in foods, the results obtained suggest features which may be of relevance to foods. A comparison of partition coefficients measured in the emulsion with those in macroscopic two-phase systems shows that they are closely correlated but, surprisingly, those measured in macroscopic two-phase systems are, on average, larger by a factor of 2.5. This implies that the aqueous dispersion medium appears to have a higher affinity for the solutes than the aqueous phase in macroscopic two-phase studies. About 70% of this factor was accounted for by the presence of gelatin in the emulsion. The affinity of the aqueous phase was increased by the presence of gelatin.

The presence of surfactant needs to be taken into account. Non-polar solutes are known to partition into the micellar environment of surfactants above their critical micelle concentration, cmc.[4,5] The distribution of solute between the aqueous and micellar environments may be expressed similarly to the partitioning between discrete phases defined by equation (1), but in terms of the micellar partition coefficient, $P^x{}_{mic}$:

$$P^x{}_{mic} = x_i^{mic}/x_i^{water} \tag{4}$$

The molar fraction micellar partition coefficient is more useful than that in terms of concentrations, as the volume of micelles of particular surfactants is not known with a high degree of certainty and it is not possible to determine the concentration of solute within micelles directly. For surfactants with a low cmc, x_i^{mic} would normally be taken as n_i^{mic}/n_{surf}, where n_i^{mic} and n_{surf} are the numbers of moles of component *i* associated with micelles and the number of moles of surfactant in the system, respectively.

Partitioning into micelles is widely used to explain why the solubility of sparingly soluble organic compounds is often greatly increased by the addition of surfactant.[5] For food emulsions, the concentration of emulsifier may be in the range 0.2–2 wt.-% and, therefore, in amounts comparable to that of an added preservative. In any system consisting of solute, surfactant, and dispersed phase, it is expected that the solute will distribute itself between water, the non-aqueous phase, and the micelles.

There are few published measurements of the distribution of food preservatives between aqueous and non-aqueous components of foods and no report of surfactant micelle–food preservative interactions. It is, however, now becoming possible to estimate the micellar[6,7] and two-phase[1,8,9] partition coefficients for particular solutes using correlations between these quantities and the partition coefficient of the solute in the octanol–water system. The latter has been adopted as a 'standard' model for the evaluation of the biological activity of various solutes which is related to the partition coefficient; published data on octanol–water partitioning are very extensive.[1,10] The most useful correlations between oil–water and surfactant–water systems and the octanol–water system are empirical and take the form[6,7,10]

$$\log P_A = a + b \log P_B \tag{5}$$

where P_A and P_B represent the partition coefficients of the same solute in two different solvent pairs, or P_A may represent a micellar partition coefficient with P_B a two-phase partition coefficient. The constants *a* and *b* are known for particular solvent pairs and often found to give good predictions for groups of solutes.

In order to illustrate the relative magnitudes of micellar and oil–water partition coefficients, measured (where available) and calculated values are shown in Table 1 for four solutes including two acids (benzoic and sorbic) used as food preservatives. Owing to the limited data on surfactant systems, the results are shown for dodecyltrimethylammonium bromide[7] (DTAB). Some data for partitioning in sodium dodecyl sulphate (SDS) are available and the coefficients are found to be comparable to those in DTAB for non-ionic solutes such as alcohols, hydrocarbons, and halohydrocarbons. Whether the positive charge on DTAB micelles affects the partition coefficient of the carboxylic acids by promoting dissociation of the acid and holding the carboxylate ion close to the micelle surface is unclear. Most of the surfactants used in food applications are non-ionic but corresponding partition data are not available. The values of the partition coefficients shown for the oil–water system apply, to a first approximation, to most vegetable oils, although the conversion of published conventional partition coefficient data to values in terms of molar fractions has been carried out assuming corn oil. It is striking that, in terms of molar fractions, the partition

Table 1 *Micellar (dodecyltrimethylammonium bromide) partition coefficients* (P^{x}_{mic}) *and two-phase (oil–water) partition coefficients* (P^{x}_{oil}), *in terms of molar fractions, for four solutes including two food preservatives (benzoic and sorbic acids), benzaldehyde to show the behaviour of a non-ionic compound and diethylamine to show the behaviour of a base*[a]

Solute	*Log* P^{x}_{mic}	*Log* P^{x}_{oil}
Benzaldehyde	2.9 (M)	−0.4 (C)
Benzoic acid	3.2 (C)	−1.1 (M)
Sorbic acid	3.1 (C)	−1.2 (M)
Diethylamine	1.9 (M)	−1.4 (C)

[a] (M) denotes measured values; (C) denotes calculated values using

$$\text{Log } P^{x}_{mic} = 0.91 \text{ Log } P^{c}_{oct} + 1.51$$
$$\text{Log } P^{c}_{oil} = 1.12 \text{ Log } P^{c}_{oct} - 0.33$$
$$P^{x}_{oil} = P^{c}_{oil}\, V^{o}_{water}/V^{o}_{oil}$$

where P^{c}_{oct} and P^{c}_{oil} are partition coefficients in terms of concentrations for the systems octanol–water and corn oil–water, respectively, and V^{o}_{water} and V^{o}_{oil} are the molar volumes of water and corn oil, respectively. Data and constants are from references 7 and 10.

coefficients of the solutes shown are more than three orders of magnitude greater in the micellar systems. Hence it is possible for a large proportion of the solute to become associated with the surfactant in solute–surfactant–oil–water systems.

3 Kinetic Effects

General Considerations—Intuitively, it can be seen that in systems where reactants partition between two or more phases, the rates of reactions involving those reactants will be modified. Therefore, where a reaction takes place exclusively in the aqueous phase, *e.g.* involving, say, an ionic and a non-ionic species, the addition of a non-aqueous phase into which the non-ionic reactant will partition, will tend to reduce the rate of reaction, provided that the concentration of that reactant is rate determining. On the other hand, the rate of reaction involving two non-ionic reactants in water may be increased by the addition of a small amount of non-aqueous phase; a tendency for the reactants to partition into the non-aqueous phase may concentrate the reactants or some desirable and reactive form of the species in question. In some instances the less polar environment may also be more conducive to reaction. In cases where the solute is required to pass from one phase to another before reaction will take place, the kinetics of the transport need to be considered; a situation may be envisaged where the rate of reaction is transport controlled.

Many of the published data on interphase transport concern the use of two macroscopic phases and application of the stagnant film model.[11,12] In most instances diffusion is slow and the rate-determining step is transport through one of the phases. In the case of diffusion out of emulsion droplets, the thickness of the diffusion layer within the droplet, in an unstirred system, will equal the radius of the droplet, whereas that in the aqueous phase will equal the dimensions of the system. The problem reduces to one of diffusion alone. On the other hand, in stirred systems, the thickness of the diffusion layer in the aqueous phase will be reduced. If a spherical, solute-containing droplet of radius r is in contact with a medium whose solute concentration may be regarded as constant while diffusion is taking place, the order of the time scale over which diffusion takes place is $r^2 D\pi^2$, where D is the diffusion coefficient.[13] If one considers 2-μm droplets and $D = 5 \times 10^{-10}\ \mathrm{m^2\,s^{-1}}$, the time scale is of the order of milliseconds. This is presumably why flavour components are quickly released from emulsions when they are diluted in the mouth. Most reactions of low molecular weight food additives which have been studied in foods or model systems are several orders of magnitude slower than the estimate made here. The rate of transport of solute through the diffusion layer may be investigated by rotating disc techniques,[14] but evidence suggests that this is unlikely to be rate limiting in stirred emulsions in which food additive reactions are taking place. In real systems, however, it is possible that solutes adsorbed at the interface may interfere generally with interfacial transport or facilitate selective transport of certain molecules.

For micellar systems, the same general considerations as for dispersed systems apply. Surfactant micelles are, however, regarded as dynamic entities and surfactant molecules exchange rapidly with those in the surrounding water,[15,16] but solute diffusion coefficients may be up to one fifth of their values in water alone.[17] This will only matter for reactions which are diffusion controlled in the absence of surfactant; this is not the case for known food preservative–food component reactions but may be relevant to free-radical chain propagations and terminations. The best known[18] kinetic effect of surfactants is their ability to catalyse reactions involving polar and non-polar reactants. Reasons for this effect include the following:

1. Concentration of non-polar reactants within the micelle environment causing an increase in rate for reactions of high kinetic order. The selective partitioning of non-polar species into micelles may also displace acid–base equilibria in solution in favour of a non-ionic form.
2. A high charge density at the micelle surface provides sites for ion-pair interactions with oppositely charged transition-state complexes, thereby increasing their stability.
3. The charge on the micelle may also perturb acid–base equilibria by stabilizing anions or cations, perhaps increasing the concentration of a particularly reactive species.

A simple method of formulating a first-order reaction of a reactant A, catalysed by micelle M, is to imagine the reversible formation of a complex MA whose concentration is rate determining, *i.e.*

$$\mathrm{M + A \rightleftharpoons MA \xrightarrow{\text{slow}} products} \tag{6}$$

If the equilibrium for formation of MA is set up rapidly and the rate of formation of products is given by k[MA], where k is a rate constant, the initial rate of reaction, V_o, as a function of [M] and [A], is given by

$$V_o = k[\mathrm{M}][\mathrm{A}]/([\mathrm{A}] + K) \tag{7}$$

where k is the equilibrium constant for the formation of MA and the concentrations are initial concentrations. This form of expression is the same as that used to describe one-substrate enzyme kinetics with [M] as enzyme concentration. By analogy, we would expect the micelle-catalysed reaction to show 'saturation' kinetics. Hence, when the micelle holds as much reactant as possible, the rate of reaction would reach the maximum rate, V_{max}, given by

$$V_{max} = k[M] \tag{8}$$

and equation (7) becomes

$$V_o = V_{max}[A]/([A] + K) \tag{9}$$

The same approach is suitable for reactions which are pseudo-first order where the micelle is regarded as containing a constant concentration of the other reactants. This requires the reactants to partition into micelles independently of one another; although such an assumption is reasonable at low concentrations within the micelle, it is probably not justified when the molar fraction of the solute becomes significant. Specific examples of heterogeneous and micelle catalysis of food preservative–food component reactions will now be considered.

Sorbic Acid–Thiol Reactions—Sorbic acid (hexadienoic acid) is an antimicrobial food preservative which is becoming used more widely because of its low toxicity. The polarized diene structure of this compound, shown in Figure 1, renders it susceptible to nucleophilic attack and although it is relatively unreactive towards amines, it will undergo slow attack by thiol groups to form the 5-substituted product.[19] In foods, such thiols could be relatively non-polar flavour components or include the thiol groups of cysteine as the free amino acid or in peptides or

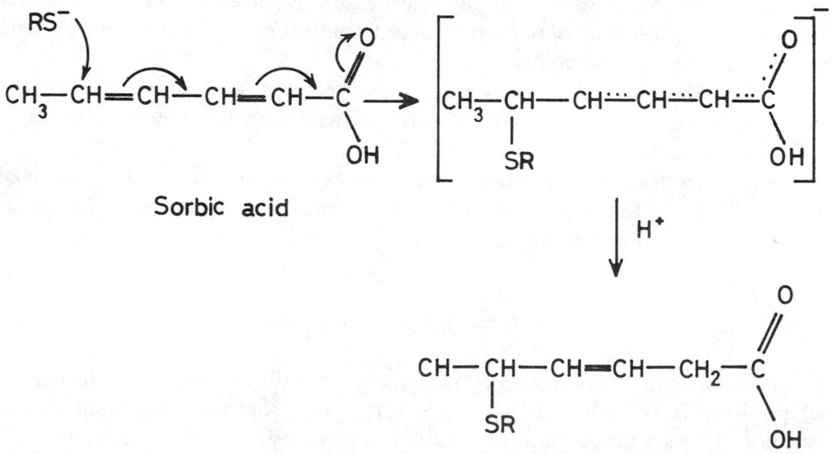

Figure 1 *Mechanism of reaction between sorbic acid and a thiol showing nucleophilic attack at position 5 and the formation of anionic intermediate I*

proteins. The reaction is of first order with respect to thiolate anion and sorbic acid and its rate depends on the presence of cationic or non-ionic surfactants.[20] For DTAB, as an example of a cationic surfactant, the effect of the surfactant concentration on the relative rate of reaction of sorbic acid with glutathione (glutamylcysteinylglycine), an example of a low molecular weight peptide, is illustrated in Figure 2. The result is typical of micellar catalysis, showing a considerable increase in rate when surfactant concentration exceeds the cmc (0.015 mol dm^{-3} for DTAB) and approaching a constant value at high concentrations. A comparison of the ability of DTAB to catalyse the reactions of different thiols with sorbic acid shows relative initial rates of reaction of 2.7, 4.6, 10.1, and 15.5 for glutathione, cysteine, mercaptoethanol, and mercaptoacetic acid, respectively, at 0.5 M DTAB ([sorbic acid] = [thiol] = 0.01 mol dm^{-3}; 0.2 M acetate buffer, pH 5.0; 80 °C), and is indicative of the importance of non-polar interactions between the thiol in question and the micelles; glutathione and cysteine are considerably more polar than mercaptoethanol and mercaptoacetic acid. On the other hand, SDS shows little or no catalytic effect, relative initial rates of 0.94, 1.78, 1.08, and 2.39, respectively, at 0.5 M SDS ([sorbic acid] = [thiol] = 0.01 M; 0.2 M acetate buffer, pH 5.0; 80 °C) being observed. It is reasonable to suggest that SDS is ineffective owing to electrostatic repulsion between the

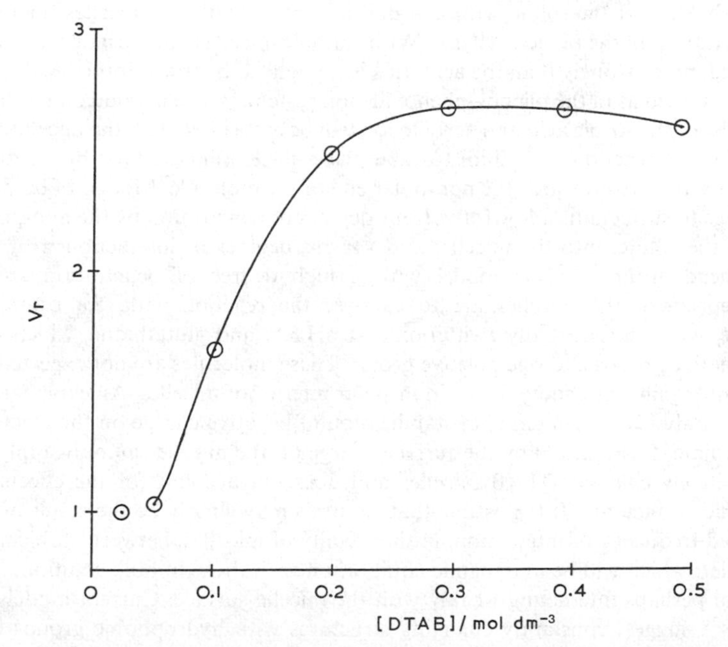

Figure 2 *Effect of dodecyltrimethylammonium bromide on the relative initial rate,* v_r, *of reaction of glutathione with sorbic acid. Reaction conditions: [sorbic acid] = [thiol] =* 0.01 mol dm^{-3}; 0.2 M *acetate buffer, pH* 5.0, 80 °C. v_r *is given by (initial rate in presence of DTAB) / (initial rate in absence of DTAB). Initial rates measured as* $-d[thiol]/dt$

Table 2 *Effect of surfactants on the relative initial rate, v_r, of the reaction of glutathione and cysteine with sorbic acid*[a]

	v_r	
Surfactant	*Glutathione*	*Cysteine*
Lecithin	1.77	3.91
Datem ester	1.48	3.44
Tween 80	2.50	3.21

[a] v_r is given by (initial rate in presence of surfactant)/(initial rate in absence of surfactant). Initial conditions: [sorbic acid] = 0.01 mol dm^{-3}; [thiol] = 0.01 M; [surfactant] = 0.8 wt.-%; acetate buffer, 0.2 mol dm^{-3}, pH 5.0; 80 °C.

thiolate anion and the negatively charged micelle. Most food surfactants are non-ionic and Table 2 shows the effects of lecithin (soya), datem ester (diacetyltartaric acid ester of C_{18} monoglyceride) and Tween 80. Whereas the cysteine reaction is catalysed nearly as effectively as by DTAB, non-ionic surfactants are less able to catalyse the glutathione reaction.

At pH 5.5, 90% of the sorbic acid will be in the form of sorbate ion. If we take the partition coefficient of sorbic acid in micelles to be in the region of 10^3 (Table 1), nearly 90% of the solute will be found associated with the micelles for total concentrations in the range 1–10 mM. With cationic micelles, the sorbate ion could well bind more strongly than the acid: this is thought to be true of other acids and their salts, such as in the phenol–phenoxide ion system.[21] The net outcome will be that most of the sorbic acid and sorbate ion will be associated with the micelles. In order for the reaction with thiol to take place, nucleophilic attack by an ionic reactant is required towards the non-polar end of the molecule. Models of binding of solutes to surfactants allow for various degrees of penetration by the non-polar part of the solute, into the micelle, and various degrees of interaction with the polar head group.[21–25] A model with a high degree of penetration seems inappropriate if the micelles are to catalyse the reaction with, for example, cysteine, which is essentially zwitterionic at pH 5.5, and glutathione, which has two negative groups and one positive group. These molecules are not expected to show any significant affinity for the non-polar interior of micelles. An explanation for the catalytic effect in terms of stabilization of negative charge on the reaction intermediate, I, (Figure 1) by the surface charge on the micelle can only apply to the positively charged DTAB micelles and does not account for the effects of non-ionic surfactants. It is possible that catalysis may simply be the result of an increased frequency of interaction, in the vicinity of micelles, between sorbic acid molecules, which will be in dynamic equilibrium with those in bulk solution, and the thiol perhaps interacting weakly with the micelle surface. Current models of micelles[26] suggest constantly changing structures with hydrophobic groups frequently being exposed to the aqueous environment. Hence the non-polar reactive end of the sorbic acid molecule will spend some time at the micelle surface exposed to attack by thiolate ion.

Preliminary experiments[20] show that the catalysis by DTAB exhibits saturation kinetics as described by equation (9) and that the stoicheiometry of the surfactant–thiol complex is in the region of 5:1 to 10:1 on a molar fraction basis. Hence it is clear that the micelles accommodate only a small number of thiol molecules consistent with micelle–thiol interactions at the micelle surface.

N-Nitrosation by Nitrite Ion—Nitrite ion is added to food mainly as an antimicrobial agent and in meats it serves also to produce characteristic nitrosylhaemoprotein pigments. At the pH of food, nitrite ion is present to a small extent as nitrous acid ($pK_a = 3.4$) and oxides of nitrogen, but mainly as the ionic species. The main reaction of this additive which has attracted interest is its nitrosation of secondary amines to N-nitrosamines, in which the nitrite-derived reactant is the nitrosonium (NO^+) ion in acidic solution or the species N_2O_3 in neutral solution. At the pH of meat, the non-ionic reagent is the most important. When dihexylamine is nitrosated in a system consisting of a dispersion of decane in aqueous buffer containing soya protein and carboxymethylcellulose,[27] and in the absence of decane, the yields of N-nitrosodihexylamine are 31.0 and 1.5%, respectively. On the other hand, if dibutylamine is the amine, the yields are 1.4 and 0.5%, respectively. Hence the dispersed phase acts to increase the yield of product, the dihexylamine reaction showing a greater effect. At pH 5.25, the apparent partition coefficients of dihexylamine and dibutylamine in decane–water are 0.061 and 0.0022, respectively. The higher yield of the nitroso compound has been linked to the partitioning of N_2O_3 and amine into the non-polar phase where nitrosation takes place and the observed partition coefficients are consistent with the observed yields. The involvement of oxides of nitrogen in nitrosation in non-aqueous phases is supported by the observation[28] that the extent of nitrosation of pyrrolidine and dipropylamine in water–benzene and water–corn oil is increased by the addition of ascorbic acid, which reduces nitrous acid to nitric oxide. One of the functions of the non-aqueous phases could be to discourage the hydrolysis of N_2O_3 which takes place in water (the equilibrium constant for the reaction

$$2\,HNO_2 \rightleftharpoons N_2O_3 + H_2O \tag{10}$$

is only of the order of 0.2 $dm^3\,mol^{-1}$) and considerable amounts of N_2O_3 may accumulate in the non-aqueous phase. The effect of ascorbic acid on heterogeneous nitrosation contrasts with its effect in homogeneous systems free from surfactant in which it acts as an inhibitor of nitrosation.

An interesting variant on experiments using liquid non-aqueous phases is the observation[29] that the rate of nitrosation of dihexylamine at pH 3.5 is over 70 times faster when the reaction is carried out in the presence of bacteria and yeast cells. The enhancement in rate varies almost linearly with cell concentration up to 11.4 $mg\,cm^{-3}$ and is regarded as being non-enzymic because if cells are heated to 100 °C, where enzymes will be denatured, the suspension retains most of its catalytic activity. On the other hand, if cells are disrupted, only 20% of the activity is associated with the aqueous fraction. It is reasonable to suppose that catalysis is due to adsorption on, or partitioning of reactants into, insoluble cell wall components such as lipid bilayers. This idea is strengthened by the observation that the catalytic effect is greatest for the least polar amines.

Table 3 *Effect of decyltrimethylammonium bromide on the relative initial rate,* v_r*, of nitrosation of secondary amines*[a]

Amine	v_r	*Amine*	v_r
Diethyl	3.2	Dihexyl	800
Di-*sec*-butyl	7.3	Morpholine	1
Methylbutyl	11	Pyrrolidine	3.8
Dibutyl	53	Piperidine	4.8

[a] v_r is given by (initial rate in presence of surfactant)/(initial rate in absence of surfactant). Reaction conditions: [amine] = 0.02 mol dm^{-3}; [NO_2^-] = 0.02 mol dm^{-3}; [surfactant] = 0.4 mol dm^{-3}; $[Br^-]_{total}$ = 0.5 mol dm^{-3}; pH 3.5; 25 °C. Data taken from reference 30.

Surfactants are also able to exert considerable kinetic effects on nitrosation reactions, as shown in Table 3 for decyltrimethylammonium bromide.[30] Kinetic data for dihexylamine show that the effect of the concentration of surfactant follows the same general behaviour as shown in Figure 2 for the sorbic acid–thiol reaction, and the observation that the enhancement in rate is greatest for the longer chain amines is indicative of the fact that non-ionic nitrosating species and amine are brought together in the micellar environment. The only food surfactant for which kinetic data are available is lecithin; a 0.8 wt.-% dispersion causes a 220-fold enhancement of the rate of nitrosation of dihexylamine. It is evident, however, that although dihexylamine has been used widely to illustrate heterogeneous and micellar catalysis of nitrosation, it has very large effects and may not be representative of the type of enhancement in rate expected for a wide range of food components. An interesting point about the reaction of dihexylamine is that it is autocatalytic. The *N*-nitroso derivative is less soluble than the amine and forms a dispersion as the reaction proceeds. A sharp increase in rate is seen[31] as soon as the solution becomes cloudy, corresponding to a product concentration of 20 μmol dm^{-3}.

Reactions of Sulphur Dioxide—Reactions in which one of the reactants is insoluble in water and forms a separate phase whereas the other can exist only in the aqueous medium need also to be considered in the context of food additive–food component reactions. Such reactions will take place at the interface. A good example is the heterogeneous sulphonation of unsaturated organic compounds, which may include fats[32] or hydrocarbons.[33] Whereas the reactions of sulphur dioxide in foods are almost entirely heterolytic reactions of sulphite ion, the autoxidation of the anion to sulphate is a free radical process[34] which involves intermediates such as $SO_5^{\cdot -}$, $SO_4^{\cdot -}$, $SO_3^{\cdot -}$, $^{\cdot}OH$ and $^{\cdot}O_2^-$. The species $^{\cdot}OH$ and $^{\cdot}O_2^-$ are excellent oxidizing agents whereas $SO_3^{\cdot -}$ is known for its ability to add to carbon—carbon double bonds. However, for the reaction to proceed most effectively a surfactant is required, presumably to allow the polar radicals to interact with the alkenyl groups. Thus, long-chain fatty acids need to be dispersed with the help of Tween to permit radical sulphonation.[32] On the other hand, although it is possible to oxidize β-carotene in homogeneous mixtures containing sulphite ion undergoing autoxidation, in an organic solvent–water mixture,[35,36]

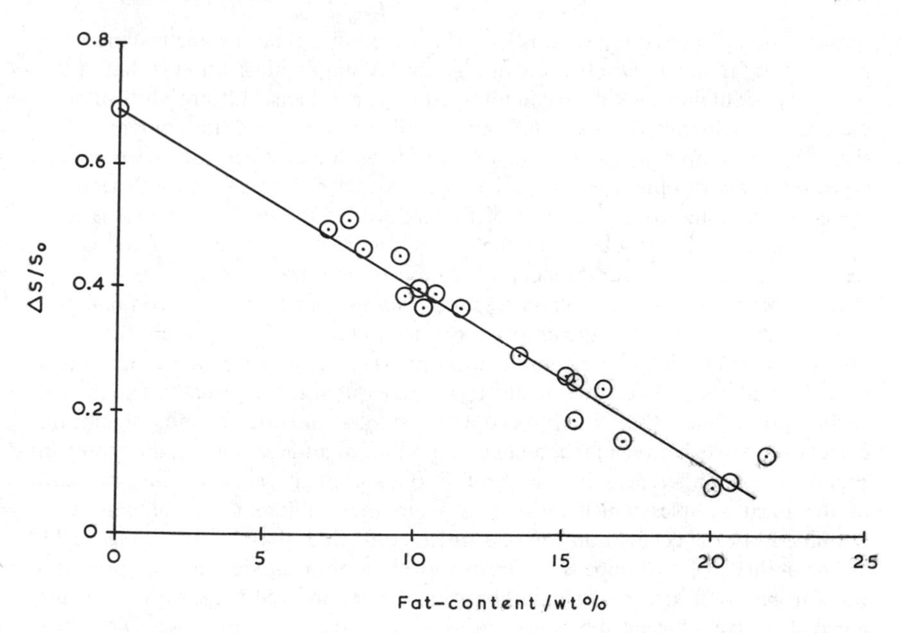

Figure 3 *Relationship between the relative loss of sulphur dioxide, $\Delta S/S_o$, and fat content for the addition of sulphur dioxide to various cuts of beef and pork. Initial sulphur dioxide content, S_o = 600 mg/kg meat. The zero fat-content measurement was obtained using pork from which fat had been extracted with acetone*

dispersions of the carotenoid in water are stable to such oxidation. Dispersions of Vitamin A, which has half the number of carbon atoms (20) as β-carotene and an OH group, in water, are oxidized quickly under these conditions. On the other hand, oxidation of dispersions of β-carotene in water is possible if a small amount of chloroform is added; it is thought that this may provide some new radical intermediate which is less polar and able to interact better with the non-polar hydrocarbon.

An interesting variant of this type of reaction is observed in the sulphonation of dodecene.[37] If sulphonation is attempted in a solution containing *tert*-butanol–water (1:1) using γ-radiation as the radical initiator, it seems that no chain sulphonation takes place until some of the reaction product, 1-dodecanesulphonate, has accumulated; the addition of this product allows the reaction to proceed correctly. It is likely that the hydrocarbon will exist in some kind of aggregate, access to which by the polar free radical is promoted by the addition of surfactant, such as the sulphonated product. Sulphonation takes place at the surface of this aggregate and the hydrophilic region so formed, with its associated secondary free radical, will interact more closely with the aqueous phase. It is not surprising, therefore, that a termination reaction involves the formation of disulphonate, with the sulphonate groups located on positions 1 and 2 of the hydrocarbon.

All examples given so far are of reactions whose rate is increased by the presence of a dispersed phase. It is important to realize that in some instances the presence of fat may *reduce* the rate of reaction. A simple illustration is that of the reactivity of sulphur dioxide in comminuted meat products.[38] If one adds sulphur dioxide to comminuted pork at 600 mg per kilogram of meat (fat content 10%), there is an instantaneous loss of *ca.* 40% of the additive. When the experiment is repeated using the same meat which has been extracted exhaustively with acetone, the corresponding loss is *ca.* 70% of that added. If the acetone extract is now added back to the residue from the extraction and the mixture is tested for reactivity towards sulphur dioxide, it is found to be unreactive.

It has been shown that the main reaction leading to instantaneous combination of the additive in meat is the cleavage of disulphide bonds of proteins and it is possible that fat released on comminution provides a barrier around the protein particles and fibres. Therefore, whilst trying to explain the different reactivities of various cuts of meat (beef and pork) towards sulphur dioxide, the most significant correlation exists between fat content and amount of additive lost, as illustrated in Figure 3. The differences in the non-fat solids content (*i.e.* including protein) of the meat samples which must arise from their different fat contents were evaluated but the correlation was less satisfactory than that shown in Figure 3.

The ability of fat to impede the transport of sulphur dioxide in meat products was further confirmed by staining fibres in defatted minced pork with malachite green. This dye is bleached by sulphur dioxide, but the extent to which colour was lost on treatment of the defatted pork sample with the additive depended linearly on the amount of fat added back to the sample.

4 Conclusion

There are numerous instances where the addition of non-aqueous components or surfactants to a reacting system has a profound effect on its chemical reactivity. When considering model systems for reactions in foods, it is important to realize the role of otherwise inert components, including particulate matter, and that their effects are exerted on both ionic and non-ionic reactions. Environments such as those within micelles may also exist at the surface of soluble protein molecules and it is important to include macromolecules as potential catalysts in food systems. An intriguing example of such a possibility concerns the ability of sorbic acid to inhibit a range of enzyme reactions. The best explanation of this action is in terms of its reactivity towards thiol groups on the enzyme molecule, but current data[19] on the reactivity of sorbic acid with thiols show the reaction to be slow (over many hours) at 80 °C. It is possible that the specific environment close to the sulphydryl groups on the enzymes in question may be conducive to much greater reactivity than measured so far.

Those concerned with studying the mechanisms of ionic reactions are well aware of the roles played by ions and the specific roles of acids and bases as catalysts, but the possibility that there may be a change in kinetics or mechanism arising from non-polar components is usually disregarded. Account of these factors seems to be the next step towards being able to understand properly chemical reactivity in the food environment.

Acknowledgements. One of the authors (B.L.W.) is indebted to AFRC for generous support of his work on the reactivity of sulphur dioxide in foods, to MAFF for generous support of research on sorbic acid, and particularly to Dr. D. J. McWeeny for continuing interest.

References

1. C. Hansch and A. Leo, 'Substituent Constants for Correlation Analysis in Chemistry and Biology,' Wiley, New York, 1979.
2. T. Matsubara and J. Texter, *J. Colloid Interface Sci.*, 1986, **112**, 421.
3. J. Texter, T. Beverly, S. R. Templar, and T. Matsubara, *J. Colloid Interface Sci.*, 1987, **120**, 389.
4. P. H. Elworthy, A. T. Florence, and C. B. MacFarlane, 'Solubilisation by Surface Active Agents,' Chapman and Hall, London, 1968.
5. A. M. Blokhus, H. Hoiland, and S. Backlund, *J. Colloid Interface Sci.*, 1986, **114**, 9.
6. C. Treiner and A. K. Chattopadhyay, *J. Colloid Interface Sci.*, 1986, **109**, 101.
7. C. Treiner and M.-H. Mannebach, *J. Colloid Interface Sci.*, 1987, **118**, 243.
8. R. D. Cramer, III, *J. Am. Chem. Soc.*, 1980, **102**, 1837.
9. R. D. Cramer, III, *J. Am. Chem. Soc.*, 1980, **102**, 1849.
10. A. Leo, C. Hansch, and D. Elkins, *Chem. Rev.*, 1971, **71**, 525.
11. H. van de Waterbeemd, S. van Boeckel, A. Jansen, and K. Gerritsma, *Eur. J. Med. Chem. Chim. Ther.*, 1980, **15**, 279.
12. H. van de Waterbeemd, P. van Bakel, and A. Jansen, *J. Pharm. Sci.*, 1981, **70**, 1081.
13. J. Crank, 'The Mathematics of Diffusion,' Clarendon Press, Oxford, 1957, p. 84.
14. W. J. Albery, J. F. Burke, E. B. Leffler, and J. Hadgraft, *J. Chem. Soc., Faraday Trans. I*, 1976, **72**, 1618.
15. K. K. Fox, *Trans. Faraday Soc.*, 1971, **67**, 2802.
16. D. J. Jobe, V. C. Reinsborough, and P. J. White, *Can. J. Chem.*, 1982, **60**, 279.
17. P. Stilbs, *J. Colloid Interface Sci.*, 1983, **94**, 463.
18. E. Cordes (ed), 'Reaction Kinetics in Micelles,' Plenum Press, New York, 1973.
19. B. L. Wedzicha and M. A. Brook, *Food Chem.*, 1988, **31**, 29.
20. B. L. Wedzicha and A. Zeb, *Int. J. Food Sci. Technol.*, 1989, in press.
21. C. A. Bunton and C. P. Cowell, *J. Colloid Interface Sci.*, 1988, **122**, 154.
22. I. Johnson and G. Olofsson *J. Colloid Interface Sci.*, 1987, **115**, 56.
23. S. D. Christian, E. Tucker, G. A. Smith, and D. S. Bushong, *J. Colloid Interface Sci.*, 1986, **113**, 439.
24. J. R. Cardinal and P. Mukerjee, *J. Phys. Chem.*, 1978, **82**, 1614.
25. P. Mukerjee and J. R. Cardinal, *J. Phys. Chem.*, **82**, 1620.
26. M. C. Woods, J. M. Haile, and J. P. O'Connell, *J. Phys. Chem.*, 1986, **90**, 1875.
27. R. C. Massey, C. Crews, R. Davies, and D. J. McWeeny, *J. Sci. Food Agric.*, 1979, **30**, 211.
28. D. S. Mottram and R. L. S. Patterson, *J. Sci. Food Agric.*, 1977, **28**, 352.
29. H. S. Yang, J. D. Okun, and M. C. Archer, *J. Agric. Food Chem.*, 1977, **25**, 1181.
30. J. D. Okun and M. C. Archer, *J. Natl. Cancer Inst.*, 1977, **58**, 409.
31. J. D. Okun and M. C. Archer, *J. Org. Chem.*, 1977, **42**, 391.
32. M. C. C. Lizada and S. F. Yang, *Lipids*, 1981, **16**, 189.
33. M. S. Kharasch, E. M. May, and F. R. Mayo, *J. Org. Chem.*, 1939, **3**, 175.
34. B. L. Wedzicha, 'Chemistry of Sulphur Dioxide in Foods,' Elsevier Applied Science, Barking, 1984, pp. 25–31.
35. G. D. Peiser and S. F. Yang, *J. Agric. Food Chem.*, 1979, **27**, 446.
36. B. L. Wedzicha and O. Lamikanra *Food Chem.*, 1983, **10**, 275.
37. A. Sakumoto, T. Miyata, and M. Washino, *Chem. Lett.*, 1975, 563.
38. B. L. Wedzicha and K. A. Harrison, *Int. J. Food Sci. Technol.*, 1989, in press.

Weak Particle Networks

By T. Van Vliet and P. Walstra

DEPARTMENT OF FOOD SCIENCE, WAGENINGEN AGRICULTURAL UNIVERSITY, WAGENINGEN, THE NETHERLANDS

1 Introduction

Liquid dispersions may become inhomogeneous or 'demix' during storage, *e.g.* by sedimentation or creaming of suspended particles; this is usually undesirable for the consumer. Examples of sedimenting or creaming particles in liquid dispersions are phosphate crystals in liquid detergents, cocoa particles and denatured protein particles in chocolate drinks, protein particles in buttermilk, fat globules in evaporated milk, and oil droplets and cellular fragments in fruit juices and nectars. Sedimentation or creaming is normally visually unattractive, and it may adversely affect the taste of beverages. In freshly pressed orange juice, the cellular fragments contain the antioxidants. Therefore, fast sedimentation soon causes an off-taste to develop in the upper layer due to oxidation. In chocolate milk the cocoa particles may sediment, which results in a light-coloured upper layer with a flat taste and a dark-brown, bitter-tasting substratum. In some beverages sedimented particles may form a kind of cake that cannot be fully redispersed on shaking.

In general, different types of demixing of liquid dispersions may be distinguished,[1,2] *viz.*:

—sedimentation or creaming;
—syneresis (formation of a clear liquid layer);
—segregation (layer formation in another way than by sedimentation, creaming or syneresis);
—formation of visible flocs.

The sedimentation of an isolated rigid spherical particle in a homogeneous medium may in some conditions be described by the Stokes equation:[3]

$$V_s = \frac{g(\Delta\rho)d^2}{18\eta_c} \tag{1}$$

where V_s is the sedimentation or creaming velocity, g the acceleration due to gravity, $\Delta\rho$ the density difference $= \rho_d - \rho_c$, ρ_d and ρ_c being the density of the particle and the continuous phase, respectively, d the particle diameter, and η_c the viscosity of the continuous phase. The most important factors causing deviations from equation (1) are Brownian motion, convection currents, and hindered separation due to neighbouring particles. The mean displacement x due to Brownian motion is given approximately by[3]

$$x = \left(\frac{kTt}{3\pi\eta_c d}\right)^{\frac{1}{2}} \tag{2}$$

where k is the Boltzmann constant, T the absolute temperature, and t time. If we determine a Peclet number (Pe) as the ratio of the calculated time needed for a particle to move over its own diameter by Brownian motion to that by sedimentation, we obtain

$$Pe = \pi d^4 g\Delta\rho/6kT \approx 10^{21} d^4 \Delta\rho \tag{3}$$

in SI units. For $Pe > 1$, disturbance of sedimentation due to Brownian motion may be neglected; this implies $d \gtrsim 1\ \mu\text{m}$ for $\Delta\rho = 100\ \text{kg m}^{-3}$. Convection currents may disturb the sedimentation of even larger particles. Particles smaller than *ca.* 0.1 μm in diameter will remain in suspension because of Brownian motion and convection currents almost indefinitely.

For the hindering effect due to neighbouring particles, different empirical or semi-empirical corrections are available of the form[4,5]

$$V_h \approx V_s(1 - \varphi)^n \tag{4}$$

where V_h is the hindered sedimentation velocity, φ the volume fraction of dispersed particles, and n depends on such conditions as Reynolds number, Peclet number, and spread in particle size. For instance, Richardson and Zaki[6] gave $n = 4.65$, but Walstra and Oortwijn[5] found $n = 8.6$ for fat globules in pasteurized milk.

In most food systems, η_c depends on the shear rate $\dot{\gamma}$ applied. The value of η_c to be inserted in equation (1) depends on V because V and the local value of $\dot{\gamma}$ are directly related. The tangential velocity component for creep flow around a sphere of radius R in a Newtonian fluid in spherical coordinates is[3]

$$V_\theta = V_\infty\left[1 - \tfrac{3}{4}\left(\frac{\mathrm{R}}{\mathrm{r}}\right) - \tfrac{1}{4}\left(\frac{\mathrm{R}}{\mathrm{r}}\right)^3\right]\sin\theta \tag{5}$$

where V_∞ is the velocity of the sphere relative to the fluid at $r \rightarrow \infty$, r is the distance from the centre of the sphere, and θ the angle between the average direction at which the fluid approaches the sphere and the line connecting the centre of the sphere and the considered point. We must consider the shear rate at the surface of the sphere, *i.e.* at $r = R$. This turns out to be

$$\dot{\gamma}_{\theta,s} = \frac{\mathrm{d}V_\theta}{\mathrm{d}r} = V_\infty \cdot \frac{3}{2R} \cdot \sin\theta \tag{6}$$

and after integration over θ and the circumference

$$\bar{\dot{\gamma}}_s = \frac{3V_\infty}{2R} = \frac{3V_\infty}{d} \tag{7}$$

For non-Newtonian fluids, the calculation of the effective $\dot{\gamma}_s$ is more complicated and, in fact, the exact relationship is not known. In general, the expression for the sum of friction and form drag on a sphere is[3]

$$F_k = (\tfrac{1}{4}\pi d^2)(\tfrac{1}{2}\rho V_\infty^2)f_c \tag{8}$$

For creeping flow of a Newtonian fluid around a sphere, $f_c = 24Re$,[7] where Re is the Reynolds number ($dV_\infty\rho/\eta$). Assuming that the shear thinning behaviour of the system may be described by a power law ($\sigma = k\dot{\gamma}^n$) and the relation $f_c = 24/Re$ still holds for $Re < 0.1$,[7] we obtain

$$f_c = \frac{24k}{\rho d^n V_\infty^{2-n}} \cdot x \tag{9}$$

where x is an unknown factor that probably will not be far from unity and that may depend on n.

Inserting equation (9) into equation (8) yields

$$F_k = 3\pi k d^{2-n} V_\infty^n x$$

For a steady-state fall of a sphere in a fluid, the force F_k is just counterbalanced by the gravitational force on a sphere less the buoyancy force. This results in

$$V_\infty = \left(\frac{1}{18} \cdot \frac{d^{1+n}\Delta\rho g}{kx}\right)^{\frac{1}{n}} \tag{10}$$

and

$$\dot{\gamma}_{\text{eff}} = \left(\frac{\sigma}{k}\right)^{\frac{n}{1}} = (3x)^{\frac{1}{n}} \frac{V_\infty}{d} \tag{11}$$

It should be stressed that these equations apply only if the fluid exhibits a power law behaviour even at shear rates far lower than the $\dot{\gamma}_{\text{eff}}$ calculated from equation (11) (the behaviour at higher $\dot{\gamma}$ is irrelevant in this respect). This implies that these equations are generally not applicable to macromolecular solutions, which normally exhibit a kind of Newtonian behaviour at low $\dot{\gamma}$; a so-called Ellis fluid then is a better model on which to base the calculations of $\dot{\gamma}$.[7] For many particle networks, a model based on a power law combined with a yield stress probably yields reasonable results, although a more refined model may be better; however as far as we know, such a treatment is not available to the extent that we can write equations analogous to equations (10) and (11).

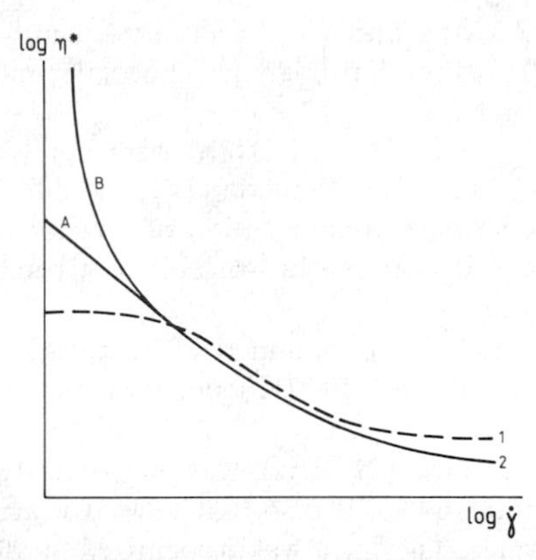

Figure 1 *Schematic representation of the shear thinning behaviour of (1) high molecular weight polymer solution and (2) weak particle 'network'; (A) without a yield stress and (B) with a yield stress*

The value of $\dot{\gamma}_{eff}$ obtained by equation (7) or (11) may differ substantially from that during usage (pouring, drinking, etc.); even for particles of 100 μm and $\Delta\rho$ = 500 kg m^{-3} in a low-viscosity fluid such as water $\dot{\gamma}_{eff}$ is already orders of magnitude lower than that during drinking while V_s would be unacceptably high. To obtain a V_s below 1 mm per day, even for particles as small as 10 μm, a high η^*_c is necessary, which implies that the product must be very shear thinning to obtain a sufficiently thin liquid at the relevant $\dot{\gamma}$ during usage. It is often difficult to fulfil these requirements merely by adding a thickening agent. Dispersions with a very high volume fraction of weakly aggregated particles are mostly better suited (see also Figure 1).

This difficulty can be circumvented completely by the formation of a weak network with a yield stress high enough to prevent sedimentation. The fracture or yield stress σ_y of such a network must be equal to or larger than the sedimentation stress exhibited by the particle, which implies that[2,8]

$$\sigma_y > 2dg(\Delta\rho)/3 \tag{12}$$

Normally a σ_y below 1 Pa, which is equivalent to the pressure exerted by a 10^{-4} m water column, suffices to prevent sedimentation.

However, the formation of such a weak network may cause various problems, such as syneresis, segregation due to uniaxial compression of the network under its own weight, or the formation of visible flocs.

In this paper we discuss some of the 'requirements' that weak networks in beverages should fulfil in order to give a satisfactory result in preventing sedimentation, while avoiding other problems.

2 Materials and Methods

Standard chocolate milk was prepared by mixing skim milk with cocoa (1.5% w/w), sucrose (7% w/w) and κ-carrageenan (mostly 0.025% w/w). The mixture was homogenized at 17.5 MPa and 75 °C and subsequently heated by indirect UHT at 135 °C for 10 s. After bottling, the chocolate milk was sterilized by autoclaving at 118 °C for 15 min.

The cocoa powders (type D11A and D11S) were kindly supplied by Cacao de Zaan (Koog aan de Zaan, The Netherlands). The difference between these two types is due to the stronger alkalinization of powder D11S, resulting in a higher salt content and a deeper red-brown colour. Chemical analyses of the powders can be found elsewhere.[1]

The κ-carrageenans used were commercial samples of Aubigel X120/A (Aubigel, France) and Genulacta K100 (Kobenhavns Pectinfabrik, Copenhagen, Denmark).

Single-strength orange juice (11° Brix) was prepared by diluting a Brasil Cultrale orange concentrate (64.95° Brix, 6.18% w/w citric acid, supplied by Coca Cola) with deionized water. The juice was pasteurized at 90 °C for 6 s. Orange juice without pulp particles was prepared by centrifugation at 360 *g* for 10 min. Serum, *i.e.* juice without pulp and cloud particles, was obtained by centrifugation at 19 200 *g* for 20 min.

Enzymic liquefaction of the pulp particles and of the serum polysaccharide and protein was achieved by adding an enzymic preparation, Rapidase C-40 (Gist-Brocades, Delft, The Netherlands). It contains polygalacturonase, pectin esterase, pectin lyase, hemicellulase and Cx-cellulase. Liquefaction was carried out by adding 50 mg l^{-1} of enzyme preparation and keeping the mixture at 30 °C for 2.5 h. Subsequently, the enzymes were inactivated by heating at 90 °C for 6 min.

Apricot nectar was prepared from canned apricots. After blanching at 125 °C for 5 min the apricots were mechanically macerated by pressing them through a sieve of mesh width 0.5 mm. Part of the purée obtained was treated with Ultrazyme 100 for varying times, then both the treated and untreated purées were heated for 10 min at 100 °C and diluted 1 : 1 with 20% sucrose solution. After dilution, the nectar was homogenized with an Ultra Turrax at 20 000 rev min^{-1} for 25 s and de-aerated. As a preservative, 0.2% w/w of sodium benzoate was added. A more extensive description is given elsewhere.[9]

Rheological properties were determined by means of two constant-stress viscometers, mostly at 20 °C. For the chocolate milk and apricot nectar a Deer Rheometer PDR 81, described more extensively by Roefs,[10] with a coaxial cylinder geometry was used. The diameter of the outer cylinder was 58 mm and that of the inner cylinder 56 mm for chocolate milk and 50 mm for apricot nectar. For this beverage the surfaces were roughened with emery paper. The applied shear stress could be varied between 10^{-2} and 10 Pa. For the orange juice a parallel-plate viscometer without bearings was used, as described previously.[11] Essentially, it consists of a ferromagnetic disk which floats on the liquid (layer thickness 5 mm, disk radius about 40 mm). It is driven and centred by a rapidly rotating magnetic field. The shear stress could be varied between 5×10^{-6} and 10^{-1} Pa by changing the distance between the disk and the rotating magnet.

A yield stress was defined as the stress below which no lasting motion of the inner cylinder or of the disk could be detected within 30 min. This implies that the shear rate was below $5 \times 10^{-6}\ s^{-1}$.

3 Results and Discussion

Required Extent of Shear Thinning—During drinking many beverages must have a viscosity of, say, ten times that of water while $\dot{\gamma}$ will be about $10^2\ s^{-1}$.[12] If one accepts a sedimentation of 1 cm during 6 months, V_s must be below $6 \times 10^{-10}\ m\ s^{-1}$. For a particle of 10 µm this implies a $\dot{\gamma}$ of about $2 \times 10^{-4}\ s^{-1}$ for a Newtonian fluid and somewhat higher for a shear thinning fluid. For a $\Delta\rho$ of $450\ kg\ m^{-3}$, η^*_c must be larger than 40 Pa s if calculated by equation (1). Comparison of these viscosity data with apparent viscosity *versus* $\dot{\gamma}$ published for macromolecular solutions[13,14] show that such systems are mostly unsuitable because they are insufficiently shear thinning or η^*_c is too high at $\dot{\gamma} = 10^2\ s^{-1}$. In principle, the calculations must be carried out with equations (10) and (11). However, most macromolecular solutions are semi-Newtonian at low shear rates and equations (1) and (7) yield a reasonable approximation.

The requirements are often better satisfied by systems which are essentially particle gels, even if macromolecules are involved. Examples of the shear thinning behaviour of some beverages are shown in Figure 2, together with that of a 0.2% w/w xanthan gum dispersion, The orange juice was non-stable. Xanthan

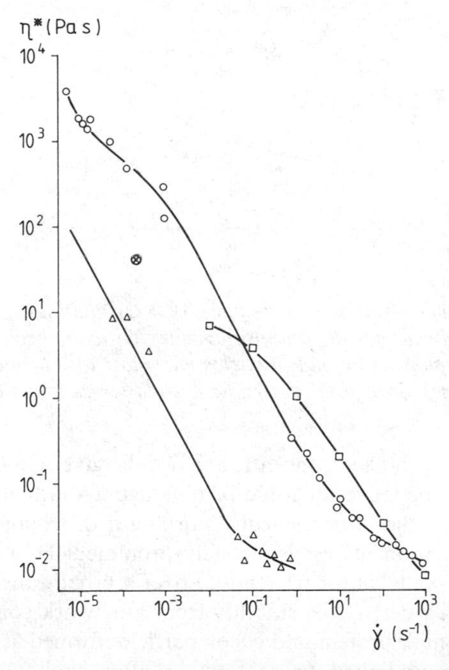

Figure 2 *Apparent viscosity* η^* *as a function of the shear rate* $\dot{\gamma}$ *for a standard chocolate milk* (○), *orange juice (Riedel Ltd., Ede)* (△) *and xanthan gum* (□). T = 20 °C. ⊗ $\eta^* = 40$ Pa s *at* $\dot{\gamma} = 2 \times 10^{-4}\ s^{-1}$ *(see text).* (Xanthan data from 'Gums and Stabilizers for the Food Industry', Elsevier Applied Science, 1986, p. 637)

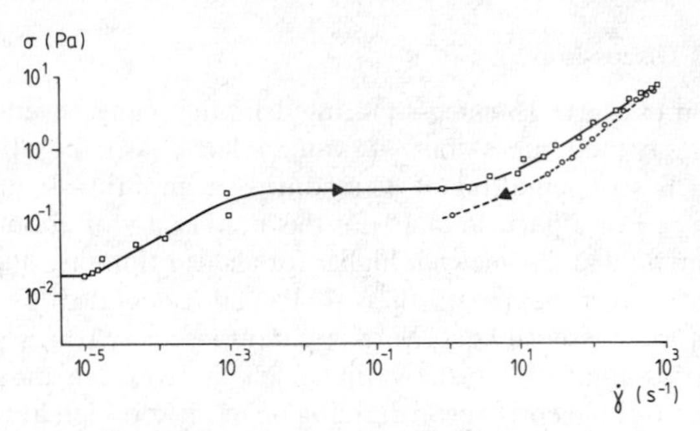

Figure 3 *Flow properties of a standard chocolate milk. The shear rate $\dot{\gamma}$ was measured as a function of the applied shear stress σ. Full curve, increasing σ; dotted curve, decreasing σ*

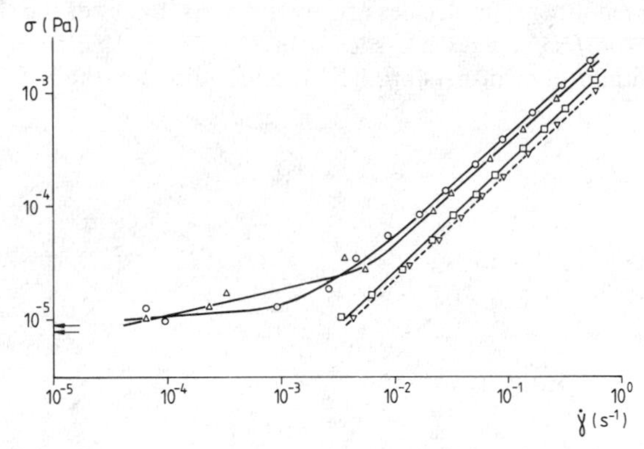

Figure 4 *Flow properties of an orange juice. (○) Original juice; (△) juice with the serum polysaccharides and protein enzymatically broken down; (□) juice with the pulp and cloud particles enzymatically broken down; and (▽, dotted line) orange juice serum. A yield stress is indicated by ←*

gum solutions with a higher concentration would give a too high η^* at $\dot{\gamma} = 10^2\,s^{-1}$. Probably, a better result could be obtained if xanthan gum were used in combination with another synergistically working hydrocolloid (*e.g.* locust bean gum).[16,17] Then gelation at rest is strongly promoted. However, so far as we know, no data are available for η^* *versus* $\dot{\gamma}$ over a broad range of $\dot{\gamma}$.

The chocolate milk derived its stability from a network consisting of voluminous aggregates of milk protein and cocoa particles formed at elevated temperatures, which are cross-linked to each other after cooling by κ-carrageenan molecules adsorbed on these aggregates.[1] In this way essentially a particle gel is formed with a yield stress of a few times 10^{-2} Pa (Figure 3).[8] Between this yield stress and a stress of *ca.* 0.2 Pa (the actual values depend on the sample of chocolate milk), a viscous component was also found. However, the apparent viscosity is very high. Presumably the bond strengths in the network are such that stresses in this range cause breaking of bonds from time to time but the rate of this process is so slow that the bonds can be restored fast enough to prevent macroscopic breakdown of the network. The latter occurs at stresses of about 0.2 Pa, resulting in a decrease in η^* by a factor of *ca.* 10^3. The cocoa particles are essential for the formation of the protein–cocoa aggregates and hence for the formation of the network. With only the extract of cocoa present (*i.e.* without the particles), the network formed is much weaker. Both the breaking stress of the network and η^* at $\dot{\gamma} = 10\ s^{-1}$ are about an order of magnitude lower.

In orange juice it is the pulp particles that are responsible for the formation of a yield stress and not the dissolved polysaccharides and protein (Figure 4).[8] Breaking down only the serum polysaccharides and protein with enzymes resulted in the same curve as for untreated juice. If, on the other hand, the pulp particles were destroyed by enzymic action, the flow properties of the resulting juice were not significantly different from those of the serum. The essential role of the pulp particles also follows from the observation that the flow properties of various juices without pulp particles were virtually identical, whereas the flow behaviour of the original juices showed great variations.[8]

The observations mentioned above all point to the remarkable fact that in these beverages the suspended particles, which in absence of a network would sediment, directly contribute to the formation of the network that retards or stops their sedimentation. Requirements other than those of a minimum σ_y or η^* at low $\dot{\gamma}$ and a maximum η^* at high $\dot{\gamma}$ that the weak particle networks must fulfil are in general as follows:

—after disturbing the gel, it must restore itself in a relatively short time;
—the formation of visible flocs or syneresis of the gel must not occur;
—the weak network must adhere to the wall of the container in order to avoid uniaxial compression under its own weight.

Reversibility of Network Structure—After disturbing the gel, the network structure must restore itself fast enough to prevent visible sedimentation or creaming during that time.

A rough estimate of the time t in which the network structure must restore itself can be obtained from the Stokes equation. Because in practice sedimentation or creaming will be hindered, a safe limit is obtained by calculating V, according to Stokes. For instance, for particles of $10\ \mu m$ with $\Delta\rho = 450\ kg\ m^{-3}$, $\eta^* = 5 \times 10^{-2}$ Pa s, and accepting a sedimentation of at most 0.5 mm, it is found that t must be $< 10^3$ s. This time can be determined experimentally by measuring the yield stress after different waiting times. For chocolate milk (Figure 3), it follows immediately from the flow curve that over time scales longer than equivalent to $\dot{\gamma} = 10^{-3}\ s^{-1}$ (*i.e.* $t > 10^3$ s), the network is able to keep (which implies to restore) its integrity. If σ had been lowered to a value below that indicated by the dotted curve, η^* increased at first and after some time the original curve was obtained.

Formation of Visible Flocs or Syneresis—Ongoing cross-linking of the structural elements forming the network may induce syneresis, which can be observed as a clear liquid layer on the top or at the bottom of the gel. Mostly it is not a serious problem in beverages with a weak network. Probably it plays a part in apricot nectar. In order to improve the stability against layer formation, apricot nectar may be given an enzyme treatment.[9] An explanation given in the literature is that such a treatment would increase the yield stress. However, by carefully measuring the rheological properties in a concentric cylinder system with roughened surfaces, we found that the enzyme treatment did not influence the yield stress whereas at higher $\dot{\gamma}$ a lower η^* was found. In both instances a σ_y of about 1 Pa was found, which in principle is high enough to prevent sedimentation. With smooth cylinders, slip was observed for the untreated apricot nectar. If the samples were stored in slowly rotating bottles, in the enzyme-treated nectar a gel surrounded by a clear liquid was obtained, whereas the untreated nectar remained homogeneous. Probably syneresis occurs in the treated apricot nectar if the network cannot stick to the wall because of continuing liquid motion. In standing bottles the enzyme treatment probably increases the extent of sticking of the network to the wall. We consider the importance of this point below.

Formation of visible flocs may occur if the weak flocs formed contract. This is a serious problem in chocolate milk where visible flocs may be formed if the heat treatment is too intense, the pH is too low, or the wrong cocoa is used. In chocolate milk made with cocoa powder D11S much greater formation of visible flocs was observed than in those made with the less alkalinized powder D11A.

Uniaxial Compression of a Network Under its Own Weight—It has been shown that for chocolate milk segregation, *i.e.* partial collapse of the weak network under its own weight, may occur.[1] This process may be described as uniaxial compression of the total network by gravitational forces until the latter are counterbalanced by the product of the uniaxial compression modulus E of the network and the strain ε [$\sim \Delta h/h$, where h is the height of the network and Δh (= $h_0 - h$) the decrease in height; h_0 is the initial height]. If there is no interaction between the network and the container, both stresses must be equal, which gives

$$E(\varepsilon,t)\varepsilon = \varphi_p \Delta\rho g h \tag{13}$$

where φ_p is the volume fraction of suspended particles including those forming the network.

As most weak gels are to some extent viscoelastic and as large deformations occur (about 1), E will depend on ε (non-linear behaviour) and on the time scale of the observation. Moreover, during uniaxial compression, φ_p of the network compounds increases. Generally, in inhomogeneous systems E increases more than in proportion to the concentration of the network compounds.[10] Therefore, E increases more than in proportion to h_0/h, resulting in a much smaller ε than calculated with equation (13).

In chocolate milk, the gravitational force due to the weight of the network is approximately 150 h Pa ($\varphi_{protein}$ and φ_{cocoa} are estimated to be 0.02 and 0.013, respectively, and $\Delta\rho$ is taken as 450 kg m^{-3}).[1] For $h = 0.1$ m this implies that $E(\varepsilon,t)\varepsilon$ must be 15 Pa to prevent uniaxial compression. Because $E \approx 0.1$–

$0.5\ \mathrm{N\,m^{-2}}$,[1] segregation will occur unless h is small. The only way to obtain such a situation is by the network adhering to the wall. Segregation also occurs in non-enzyme-treated apricot nectar, where $E = 2\text{–}3\ \mathrm{N\,m^{-2}}$, as derived from a plot of the shear stress against the shear strain. The gravitational force causes a stress of about 50 h (in SI units), assuming $\varphi_p \Delta\rho = 5\ \mathrm{kg\,m^{-3}}$.[9]

Equation (13) tells us nothing about the velocity of segregation. Owing to the high φ_f of the network elements, which are, *e.g.* voluminous aggregates or cellular fragments, hindered sedimentation will occur. A calculation according to equation (4) is questionable because no separate particles are sedimenting, but rather a network as a whole. Such a network is very inhomogeneous, containing dense particles flocculated into less dense aggregates, which in turn form the network. These aggregates contain much water. The velocity of segregation would then be determined more by the velocity of the liquid flow through the network than by pure hindered sedimentation. According to Darcy's law, the overall liquid flux v in one direction through a porous medium is[18]

$$v = -(B/\eta_c)\ \nabla P \tag{14}$$

where B is the permeability coefficient and ∇P the pressure gradient, which in this instance has only one component. If we consider the network to be built of more or less spherical particles, an assumption of B may be made using a modified Kozeny–Carmen equation:[10,19]

$$B = \frac{(1 - \varphi_f)^3 {d_{32}}^2}{180{\varphi_f}^2} \tag{15}$$

where d_{32} is the volume-surface diameter of the particles.

The permeability of the aggregates will be effectively zero compared with B for the whole network. Therefore, the aggregates should be taken as the particles in the considered model. For chocolate milk the diameter of the aggregates is about 50 μm as judged under a microscope. φ_f will be initially around 0.5 ($\pm$0.15) as follows from the final height of the segregated layer after a very long storage period. Inserting $\varphi_f = 0.5$ and $d_{32} = 50\ \mu\mathrm{m}$ in equation (15) gives for $B \approx 7 \times 10^{-12}\ \mathrm{m^2\,s^{-1}}$. In view of the uncertainty in φ_f and d_{32}, this value has at best an order of magnitude accuracy. The viscosity of a 12% sugar solution with some dissolved salts and about 1% of protein will be *ca.* 2×10^{-3} Pa s. ∇P will be caused by the weight of the network. Because only the aggregated milk protein and the cocoa particles contribute to the weight, one may set ∇P equal to $\varphi_p \Delta\rho g$. A rough estimate then gives $\nabla P = 150\ \mathrm{Pa\,m^{-1}}$ (see above). Inserting the values for B, η_c and ∇P in equation (10) gives $v = 5 \times 10^{-7}\ \mathrm{ms^{-1}}$. This value for v implies that the segregation velocity would be initially a few centimetres per day. For some chocolate milks that are rather unstable against segregation, values of up to 3 cm day^{-1} have been observed. However, in most instances the segregation velocity is about a factor of 10 lower (Figure 5). In agreement with the reasoning given above, the segregation velocity decreased as a function of time, because φ_f increases steadily which causes B to decrease, $E\varepsilon$ to increase and hence v to decrease considerably.

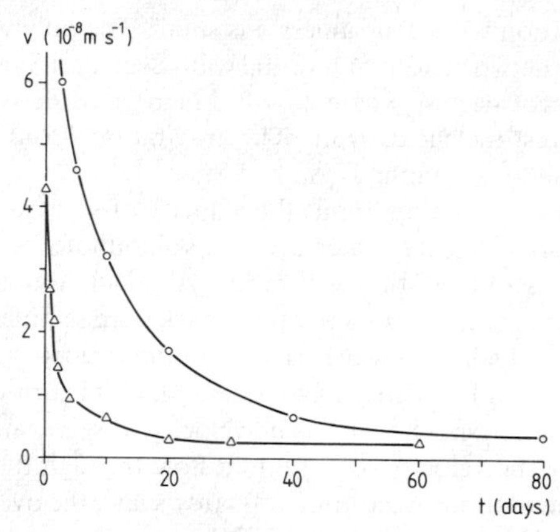

Figure 5 *Segregation velocity as function of storage time for two chocolate milk samples (pH = 6.6). Different milk was used:* (○) *UHT and bottle sterilized.* 0.03% κ-*carrageenan;* (△) *UHT sterilized,* 0.025% κ-*carrageenan*
(Data from reference 20)

An important factor determining whether segregation will occur is adherence of the network to the wall of the container. Good adherence may drastically reduce the value of the effective h in equation (13). This implies that the shape and surface properties of the container may markedly influence the occurrence of segregation. In chocolate milk this adherence phenomenon depends on, among other things, the type of κ-carrageenan used.[1] In apricot nectar the adherence is probably improved by an enzyme treatment as described above.

References

1. Th. van den Boomgaard, T. van Vliet and A. C. M. van Hooydonk, *Int. J. Food Sci. Technol.*, 1987, **22**, 279.
2. E. Dickinson, in 'Gums and Stabilizers for the Food Industry,' eds. G. O. Phillips, P. A. Williams, and D. J. Wedlock, Vol. 4, IRL Press, Oxford, 1988, 249.
3. R. B. Bird, W. E. Stewart, and E. W. Lightfoot, 'Transport Phenomena,' Wiley, New York, 1960.
4. D. F. Darling, in 'Food Structure and Behaviour.' eds. J. M. V. Blanshard and P. Lillford, Academic Press, London, 1987, p. 107.
5. P. Walstra and H. Oortwijn, *Neth. Milk Dairy J.*, 1975, **29**, 263.
6. J. F. Richardson and W. N. Zaki, *Trans. Inst. Chem. Eng.*, 1954, **32**, 35.
7. J. C. Slattery and R. B. Bird, *Chem. Eng. Sci.*, 1961, **16**, 231.
8. T. van Vliet and A. C. M. van Hooydonk, in Proceedings of the 9th International Congress on Rheology, Mexico, eds. B. Mena, A. Garcia-Rejon, and C. Rangel-Nafaile, Universidad Nacional Autónoma de, 1984, Vol. 4, p. 115.
9. H. A. I. Siliha, *PhD Thesis*, Wageningen Agricultural University, 1985.
10. S. P. F. M. Roefs, *PhD Thesis*, Wageningen Agricultural University, 1986.
11. T. van Vliet, A. E. A. de Groot-Mostert, and A. Prins, *J. Phys. E*, 1981, **14**, 745.
12. F. Sharma and P. Sherman. *J. Texture Stud.*, 1973, **4**, 111.

13. E. R. Morris, in 'Gums and Stabilizers for the Food Industry,' eds. G. O. Phillips, D. J. Wedlock, and P. A. Williams, Vol. 2, Pergamon Press, Oxford, 1984, 57.
14. B. Launay, J. L. Doublier, and G. Cuvelier, in 'Functional Properties of Food Macromolecules,' eds. J. R. Mitchell and D. A. Ledward, Elsevier Applied Science, Barking, 1986, p. 1.
15. M. Milas, M. Rinardo, and B. Tinland, in 'Gums and Stabilizers for the Food Industry,' eds. G. O. Phillips, D. J. Wedlock, and P. A. Williams, Vol. 3, Elsevier Applied Science, Barking, 1986, p. 637.
16. I. C. M. Dea and E. R. Morris, in 'Extracellular Microbia of Polysaccharides,' eds. P. A. Sanford and A. Laskin, ACS Symposium Series, No. 45, American Chemical Society, Washington, DC, 1977, p. 174.
17. E. Dickinson and G. Stainsby, 'Colloids in Food,' Elsevier Applied Science, Barking, 1987.
18. A. E. Scheidegger, 'The Physics of Flow through Porous Media,' Oxford University Press, Oxford, 1960.
19. H. J. M. van Dijk and P. Walstra, *Neth. Milk Dairy J.*, 1986, **40**, 3.
20. H. A. Deckers and A. C. M. van Hooydonk, to be published.

Influence of the Emulsifier on the Sedimentation of Water-in-Oil Emulsions

By F. van Voorst Vader and F. Groeneweg

UNILEVER RESEARCH LABORATORIUM VLAARDINGEN, P.O. BOX 114, 3130 AC VLAARDINGEN, THE NETHERLANDS

1 Introduction

A comparison of the properties of various commercial water-in-oil (W/O) emulsifiers showed that W/O emulsions having the same water content but prepared with different emulsifiers differed widely in sedimentation behaviour, although the water droplets were of comparable size (Figure 1). For instance, 30% W/O emulsions stabilized with Admul WOL (polyricinoleic acid esterified with polyglycerol) had separated after 6 h into a supernatant oil layer and a stable

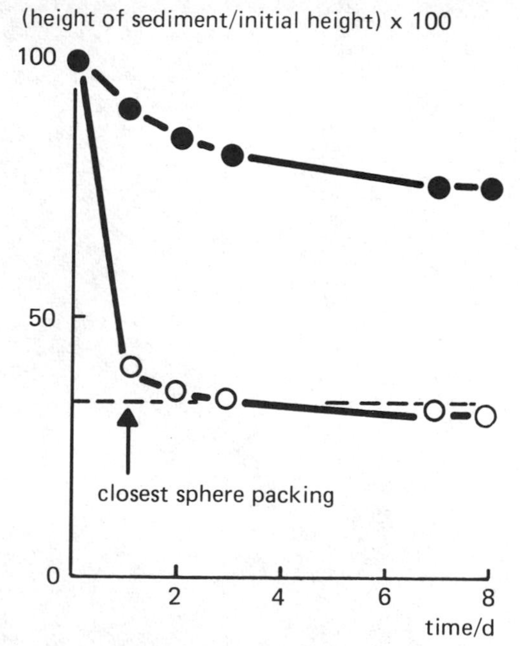

Figure 1 *Sedimentation of* 30% *W/O emulsions (water in paraffin oil). Emulsifiers:* (○) 1% *Admul WOL;* (●) 1% *DK-F10*

78% W/O emulsion, whereas in emulsions stabilized with DK-F10 (a mixture of di-, tri-, and polyesters of hydrogenated tallow fatty acids with sucrose) only slight sedimentation occurred during this period. We set out to establish the emulsifier property that was responsible for this difference.

Owing to van der Waals interactions, W/O emulsions will always flocculate, resulting in the formation of a droplet network. For dispersions of solids it has been shown[1] that such a network will be stable against sedimentation provided that its shear storage modulus (G') is sufficiently high to resist the gravitational force exerted on the droplets. On increasing the volume fraction of the dispersed phase, a 'gel point' is reached at which the system shows consolidation. For strongly flocculated systems, the boundary sedimentation rate just above this gel point is negligible. The behaviour of the dispersion in a force field can be modelled if the yield stress of the particle network formed is assumed to increase with the volume fraction of the dispersed phase, φ. This yield stress counteracts the compressive load $P(\varphi)$ at the bottom of a sediment with critical height H_s (Figure 2).

$$P(\varphi) = g\Delta\rho \int_0^{H_s} \varphi(z)\mathrm{d}z \qquad (1)$$

where $\Delta\rho$ = density difference between the phases, g = gravitational constant, and z = vertical coordinate. $P(\varphi)$ is related to the compressive modulus $K(\varphi)$ of the sediment by

$$\mathrm{d}P = K(\varphi)\mathrm{d}\ln\varphi \qquad (2)$$

Experimental results[2] suggest that $K(\varphi) \approx \frac{2}{3}G'(\varphi)$, leading to

$$H_s = \frac{2}{3g\Delta\rho\varphi} \int_{\varphi \text{ at } G'=0}^{\varphi} G'(\varphi)\mathrm{d}\ln\varphi \qquad (3)$$

where φ is assumed to be independent of z.

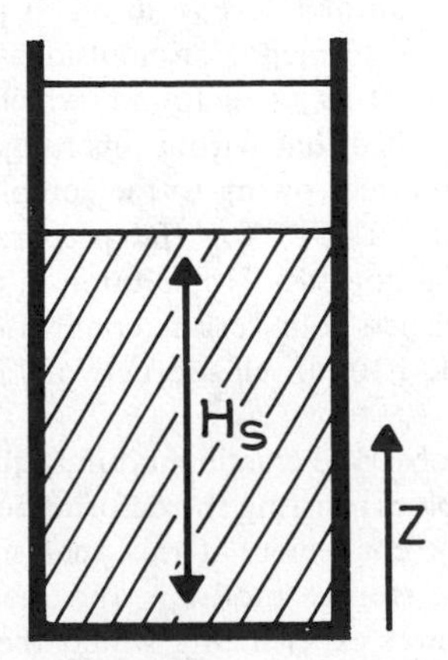

Figure 2 *Analysis of sediment height* (H_s)

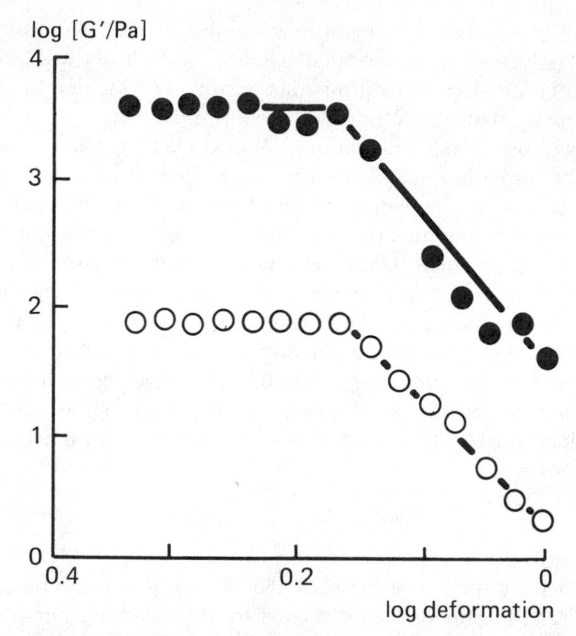

Figure 3 *Storage modulus* (G′) *at* $\varphi = 0.25$ (○) *and* 0.4 (●) *of flocculated W/O emulsions stabilized with DK-F10. Frequency:* 0.63 cycles s^{-1}

If the relationship between G' and φ is known, the maximum height of a sediment at a certain value of φ can be calculated. This theory has been confirmed by experimental data.[1]

We wished to investigate whether the above theory was also relevant to emulsions stabilized with DK-F10. To this end, the storage moduli of W/O emulsions stabilized with 0.5% w/w of DK-F10 were measured. However, as DK-F10 is virtually insoluble in oil at room temperature, we first measured the storage modulus of a 1% dispersion of DK-F10 in oil. It proved to be zero, indicating that no crystal network was formed. For emulsions containing 25–45% of water, storage moduli of 100–5000 Pa were found, which slowly increased with time (Figure 3). It should be noted that within this range, the results of the measurements are difficult to reproduce owing to the complicated method recommended by the manufacturer of DK-F10 for the preparation of these emulsions. It involves the addition of a hot DK-F10 solution in oil to cold water followed by phase inversion. Nevertheless, our results confirm our hypothesis that in emulsions stabilized with DK-F10, an elastic network is formed between the water droplets.

In order to relate the observed elastic moduli to the properties of the emulsifier used, the nature of the forces resisting the deformation of the elastic network must be analysed. It would be convenient if this analysis also revealed an analytical relationship between the storage modulus and the volume fraction of the dispersed phase because fewer experiments would then have to be carried out to evaluate the right-hand side of equation (3).

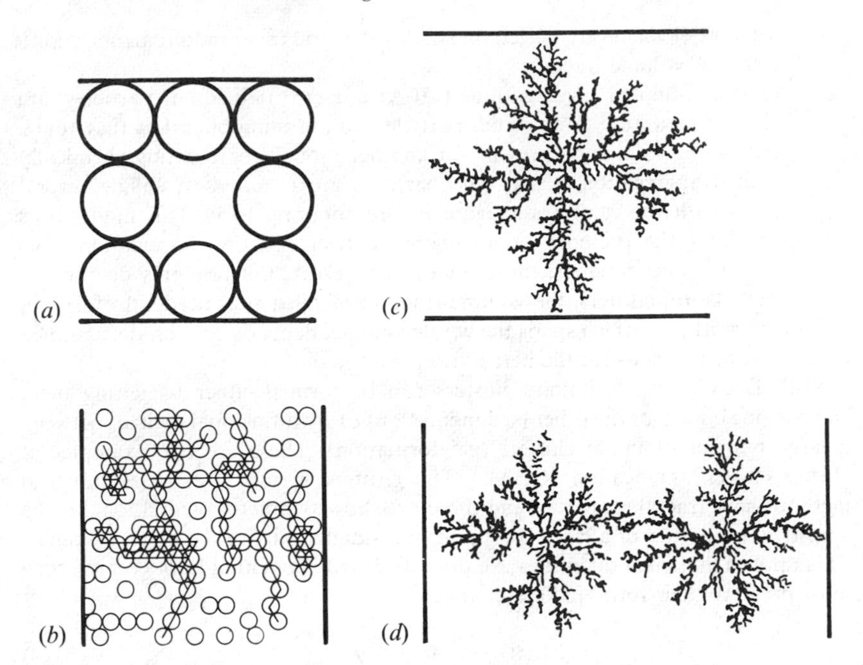

Figure 4 *Examples of network models:* (a) *linear network,* (b–d) *stochastic networks* (b) *Percolation network; infinite aggregates exist only if* $\varphi \geqslant \varphi_c (\varphi_c =$ *percolation limit);* (c) *diffusion-limited aggregate or chemically limited aggregate;* (d) *cluster–cluster aggregate*

2 Microstructural Models

The nature of the interaction forces expected to resist the macroscopic deformation of a flocculated network depends on the choice of the flocculate network geometry. Older models aim at geometrical simplicity rather than at realism. Thus, models have been proposed[3,4] which assume that all particles or droplets are arranged in linear chains of equal stiffness. One third of these chains are assumed to be oriented parallel to the extension or compression and the others perpendicularly to the deformation. Only normal interactions between the chain elements are considered. Models based on such homogeneous structures always predict that G' will be proportional to φ. Introduction of such a relationship into equation (3) leads to the conclusion that the maximum height of the sediment is independent of the volume fraction of the dispersed phase. This is contrary to the experimental results, which indicate a strong influence of φ.

Recently, novel methods of modelling flocculate networks have been developed on the basis of stochastic models. They all have in common that the network geometry is generated by a random process, which may be chosen in such a way that it approaches the real situation more closely than the simple models described above (Figure 4). Examples are as follows:

1. Percolation networks. Here the available space is filled with a grid, and

network elements are placed on sites of this grid in a random manner up to a certain volume fraction.[5]

2. Diffusion-limited aggregation (DLA)[6]. Here individual particles are allowed to diffuse to an initial particle and are immobilized as they touch the aggregate already present. In another model, representing chemically limited aggregation (CLA), the particles are required to collide several times with the central aggregate before adhering to it. This model thus simulates the presence of an energy barrier resisting flocculation. The resulting flocs resemble those obtained by DLA; they are only denser.

For percolation clusters, the volume fraction of filled sites needed to form an infinite network, φ_c, which spans the whole volume, depends only on the assumed grid geometry; φ_c is called the percolation threshold.

With DLA or CLA, infinite clusters can be formed either by letting them subside one on top of the other (sedimentation) or by diffusional contact between a large number of initial clusters (gel formation). These are two examples of cluster–cluster aggregation (CCA).[7] The clusters formed by these statistical methods have fractal properties and therefore show symmetry on dilation, *i.e.* the statistical properties of a larger part and of a smaller part of the floc are similar. This implies that such quantities are described by exponential laws because they must preserve their form under a change of scale:

$$A(\lambda\chi) = \mathrm{f}(\lambda)\,\mathrm{A}(\chi) \tag{4}$$

It follows that

$$A(\chi) = C\chi^{g} \tag{5}$$

The exponents involved in the description of the various parameters can be mutually related by scaling laws. Their values depend mainly on the dimension of the cluster and not on the detailed network geometry. In some instances, they can be calculated by mathematical derivation. In most instances, however, computer simulation must be used.

This approach can be applied to describe the storage moduli of such flocculated networks. External forces exerted on such a network are resisted by its 'backbone,' *i.e.* the shortest continuous path across the network between opposite sides (Figure 5). For a percolation network, such a path consists mainly of tortuous single chains of particles and for a minor part of dense aggregates. For CCA-generated aggregates, the single chains in the backbone are even more predominant. The tortuous chains react to an external force by bending; their resistance to deformation is therefore determined by the resistance to shear deformation of the particle–particle contacts. Multiple connected, dense aggregates react to an external force by compression or extension. The resistance to deformation is then determined by the resistance of the particles to compression. This is the only type of deformation considered in the older models discussed above. However, at lower values of φ, bending predominates. If a percolation network is assumed, the following scaling equation can be derived:[8]

$$G'(\varphi) = G'_0(\varphi - \varphi_c)^{\tau_p} \tag{6}$$

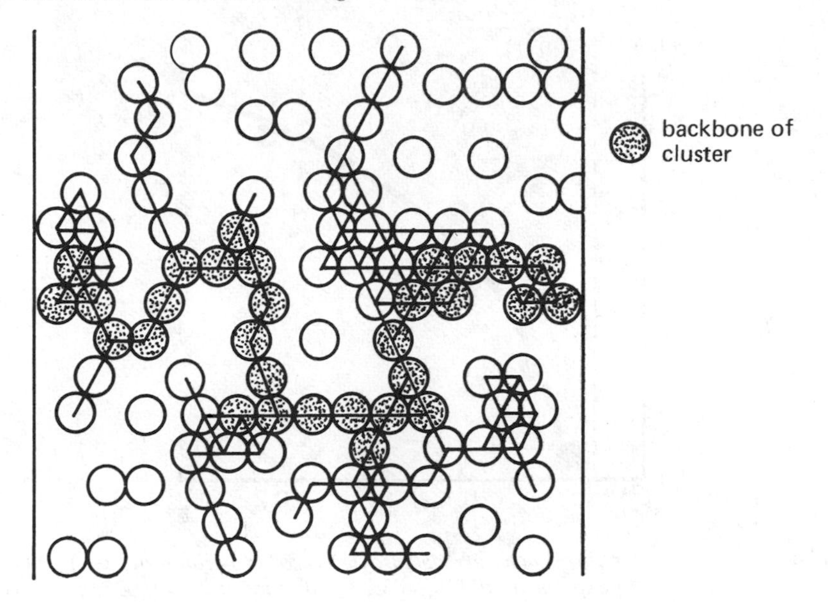

Figure 5 *Backbone of a stochastic cluster*

where G'_0 is a scaling constant. Close to the percolation threshold, the calculated value of the critical exponent $\tau_p = 3.55$; τ_p approaches 1 at the limit of closest packing of particles. For a grid allowing the closest packing of spheres, $\varphi_c \geqslant 0.144$. For CCA, the scaling equation is

$$G'(\varphi) = G'_0\varphi^{\tau} \tag{7}$$

where $\tau = 3.5$ for diffusion-limited aggregation and 4.5 for chemically limited aggregation.[9] Hence the relationship between the storage modulus and the volume fraction of the disperse phase for a stochastic network can generally be described by equation (6), where, depending on the nature of the stochastic process chosen to generate the network, φ_c ranges from 0 to 0.15 and τ from 1 to 4.5. It has indeed been found that equation (6) can be used to describe the relationship between G' and φ for flocculated dispersions. Examples are flocculated aqueous dispersions of latex, attapulgite,[1,2] and dispersions of tristearoylglycerol in paraffin oil[10] (Figure 6). It has not yet been well established whether these experimental values of φ_c and τ can be used to obtain information on the nature of the stochastic clusters present in these dispersions. For emulsions, the older theories assume that G' is determined by compression of the droplets. The resulting storage modulus is.[11]

$$G = \frac{\sigma + \varepsilon_d}{8a} \cdot \varphi \tag{8}$$

where σ = interfacial tension, ε_d = surface dilational modulus, and a = droplet radius.

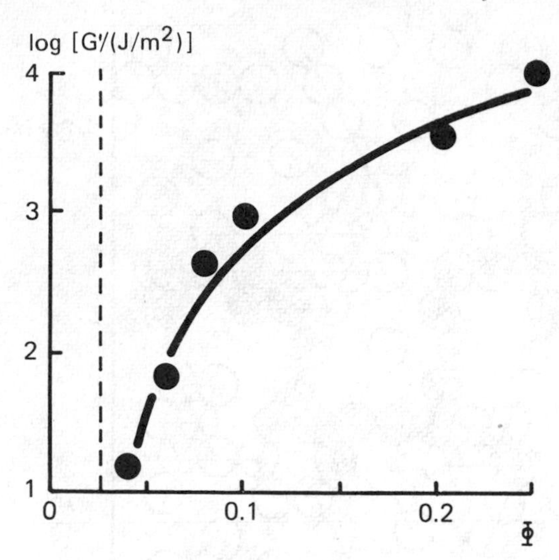

Figure 6 *G′ (φ) for dispersions of tristearoylglycerol in paraffin oil:* (—) *calculated;* (●) *experimental.* $G'_0 = 3.1 \times 10^5$ J m^{-2}; $\varphi_c = 0.025$; $\tau = 2.41$

Different results are obtained for stochastic networks where bending of the droplet chains is assumed. On contact between droplets stabilized by a monomolecular adsorption layer of emulsifier, a thin film will often be formed at an angle φ to the adjoining meniscus. The radius of the film is

$$r_a = a\sin\theta \tag{9}$$

For W/O emulsions, the value of φ has been related to the van der Waals attraction between the droplets:[12]

$$\cos\theta = 1 - \frac{A_H}{24\pi\sigma d^2} \tag{10}$$

where A_H = Hamaker constant, σ = interfacial tension and d = film thickness. However, contact angles between films and their adjoining menisci have also been observed for droplets joined by polymer bridging[13] or droplets that form a liquid crystalline phase at the W/O interface.[14] In these instances, their values are determined by other interaction forces.

To quantify the resistance to bending of a chain of droplets, we consider three droplets (Figure 7). The lines connecting their centres join at an angle α. As a first approximation, the resistance to bending, *i.e.* to a change in α, can be modelled by replacing the circular contact areas by two squares of edge length $2a\sin\theta$, which are wrapped around a cylinder with radius a and joined by two surface layers around the cylinder. On changing the angle α, one of the surface layers will suffer

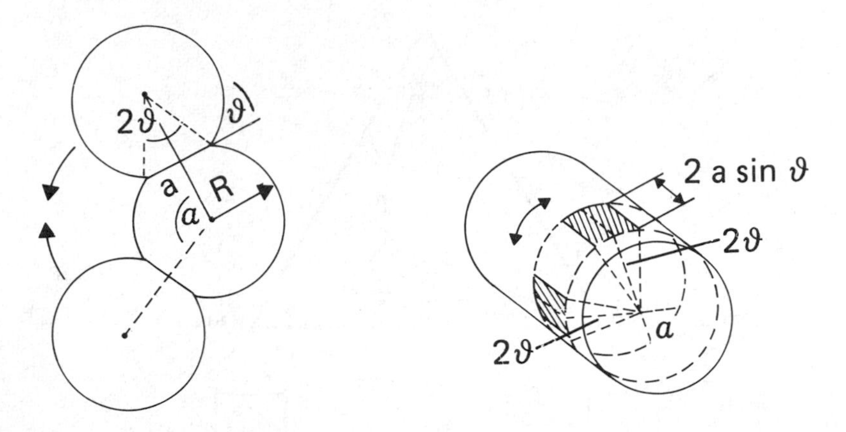

Figure 7 *Bending of droplet chains.* θ = *contact angle between film and meniscus;* a = R*sin*θ

simple elongation and the other simple compression, while the sum of their surface areas remains constant. The microscopic force ΔF resisting displacement of one of the squares through an angle $\mathrm{d}\alpha$ is

$$\Delta F = E_s \cdot 2a\sin\theta \left(\frac{1}{\alpha - 2\theta} + \frac{1}{2\pi - \alpha - 2\theta} \right) \mathrm{d}\alpha \qquad (11)$$

where E_s is the elastic surface modulus. On replacing $\mathrm{d}\alpha$ by the change in length of the droplet chain, $\mathrm{d}x$, we obtain

$$\mathrm{d}F = E_s \cdot \frac{\sin\theta}{\cos\alpha} \left(\frac{1}{\alpha - 2\theta} + \frac{1}{2\pi - \alpha - 2\theta} \right) \mathrm{d}x \qquad (12)$$

Permanent inhibition of sedimentation is possible only if E_s does not vanish with time, *i.e.* if the surface layers behave as elastic strips. For small deformations of such strips:

$$E_s = 3\,\mu'_s \qquad (13)$$

where μ'_s is the surface shear storage modulus. It follows that sedimentation can be arrested only if the O/W interface has a surface shear storage modulus $\mu'_s > 0$. The storage modulus of the stochastic network will then be proportional to μ'_s. To obtain dimensional similarity between G' and the pre-exponential factor in (7), a length parameter has to be introduced. The only linear quantity relevant to our problem is the droplet radius, a, indicating that

$$G'(\varphi) \approx \frac{\mu'_s}{a} (\varphi - \varphi_c)^{\tau} \qquad (14)$$

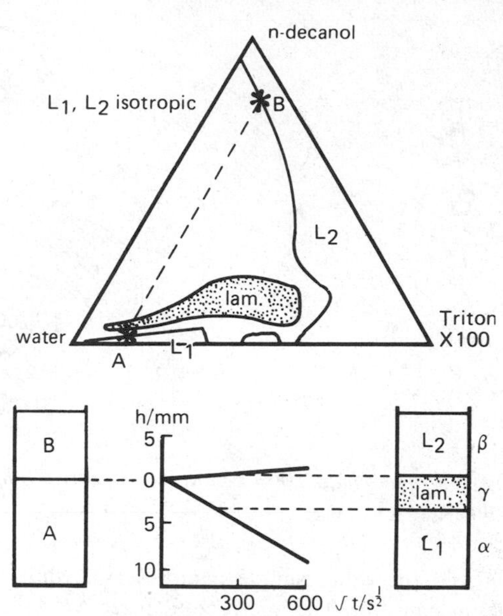

Figure 8 *Physical formation of novel phase*

3 Application to Experimental Data

The values of μ'_s for interfaces at which monolayers of low molecular weight emulsifiers are adsorbed are generally very low. The following interfaces are known to show appreciable surface shear storage moduli:

1. Entangled polymer layers that are adsorbed at an interface.[15] Thus the present theory explains why the presence of an adsorbed polymer at the W/O interface of an emulsion of toluene in water stabilized with Span 60 strongly decreases the rate of sedimentation.[16]
2. O/W interfaces at which a crystalline or liquid crystalline layer has been formed.[17] This situation has been investigated on systems in which a liquid crystalline phase is formed at the W/O interface on contact between the bulk phases. In the system decanol–water–Triton X-100 (Figure 8), a liquid crystalline layer is formed at the decanol–aqueous solution interface. The thickness of the layer increases with time.

The growth is determined by the rate of diffusion of the bulk phase components through the initial liquid crystalline layer so that the thickness of this layer, h, is proportional to the square root of time:

$$h = K\sqrt{t} \tag{15}$$

The shear moduli of the liquid crystalline layers formed can be measured with a double ring surface shear rheometer developed by de Feijter (Figure 9). In this

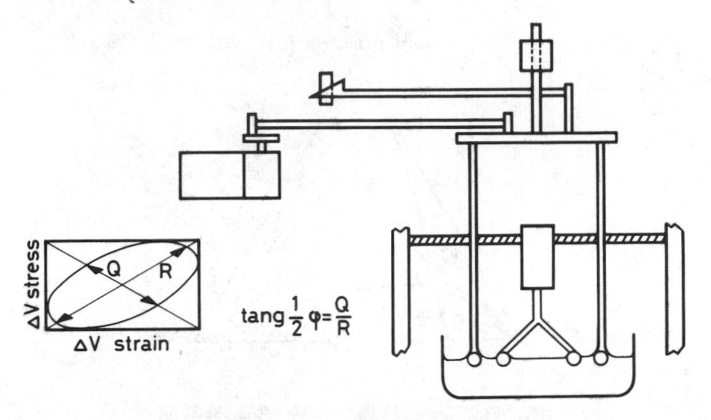

Figure 9 *Surface rheometer.* $\mu_s = \mu'_s + i\mu''_s = |\mu_s|(\cos\varphi + i\sin\varphi)$; μ'_s = *surface shear storage modulus*; μ''_s = *surface shear loss modulus*; φ = *loss angle*

apparatus two rings are placed concentrically at the interface. The outer ring performs a sinusoidal motion and the torque exerted on the stationary inner cylinder is measured. The theory corrects for the hydrodynamic energy dissipation in the adjoining liquid phases so that the rheological properties of the interface layer itself can be evaluated. It is found that

$$\mu'_s = hG' \tag{16}$$

$$\mu''_s = hG'' \tag{17}$$

where G' and G'' are the shear bulk moduli of the liquid crystalline layer and μ''_s is the surface shear loss modulus. Formation of liquid crystalline layers at the interface is also observed for W/O emulsions prepared from a solution of α-monoisostearyl glyceryl ether in paraffin oil and water.[18] From the phase diagram of this system (Figure 10a) it is clear that above a certain initial concentration of emulsifier in the oil phase a liquid crystalline phase will be formed on contact between such a solution and water. As expected on the basis of the above theory, W/O emulsions prepared with sufficiently concentrated solutions of emulsifier in paraffin oil do not show sedimentation. The presence of liquid crystalline material can be readily detected by electron microscopy (Figure 10b).

An analogous phenomenon is observed for our DK-F10 emulsions. On contact between the hot solution of DK-F10 in oil and the cooled aqueous phase, thin shells of emulsifier are deposited on the droplets and form mutual bonds by sintering. Assuming the thickness of these shells to be constant, its value, h, can be calculated from the equation

$$h = \frac{\varphi_E}{3\varphi} \cdot a \tag{18}$$

where φ_E = volume fraction of solid emulsifier in the system. For the systems

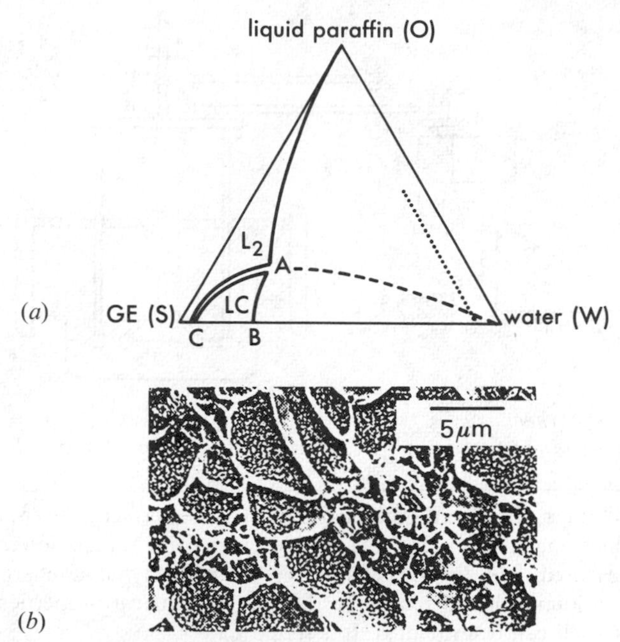

Figure 10 (a) *Ternary phase diagram of* α-*monoisostearyl glyceryl ether (GE)–water–liquid paraffin system at* 25 °C; (b) *scanning electron micrograph of GE emulsion*

(Reproduced with permission from *Yukagaku*, 1987, **36**, 588)

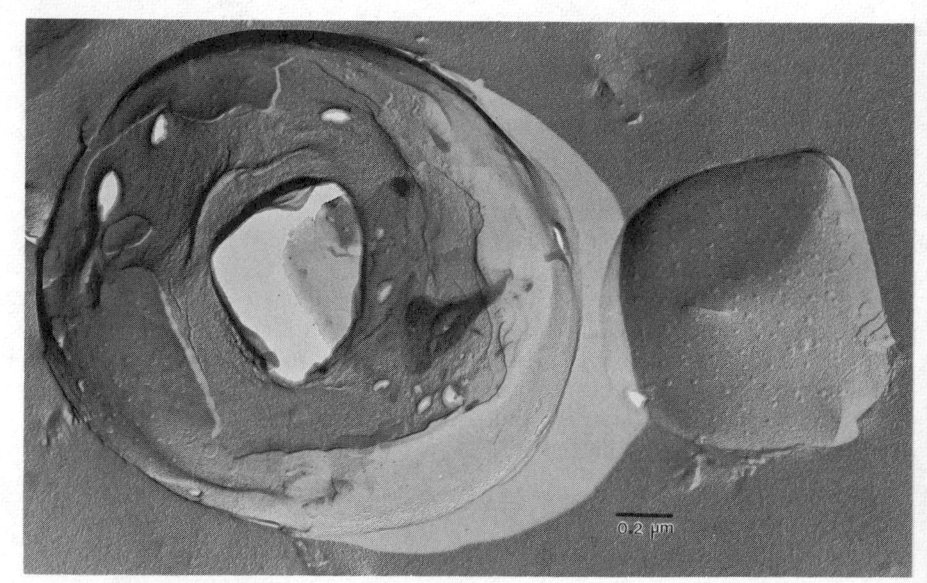

Figure 11 *DK-F*10 *scales on W/O emulsion droplets*

investigated the expected values of h ranged from 0.1 to 0.01 μm. In fact, the shells were easy to observe by electron microscopy (Figure 11).

Hence the difference in sedimentation behaviour between W/O emulsions stabilized with Admul-WOL and those stabilized with DK-F10 can now be explained. The DK-F10 shells around the water droplets sinter together leading to a high μ'_s and a strong resistance to chain bending in the flocculate, whereas Admul-WOL stabilizes W/O emulsions by monolayer adsorption. The value of μ'_s at the W/O interface of the latter emulsions is very low, resulting in a high rate of sedimentation.

4 Conclusions

Sedimentation of flocculated W/O emulsions is determined by the surface shear moduli of the droplet surfaces and by the volume fractions of the dispersed phases.

The theory of stochastic networks as developed by computer simulation offers a promising route to describe the rheological properties of flocculated emulsions.

References

1. R. Buscall, *Colloids Surf.*, 1982, **5**, 269.
2. R. Buscall and L. R. White, *J. Chem. Soc.,, Faraday Trans 1*, 1987, **83**, 873.
3. M. van den Tempel, *J. Colloid Sci.*, 1961, **16**, 281.
4. M. van den Tempel, *J. Colloid Interface Sci.*, 1979, **71**, 18.
5. J. P. Clerq, G. Giraud, J. Roussenq, R. Blanc, J. P. Carton, E. Guyon, H. Ottavi, and D. Stauffer, *Ann. Phys. (Paris)*, 1983, **8**, 5.
6. T. A. Witten and L. M. Sander, *Phys. Rev. Lett.*, 1981, **47**, 1400.
7. W. D. Brown and R. C. Ball, *J. Phys. A*, 1985, **18**, L517.
8. Y. Kantor and I. Webman, *Phys. Rev. Lett.*, 1984, **52**, 1891.
9. W. D. Brown, *Thesis*, Cambridge University, 1987.
10. J. P. M. Papenhuyzen, *Thesis*, Eindhoven University, 1970.
11. J. Kamphuis, *Thesis*, Twente University, 1984.
12. J. Requena, D. F. Billet, and D. A. Haydon, *Proc. R. Soc. London, Ser. A*, 1975, **347**, 141.
13. F. van Voorst Vader and H. Dekker, *J. Colloid Interface Sci.*, 1981, **83**, 377
14. J. A. Sohara, *Dissert ıtion*, Lehigh University, 1986.
15. J. A. de Feijter and J. Benjamins, *J. Colloid Interface Sci.*, 1979, **70**, 375.
16. P. Becher, in 'Interfacial Phenomena in Apolar Media (Surfactant Science Series, Vol. 21),' eds. H. F. Eicke and G. D. Parfitt, Marcel Dekker, New York, 1987, p. 257.
17. H. S. Kielman and P. J. F. van Steen, *J. Phys. (Paris)*, 1979, **40**, C3-447.
18. Y. Suzuki and H. Tsutsumi, *Yukagaku*, 1987, **36**, 588.

Towards a Comprehensive Theory for Sedimentation in Colloidal Suspension

By G. C. Barker and M. J. Grimson

THEORY AND COMPUTATIONAL SCIENCE GROUP, AFRC INSTITUTE OF FOOD RESEARCH, NORWICH LABORATORY, COLNEY LANE, NORWICH NR4 7UA, UK

1 Introduction

There are many important situations in which a two-phase material is separated under the action of an external field, *e.g.* blood purification, food emulsion stability, oil recovery, and industrial waste management. The process of physical interest is described as batch separation, thickening, creaming, or sedimentation and is one of the most common features of design procedures. The simplest observations of sedimentation show, for an initially homogeneous sample, the formation and propagation of one or more menisci. Many experiments are restricted to measurements of the meniscus kinematics but, more recently, non-intrusive X-ray, γ-ray, and ultrasonic techniques have been developed which allow a full determination of the distribution of dispersed phase material at several times. A detailed understanding of these experiments requires a working description of sedimentation at the continuum level which also incorporates some of the colloidal aspects of the process. For a large part of the experiment the interface(s) travel parallel or antiparallel to the external field and apparently without change of form. Finally, there is a single interface at rest separating essentially pure solvent from a concentrated sediment which has an amorphous structure. These observations suggest a one-dimensional solitary wave formulation of sedimentation with all pertinent features being included in a simple non-linear partial differential equation (Burgers' equation) which is derived below. This formulation can be supplemented by computer simulations which model the process at a particulate level.

The detailed dynamics of colloidally sized objects supported in a viscous fluid and placed in an external field have received extensive theoretical study. The microscopic framework for evaluating the mobility tensors and settling velocities, developed by Batchelor,[1,2] is sufficiently complex to limit all results to low-order series expansions in the volume fraction φ of the dispersed phase. Therefore, applications are restricted to dilute systems. Previous continuum models of sedimentation, based on the continuity equation for the dispersed phase volume fraction (*e.g.* references 3 and 4) ignore effects which arise from the particulate nature of the dispersion and depend strongly on numerical or graphical solutions of partial differential equations which require empirical input. Consequently, these models suffer from a lack of generality and are often restricted to one specific application.

We may identify several aspects of real sedimenting colloids which hinder comparisons with rigorous theoretical treatments. (i) Aggregation processes, both reversible and irreversible (coagulation), lead to the formation of time-dependent cluster and size distributions even for inherently monodisperse systems. The formation of aggregates lead to a build-up of interparticle stresses which must be included in the force balance with gravitational and viscous drag forces. The aggregation processes are strongly influenced by the details of the interparticle interactions, including the attractive portions, and it is precisely these details which remain unknown in most instances and which cannot be incorporated in the Batchelor formalism, thereby limiting the range of application for fully self-contained microscopic calculations. (ii) The ability of many colloidal droplets to deform introduces significant changes in the pair distribution of the drops[5] and therefore changes in the effect of interparticle forces, together with unknown modifications of the direct and indirect (hydrodynamic) interparticle interactions themselves. The microscopic framework depends on rigid particles. (iii) The two-component picture is rarely realized in practice. Often the continuous phase contains a dissolved third component (polymer, electrolyte, polysaccharide) and may be inhomogeneous; alternatively, surface-active material may be present or the dispersed objects may be complex (irregular shapes, capsular, *etc*). In all practical instances the dispersed particles have polydisperse sizes and the effects of sedimentation cannot be separated from size segregation effects which occur simultaneously.

We have recently presented some simple expressions for the variation of concentration profile with time in a model colloidal dispersion undergoing sedimentation.[6] The results, which are exact within the model and which can be expressed in terms of a small number of easily interpreted parameters, provide a representation of sedimentation phenomena which facilitates immediate comparison with experiment. Here we shall explore the versatility of this model and, using comparison with simple hard-particle computer simulations, we shall examine the effect of particle properties on the observed wave motions.

We initially consider monodisperse particles sedimenting in an external field of strength g which varies in only one spatial direction z. The force balance equation for a small volume element of suspension can be written as

$$\Delta\rho g\varphi u/v(\varphi) - \Delta\rho g\varphi - (\partial P/\partial\varphi)(\partial\varphi/\partial z) = 0 \quad (1)$$

where we have neglected inertial effects and included interparticle stresses in the form of a particle pressure, P.[7] We have assumed that P is a function of volume fraction only. The second term in this equation represents the effect of the external field on the non-equilibrium matter distribution and the first term arises from the viscous drag on monodisperse particles of excess density $\Delta\rho$ falling with velocity u. The velocity function $v(\varphi)$, which is known accurately only for dilute systems of hard spheres,[2] is the single particle sedimentation velocity in the absence of interparticle stress. We assume that $v(\varphi)$ is a function of φ alone.

Equation (1) leads to a particle flux, $J(\varphi)$, of the general form

$$J(\varphi) = v(\varphi)\varphi + \varepsilon(\varphi)(\partial\varphi/\partial z) \tag{2}$$

In practice, the origin of diffusive flux in equation (2) may be very complex and therefore we cannot be precise about the form of the coefficient $\varepsilon(\varphi)$. The classical theory of Brownian motion, extended to non-equilibrium systems and therefore including both hydrodynamic and direct particle–particle interactions, leads to a generalized force which accounts for diffusive transport in inhomogeneous systems and which therefore makes an important contribution to the second term in equation (2). Also, we note that the particulate environment changes very rapidly as we move through the interface, causing relevant length scales to be shorter. This may invalidate continuum approximations and necessitate a full multi-component microscopic treatment in the region of the menisci.

We present equation (2) as two terms of a generalized 'gradient' expansion for the dispersed phase flux and we regard $v(\varphi)$ and $\varepsilon(\varphi)$ as undetermined coefficients. Specification of the sedimentation problem is completed by the continuity equation for dispersed phase volume fraction:

$$\partial\varphi(z,t)/\partial t + \varepsilon J(\varphi,\partial\varphi/\partial z)/\partial z = 0 \tag{3}$$

combined with an initial distribution, $\varphi(z,0)$, of dispersed phase material and appropriate boundary conditions.

2 Non-linear Wave Solutions

The complex sedimentation problem presented in the Introduction may be reduced, by appropriate choice of the coefficients v and ε, to a non-linear wave equation which has analytical solutions.[6] We first approximate $\varepsilon(\varphi)$ by a constant, ε_o, so that the dominant source of non-linearity in J arises from the external field. Equation (3) becomes

$$\partial\varphi(z,t)/\partial t - [v(\varphi) + \varphi\partial v(\varphi)/\partial\varphi]\partial\varphi/\partial z - \varepsilon_o\partial^2\varphi/\partial z^2 = 0 \tag{4}$$

It is clear that $v(\varphi)$ should reduce to the single-particle velocity in unbounded fluid, u_o (Stokes velocity), as $\varphi \to 0$ and should become zero as φ approaches the maximum packing fraction φ_m. The simplest function satisfying these conditions is the linear velocity function

$$v(\varphi) = u_o(1 - \varphi/\varphi_m) \tag{5}$$

The approximations made above are appropriate to dilute systems and may be regarded as a revealing first approximation to the complex behaviour contained in equation (2). Boundary conditions are chosen such that

$$\varphi(0,t) = \varphi_m$$

$$\varphi(L,t) = 0$$

The model will give the variation of the volume fraction, $\varphi(z,t)$, of a plug of dispersed phase material with length L and known initial distribution located in an infinite tube between close-packed sediment and pure solvent. When $\varepsilon_o \ll 1$ this can be used as an approximation to the behaviour observed in a finite tube where strictly the boundaries are fluxless.

In dimensionless form we may now write the continuity equation for the dispersed phase material as

$$\partial c/\partial t - (1 - 2c)\partial c/\partial z - \varepsilon_o \partial^2 c/\partial z^2 = 0 \tag{6}$$

where the concentration $c(z,t)$ represents the volume fraction expressed in units of φ_m. All lengths are scaled by the length L of the inserted plug of non-equilibrium material (the origin of z is taken as the base of the inserted plug and the external field points along the negative z axis) and t is time scaled by $|u_o|/L$. The small parameter in dimensionless form is $\partial = |\varepsilon/u_o L|$. Equation (6) is Burgers' equation. A particular transformation due to Hopf–Cole linearizes equation (6) and leads to a formal solution:[8]

$$c(z,t) = \int_{-\infty}^{\infty} (y,0)\exp[\mathrm{f}(z,t,y)]\mathrm{d}y \Big/ \int_{-\infty}^{\infty} [\mathrm{f}(z,t,y)]\mathrm{d}y \tag{7}$$

where

$$\mathrm{f}(z,t,y) = -(z - y + t)^2/4\delta t + (1/\delta) \int_{y}^{\infty} c(z',0)\mathrm{d}z' \tag{8}$$

For a piecewise linear initial distribution of dispersed material, the results of integrations (7) and (8) have analytical forms and therefore the solutions of equation (6) may be written as closed expressions.[6] The dominant features of these solutions are travelling steps of concentration or solitary waves. Each wave has a tanh-like shape with width *ca.* δ and travels, without change of form, from the position of a discontinuity in the initial distribution function towards the position of the equilibrium meniscus.

The most prominent solitary waves are associated with the initial boundaries of the inserted material and, for uniform initial distribution with concentration c_0, they travel with constant velocities $c_0 \mathrm{u}_0$ and $(c_0 - 1)u_0$, independent of δ, for a major portion of the sedimentation experiment. When the two approaching waves interfere, in the region of the final step, the velocities show perturbations before becoming zero. For $c_0 = 0.25$ and $\delta_0 = 0.001$ the evolving concentration distribution is shown in Figure 1.

Analytical solutions of the simple equation (6) form a valuable starting point for a discussion of the complex sedimentation behaviour contained in equations (2) and (3) with zero flux boundary conditions. Numerical, non-linear wave-like solutions of this problem can be obtained and understood because of their close association with analytical solitary wave forms.

Experimental evidence[9,10] suggests the Richardson–Zaki form of velocity function given by

$$v(c) = u_0(1 - c)^p \tag{9}$$

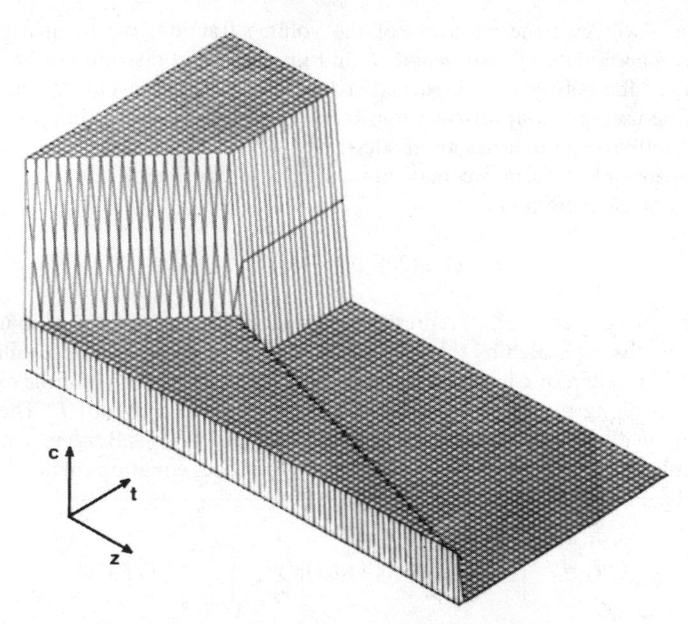

Figure 1 *The time-dependent distribution function of dispersed phase material for a solitary wave model of sedimentation with* v(c) = (1 − c). *The initial uniform concentration is* $c_o = 0.25$ *and* $\delta = 0.001$ $(0 < t < 2)$

where $p = 5$–7 is appropriate for non-flocculated suspensions. For an initially uniform distribution with $c_0 = 0.25$ we have solved equation (4) using equation (9) with $p = 6$ and $\delta = \delta_0 = 0.001$. The time-dependent distribution of dispersed phase material is shown in Figure 2. The sedimentation behaviour can be separated into two regimes. Initially, wavefronts propagate from the fluxless boundaries with constant velocities. The transition region between sediment and uniform suspension is much more diffuse than that between suspension and pure solvent and the sediment does not immediately obtain the maximum packing density at the tube end. This regime shows essentially solitary wave behaviour. After coalescence of the two fronts, the single interface moves with steadily decreasing speed as the whole sediment slowly thickens. The two-stage character arises from the strong asymmetry of the velocity function (9). The slow thickening region corresponds to a region of the velocity curve, at moderate and high volume fractions, in which suspended particle velocities are small and largely insensitive to volume fraction.

Further modifications to the evolving concentration profile are made by incorporating a functional dependence on φ into the generalized diffusion coefficient. A precise functional form for $\delta(\varphi)$ based on rigorous physical arguments is unknown for real sedimenting dispersions. For stable colloidal dispersions the particle pressure may be identified with the particle osmotic pressure and expressed by

$$P \propto (1 - c)^{-1} \tag{10}$$

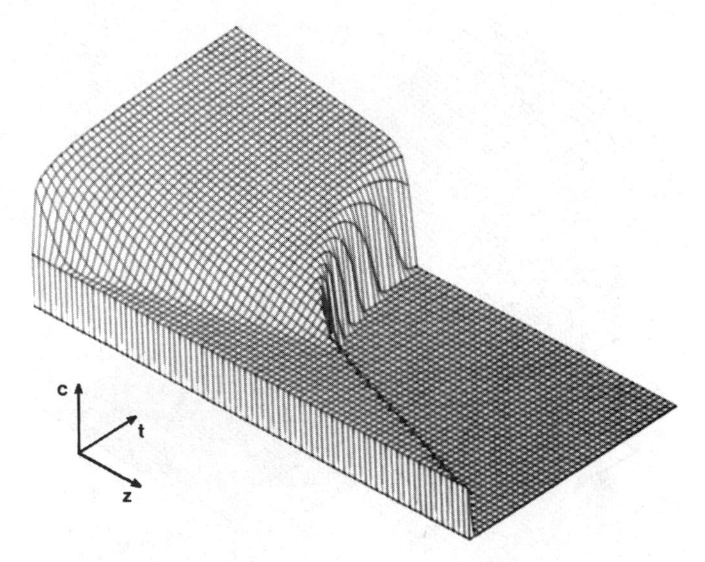

Figure 2 *The time-dependent distribution function of dispersed phase material for a sedimentation model with* $v(c) = (1 - c)^6$ *and* $\delta(c) = 0.001$. *The initial uniform concentration is* $c_o = 0.25$ $(0 < t < 8)$

In this instance, using equation (8), we may write

$$\delta(c) = \delta_0(1 - c)^{p-2} \tag{11}$$

Aggregation effects, particle deformability, and the details of the thermodynamic state all contribute to the form of equation (11). Other constitutive equations have been suggested for $P(\varphi)$ largely through comparison with experimental and computer simulation data. All these functions rise sharply as volume fraction approaches close packing (*e.g.* reference 4), so that equation (11) is sufficient to represent qualitatively the effects introduced by $\delta(\varphi)$.

Using equations (9) and (11) with $\delta_0 = 0.001$ and $p = 6$ we have solved the general equation (3) for an initially uniform distribution of dispersed material with $c_0 = 0.25$ and zero flux boundary conditions. The results, shown in Figure 3, display the two-stage sedimentation behaviour of Figure 2, but associated with the large interparticle stress at high volume fraction there is now an additional 'consolidation' at the bottom of the sediment which appears as a sharply rising section of the concentration profile. This consolidation effect is apparent also in the increased velocity of the final, unified, meniscus. The velocity of the upper meniscus in the first phase of sedimentation is unaltered by the φ dependence introduced in δ. A further decrease in δ causes a conflict between the rising consolidation front and the maximum packing criterion. In this event it is possible for the consolidation front to resemble an additional shock.

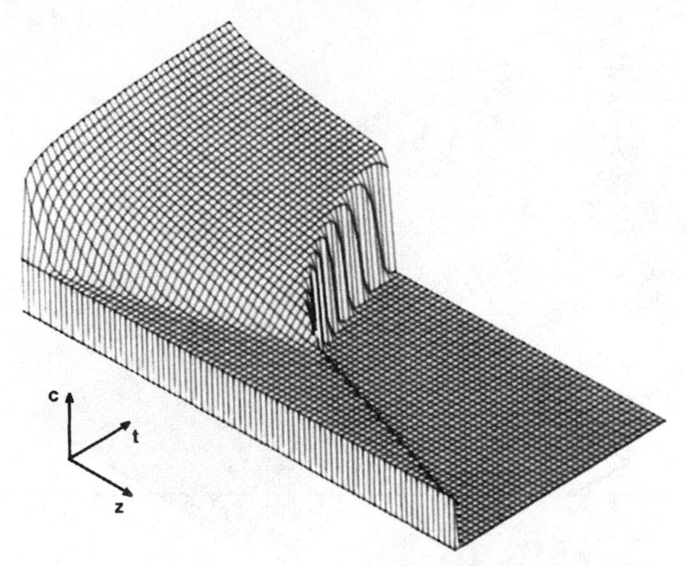

Figure 3 *The time-dependent distribution function of dispersed phase material for a sedimentation model with* $v(c) = (1 - c)^6$ *and* $\delta(c) = 0.001(1 - c)^4$. *The initial uniform concentration is* $c_o = 0.25$ $(0 < t < 8)$

3 Computer Simulations of Unstable Particle Dispersions

Many significant features of sedimentation behaviour are associated with those particles located around sharp steps in particle concentration. Pusey and Tough[11] have pointed out that Batchelor's theory of the sedimentation velocity is only relevant for time scales which are much longer than velocity fluctuation times caused by fluid particle impacts but much smaller than any 'structural relaxation' time. In the vicinity of the interface the particle environment changes rapidly and relaxation times can vary so that the statistical properties of the particle velocities may be a consequence of spatial configuration. Furthermore, collective processes on long time scales cannot be neglected and a continuum approach may be inappropriate. In order to see the role of simple particle exclusions in collections of hard particles acted upon by an external field, we have performed simulations in which the effect of the supporting fluid is reduced to a minimum. The fluid drag appears only as a steady average velocity imparted to isolated particles.

The simulations were performed using a dynamic hard-particle Monte Carlo algorithm. A system of N hard particles was placed in a tube with hard boundaries at the tube ends and periodic boundary conditions in other directions. Uniform initial configurations, of fixed concentration and size distribution, were generated from conventional random sequential adsorption algorithms.

The simulation proceeds by choosing a particle, i, at random and moving it a distance $m_i g\Delta$ in the direction of the gravitational field and a distance $dr\Delta$ in a random direction to simulate the diffusional motion, where r is a random number in the interval $[0,1]$, d the diffusion strength and m_i the relative size of particle i.

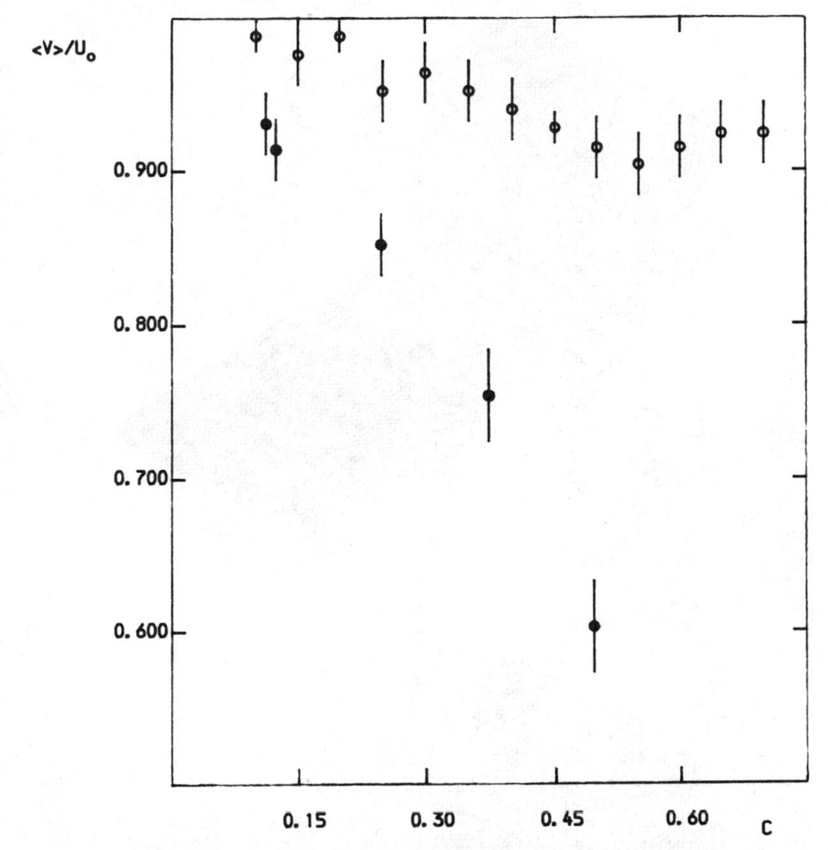

Figure 4 *Plot of the upper meniscus velocity against the initial concentration for hard-particle computer simulations of sedimentation.* (○) *One-dimensional results;* (●) *two-dimensional results*

The Monte Carlo step size Δ is chosen to be sufficiently small to ensure that the sediment at the base of the tube has a volume fraction within 1% of the accepted close-packed value. The random move was accepted with a probability that was inversely proportional to m_i. If the move led to an overlap of two or more particles, it was rejected; otherwise, the move was accepted and the particle coordinates updated. Another particle was then chosen at random. The random move has only a small role in the systematics of sedimentation. Its principal function lies in determining the sedimentation velocities, but as we only measure meniscus velocities with respect to single-particle settling velocities, the magnitude of d is not critical and is chosen to facilitate data collection best.

We have performed simulations in one dimension, using hard rods, and two dimensions, using hard disks, with $N = 1000$. In this simple model the average settling behaviour of collections of hard particles closely follows the evolution

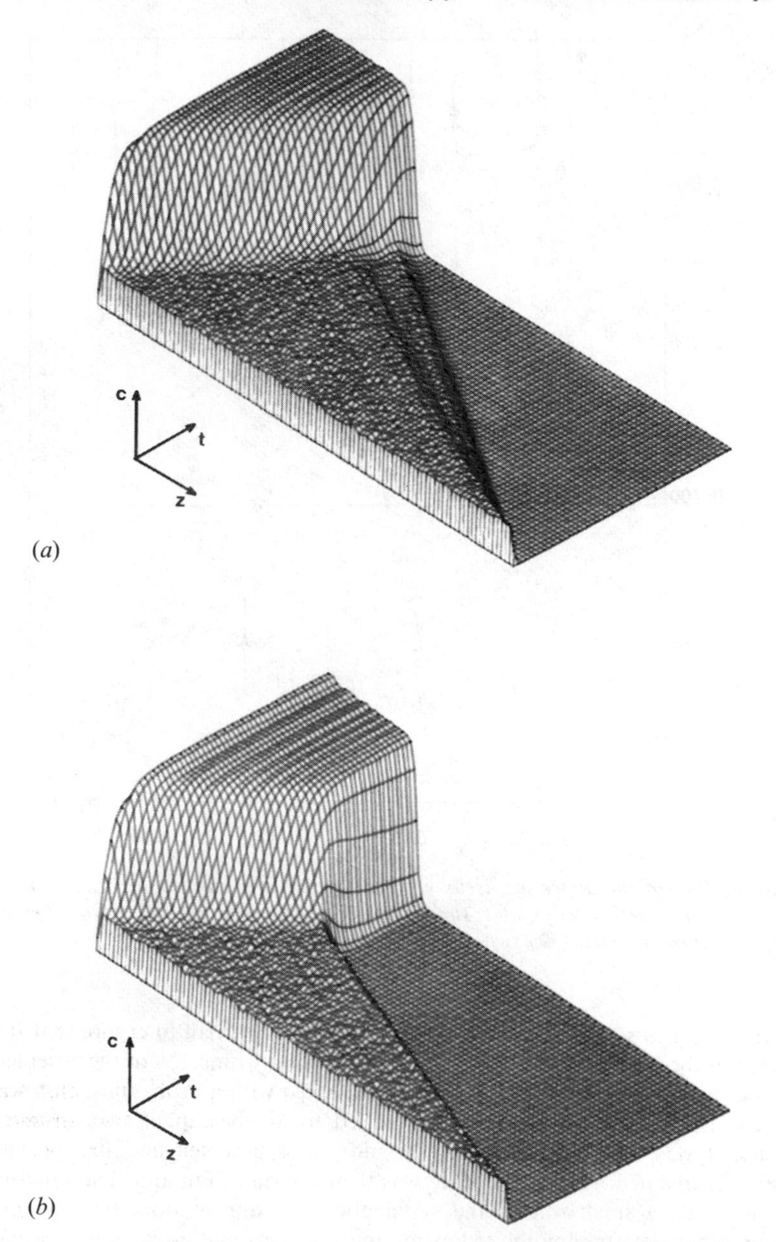

Figure 5 *The average time-dependent distribution function of dispersed particles in a two-dimensional computer simulation of sedimentation. The initial uniform concentration is* $c_o = 0.18$. (a) *A* 50 : 50 *by weight bidisperse system of particles with size ratio* 3 : 4. (b) *Monodisperse system of the larger particles*

of the distribution of dispersed phase material previously obtained from a continuum sedimentation model (*e.g.* Figure 5b). The propagating steps of concentration are typically 5–20 particles wide. We have evaluated the mean meniscus velocity from averages over ten simulations performed with different initial configurations and using independent sequences of pseudo random numbers. The upper meniscus velocity is plotted against the initial, uniform, concentration, c_0, in Figure 4.

For small values of c_0 the upper meniscus falls with a speed very close to the speed, u_0, of a single isolated particle. At larger concentrations the meniscus moves more slowly. The retardation is due to hard-particle exclusion effects and is most prominent for the two-dimensional simulations. The results are easily extended to three dimensions. In real suspensions the particle collisions are screened by the indirect forces. However, the results above indicate that, at moderate and high particle concentrations where hydrodynamics become inefficient, the resistance from particle–particle interactions should be included in evaluations of drag and sedimentation coefficients.

Snapshot pictures of single *N*-particle configurations taken from the two-dimensional simulations show short-range correlations between the particle positions so that the particle volume fraction is not a full expression of the local environment. To investigate this we have performed two-dimensional simulations, as above, using a bidisperse mixture of disks. The radii of the two disk species are in the ratio 3:4 and they are a 50:50 mixture by weight with a total concentration $c_0 = 0.18$. We have plotted the evolution of the average particle distribution function in Figure 5a. In Figure 5b we have plotted, using identical axes, the average distribution function for a monodisperse collection of the large particles which have the same overall concentration. In Figure 5a the most rapidly moving meniscus corresponds to the settling of larger particles in a polydisperse environment with $c_0 = 0.18$. The falling meniscus in Figure 5b corresponds to the same particles falling through a monodisperse environment with $c_0 = 0.18$. Clearly, the presence of the smaller particles gives added resistance to the downward progress of the larger particles. The mean velocities, in units of the speed of an isolated large particle, are 0.56 and 0.82, respectively.

We may explain this retardation in terms of the occupation of voids, which appear below large particles, by lateral moves of smaller particles. Similar moves by larger particles have a high probability of rejection. Another clear demonstration of this effect, *i.e.* the preferential occupation of voids by small particles, is given in the following section.

4 Size Segregation in Particulate Dispersions

The computer experiments discussed in the previous section are single, unidirectional processes representing the unperturbed settling of hard particles in an external field. However, many real sedimenting systems are subject, either constantly or periodically, to external excitation or stimuli in the form of vibrations, stirring, or shaking, *etc.* These highly non-equilibrium processes lead to driving forces which cannot easily be incorporated into force balance equations of the form of equation (1) but whose effects on concentrated sediments of polydisperse

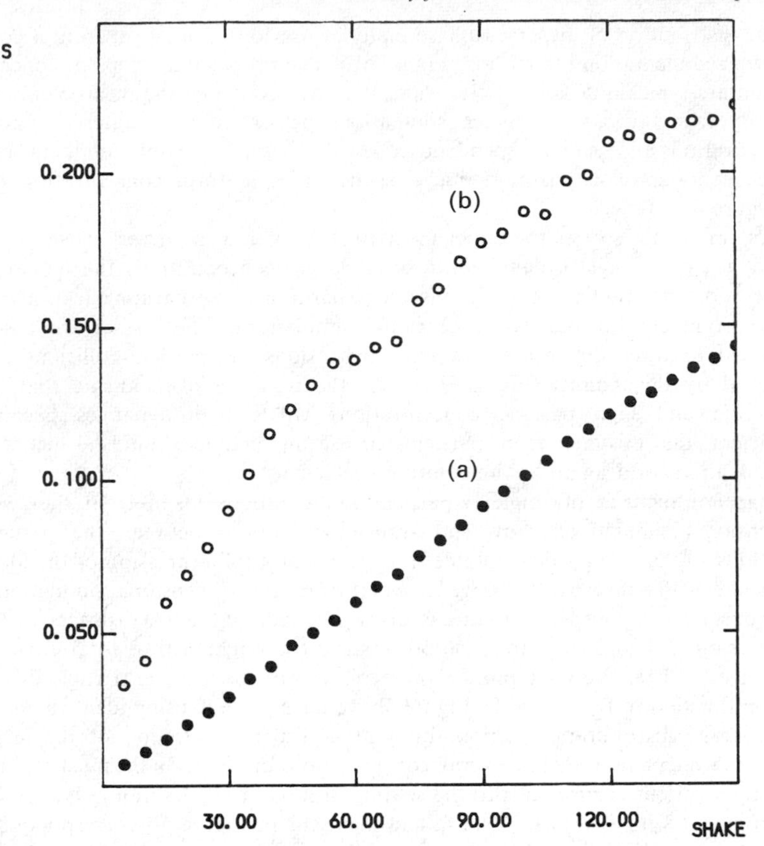

Figure 6 *Separation coefficient* versus *cycle number for a computer simulation of sediment shaking with* (a) *polydisperse disks and* (b) *polydisperse random dimers*

particles can clearly be illustrated by two-dimensional computer simulations.

We have performed athermal shaking simulations for polydisperse systems of hard particles using a modified Monte Carlo procedure.[12] Sediments formed during the simulations in the previous section were repeatedly disturbed or 'shaken.' A single shake consists of two parts. First, the particles are uniformly raised through a distance equal to the diameter of the smallest particle, relative particle positions remaining unaltered. Second, the particles are moved individually by a series of random steps. Steps which increase the system potential energy or which cause particles to overlap are disallowed. A shake is terminated, and a new sediment thereby formed, when the average rate of change of system potential energy falls below 0.1%. Typical shakes represent 2500 Monte Carlo steps per particle. For the small systems we have considered, shaking does not produce noticeable changes in the particle volume fraction of the sediment. After

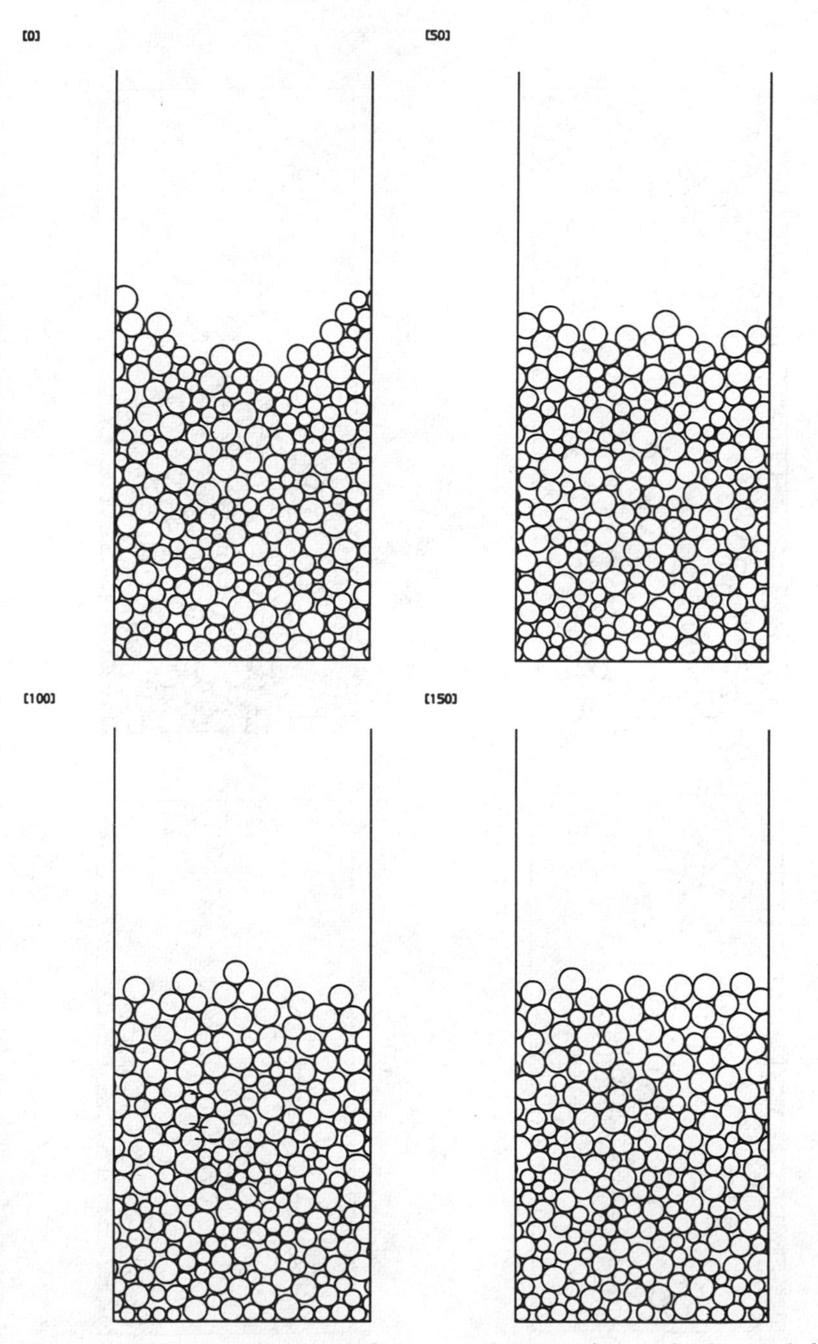

Figure 7 *Typical configurations from a computer simulation of sediment shaking with polydisperse disks after* 0, 50, 100 *and* 150 *cycles*

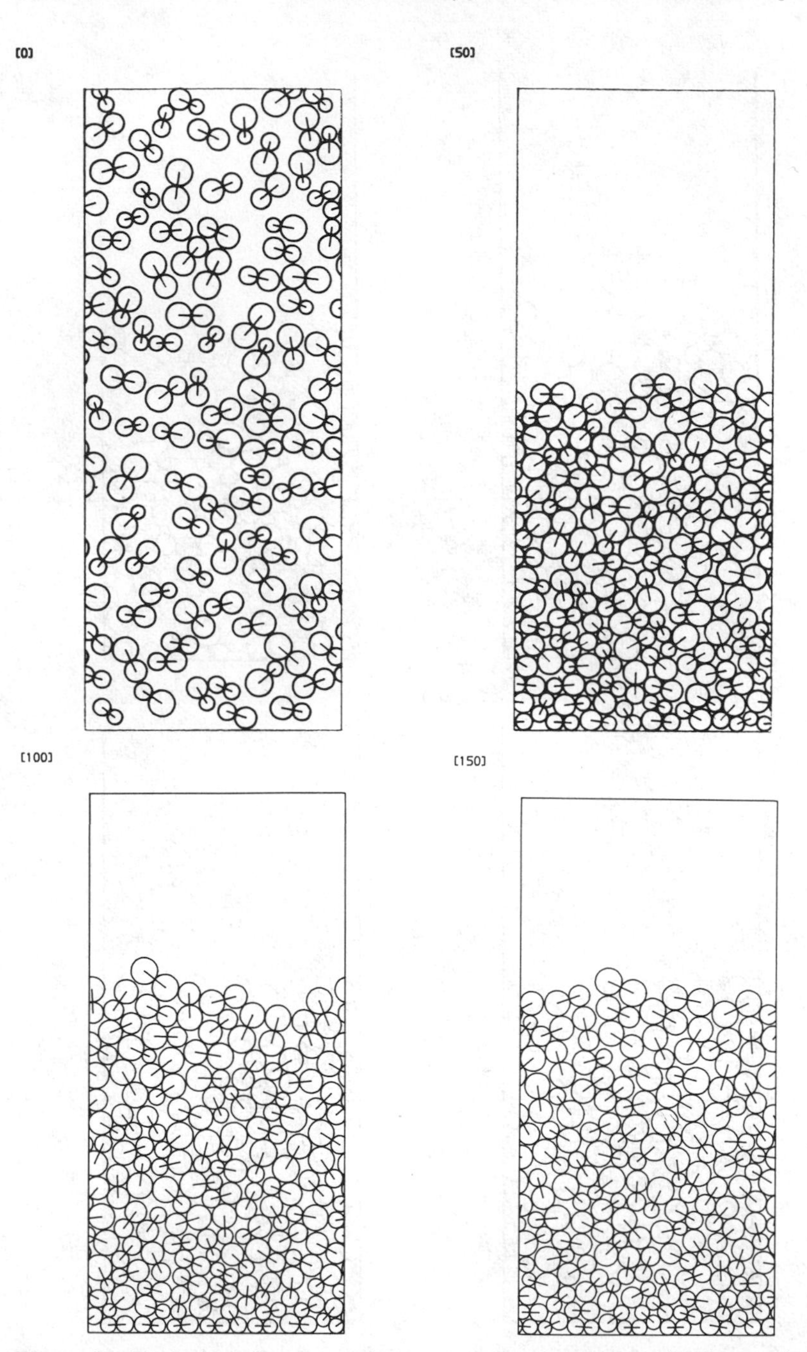

Figure 8 *Typical configurations from a computer simulation of sediment shaking with polydisperse random dimers after* 0, 50, 100 *and* 150 *cycles*

each shake we may measure the size segregation of a polydisperse set of particles by the segregation parameter, s, given by

$$s = \sum[(R_i - R_{\text{min}})z_i] \Big/ \bar{z} \sum_i (R_i - R_{\text{min}}) - 1 \tag{12}$$

where N is the total number of particles, R_i the size of the ith particle which is at a height z_i, R_{min} the minimum disk radius and $\bar{z}$ the mean height of the disks.

To illustrate the size separation which occurs during shaking we have performed simulations using 200 polydisperse hard disks in the form of (a) pure monomers and (b) random dimers. The disk sizes, R_i, are uniformly distributed in the range $R_0 < R_i < 2R_0$ and the strip width is $L = 20R_0$. In case (a) random moves are downward and sideways translations whereas in case (b) there are additionally random rotations around the centre of mass of the dimers. Each simulation was repeated ten times, using independent sets of disks and independent sequences of random numbers, and average results for the evolution of the separation coefficient, s, are presented in Figure 6. Typical configurations, reached after 0, 50, 100, and 150 shakes, are illustrated in Figures 7 and 8 for monomers and dimers, respectively. (Note that the zeroth configuration is a random close-packed configuration for the monomers and a random sequentially adsorbed configuration for the dimers.)

Symmetric and anisotropic particulate systems exhibit clear size gradation after 150 shakes. The initial growth of the separation coefficient is approximately linear in both examples. Rosato *et al.*[12] presented purely geometric arguments to explain size segregation within bidisperse disk systems. Our results for polydisperse and asymmetric particles, in which the initial separation rate is proportional to the size of the 'shaking' displacement from the bottom of the container, confirm this general hypothesis. In the shaking process, voids created beneath large particles are easily filled by the movement of a single small particle, thereby preventing downward progess of the larger species. In contrast, several small particles must move cooperatively to create a void large enough for a large particle to enter, hence making large-particle settling unlikely. Angular correlations of the rising anisotropic particles, which will be discussed fully elsewhere, support the geometric interpretation.

This two-dimensional model of size segregation provides valuable insights into the discussion of vibrationally induced separations of particulate matter. A full description will lead to the formulation of counter intuitive driving forces which must be included in representative force balance equations.

5 Discussion

We cannot give quantitative comparison with experimental data in this paper but we can give support for, and show the versatility of, the general flux in equation (2) by making qualitative assessments of published sedimentation profiles. For the sedimentation of suspensions of large glass beads, gravitational forces dominate interparticle forces and we therefore expect non-consolidatory sedimentation behaviour. Dunand and Soucemarianadin[13] have given experimental results for glass beads (10^{-4}m) suspended in corn syrup solution and in hydroxypropyl guar. For $c_0 = 0.3$ the results show simple solitary wave behaviour in the first case and sediment thickening in the second case, where the resistive forces are increased by the presence of the guar. Similar results can be seen in a variety of experimental measurements, *e.g.* erythrocyte sedimentation[14] and 6×10^{-5}m illite particles in toluene.[15] Carter *et al*[16] have given concentration profiles of sedimenting n-alkanes in water emulsions. In this instance the particles are small (*ca.* 10^{-6}m) and have long-range colloidal interactions so that the particle pressure cannot be ignored. The experimental profiles show consolidation behaviour as seen in Figure 3.

Coupled equations of the form of equation (4) may be solved simultaneously to provide a description of sedimenting polydisperse suspensions which often appear as superpositions of solitary waves.[15] Further extensions of the model may include the examination of sedimentation behaviour in irreversibly flocculated systems. Buscall and White[7] have shown that, if consolidation is not the rate-determining process, the sedimentation particle flux in a flocculated network can be written in the form of equation (2), where $P(\varphi)$ represents a concentration-dependent yield stress. The continuity equation must then be solved with a moving boundary condition.

Sedimentation leads to the build-up of dense, disordered, and stressed aggregates. Under these conditions a continuous description does not give a full picture of the non-equilibrium behaviour. The collective modes of the particulate system have a microstructural dependence which appears as an explicit size dependence for the relative particle motions in polydisperse systems. Voids created beneath large particles are preferentially filled by small particles, which leads to a size-dependent driving force. The effect occurs for mixtures of particles which have comparable sizes and is distinct from 'sifting', which occurs in mixtures of particles which have disparate sizes. Size segregation effects are observed most easily in shaken collections of non-colloidal particles (*e.g.* sand grains, mixed nuts). However, the origin of this process is geometric so that the qualitative features should be independent of particle size, shape, and dimensions.

Acknowledgements. We thank Mr. B. Wright for encoding the initial differential equation programs and we acknowledge support from the Ministry of Agriculture, Fisheries, and Food.

References

1. G. K. Batchelor, *J. Fluid. Mech.*, 1976, **74**, 1.
2. G. K. Batchelor, *J. Fluid. Mech.*, 1982, **119**, 379.
3. G. J. Kynch, *Trans. Faraday Soc.*, 1952, **48**, 166.
4. F. Concha and M. C. Bustos, *AIChE J.*, 1987, **33**, 312.
5. G. C. Barker and M. J. Grimson, *Mol. Phys.*, 1987, **62**, 269.
6. G. C. Barker and M. J. Grimson, *J. Phys. A*, 1987, **20**, 305.
7. R. Buscall, J. R. White, *J. Chem. Soc., Faraday Trans. 1*, 1987, **83**, 873.
8. G. B. Whitam, 'Linear and Non-linear Waves', Wiley, New York, 1974.
9. R. H. Davis and K. H. Birdsell, *AIChE J.* 1988, **34**, 123.
10. R. Buscall, J. W. Goodwin, R. H. Ottewill, and T. F. Tadros, *J. Colloid Interface Sci.*, 1982, **85**, 78.

11. P. N. Pusey and R. J. Tough, *J. Phys. A.*, 1982, **15**, 1291.
12. A. Rosato, K. J. Strandburg, F. Prinz, and R. H. Swendson, *Phys. Rev. Lett.*, 1987, **58**, 1038.
13. A. Dunand and A. Soucemarianadin, in 'Proceedings, 60th Annual Technical Conference and Exhibition of the Society of Petroleum Engineers, Las Vegas,' 1985, SPE 14259.
14. C. D. Hill, A. Bedford, and D. S. Drumheller, *J. Appl. Mech.* 1980, **7**, 261.
15. Y. T. Shih, D. Gidaspow, and D. T. Wasan, *Colloids Surf.*, 1986, **21**, 393.
16. C. Carter, D. J. Hibberd, A. M. Howe, A. R. Mackie, M. M. Robins, personal communication (1988).

Mechanisms of Fracture in Meat and Meat Products

By Peter Purslow

AFRC INSTITUTE OF FOOD RESEARCH, BRISTOL LABORATORY, LANGFORD, BRISTOL BS18 7DY, UK

1 Introduction

Fracture is simply the breakdown or separation of a material or structure into two or more parts. The fracture behaviour of a food is an important aspect of its mechanical properties in two ways:

1. When we chew food, it is broken down (*i.e.* fractured) in the mouth. The ease or difficulty with which this breakdown can be achieved forms the basis of our sensory perception of some textural qualities of the food, such as toughness.
2. Many processing operations on foods involve fracture. For example, the meat used as the raw material for restructured meat products is broken into relatively large pieces (pre-broken), and then cut into flakes. Wheat is fractured in the milling process to produce flour, and the manufacture of potato crisps involves slicing the potatoes. In many of these processes, the way in which the raw material breaks or cuts affects the quality of the size-reduced piece and hence end-product quality.

Given the importance of the fracture properties of food to their textural quality and processing characteristics, it is important that we understand the basis of these properties at a cause-and-effect, non-empirical level so as to be able to control and manipulate food texture and processing operations to the best advantage.

In this paper, some of the basic concepts of fracture and fracture mechanics will be outlined, to show in general terms how we go about quantifying fracture behaviour, and also some of the general mechanisms of fracture that need to be recognized when looking at the properties of composite foodstuffs. Applications of these concepts will then be demonstrated by considering the very different fracture behaviour of cooked meat, frozen meat, and meat products.

2 Measurement of Fracture Properties

The simplest and most obvious measurements that can be taken to quantify the point of fracture of a material are shown in Figure 1. Application of increasing strains to a sample of material results in a concomitant increase in stress. At some

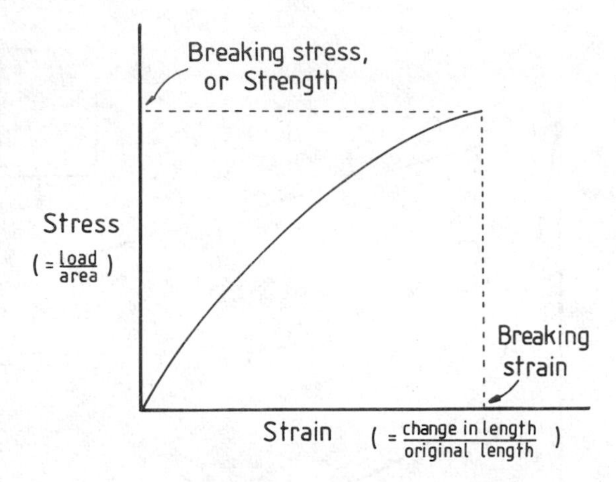

Figure 1 *Definition of breaking stress and breaking strain*

point the material will break, and the breaking stress (*i.e.* strength) and breaking strain can be taken as measurements of the fracture properties of the materials. Although strength is a conceptually adequate measurement, it may not in practice give us a complete picture of the fracture behaviour of materials. Some materials are very sensitive to the presence of even small defects or cracks, with the result that the material breaks at much lower than expected loads. For example, we know from everyday experience that the easiest way to break a sheet of glass is to mark it with a small scratch; it thus will break along the scratch line, at a lower load. Cream cracker biscuits also demonstrate this; hair-line cracks in a biscuit that are too small to be easily seen nevertheless can reduce its strength to a point where the loads normally exerted by a knife spreading on butter are too great, and the biscuit crumbles in the hand. Such materials are described as notch sensitive.[1]

How do we tell if materials are notch sensitive? Figure 2 demonstrates the variation in strength of a specimen with increasing defect size for both completely notch-sensitive and notch-insensitive cases. By comparison of the behaviour of any material under investigation with these two boundary cases, the degree of its notch sensitivity can be assessed. With a completely notch-insensitive material, the breaking stress falls away linearly with increasing crack length a, expressed as a proportion of the overall width of the specimen W, strictly in proportion with the remaining amount of load-bearing cross-section. In a notch-sensitive material, the presence of a notch or crack causes a stress concentration at the tip of the crack which causes fracture to occur at applied stresses which are very much lower than that required to break an unnotched specimen. This stress concentration is a function of the size of the crack, and this results in the fracture strength decreasing rapidly with increasing crack length for even very small cracks.

As a first approximation, we can say that if a material is completely notch insensitive, then strength is a fair measurement of its fracture properties; if there are small cracks or defects in the material, they would have to be noticeably large before the apparent strength of the material had fallen appreciably. However,

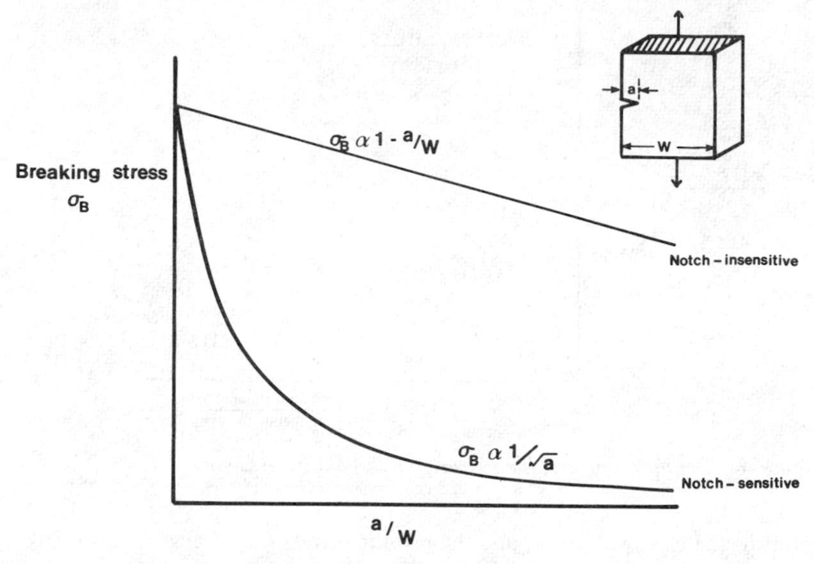

Figure 2 *Diagram showing notch-insensitive (top line) and notch-sensitive (bottom line) behaviour. Breaking stress,* i.e. *maximum load/nominal cross-sectional area (shaded area), is plotted against crack length* (a)/*specimen width* (W)

with notch-sensitive materials, the presence of even microscopic cracks may decrease the apparent strength by several orders of magnitude, which means in practice that strength will not be a unique value characteristic of the material. It is therefore necessary to adopt a different approach to quantify the resistance of notch-sensitive materials to fracture, which uses the simplest of concepts from fracture mechanics.[2-4]

3 Fracture Toughness

Fracture mechanics define the toughness of a material as the resistance provided by the material to the propagation of cracks through it. We can simply define the energy required to propagate fracture in the following way:

$$\text{Toughness, } R = \frac{\text{energy used in propagation } (U)}{\text{area through which fracture progagated } (A)}$$

and, in the limit of infinitesimal extensions of the crack propagation area (A), $R = \mathrm{d}U/\mathrm{d}A$, *i.e.* the rate of energy consumption with respect to crack growth.[4]

Figure 3 shows a basic practical method of measuring R. We take a test specimen of an elastic material with some length of 'starter' crack, a, and load it up obtaining the load-deflection curve 0B which describes its behaviour until the crack starts to propagate, at B. Crack propagation by an amount Δa occurs at B

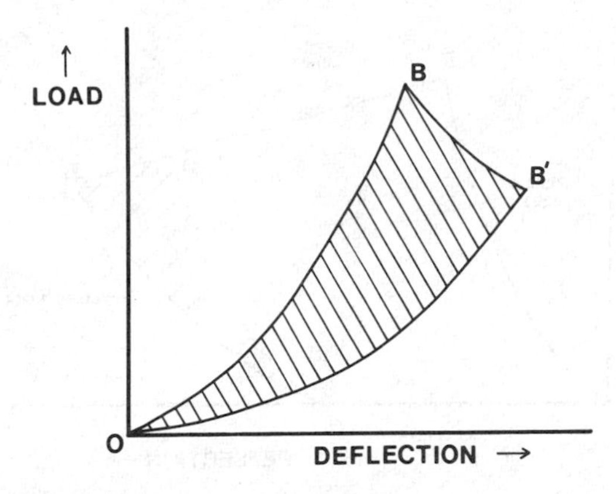

Figure 3 *Load-deflection diagram for elastic fracture*

to B′; B′0 gives the load-deflection curve when the specimen is unloaded. The shaded area 0BB′ represents the energy used in propagating the crack by Δa. The area of crack propagation is Δa times the thickness of the specimen, and R can now simply be calculated as energy used/area of propagation.

In elastic fracture, this energy required to propagate a crack (R) is supplied by the release of strain energy elastically stored in the body of the specimen; a specimen with a longer crack can store less strain energy for a given extension than the same specimen with a shorter crack. For *slow* crack propagation, where the strain energy release is just enough to feed crack propagation, $R = \mathrm{d}U/\mathrm{d}A = J_c$, the critical strain energy release rate (J_c is usually reserved for non-linear elastic materials, with the symbol G_c being used for the critical strain energy release rate of linear elastic materials). J_c(or G_c) is ideally a property characteristic of that material, whose value should be independent of the testing situation.

4 Elastic–Plastic Fracture

In elastic fracture, all the strain energy put into a loaded specimen can be released again on unloading the specimen, or can be used up irreversibly to drive crack propagation. All the energy dissipation is localized at the crack tip and no other dissipation of energy occurs. However, some materials may also deform plastically, where energy is dissipated in yielding. Irreversible dissipation of energy can also occur generally, at all points within a specimen, by viscous-type processes in viscoelastic materials[5] (most hydrated biological materials are noticeably viscoelastic). Materials that deform and fracture plastically do not have unloading curves which go back to the origin of the load *versus* deformation curve; there is some irreversible deformation, or 'permanent set.'

In materials that undergo an appreciable large-scale yielding during fracture, the total energy required to propagate fracture clearly has two components: the energy dissipated plastically and the work intrinsically used in propagating the fracture, *i.e.* $\mathrm{d}U_{\text{total}}/\mathrm{d}A = R_f + R_p$, where R_f is the (intrinsic) fracture toughness

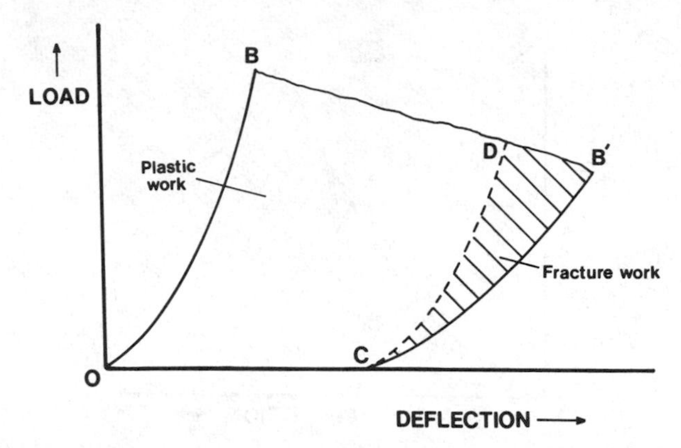

Figure 4 *Load-deflection diagram for elastic–plastic fracture, showing partitioning of energy into plastic and fracture work components*

and R_p the plastic work. In practical terms, we can 'partition' the total energy required to fracture such a material into plastic work and fracture work by use of the approach shown schematically in Figure 4. A specimen with a shorter crack is loaded (line 0B) until propagation starts at B and continues to B′. After propagation of the fracture by a known amount, the material is unloaded (line B′C). When unloaded, *i.e.* at zero load, there is some permanent extension of the material, represented by the distance 0C. Atkins and Mai[4] argue that, if a line parallel to the original loading line 0B is drawn through the point C (*i.e.* the constructed line CD), then the shaded area B′CD again represents the (elastic) energy used in fracture, in a manner analogous to 0BB′ in Figure 3, and that the area 0BDC represents the plastic work done. This graphical approach therefore allows the total work to be partitioned into R_f and R_p, the fracture and plastic work components.

These very basic considerations of strength, notch sensitivity, and elastic and elastic–plastic fracture toughness should serve as a minimum requirement to show how the fracture properties of foods can be quantified. Before going on to consider examples of the application of these concepts, it is pertinent to discuss some of the structural processes or mechanisms that can occur during the fracture of a material, in order that we can recognize these mechanisms when looking at the fracture behaviour of foods.

5 Fracture Mechanisms

The most obvious structural 'event' during fracture is the splitting apart of two planes of atoms so as to create two new surfaces. In brittle homogeneous materials such as glass, fracture does indeed result in two clean, plane surfaces, and the energy required to make such surfaces is merely the free surface energy of the material. This is relatively small (1–10 J m^{-2} for glass[6]), and so these materials are

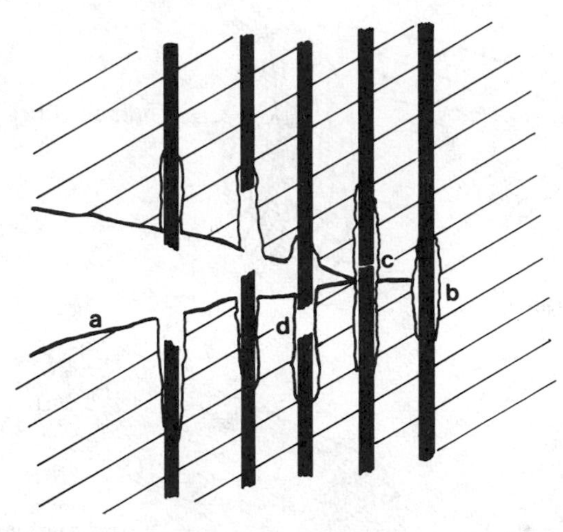

Figure 5 *Fracture mechanisms in a fibrous composite.* (a) *Matrix cracking;* (b) *fibre debonding;* (c) *fibre cracking;* (d) *fibre pull-out*

very easy to fracture. Tougher biological materials, such as wood and skin, require approximately $10^4\,\mathrm{J\,m^{-2}}$ to fracture.[6,7] We are therefore looking for structural changes or processes which occur along or near the path of the fracture which will irreversibly absorb these much higher quantities of energy, so making the material tougher.

In composite materials, the interface between the two phases provides opportunities for energy to be absorbed by processes such as separation of the interface ('debonding') and slippage of one phase past another.[8-10] If the dispersed phase is finely divided and well distributed in the continuous phase, these mechanisms provide a capacity for a great deal of energy dissipation because the total area of interface available is very large. Figure 5 depicts schematically such fracture mechanisms in a fibrous composite. Propagation of a crack through such a composite involves (a) simple cracking of the matrix material, followed by (b) debonding of the fibre–matrix interface as the crack runs into the proximity of a fibre. This is then followed by (c) the fracture of the debonded fibre, which would occur at any position along its debonded length. Finally, there is further dissipation of energy as the debonded and fractured fibre is pulled out of the matrix (d). The energies involved in debonding and fibre pull-out can be high in conventional fibre composites.[8] Such mechanisms may be revealed by close examination of the surfaces produced by fracture; if there is a great deal of fibre pull-out, for example, fractured surfaces will contain a jagged mass of separated and sticking-out broken fibre ends.

Having covered some of the basic concepts of fracture and some of the structural mechanisms generally involved, let us now consider the fracture properties of meat to see some examples of these points.

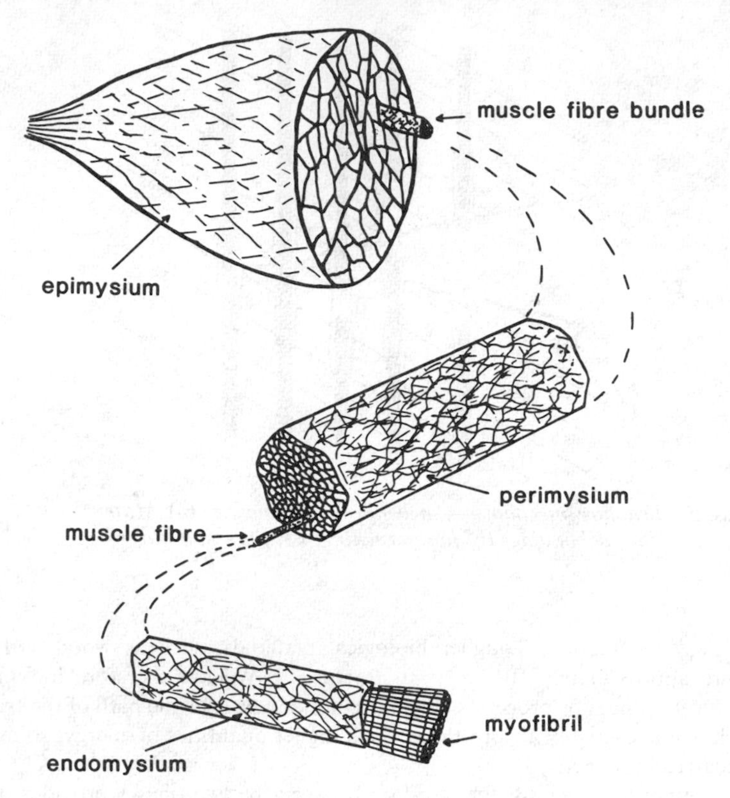

Figure 6 *The general structure of meat*
(Reproduced with permission from 'Advances in Meat Research', Vol. 4, Van Nostrand Reinhold, New York, 1987)

6 Meat Structure

The general structure of meat is outlined in Figure 6. It is a fibrous composite material with hierarchical organization. Each individual muscle fibre is *ca.* 10–100 µm in diameter and contains approximately 1000 myofibrils, the contractile organelles of the cell.[11] Surrounding each muscle fibre is a connective tissue 'phase,' the endomysium, itself a fibrous composite structure. Individual muscle fibres are aggregated into fibre bundles, 1–10 mm in diameter, which are enveloped in their own connective tissue phase, the perimysium. The whole muscle is made up of many such bundles of muscle fibres, and is ensheathed in another connective tissue, the epimysium (a fuller description of the structure of muscle and its components can be found elsewhere[12]). The structure of meat is therefore complex with many levels of organization and many interfaces between these different sub-structures, so providing a large number of possible structural mechanisms that could conceivably be involved in the fracture process.

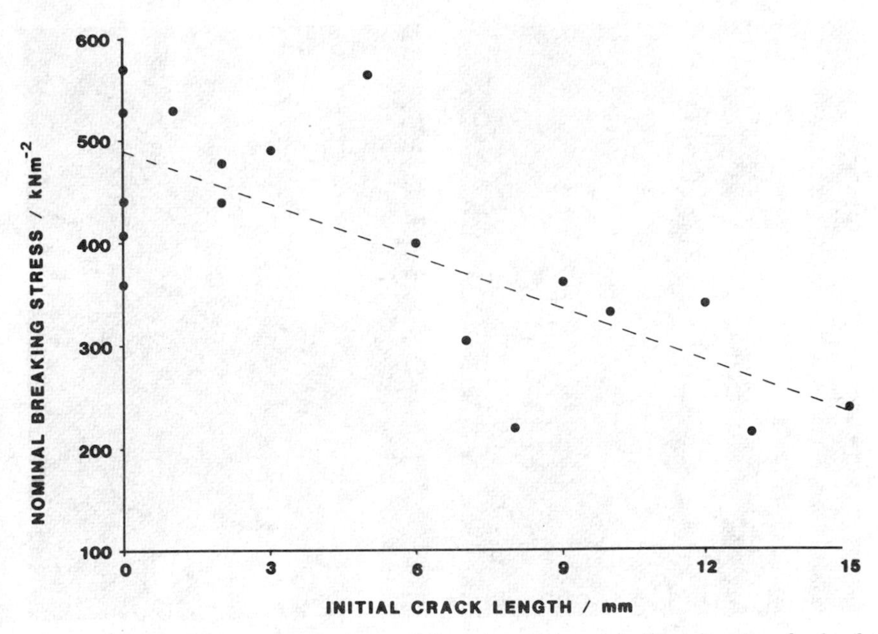

Figure 7 *Breaking stress* versus *crack length across muscle fibre direction for beef semitendinosus cooked to* 80 °C *and pulled along the fibre direction. The dotted line is the theoretical fit for complete notch-insensitivity*
(Reproduced with permission from *Meat Sci.*, 1985, **12**, 39)

7 Fracture of Cooked Meat

The study of the fracture behaviour of meat is perhaps most important in relation to cooked meat, as meat is cooked before we eat it. In looking at the fracture of cooked meat, the aim is to understand which of the many sub-structures of the material outlined above are principally involved in the fracture process, as these will be the determinants of the textural qualities of the material.

If we reconsider our basic fracture concepts, perhaps the first question we should ask is how we can quantify the fracture properties of cooked meat—is the material notch sensitive? Figure 7 shows the result of an experiment to assess the notch sensitivity of beef semitendinosus muscle cooked to 80 °C, by measuring the breaking stress of strips of muscle pulled along the muscle fibre direction as a function of the length of notches cut into one edge of the specimen.[13] The breaking stress of unnotched specimens was approximately 500 kPa. With an edge-notch of 15 mm (which corresponded to half the width of the 30 mm wide samples used), the breaking stress had fallen to approximately 250 kPa, *i.e.* the breaking stress was halved when notches were cut half-way across the specimen width. This is in agreement with completely notch-insensitive behaviour, and therefore strength is a reasonably adequate quantity to use to describe the fracture properties of cooked meat.

(*a*) (*b*)

Figure 8 (a) *Initial fracture across the muscle fibre direction;* (b) *subsequent fracture of separated bundles of muscle fibres.*
(Reproduced with permission from *Meat Sci.*, 1985, **12**, 39)

Measurements of the tensile strength along and across the muscle fibre direction show cooked meat to be highly anisotropic; the strength of (unnotched) longitudinal slices of cooked beef semitendinosus muscle pulled along the muscle fibre direction was found to be 300–400 kPa, whereas the strength of the same material pulled perpendicular to the fibre direction was only 20–30 kPa, *i.e.* an order of magnitude lower.[13] It has been shown that fracturing the cooked meat by pulling across the muscle fibre direction involves the separation of bundles of muscle fibres, by debonding of perimysial connective tissue from the endomysium on the surface of the muscle fibres at the periphery of the bundles.[13,14] Fracture of the material by pulling along the fibre direction will obviously have to involve fracture of all the fibre bundles and fibres across the specimen in order for the specimen to be separated into two parts, and the strength results show that this is far more difficult to achieve than lateral separation of the bundles.

Figure 8a shows the appearance of a pre-notched piece of cooked meat being pulled along the muscle direction as fracture has just begun. The first event is the debonding of interfaces between the muscle bundles at right-angles to the pre-cut notch, *i.e.* along the muscle fibre direction. Figure 8b shows the final appearance of the specimen after it has completely fractured. Following the separation of the material into debonded, independent bundles, on further extension each bundle snaps cleanly across as a single unit. Closer examination[14] of the ends of these bundles shows that there is little or no debonding or pull-out of individual fibres within each bundle, and the fracture surface across the whole bundle is fairly smooth. This suggests that fracture of the bundles is dominated by the stress required to fracture the myofibrillar material within the fibres. Any perimysial sheaths bridging the gap between broken ends of the fibre bundles then have to be fractured.

These simple tests have therefore identified the debonding of muscle fibre bundles from their surrounding perimysial connective tissue and from each other, the fracture of myofibrils within muscle fibres, and the fracture of debonded perimysial strands as the dominant structural mechanisms in the fracture behaviour of cooked meat. These observations are important, as they allow us to focus our attention on these structures as determinants of meat toughness in a much more precise way.

8 Fracture of Frozen Meat

A knowledge of the fracture properties of frozen meat is important in relation to meat processing; most reformed steaks manufactured in the UK are produced from material that has been comminuted in a semi-frozen state.

The work of Dobraszczk *et al.*[15] has shown that frozen meat has a very different fracture behaviour to that of cooked meat. In the frozen state, meat is more notch sensitive, and so an energy-based measurement of fracture toughness is more appropriate here.

In general, the work required to fracture frozen meat shows a strong dependence on temperature. Figure 9 shows data on the toughness (work to fracture) of frozen meat as a function of temperature, from a variety of test configurations and in two orientations, along and across the fibre direction. In all instances there is a very obvious peak in the energy required to fracture the material in the region of -10 to -15 °C, and the anisotropy of toughness is not so marked as in cooked meat. The fracture of frozen meat in this temperatue range is in fact elastic–plastic, and the approach outlined above of partitioning the total energy consumed in fracture into plastic work and fracture work components yields the result shown in Figure 10, for cracks travelling along the muscle fibre direction (the same form of result is also seen for an elastic–plastic analysis of work to fracture across the muscle fibre direction). As Figure 10 shows, the total work to fracture is dominated by the plastic work term, with the fracture work component being relatively small and almost constant with temperature. It is a peak in the plastic work at intermediate temperatures which is responsible for the peaks in total work shown in Figure 9. At -30 °C and below, the fracture behaviour of frozen meat is then a brittle one. Scanning electron microscopy of the fractured surfaces[15] reveals that cracks can travel in an almost smooth path straight across the muscle fibres and fibre bundles at this temperature. At higher temperatures (-20 °C and above), there is a transition in fracture behaviour to a plastic type of fracture, and scanning electron microscopy shows a great deal of separation and pull-out of individual fibres.[15] The decrease in plastic work above -15 °C or so does not reflect a decrease in the plastic behaviour of the material, rather that it just takes less energy to deform plastically at higher temperatures.

Dobraszcak *et al.*[15] concluded that the properties of frozen meat are, not

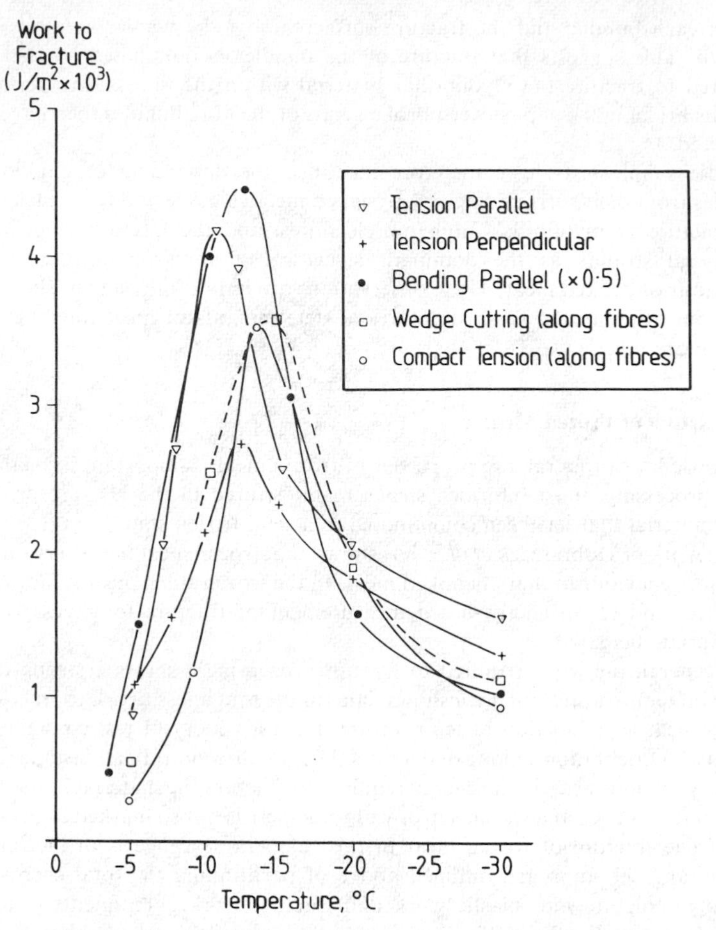

Figure 9 *Work to fracture* versus *temperature of frozen meat for several test geometries*
(Redrawn with permission from *Meat Sci.*, 1987, **21**, 25)

surprisingly, dominated by those of ice, and that the softening and plasticizing of frozen meat in the temperature region of −10 to −15 °C is analogous with peaks in irreversible energy dissipation in ice found by other workers at −10 to −13 °C.[16–18] Ice undergoes viscoelastic deformation at these temperatures, owing to the ability of ice crystals to slide past each other as a result of the formation of unfrozen water at grain boundaries.[16] The columnar ice crystals that occur in meat frozen at commercially relevant (*i.e.* slow) rates of cooling are anisotropic[19] and align along the muscle fibre direction. The anisotropy in toughness along and across the muscle fibre direction seen in Figure 9 is consistent with this anisotropy in columnar ice properties.[15]

An extension of this approach indicates how the elastic to plastic transition in

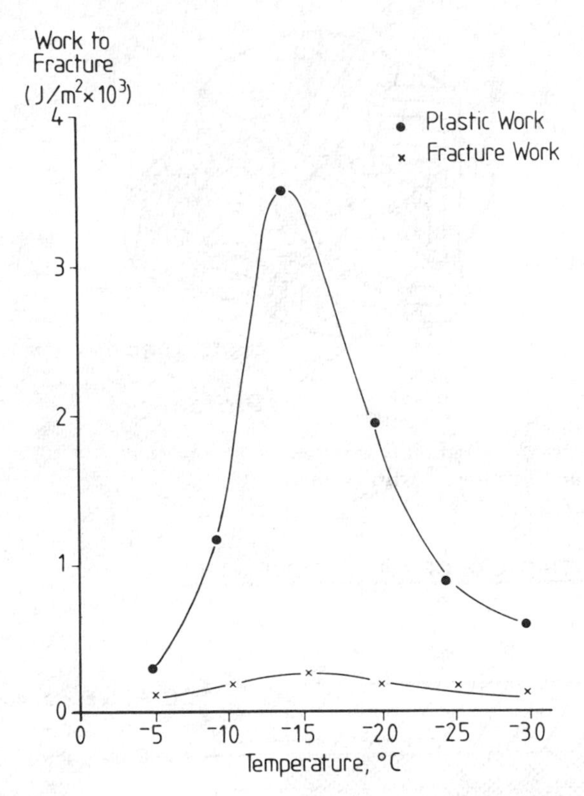

Figure 10 *Plastic work and fracture work components of work to fracture* versus *temperature for crack propagation along the fibre direction in frozen meat*
(Reproduced with permission from *Meat Sci.*, 1987, **21**, 25)

the fracture behaviour of frozen meat occurs as a function not just of temperature but also of rate of fracture and the size of the specimen.[15] All of these are relevant if we wish to understand the response of the frozen material to the fracture conditions that may occur in a variety of industrial comminution processes. It is important to reach such an understanding if we want to control and manipulate the cutting process on anything more than an empirical, trial-and-error basis.

9 Fracture of Adhesive Junctions in Meat Products

As a final example of the application of some basic concepts to food fracture, let us briefly consider adhesive fracture in reformed meats.

Figure 11 is a schematic diagram of a reformed meat product, which consists of pieces of meat in a matrix of myofibrillar proteins (mainly myosin) that has been extracted from the meat pieces themselves by the action of added salts and mechanical agitation. These extracted proteins then heat-set on cooking to form a gel that 'binds' the meat pieces together into a coherent solid. The way in which

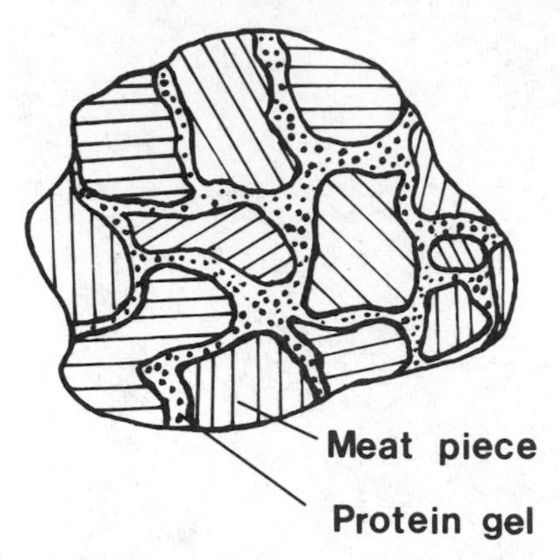

Figure 11 *Schematic diagram of a reformed meat product*
(From Jolley and Purslow,[25] with permission)

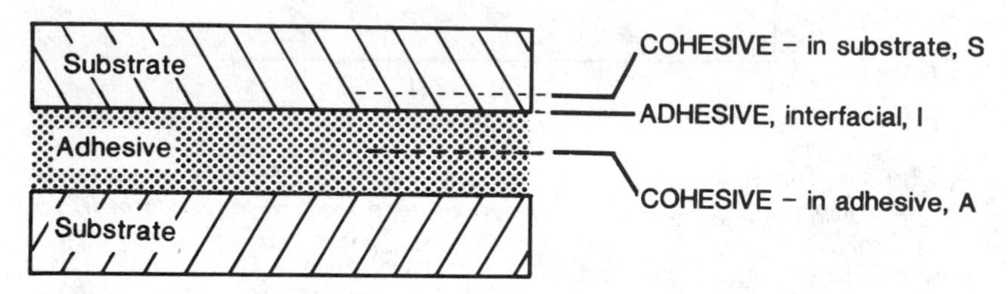

Figure 12 *Schematic diagram of an adhesive junction showing three possible fracture modes*

the meat–gel–meat adhesive junctions then come apart in the mouth as the product is chewed will play a major role in our sensory appreciation of the eating quality of the product, and hence the fracture properties of these junctions are important to understand.

At the simplest level, an adhesive joint can fracture by three different mechanisms (three modes of fracture),[19] as outlined in Figure 12, which shows schematically a junction between two pieces of substrate (meat pieces) stuck with an adhesive (protein gel). The junction can fail by (a) the gel failing cohesively, *i.e.* fracture occurs totally within the gel phase (cohesive-in-adhesive failure); (b) fracture running totally within the meat piece (cohesive-in-substrate failure); or (c) true interfacial or adhesive failure—the junction between meat and gel separates. There will be a characteristic energy required to propagate each type of failure mode (designated A, S, and I, respectively) and these can be expected to be

(*a*)

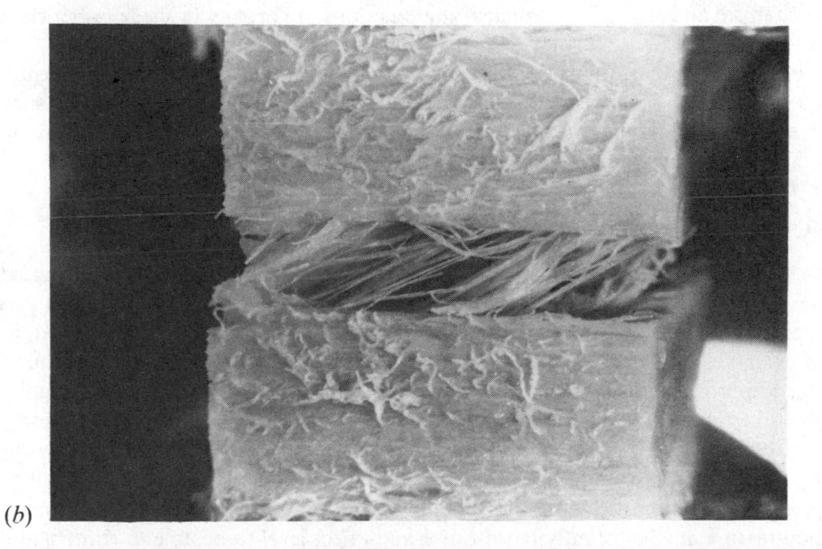

(*b*)

Figure 13 *Fracture behaviour of meat–myosin gel–meat junctions subject to tensile forces.* (a) *A junction with muscle fibres in both meat pieces perpendicular to the junction.* (b) *Junction with muscle fibres in both pieces parallel to the junction plane*

(From Purslow *et al.*,[20] with permission)

different from each other. The mode of fracture that requires the least energy will naturally be the one that is most likely to occur.

Work in our laboratory[20] has demonstrated that different modes of fracture of meat–myosin gel–meat junctions occur according to the orientation of muscle fibres in the surface of the meat pieces. Figure 13a shows a junction, made by cooking two pieces of beef semitendinosus muscle stuck together with a purified myosin solution, on the point of being pulled apart in tension. The muscle fibres in both meat pieces are running vertically, at right-angles to the horizontal junction plane. The assembly is failing at the glue line, and our best evidence to date is that the myosin gel is largely remaining intact so that the fracture mechanism is one of interfacial failure between gel and meat piece. However, if, as in Figure 13b, the muscle fibres in both meat pieces run parallel to the (horizontal) plane of the junction, the material fails differently; cohesive-in-(meat) substrate failure occurs. Fibres are being peeled off the surface of one or both meat pieces. This difference in failure mechanism is reflected in very different tensile strengths of the two types of junctions. The type of junction shown in Figure 13a has an average tensile strength of 60 kPa, whereas the type shown in Figure 13b has an average tensile strength of 21 kPa.[20] The strength of the gel and gel–meat interface is probably the same in Figures 13a and b, but in Figure 13b these strengths are greater than the lateral cohesive strength of the meat, so that this junction fails by the latter mechanism (*i.e.* the weakest mode). Broadly, the effect of orientation on fracture mechanism and strength is similar if the junctions are broken in shear rather than in tension.[21]

It is hoped that this brief example shows how we are gaining a greater understanding of the basis of fracture properties in reformed meats. Again, such an understanding is important if we want to 'design' meat products with optimum textural characteristics.

10 Conclusions

This paper has reviewed briefly some of the more basic concepts of fracture that are useful in looking at the properties of a wide range of materials, and has attempted to demonstrate the application of these concepts to one food, meat, as an example. This is not meant to imply, however, that this fracture mechanics-based approach is limited in usefulness only to meat; such an approach has great general applicability, and has already been applied to other foods such as cheese[22] and vegetables.[23,24] A wider application is expected in future because it allows us to understand more about why foods fracture as they do, as well as enabling us to quantify fracture more satisfactorily. This is important if we want to understand food texture at a sufficiently basic cause-and-effect level to be able to control and manipulate it to best advantage.

References

1. A. Kelly, 'Strong Solids' Clarendon Press, Oxford, 1966.
2. J. F. Knott, 'Fundamentals of Fracture Mechanics,' Butterworths, London, 1973.
3. D. P. Isherwood, in 'Food Structure—its Creation and Evaluation,' eds. J. M. V. Blanshard and J. R. Mitchell, Butterworths, London, 1988, p. 93.

4. A. G. Atkins and Y. W. Mai, 'Elastic and Plastic Fracture,' Ellis Horwood, Chichester, 1985.
5. J. G. Williams, 'Fracture Mechanics of Polymers,' Ellis Horwood, Chichester, 1984.
6. J. E. Gordon, 'Structures,' Penguin, Harmondsworth, 1978.
7. P. P. Purslow, *J. Mater. Sci.*, 1983, **18**, 3591.
8. A. Kelly, *Proc. R. Soc. London, Ser. A*, 1970, **319**, 95.
9. J. K. Wells and P. W. R. Beaumont, *J. Mater. Sci.*, 1982, **17**, 397.
10. B. Harris, in 'The Mechanical Properties of Biological Materials,' eds J. F. V. Vincent and J. D. Currey, Cambridge University Press, Cambridge, 1980, p. 37.
11. C. R. Bagshaw, 'Muscle Contraction,' Chapman and Hall, London, 1982.
12. H. Schmalbruch, 'Skeletal Muscle,' Springer, Berlin, 1985.
13. P. P. Purslow, *Meat Sci.*, 1985, **12**, 39.
14. P. P. Purslow, in 'Advances in Meat Research,' eds A. M. Pearson, T. R. Dutson, and A. J. Bailey, Vol. 4, Van Nostrand Reinhold, New York, 1987, p. 187.
15. B. J. Dobraszczk, A. G. Atkins, G. Jeronimidis, and P. P. Purslow, *Meat Sci.*, 1987, **21**, 25.
16. P. Barnes, D. Tabor, and J. F. C. Walker, *Proc. R. Soc. London, Ser. A*, 1971, **324**, 127.
17. R. Vassoille, J. Perez, and J. Tatibouet, *Cryo-Lett.*, 1984, **5**, 393.
18. P. S. Belton, R. R. Jackson, and K. J. Packer, *Biochim. Biophys. Acta*, 1972, **286**, 16.
19. A. J. Kinloch and S. J. Shaw, in 'Developments in Adhesives—2', ed. A. J. Kinloch, Elsevier Applied Science, Barking, 1981, p. 83.
20. P. P. Purslow, S. M. Donnelly, and A. W. J. Savage, *Meat Sci.*, 1987, **19**, 227.
21. S. M. Donnelly and P. P. Purslow, *Meat Sci.*, 1987, **21**, 145.
22. M. L. Green, R. J. Marshall, and B. E. Brooker, *J. Texture Stud.*, 1985, **16**, 351.
23. A. G. Atkins and J. F. V. Vincent, *J. Mater. Sci. Lett.*, 1984, **3**, 310.
24. D. Schoorn and J. E. Holt, *J. Texture Stud.*, 1983, **14**, 61.
25. P. D. Jolley and P. P. Purslow, in 'Food Structure—its Creation and Evaluation,' eds J. M. V. Blanshard andJ. R. Mitchell, Butterworths, London, 1988, p. 231.

Continuous Sausage Processing

By Larry L. Borchert

OSCAR MAYER FOODS CORPORATION, MADISON, WI, USA

Sausage making traditionally has been an art, and manufacturing procedures have been developed largely on an empirical basis. The knowledge and processes thus developed have, in the past, made possible the production of sausages of excellent quality under desirable conditions. However, when complications such as fat or water separation occurred during processing, resulting in poor texture, an inadequate understanding of the factors involved made it difficult to select the proper corrective measures. Additionally, the traditional processes were highly labour intensive, resulting in relatively expensive and, because of excessive handling, short-lived products. Hence, it became desirable to initiate, in the late 1950s, a comprehensive industrial research programme designed to overcome these problems. The programme integrated many facets of science (including, for the first time, a study of emulsified sausage ultrastructure), engineering, and industrial technology. After more than 5 years of research and development, Oscar Mayer Foods perfected a unique, high-speed, continuous process for manufacturing wieners.

The aim of this paper is not to give a comprehensive review of all that has been written on the topic of sausage emulsions, water- and fat-holding, but rather to discuss the concepts which are important for understanding the continuous sausage process. The fundamental topics have been reviewed previously.[1-3]

The discussion will be limited to highly comminuted sausages, which include primarily wieners (also known in the USA as frankfurters and hot dogs) and bologna. For our purposes they all contain approximately 55% moisture (45% from meat and 10% added), 30% meat fat, 11% meat protein, 2.5% salt, spices, and curing agents. Traditionally, wieners and bologna were made by bringing lean meat, fat meat, water, salt, spices, and curing agents together in a large, rotating, metal bowl where high-speed rotating knives cut and mixed the ingredients into a highly comminuted batter. The batter was then extruded into a smoke- and moisture-permeable casing, linked into sausages of the proper length, smoked, cooked, cooled, and packaged for sale.

The Oscar Mayer continuous wiener process (Figure 1) emulates all the traditional individual steps, but dramatically reduces the amount of time and labour required to produce wieners. The process converts the raw meat ingredients into finished wieners, packaged in a vacuum-sealed film, at a rate of 37 800 wieners (1700 kg) per hour. In a period of 45 min, the batched meat ingredients

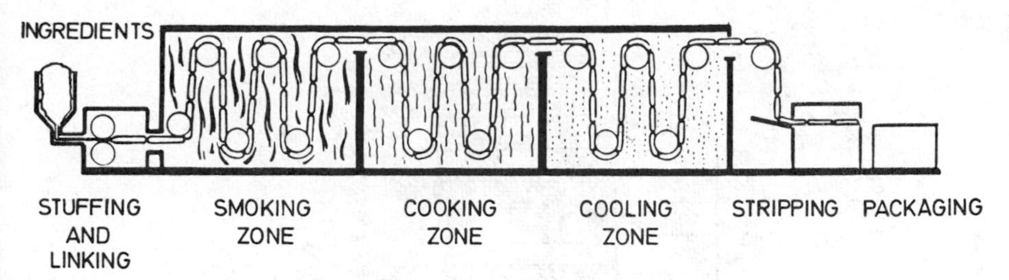

Figure 1 *Oscar Mayer Foods Corp. continuous wiener process*

are emulsified, stuffed into protective cellulose casings, smoked, cooked, chilled, removed from casings, and packaged.

Prior to the development of this process, the individual steps and various handling procedures in manufacturing required a minimum of 9 h. Now the whole process is automated and the product is untouched by human hands, ensuring a low labour intensity and a long refrigerated shelf-life. Even packaging is handled mechanically in the process. After the cellulose casings have been stripped away, the 'skinless' wieners are conveyed automatically to the vacuum-sealing packaging system. Hence the fresh flavour of the wieners, a most perishable quality in all processed meats, is captured only minutes after emerging from the last phase of the continuous process.

The continuous processing technique is one of the most significant developments in the history of the meat industry, because it is now possible to achieve the highest degree of quality control, ensuring the ultimate benefit to consumers. All of the manufacturing operations are easily controlled to produce consistently a mildly flavoured product of uniform quality. A modification of the continuous wiener process now exists for manufacturing long (2 m), large diameter (10 cm) slicing sausages, such as bologna.

Hansen[4] began the scientific understanding of sausage emulsions and their stability by microscopically examining hundreds of samples representing modifications to the emulsifying technique. His study supported the development of the Mark V emulsifier.[5] The device is powered by a 75 h.p. electric motor which drives a rotor, containing 120 crescent-shaped knives, within the vacuumized stator (Figure 2). The meat mixture is pumped into the emulsifier, where it is vacuumized and comminuted, leaving the process as a fine, very stable meat emulsion. The stability of the emulsion permits great latitude in the physical abuse it can tolerate and the variety of raw materials which can be used.

Subsequent to Hansen's work we have learned that a proper understanding of the physical properties of comminuted meat systems cannot be obtained without a knowledge of the microstructure of the material. Most methods available for determining emulsifying and water-holding capacity of the systems are empirical and the results depend on the experimental conditions used. If these methods are used without an understanding of microstructure, the results and their interpretation may be confusing. For example, good fat-holding does not necessarily imply that all the fat has been emulsified. It can simply be entrapped in voids or it can be

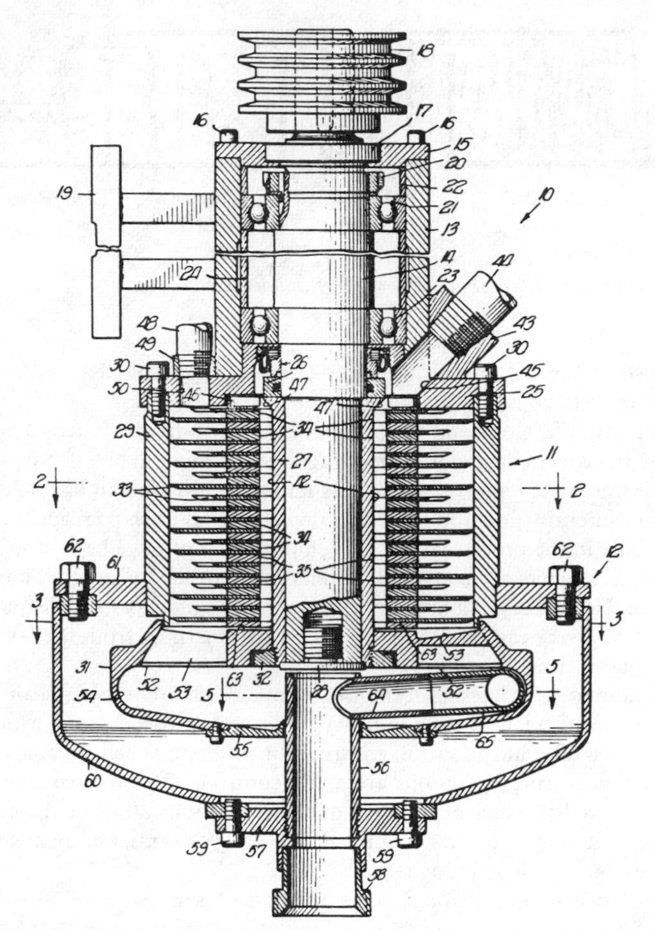

Figure 2 *Continuous chopper design*

in an intact fat cell and still make an important contribution to the structure of the product. Even if the fat is truly emulsified, it is sometimes difficult to determine whether it is the properties of the emulsifier at the interface, the properties of the surrounding network structure, or both that determine the fat-holding properties of a product.[3]

Numerous workers,[4,6-9] using light microscopy, have shown that highly comminuted meat products appear structurally similar to classical oil-in-water emulsions. In the formation of the meat emulsion, the animal fat forms the discontinuous phase, and the water, protein, and salt represent the continuous matrix. Although some of these researchers[4,6,7] have shown that a dense membranous layer surrounds each fat globule, the low resolution of light microscopy limited the examination of the membrane and the visibility of small fat globules.

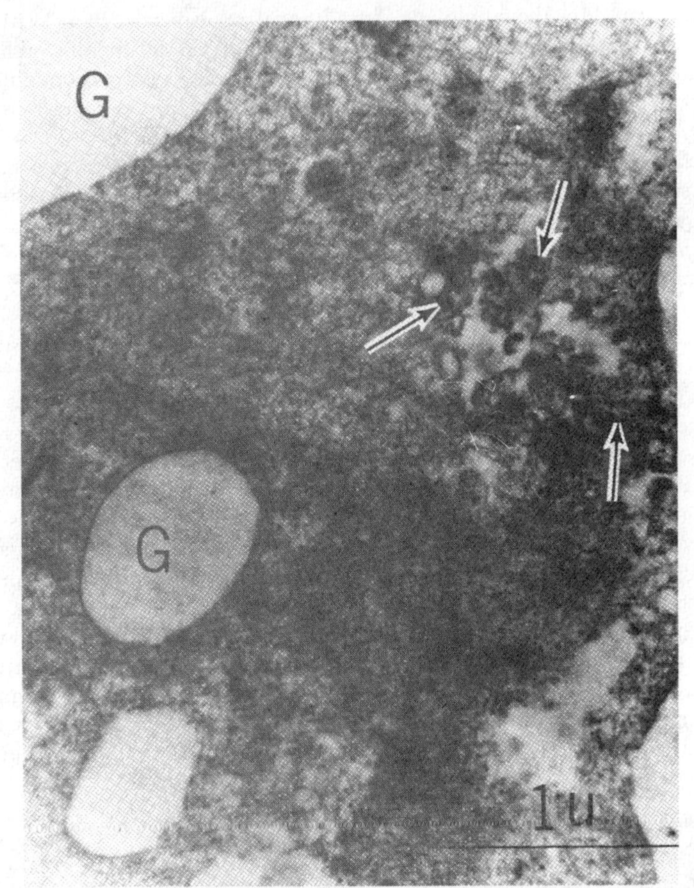

Figure 3 *An electron micrograph of an uncooked meat emulsion. G indicates fat globule surrounded by dense membrane. Arrows designate muscle cell organelles*
(Reproduced with permission from *J. Food Sci.*, 1967, **32**, 419)

Likewise, published reports have not clearly shown the effect of heat processing on the protein matrix and the globule membrane of the meat emulsion.

Borchert *et al.*[10] were the first to demonstrate, using electron microscopy, the presence of an interfacial film around the fat globules, some as small as 0.1 μm, before heat treatment (Figure 3). After heat treatment they found not only that the membrane surrounding the fat globules had been disrupted but that the previously dense protein phase had also been disrupted, as a result of coagulation, into dense, irregular zones. This was subsequently confirmed using scanning electron microscopy.[11,12]

The right to use the term 'emulsion' in conjunction with highly comminuted meat systems has been challenged because even though a proteinaceous interfacial membrane envelops the fat globule,[10] traditional emulsifiers such as lecithin and mono- and diglycerides cause a dramatic decrease in fat-holding ability.[9,13] Challenges have also come from those who have studied systems comminuted in traditional bowl choppers, where undisturbed clusters of normal fat cells are often seen.[14,15]

True emulsions are defined as two immiscible liquids where one liquid is dispersed as droplets in the other liquid and stabilized by an emulsifier or a combination of emulsifier(s) and stabilizer(s). In wieners, where fat is dispersed in a continuous matrix, fat-holding properties are influenced by more factors than the interfacial film; the whole structure has to be taken into account.[2,3] Stokes law, for example, in which the rate of emulsion separation is inversely proportional to the viscosity of the continuous phase and directly proportional to the diameter of the fat globules, prevails in the uncooked wiener. If the matrix is solid, the fat can be effectively entrapped in the network structure regardless of the presence of an interfacial film. If the fat is solid, the degree of crystallinity and the type of crystals formed will be of importance to the behaviour of the dispersed fat phase. Finally, it has to be pointed out that solid or semi-solid food systems are rarely just two-phase systems but rather multi-phase systems. Changes in the stability of one phase may determine the instability of the whole system and fat instability may not necessarily have arisen from changes in the immediate environment of the dispersed fat.

Our knowledge of meat-based emulsions is meagre when compared with what is known about many food systems. The fundamental work on meat emulsions described here was undertaken to assist in the development of a commercial continuous sausage-making process. These findings provided a foundation for the system and a base upon which a continuing flow of new information is being built.

References

1. J. Schut, in 'Food Emulsions,' ed. S. Friberg, Marcel Dekker, New York, 1976, p. 385.
2. J. C. Acton, G. R. Ziegler, and D. L. Burge, Jr., in 'Critical Reviews in Food Science and Nutrition, Vol. 18, CRC Press, Boca Raton, Florida, 1983, p. 99.
3. A.-M. Hermansson, in 'Functional Properties of Food Macromolecules,' eds. J. R. Mitchell and D. A. Ledward, Elsevier, Amsterdam, New York, 1986, p. 273.
4. L. J. Hansen, *Food Technol.*, 1960, **14**, 565.
5. E. Schmook, Jr. and A. H. Vedvik, *US Pat.*, 3, 108, 626, 1963.
6. C. E. Swift, C. Lockett, and A. J. Fryar, *Food Technol.*, 1961, **15**, 468.
7. R. L. Helmer and R. L. Saffle, *Food Technol.*, 1963, **17**, 115.
8. J. A. Carpenter and R. L. Saffle, *J. Food Sci.*, 1964, **29**, 774.
9. J. A. Meyer, W. L. Brown, N. E. Giltner, and J. R. Guinn, *Food Technol.*, 1964, **18**, 138.
10. L. L. Borchert, M. L. Greaser, J. C. Bard, R. G. Cassens, and E. J. Briskey, *J. Food Sci.*, 1967, **32**, 419.
11. D. M. Theno and G. R. Schmidt, *J. Food Sci.*, **43**, 845.
12. K. W. Jones and R. W. Mandigo, *J. Food Sci.*, 1982, **47**, 1920.
13. K. O. Honikel, *Fleischwirtschaft*, 1982, **62**, 1390.
14. A. H. A. van den Oord and P. R. Visser, *Fleischwirtshaft*, 1973, **53**, 1427.
15. G. G. Evans and M. D. Ranken, *J. Food Technol.*, 1975, **10**, 63.

Technological Problems in Margarine and Low-calorie Spreads

By J. Madsen

GRINDSTED PRODUCTS A/S, BRABAND, DENMARK

1 Introduction

Today, margarine is a well established product of high quality and, because of the highly developed know-how, it is seldom that technological problems occur. There are, however, always new personnel to be educated and new margarine factories being built (mainly overseas where special local problems may arise), so that problems tend to be specific trouble shooting rather than generic and can be solved by better communication. This is not the situation with novel low-calorie spreads, where further substantial developments are still feasible. The incorporation of greatly increased levels of water in low-calorie spreads brings many problems; these encompass both microbiological and physical stability, which themselves are not unrelated. Greater insight into structural behaviour is needed for optimum formulation and processing decisions.

Many different brands of low-calorie spread are available on the market in Europe, and the production technology and recipes have developed much since the first product was introduced in 1968. Low-calorie spreads are still being developed in Europe, new factories are starting to produce it, new recipes are being introduced, *etc*., and it is with low-calorie spreads that most problems arise. Low-calorie spreads in Europe consist of 39–41% fats and a minimum of 50% water.

2 Low-calorie Spreads—Potential Problems

The microbiological shelf-life of a low-calorie spread with milk protein is shorter than that of a corresponding water-based product. The reason for using milk protein, in the form of skimmed milk, milk powder, whey powder, or whey protein, *etc*., is that an improved flavour release is obtained in the low-fat spread, as the milk protein gives a looser emulsion. This is illustrated in Figure 1. Further, with 7–8% protein a higher nutritive value can be claimed.

The water phase is normally pasteurized at high temperature, but as burning problems occur in the pasteurizer, and there is always some post-contamination,

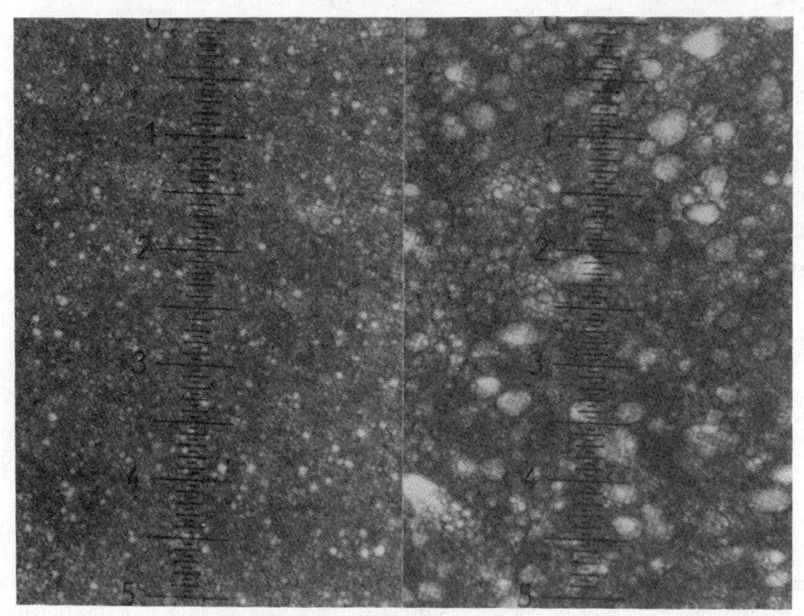

Figure 1 *Water dispersion: left, water-based; right, protein-based low-calorie spread. One scale unit is* 100 μm

microbiological problems arise. Pasteurization of the liquid emulsion is preferable, but this might cause emulsion instability. Preservatives are normally used but at relatively high pH values, as it is desirable to keep the milk protein dissolved in order to give better emulsifying properties.

With 1–2% milk dry matter calculated on the emulsion a stabilizer must also be used to give a sufficient water-binding effect.

During open-pan frying, the added fat melts and if any water droplets are present they coalesce to larger droplets which sink on to the hot frying pan surface. Here the water evaporates explosively, as illustrated in Figure 2. Fat is often thrown outside the frying pan; the effect is called 'spattering.' In margarine this is usually not a problem when emulsifiers with anti-spattering properties are used, but these are not effective in low-calorie spreads with their much higher water content. All low-calorie spreads on the market today give heavy spattering and the milk protein-containing type also undergo burning of the protein on the frying pan.

In low-calorie butter there is, as with butter, a problem of spreadability at refrigeration temperatures owing to the steep solid fat content-curve of butter fat. Some improvement can be obtained by using certain emulsifiers and an intermediate crystallizer in the tube chiller. The use of high contents of protein, (approximately 7%) seems to give improved spreadability.

Problems with emulsion instability in the tube chiller arise more frequently during low-calorie butter production because the cooling is more critical owing to the hardness of butter fat at lower temperatures.

Figure 2 *Spattering during open-pan frying*

Opening a tub of low-fat spread from the refrigerator will, at room temperature, give condensed water on the low fat spread, particularly when it is closed and put into the refrigerator again. This does not happen with tub margarine with its lower water content. Reducing the water activity of low-calorie spreads by means of hydrocolloids will only help to a certain extent, and factors such as mouthfeel and cost prevent the use of too much hydrocolloid.

Defective batches of low-fat spread cannot always be reworked owing to the risk of microbiological infection, so it is often required to re-use the fat phase only. Owing to the high water content, the water is bound more strongly than in margarine, so it is difficult to separate the spread into water and fat phase again. Regarding larger quantities, it is not practically applicable to precipitate the protein with salt or separate the emulsion by deep freezing.

Margarine and Shortening—Potential Problems

Layering of soft margarine in tubs is often in the form of a hemisphere. When the margarine is in normal use as in spreading on bread, this is not normally seen, but if the tub is pressed at the sides the hemisphere will separate and be pressed up as shown in Figure 3.

This layering is supposedly caused by breakage of the crystal lattice by the volumetric form of the filling machine. The more crystal lattice has set the greater is the problem, but owing to the consistency at filling, the resting tube cannot be too small. The formation of the crystal lattice can be delayed by using a pin mixer, but this may make the margarine too soft for filling.

Spattering in open-pan frying when using margarine can cause pollution in the kitchen and much work is being done to reduce spattering. The conditions during water evaporation have been mentioned. The problem could be analysed more satisfactorily if a more detailed explanation of the spattering could be found. Emulsifier type, water phase composition, and to a minor extent fat blend influence the spattering, but which factors give minimum spattering are found empirically, using trial-and-error methods.

Layering in puff pastry margarine gives holes in the thin layer of margarine in the puff pastry dough and thus an uneven structure and less volume in the puff pastry. The layering is often formed in the packing machine, where the feeding screws break the crystal lattice, but the pressure is not high enough to reform the crystal structure.

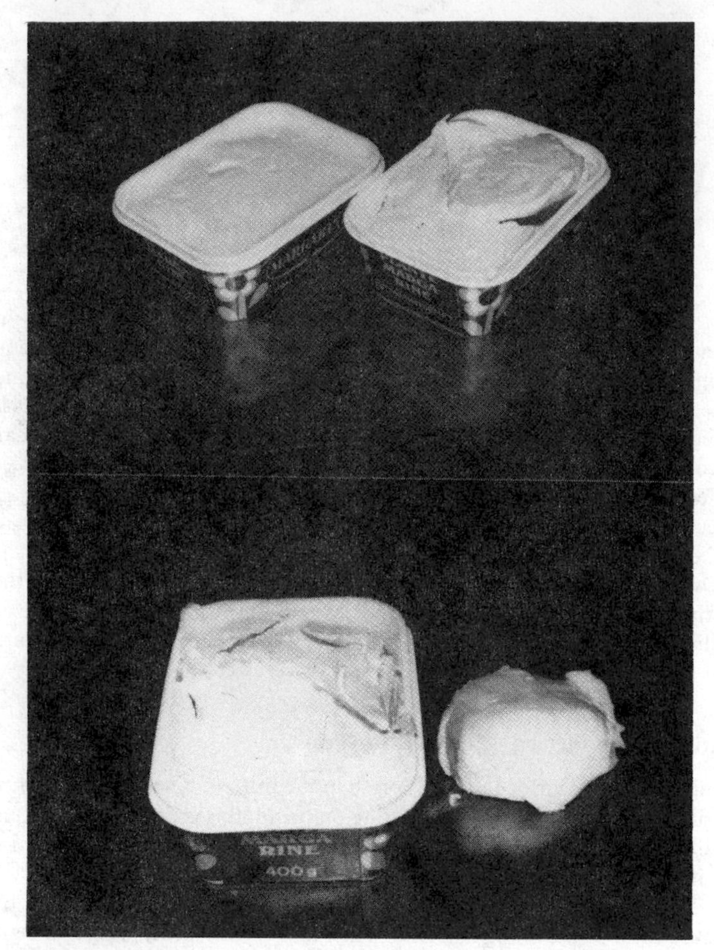

Figure 3 *Hemisphere formed in tub margarine*

Off-taste in coconut-containing products such as shortening occurs particularly in the Far East (the Philippines). The product acquires a pronounced soapy taste because of hydrolysis, as it is not possible to remove the water completely. About 0.003% of water will result in enough free fatty acids to give an off-taste in the coconut oil. Oddly enough, the same problem does not occur in margarine.

Insufficient investment and mistreatment of margarine are mainly cost/communication problems, but do certainly give margarine quality problems in many factories.

The following list indicates some of the operations and locations where problems can also arise in a margarine factory.

Raw material
- Quality
- Consistency
- Oil blend
- SFI

Preparation of solutions
- Dissolving salt, milk powder, whey powder

Hygiene
- Microorganisms
- Dirt (foreign materials)

Emulsion preparation
- Batch automatic system
- Continuous emulsification

Process control
- Temperature, ammonia, flow

Plant arrangements
- Three-way valve not cleaned, *etc.*

Channelling
- Resting tube, bend minimum 15 cm, preferably 30 cm
- Non-jacketing

Corrosion
- Salt crystals

Pipework welding
- Not smooth inside

Storage, distribution
- Non-constant temperature

The Role of Fat Crystals in Emulsion Stability

By I. J. Campbell

UNILEVER RESEARCH, COLWORTH HOUSE, SHARNBROOK, BEDFORD MK44 1LQ, UK

1 Introduction

It has been known for some time[1] that solid particles adsorbed at the oil–water interface can stabilize emulsions against coalescence. Similar crystal adsorption phenomena are employed in the ore flotation industry. Schulman and Leja[2] conducted extensive studies on the wetting of barium sulphate by surfactant solutions in both air and oil systems. The work demonstrated that the wetting of polar surfaces (in this instance barium sulphate) by water was reduced by the addition of a surfactant. This was attributed to surfactant adsorption at the surface. The work clearly demonstrated a correlation between the contact angle measured on the solid surface and the type of emulsion (oil-in-water or water-in-oil) which could be stabilized. Bascom and Singleterry[3] showed that the wetting of non-polar surfaces such as PTFE and polyethylene was increased by lowering the interfacial tension.

Contact-angle studies on food-type systems have rarely been investigated. Lucassen-Reynders[4] and Bargeman[5] are among the few workers in this area. Both showed that the contact angle in the system glycerol tristearate–surfactant solution–oil was unaffected by the addition of a surfactant (except at very low interfacial tensions). However, Lucassen-Reynders demonstrated that water-in-oil emulsions could not be stabilized in the absence of a surfactant.

Many food systems containing fatty-type materials contain a range of glycerides such as triglycerides and monoglycerides in addition to free fatty acids. This study was aimed at characterizing the different types of fat crystals with respect to surface polarity, and an attempt was made to find a correlation between emulsion stability and crystal wettability.

2 Experimental

Materials—The monoglycerides used were commercially available (PPF International) distilled monoglycerides based on vegetable or animal fat feedstocks. The fatty acid analyses of the monoglycerides were therefore the same as in the original feedstock. Three different monoglycerides were used, a saturated, a part-saturated, and an unsaturated type. The Span 80 was obtained from Atlas

Products. The lecithin was from soya bean oil and was supplied by Unimills. The fatty acids were obtained from BDH (specially pure range, >99% purity). The palm oil and bean oil were supplied by Loders and Nucoline and were further treated by passage through a silica column to remove the partial glycerides and free fatty acids. Spray-dried sodium caseinate (spray bland) was obtained from DeMelkindustrie Veghel.

Emulsion Preparation—The water-in-oil emulsions produced to examine the influence of crystals on emulsion stability were prepared as follows. A fat crystal dispersion in silica-treated bean oil was prepared by dissolving the fat crystals in the oil at 80 °C. The bean oil was then cooled to 5 °C by stirring the beaker of oil in an ice-bath. The emulsions were prepared by adding water (50% w/w) to the crystal-in-oil dispersion using a laboratory turbine-type stirrer. The emulsions were stirred for a further 1 min and then left for 1 h before being photographed.

Contact Angle Measurements—The method employed to measure contact angles was an improved version of the method by Darling described by van Boekel.[6] The redesigned contact-angle apparatus is shown schematically in Figure 1. The plug was constructed from PTFE. A hypodermic needle was inserted into the hole and the plug positioned over a plate with the recess facing upwards. A liquid sample of hardened fat (*e.g.* hardened palm oil) or paraffin wax was then pipetted into the recess until level with the lip of the recess. This 'base' fat (which does not produce the measuring surface) was then allowed to crystallize against air. When the base fat had set, the fat sample for measurement was pipetted on to the surface of the base fat as shown in Figure 1. The fat sample for measurement was not more than 2 mm deep and so set very rapidly, leaving a flat surface suitable for contact-angle measurements. Most of the contact-angle measurements were made on fat

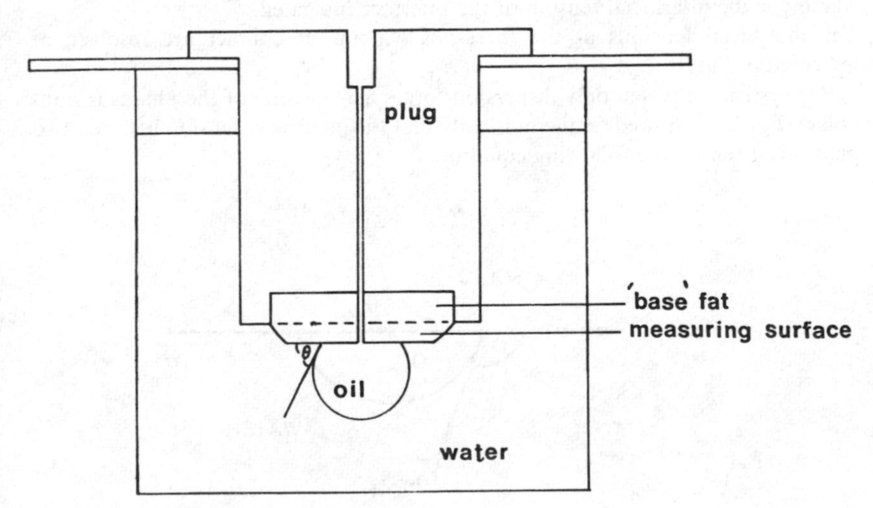

Figure 1 *Contact-angle apparatus*

surfaces set against air and therefore no further treatment of the contact angle plug was necessary. In those instances where contact angles were to be measured on surfaces set against water, the plug with the fat surface set against air was placed upturned in a beaker of the water phase at 60 °C. This had the effect of melting some of the fat, which subsequently recrystallized as the water phase cooled.

To enable an oil drop to be injected on to the oil surface, the contact-angle plug had to be further prepared as follows. A fine glass-fibre was pushed down the entrenched needle and through the fat sample. The fibre was pulled all the way through the fat sample to leave a fine capillary. The prepared contact-angle plug was then placed partially submerged in the water phase as shown in Figure 1. An Agla microsyringe was used to inject oil down the fine capillary to produce an oil drop on the fat surface. Any air present in the fine capillary could first be removed by sucking it back into the syringe.

The equilibium contact angle at the fat surface–oil drop–water phase interface was measured using a goniometer eyepiece similar to that one described by Zisman.[7] The contact angle was measured in the oil drop but quoted in the water phase according to the usual convention. Although the contact angles in this paper are for oil drops formed at the fat surface–water interface, identical results were mostly obtained for water drops formed at a fat surface–oil interface. When protein was present in the water phase, the angles were different for the oil drop and water drop systems.

Contact Angle Principles—The wetting of a solid (s) by water (w) and oil (o) is described by the Young equation

$$\gamma_{ow}\cos\theta = \gamma_{so} - \gamma_{sw} \tag{1}$$

where γ is the interfacial tension of the interface indicated.
The interfacial tensions at the three-phase point of contact are resolved as indicated in Figure 2.

For systems in which only dispersion forces act (*i.e.* one of the phases is non-polar), Fowkes[8] showed semi-empirically that the interfacial tension between two phases is given by the following equation:

$$\gamma_{1,2} = \gamma_1 + \gamma_2 - 2\sqrt{\gamma_1{}^d \gamma_2{}^d} \tag{2}$$

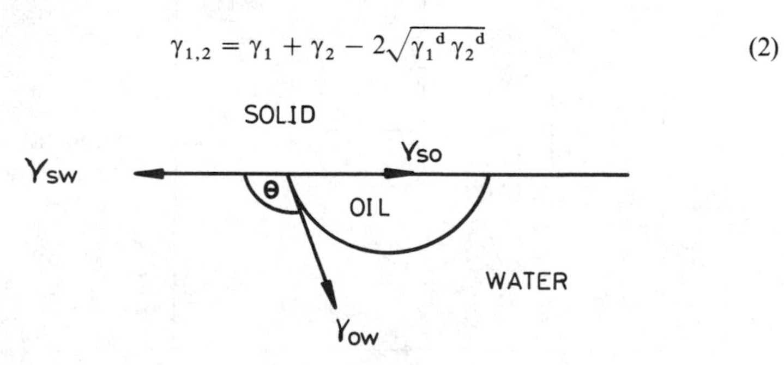

Figure 2 *The Young equation in a solid–oil–water system*

where γ_1 and γ_2 are the surface tensions of the two contacting phases and $\gamma_1{}^d$ and $\gamma_2{}^d$ are the dispersion force contributions to the surface tension (*i.e.* for water $\gamma = 71.2\,\mathrm{mN\,m^{-1}}$ and $\gamma^d = 21.8\,\mathrm{mN\,m^{-1}}$ at 20 °C).[8] For a system containing a polar phase, the Fowkes equation needs to be modified to include the polar interaction between the two phases. Tamai *et al.*[9] suggested the following modification:

$$\gamma_{1,2} = \gamma_1 + \gamma_2 - 2\sqrt{\gamma_1{}^d \gamma_2{}^d} - I_{12} \tag{3}$$

where I_{12} is the energy term due to the stabilization by polar forces.

For a solid–oil–water system, the three relevant equations are as follows:

$$\gamma_{ow} = \gamma_o + \gamma_w - 2\sqrt{\gamma_o{}^d \gamma_w{}^d} - I_{ow} \tag{4}$$

$$\gamma_{so} = \gamma_s + \gamma_o - 2\sqrt{\gamma_s{}^d \gamma_o{}^d} - I_{so} \tag{5}$$

and

$$\gamma_{sw} = \gamma_s + \gamma_w - 2\sqrt{\gamma_s{}^d \gamma_w{}^d} - I_{sw} \tag{6}$$

Combining equations (5) and (6) with the Young equation for a solid–oil–water system [*i.e.* equation (1)] gives

$$\gamma_{ow}\cos\theta = \gamma_o - \gamma_w - 2\sqrt{\gamma_s{}^d \gamma_o{}^d} + 2\sqrt{\gamma_s{}^d \gamma_w{}^d} - I_{so} + I_{sw} \tag{7}$$

If one considers the case for a non-polar oil, the $I_{so} = 0$ and therefore equation (7) reduces to

$$\gamma_{ow}\cos\theta = \gamma_o - \gamma_w - 2\sqrt{\gamma_s{}^d \gamma_o{}^d} + 2\sqrt{\gamma_s{}^d \gamma_w{}^d} + I_{sw} \tag{8}$$

By substituting γ_w in equation (8) by γ_w from equation (4) and as we are restricting ourselves to working with non-polar oils with $I_{ow} = 0$, then

$$\gamma_{ow}\cos\theta = 2\gamma_o - \gamma_{ow} - 2\sqrt{\gamma_o{}^d \gamma_w{}^d} - 2\sqrt{\gamma_s{}^d \gamma_o{}^d} + 2\sqrt{\gamma_s{}^d \gamma_w{}^d} + I_{sw} \tag{9}$$

As $\gamma_o = \gamma_o{}^d$ for non-polar oils, equation (9) can be rearranged to give

$$\gamma_{ow}\cos\theta = -\gamma_{ow} + 2(\sqrt{\gamma_o{}^d} - \sqrt{\gamma_w{}^d})(\sqrt{\gamma_o{}^d} - \sqrt{\gamma_s{}^d}) + I_{sw} \tag{10}$$

or

$$\gamma_{ow}\cos\theta = -\gamma_{ow} + C + I_{sw} \tag{11}$$

This equation is similar in form to that discussed by Lucassen-Reynders,[10] van Boekel,[6] and van Voorst Vader[11] for non-polar systems. The consequence of equation (10) for non-polar systems is that when the constant $2(\sqrt{\gamma_o{}^d} - \sqrt{\gamma_w{}^d})(\sqrt{\gamma_o{}^d} - \sqrt{\gamma_s{}^d})$ is small (which it will be in the case of a water–oil system),

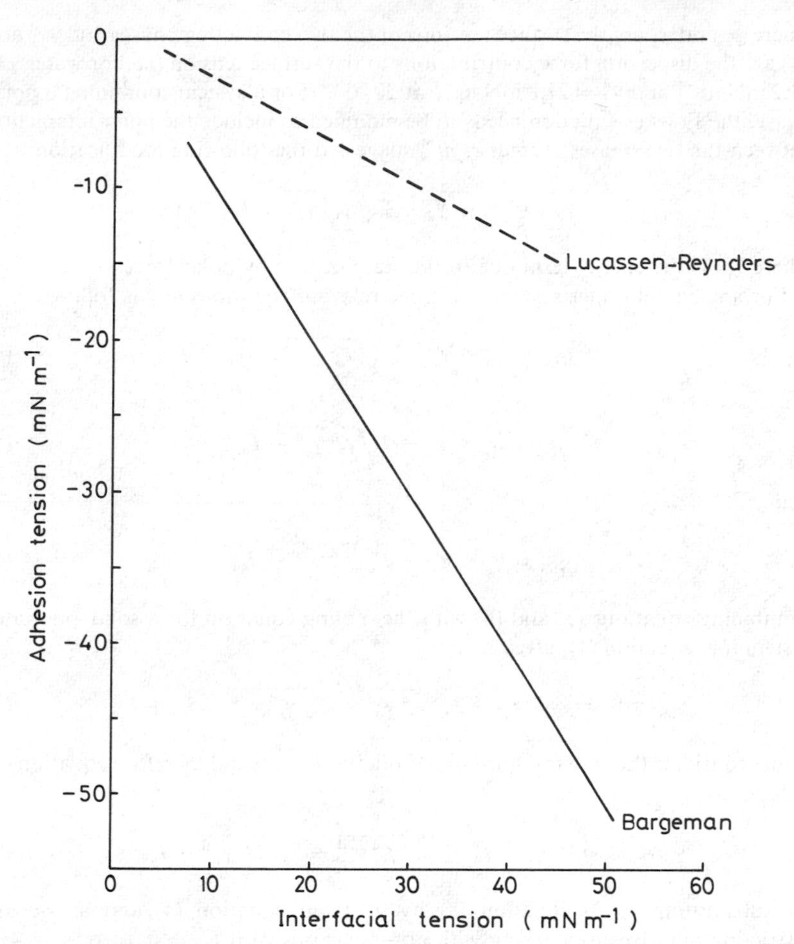

Figure 3 *Adhesion tension* versus *interfacial tension for glycerol tristerate–water–paraffin oil using the results of Lucassen-Reynders*[4] *and Bargeman*[5]

the contact angle is virtually independent of the oil–water interfacial tension until γ_{ow} approaches the value of the constant, when the contact angle will rapidly change. The results of Lucassen-Reynders and Bargeman are shown in Figure 3, in which the adhesion tension (γ_{ow} cosθ) is plotted against γ_{ow}. The differences in the measured contact angles are attributed to the different techniques (porous plug and sessile drop, respectively). A constant contact angle is indicated by a straight line passing through the origin; the data in Figure 3 indicate that $C + I_{sw}$ is small.

For polar solids in a water–non-polar oil system, equation (11) can be represented as follows:

$$\cos\theta = \frac{(I_{sw} + C)}{\gamma_{ow}} - 1 \tag{12}$$

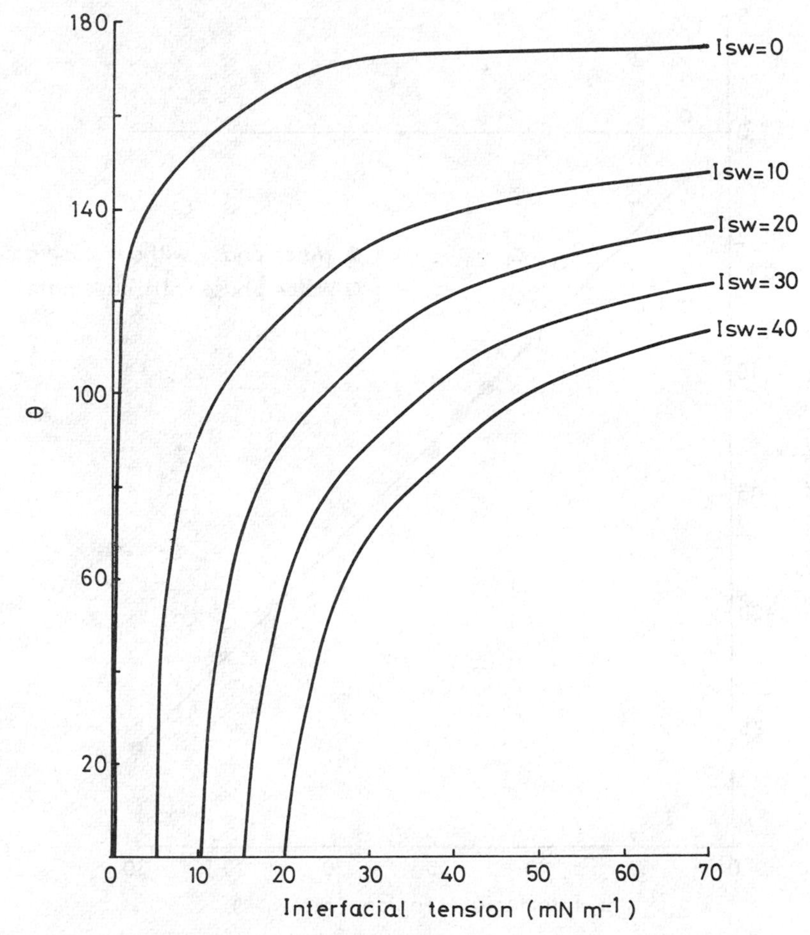

Figure 4 *Theoretical plot of contact angle* versus *interfacial tension for a range of surface polarities when* C = 0.2. *Curves* 1–5 *represent* I_{sw} = 0, 10, 20, 30 *and*, 40 mN m^{-1} *respectively*

Figure 4 shows a theoretical plot of θ against γ_{ow} for a range of I_{sw} and γ_{ow} values. It is readily apparent that polar surfaces are more easily wetted than non-polar surfaces and hence polar crystals should be able to stabilize water-in-oil emulsions at higher interfacial tensions.

3 Results and Discussion

Contact Angle Measurements—The effect of added emulsifier (at a level of 0.3% w/w) to the oil phase was examined for a hardened palm oil–water–bean oil system. The emulsifiers examined were three different monoglycerides (iodine values ranging from 0 to 100), lecithin and Span 80. A measurement was also

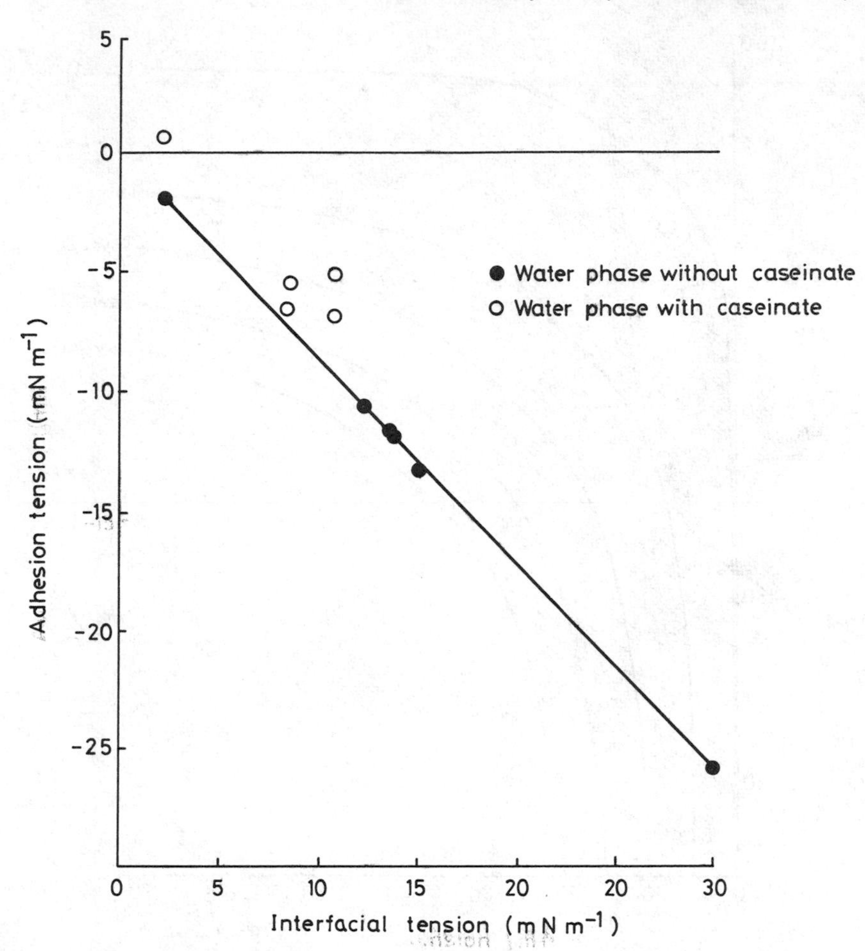

Figure 5 *Adhesion tension* versus *interfacial tension for hardened palm oil–water–bean oil + emulsifier systems with* (□) *and without* (●) 1% w/w *sodium caseinate in the water phase*

made in the absence of emulsifier. The results showed that a constant contact angle of about 150° was obtained for all the emulsifiers. A plot of $\gamma_{ow}\cos\theta$ against γ_{ow} gave a straight line almost passing through the origin. The addition of 1.0% of sodium caseinate to the water phase led to smaller contact angles in every case (see Table 1). The protein data when plotted as adhesion tension *vs.* γ_{ow} did not fall on a straight line (see Figure 5). The measured interfacial tensions at the protein–oil interface were not appreciably lower than those in the non-protein system and therefore it must be concluded that the value of I_{sw} is affected by protein adsorption. Presumably the emulsifier in the oil phase influences the degree of protein adsorption at the fat surface.

Table 1 *Effect of emulsifier (at the* 0.3% w/w *level) in the oil phase on the contact angle in a hardened palm oil–bean oil–water phase system*

Emulsifier	*Contact angle measured against water* (°)	*Contact angle measured against* 1% *protein solution* (°)
No added emulsifier	150	
Saturated monoglyceride	150	148
Part-saturated monoglyceride	151	142
Unsaturated monoglyceride	150	130
Span 80	151	130
Lecithin	149	82

Table 2 *Effect of emulsifier when added to the fat phase in a hardened palm oil–bean oil–water system*

	Contact angle at various emulsifier concentrations (°)		
Emulsifier	1%	10%	100%
Saturated monoglyceride	146	142	61
Part-saturated monoglyceride	138	81	63
Unsaturated monoglyceride	125	73	66
Lecithin	132	109	
Span 80	142		

The same fat-soluble emulsifiers were examined when added to the fat phase. Lecithin and Span 80 were only examined at low addition levels as these emulsifiers are liquid in form. The contact angles obtained are shown in Table 2.

It is clearly seen that the addition of emulsifier to the fat phase decreases the contact angle (*i.e.* the surface becomes more polar). The higher the emulsifier level, the lower is the contact angle. It is interesting that all the monoglycerides show approximately the same degree of surface polarity when used alone. It is apparent that monoglyceride crystals can be considered fairly polar in nature.

The third class of fat crystals examined were fatty acids (butyric, hexanoic, palmitic and stearic). The fat samples were prepared set against air and set against water. Table 3 shows that the fatty acid crystals are more polar than triglyceride but less polar than monoglyceride; the higher the fatty acid chain length, the more polar is the fat surface. The results also demonstrate that a fat surface when crystallized against water leads to a more polar surface than one set against air; presumably greater orientation of the fatty acid molecules occurs when crystallization occurs against water. An extremely polar surface is produced when the fatty acid surface is brought into contact with sodium hydroxide. The fatty acid is clearly dissociated, effectively forming a soap of the fatty acid. One would predict that the stability of water-in-oil emulsions stabilized by fatty acid crystals will be dependent on pH.

Table 3 *Effect of fatty acid on the contact angle in a fatty acid–bean oil–water system*

Fatty acid	θ *for surface set against water measured against water* (°)	θ *for surface set against air measured against water* (°)	θ *for surface set against air measured against* 0.1 M *NaOH* (°)
Butyric acid	145	155	127
Hexanoic acid	138	153	112
Palmitic acid	124	148	44
Stearic acid	117	138	20

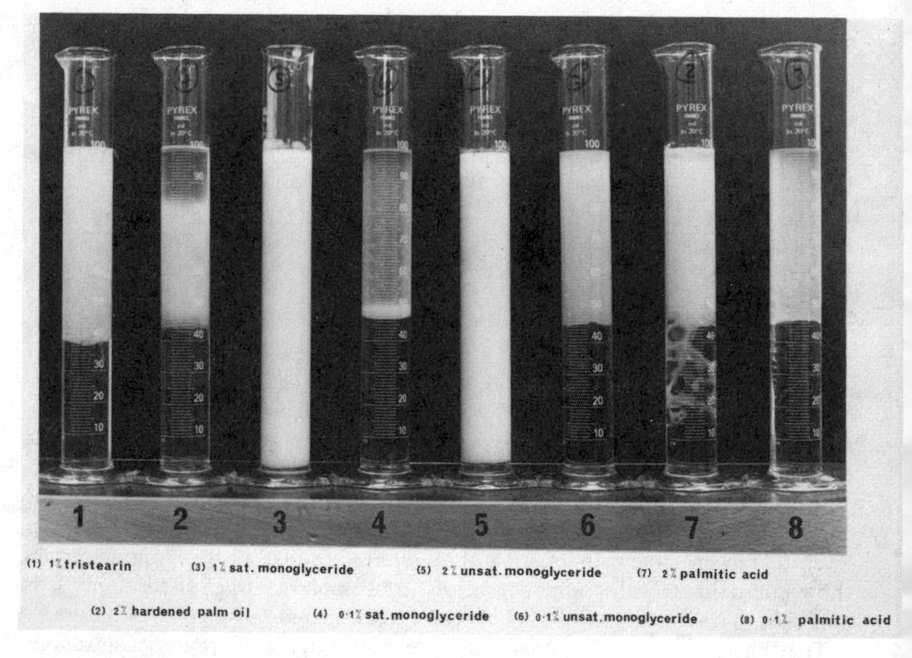

Figure 6 *Stability of water-in-oil emulsions*

Stability of Water-in-Oil Emulsions—The contact angle measurements demonstrate that triglyceride crystals are non-polar, fatty acid crystals are more polar, and monoglyceride crystals are the most polar of the fats examined. The emulsion stability experiments were conducted to examine the effect of the different polarity crystals on the stability of water-in-oil emulsions. Figure 6 shows the stability of water-in-oil emulsions prepared with crystal dispersions of triglyceride, monoglyceride, and fatty acid. It is apparent that monoglyceride crystals produce very stable emulsions whereas triglyceride and fatty acid crystals (at neutral pH) do not. Comparison of the emulsions in tubes 3 and 4 and tubes 5 and 6 indicates

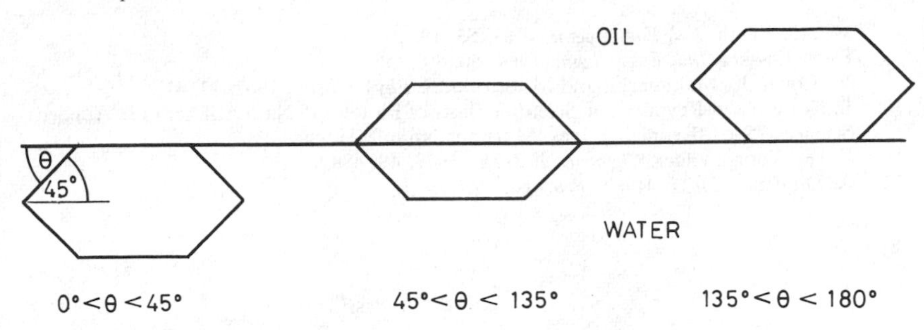

Figure 7 *Effect of the contact angle on the interfacial position of a crystal with simple geometry*

that there is an emulsifier concentration dependence on the stability of the emulsions. It was found that 0.8% of unsaturated monoglyceride was needed to give stable emulsions, whereas 0.2% of a saturated monoglyceride was sufficient. This observation is related to the solubility of the monoglyceride in the oil phase, and verifies that the fat crystals provide the stabilization mechanism for the stability of the emulsions, not the soluble monoglyceride.

Although there is a correlation between the measured contact angle and emulsion stability (more polar surfaces only produce stable emulsions), one can only conclude from this study that good emulsion stability is achieved with an angle of about 65° (*i.e.* monoglyceride). This angle is much lower than one would have predicted for water-in-oil stability. In practice, the crystal shape will also affect the stability of the emulsion, as the equilibrium position of the crystal will depend on the crystal geometry.[12] If one considers a simple crystal geometry such as that shown in Figure 7, then the crystal will have only three possible equilibrium positions, depending on the contact angle.

For the crystal geometry shown, stable oil-in-water emulsions will result only when the contact angle is less than 45° (state 1). An angle between 45° and 135° will not stabilize either water-in-oil or oil-in-water emulsions and an angle $>135°$ (state 3) will stabilize water-in-oil emulsions.

It is evident from this work that the degree of surface polarity can give a good indication of the stability of an emulsion. However, the measured contact angles need to be interpreted with care. Factors such as crystal geometry or the order in which the phases are brought into contact (causing preferential surfactant adsorption to one interface) may in practice generate a different crystal wettability to the one measured.

References

1. S. U. Pickering, *J. Chem. Soc.*, 1907, **91**, 2001.
2. J. H. Schulman and J. Leja, *Trans. Faraday Soc.*, 1954, **50**, 599.
3. W. D. Bascom and C. R. Singleterry, *J. Phys. Chem.*, 1962, **66**, 236.
4. E. H. Lucassen-Reynders, *Thesis*, University of Utrecht, 1962.
5. D. Bargeman, *J. Colloid Interface Sci.*, 1972, **40**, 311.
6. M. A. J. S. van Boekel, *Thesis*, University of Wageningen, 1980.

7. W. A. Zisman, *Ind. Eng. Chem.*, 1963, **55**, 19.
8. F. N. Fowkes, *Ind. Eng. Chem.*, 1964, **56**, 40.
9. Y. Tamai, K. Makuunchi, and M. Suzuki, *J. Phys. Chem.*, 1967, **71**, 4176.
10. E. H. Lucassen-Reynders, in 'Scientific Basis of Flotation,' Nato ASI Series E, Applied Sciences 75, ed. Kenneth J. Ives, Martinus Nijhoff, Hague, 1984.
11. F. van Voorst Vader, *Chem. Eng. Tech.*, 1977, **49**, 488.
12. A. Dippenar, *Int. J. Miner. Process.*, 1982, **9**, 1.

The Colloid Chemistry of Black Tea

By R. S. Harbron*

SCHOOL OF CHEMISTRY, UNIVERSITY OF BRISTOL, BRISTOL BS8 1TS, UK

R. H. Ottewill

SCHOOL OF CHEMISTRY, UNIVERSITY OF BRISTOL, BRISTOL BS8 1TS, UK

and

R. D. Bee

UNILEVER RESEARCH, COLWORTH HOUSE, SHARNBROOK, BEDFORD, MK44 1LQ UK

1 Introduction

When infusions of black tea, prepared using hot water, are allowed to cool a colloidal dispersion is formed which is commonly termed 'tea cream.' The composition of tea cream and the chemical factors causing its formation have been the subject of considerable investigations over several decades.[1-8] The composition is extremely complex and the particles are known to contain quantities of all the substances present in the infusion from which it separated. However, it has been established that the major constituents are theaflavins, thearubigins, and caffeine, and it has been postulated that it is the complexation amongst these species that is responsible for cream formation.[4,5] Although much attention has been given to determining the chemical composition of tea cream, relatively little work appears to have been directed towards understanding the physical nature of the tea cream particles.

In recent previous publications the morphology of tea cream particles was described.[9,10] In this paper we report an investigation of the colloidal properties of tea cream particles. As the particles appeared to be stable in the colloidal sense, it seemed appropriate to examine whether current understanding of colloid stability was of any significance in the field of tea chemistry. It was therefore of interest to determine the origin of the colloid stability of the particles and hence the first part of the work was to determine the range of concentrations and temperatures which led to the formation of stable colloidal particles. The

* Present address: FOSROC Technology Ltd., Bourne Road, Aston, Birmingham B6 7RB, UK.

particles, largely prepared in the dilute region, were then examined by microelectrophoresis in order to obtain an indication of their surface electrical properties. It was found that they were negatively charged at the natural pH of the tea infusion, 4.7, the electrophoretic mobility increasing with pH between 2 and 6 and then levelling off. The influence of various inorganic cations on the electrophoretic mobility was also examined as a means of obtaining a 'charge reversal spectrum' in order to probe the nature of the chemical groups on the surface by the method devised by Bungenberg de Jong.[11] These investigations, coupled with an investigation of the coagulation behaviour of tea cream particles, suggest that they behave, in many ways, as lyophobic colloids.

2 Experimental

Tea Extract and Tea Infusions—The tea extract used was prepared by hot water counter-current stripping of black Ceylon tea, made initially at 5% tea solids and then concentrated and spray-dried to a fine-flowing powder. All experimental samples were taken from a single bulk sample which was stored at room temperature in a sealed steel drum. The extract contained approximately 3% by weight of moisture and was contaminated with about 3% by weight of insoluble substances, which consisted of dust and cellular debris from the original plant leaf and also some crystalline material, mainly magnesium oxalate dihydrate. For infusions containing up to *ca.* 10% of tea solids the insoluble material was removed by filtering the hot infusion through a previously heated, sintered-glass funnel.

Tea infusions were prepared by refluxing the tea 'solution' at 95 °C for 30 min in a glass flask in a thermostatted bath. All the infusions were prepared using distilled water and were allowed to cool under ambient conditions. They were used as quickly as possible in order to avoid the possibility of microbiological contamination.

Electron Microscopy—A sucessful procedure for examining the tea cream particles was found to be to add 1 cm^3 of the tea infusion to 4 cm^3 of a 0.13% solution of osmium tetraoxide. After allowing the particles to fix for 30 min they were spun down in a bench centrifuge at 3000 rev min^{-1} for 5 min. The supernatant was then decanted and the particles rinsed with distilled water. Finally, they were redispersed in a small volume of distilled water and deposited on to Formvar-coated copper grids. These were examined with either a Hitachi HS7 or a Hitachi HS7S transmission electron microscope. Some of the samples were shadowed with a thin layer of chromium at an angle of 30° to the horizontal.

Dry Solids Concentration—The dry solids concentration was determined by evaporating 5–10 cm^3 of the sample, placed in a shallow glass vessel, in an oven at 70 °C for 2 days. The sample was then allowed to cool in a vacuum desiccator. Shallow vessels were used to try to reduce errors caused by the formation of tea glass. Concentrations are given as % w/w unless stated otherwise.

Electrophoretic Measurements—Measurements of electrophoretic mobility were made using either a Rank Particle Microelectrophoresis Apparatus Mark 2 or a PenKem System 3000 Automated Electrokinetic Analyser. In view of the large dilutions involved in preparing samples for electrophoresis, the most extensive measurements were carried out using dialysed samples of tea infusions as the dialysed systems appeared to be more stable to dilution. The tea infusions were dialysed against doubly distilled water at room temperature using dialysis tubing which had been well boiled in distilled water to sterilize it; over a period of 2 weeks no problems were encountered from microbiological contamination. The ratio of dialysate to sample used was 20:1.

For electrophoretic measurements, 0.3 cm^3 of the dialysed tea infusion was added to 20 cm^3 of electrolyte solution to give a final concentration of *ca.* 0.01% w/w. The resulting dispersion was then adjusted to pH 5.0 to bring the sample close to pH 4.7, the natural pH of tea infusions. In the systems where hydrolysable cations were used some measurements were also made at lower pH values.

Critical Coagulation Concentrations—These were determined by adding 0.3 cm^3 of the tea infusion to 20 cm^3 of an electrolyte solution. This was done for a number of sample tubes in order to span a range of electrolyte concentrations; most samples were adjusted to pH 5.0. The samples were shaken and allowed to stand and were subsequently examined visually at hourly intervals for coagulation by comparison with a control sample prepared by dilution with distilled water only. After 4 h the region where coagulation was occurring became apparent and hence a further set of tubes was prepared spanning this concentration range in smaller intervals of electrolyte concentration. This was repeated until the critical coagulation concentration could be estimated within narrow limits of electrolyte concentration.

3 Results

Electron Microscopy—Electron micrographs were taken of osmium tetraoxide-fixed tea cream particles prepared from whole tea infusions at different concentrations in the range 0.1–5.0%. At the lowest concentration examined (0.1%) the particles were present as irregularly shaped granules or loose aggregates (reference 9, Figure 3). The particles, which appeared to be single entities, were approximately 20–30 nm in diameter.

The particles prepared at a concentration of 0.5% were still ill-defined and there were many loose aggregates, which seemed to be made up from more compact particles of about 100 nm in diameter.

The preparations from 1% infusions appeared to consist of two different types of particles (Figure 1b). In the background there are small, lightly stained particles and in the foreground larger, more heavily stained particles. The smaller particles appeared to be granular, loosely aggregated material whereas the larger particles, *ca.* 250 nm in diameter, had an angular appearance. It appeared that the larger particles were aggregates of tightly packed smaller particles.

At a concentration of 2.5%, two types of particles were also observed. The

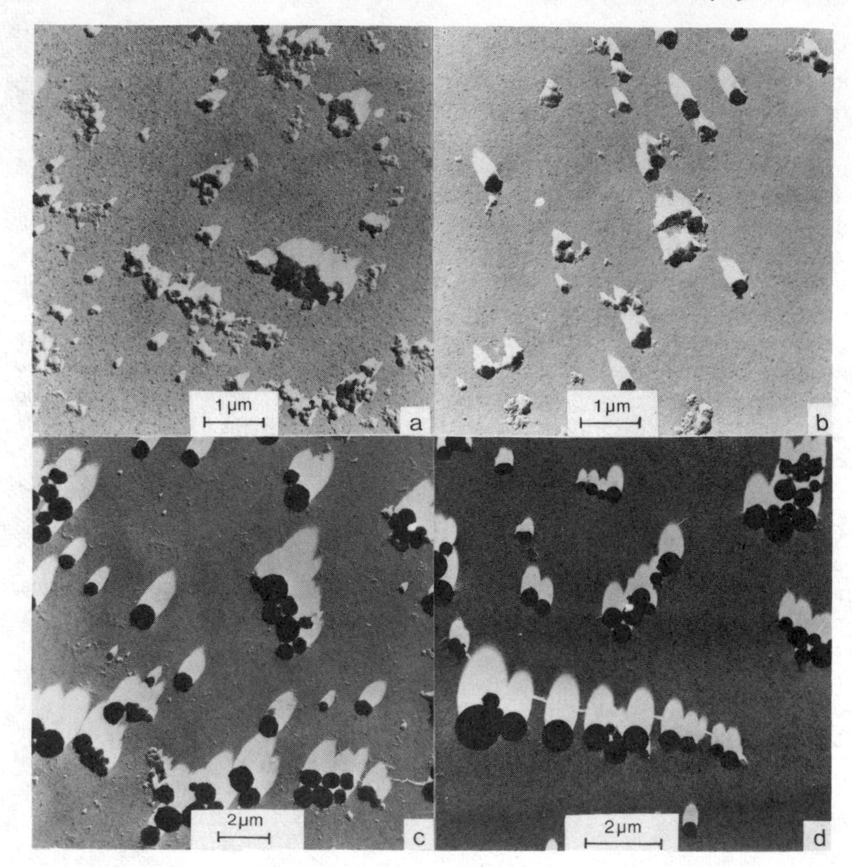

Figure 1 *Electron micrographs of tea cream particles, fixed at ambient temperature with osmium tetraoxide. The solids concentrations used for the infusions were as follows:* (a) 0.5; (b) 1.0; (c) 2.5; (d) 5.0% w/w

larger, heavily stained particles had increased in size to about 500 nm and, although the surface of these particles was not smooth, they appeared to be virtually spherical. In general, the particles were homogeneously stained (Figure 1c).

An increase in the solids concentration to 5% resulted in a substantial reduction in the number of small particles and a further increase in the size of the large particles to about 700 nm (Figure 1d). The particles appeared to be fairly smooth spheres which were homogeneously stained.

Over the concentration range 10–40% the change in particle morphology was not so evident. Highly stained spherical particles were observed and at each concentration there was a wide range of particle sizes; the mean particle size seemed to increase with increasing solids concentration.

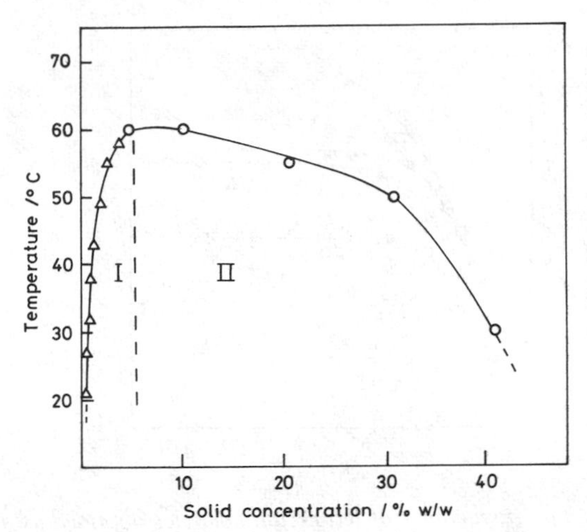

Figure 2 *Diagram to illustrate the boundary for the separation of tea cream and its dependence on temperature and composition.* (a) △, *points from Rutter;*[7] (b) ○, *this work*

Tea Cream—a Phase Separation Process—The separation of tea cream during the cooling of tea infusions can usefully be considered in terms of a phase diagram. A representation of the information obtained in this way is shown in Figure 2. The drawn curve represents the temperature above which the tea cream did not separate at any temperature. The open circles represent the temperatures of phase separation obtained by optical microscopy and the triangles represent the 'cream points' obtained by Rutter.[7] In view of the nature of the system and the methods used to obtain the separation temperatures, the phase boundary should be considered as a broad region of a few degrees rather than a sharp boundary. The diagram obtained is similar to the classical case of two liquids which mix at high temperatures and then separate into two immiscible liquid phases on cooling.[12]

The maximum in the cooling curve occurs at *ca.* 6% w/w tea solids and the type of particle obtained appeared to depend on whether the initial composition of the tea infusion was to the left or the right of the maximum. The particles obtained to the left, region I, were found to be granular in nature, whereas those to the right, region II, were found to be spherical. This conclusion was reached after examination of the particles by either electron microscopy or optical microscopy.

Electrophoretic Mobility—pH Dependence—The results obtained for the variation of the electrophoretic mobility of tea particles as a function of pH are shown in Figure 3. One set of measurements was obtained on the undialysed tea sample after dilution 1:20 with water. This sample had a specific conductance of $0.34\ \Omega^{-1}\ m^{-1}$, which for a 1:1 electrolyte is equivalent to an electrolyte concentration of *ca.* 2×10^{-3} mol dm^{-3}. The dialysed tea infusions after dilution were examined in water and in 10^{-3} and 10^{-2} mol dm^{-3} potassium chloride solutions.

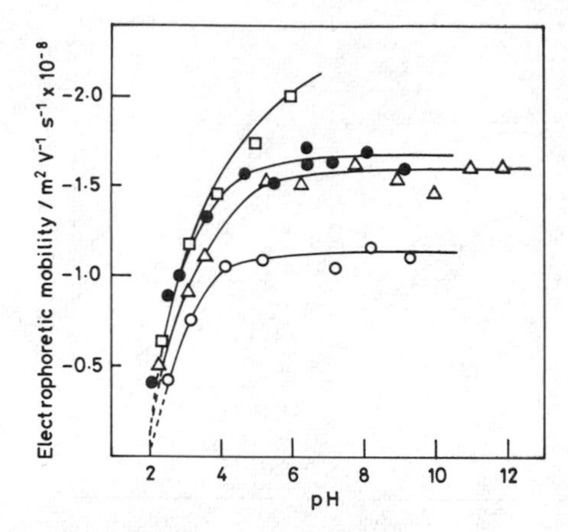

Figure 3 *Electrophoretic mobility of tea cream particles obtained as a function of pH at 25 °C. Open symbols, dialysed particles under various conditions:* □, *in distilled water;* △, *in* 10^{-3} mol dm^{-3} *potassium chloride solution;* ○, *in* 10^{-2} mol dm^{-3} *potassium chloride solution;* ●, *undialysed tea cream particles in distilled water*

The samples in electrolyte solutions appeared to reach limiting mobilities above *ca.* pH 6.0. It is of interest that the curve for the dialysed tea sample in 10^{-3} mol dm^{-3} potassium chloride solution is very similar to that of the diluted whole tea infusion particles, suggesting that dialysis does not substantially change the nature of electrokinetically determined surface properties.

It was found that considerable dissolution of the tea particles occurred at pH values greater than 9.0. The mobility values presented were therefore based on the particles which were apparently insoluble at the pH of measurement. Consequently, the experimental error was higher in this region than at the lower pH values, *i.e.* the standard deviation was of the order of 12% rather than 6%.

The mobility curves essentially reflect the changing ionization of the surface groups on the particles as a function of pH. It is clear that H^+ and OH^- ions should be regarded as potential determining ions for these systems.

Influence of Inorganic Ions on the Mobility of Tea Cream Particles—For most of the samples examined, the electrophoretic mobility was found to be essentially constant at low electrolyte concentrations and then to decrease with increasing quantities of added electrolyte. The electrolyte concentration at which the particle mobility first started to decrease depended on the valency of the cation and was 10^{-3}, 10^{-5}, and 10^{-6} mol dm^{-3} for monovalent, divalent, and trivalent cations, respectively. With thorium nitrate and aluminium nitrate solutions at pH 5, reversal of the charge on the particles from negative to positive was observed. The results presented here are those obtained on dialysed tea infusions; those obtained on whole tea infusions were very similar. Figure 4 gives the results obtained for

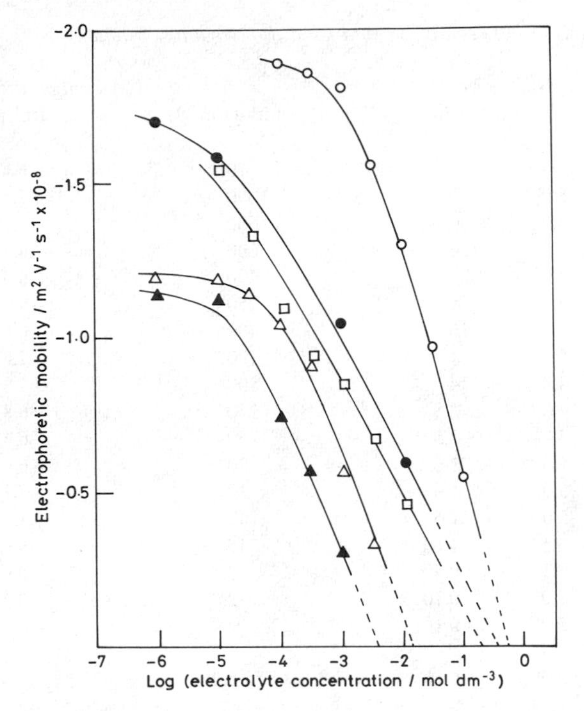

Figure 4 *Electrophoretic mobility of dialysed tea cream particles obtained as a function of electrolyte concentration in the presence of various cations at* 25 °C: ○, *Na^+ at pH* 5; ●, *Mg^{2+} at pH* 5; □, *Ca^{2+} at pH* 5; △, *UO_2^{2+} at pH* 3.5; ▲, *Al^{3+} at pH* 3

various electrolytes using non-hydrolysable cations and Figure 5 shows the results obtained in the presence of thorium and aluminium nitrates at pH 5.0.

As shown by Bungenberg de Jong[11] and others,[13,14] the sequence of concentrations at which the charge on the particles is reversed with various cations can be used as a means of obtaining information about the nature of the particle surface. This sequence has been termed a 'reversal of charge spectrum.' The experimentally determined spectrum is then compared with spectra of known surfaces. In this work, reversal of charge concentrations were obtained from the mobility data. For thorium and aluminium at pH 5 the values were obtained from the intersection point on the log(concentration) axis. For the other electrolyte systems it was obtained by extrapolating the curves to the point of zero mobility. The results obtained are shown in Figure 6.

Critical Coagulation Concentrations—The results obtained for the critical coagulation concentrations of dialysed tea infusions using a range of electrolyte concentrations are presented in Table 1. Also listed, for comparison, are the reversal of charge concentrations (r.c.c.) obtained by electrophoresis.

Table 1 *Comparison between c.c.c. and r.c.c. for various cations*

Cation	*pH*	*c.c.c.* (mmol dm^{-3})	*r.c.c.* (mmol dm^{-3})	*ζ-potential at c.c.c.* (mV)
H^+		6 (pH 2.2)	10	−6.4
Li^+	5.0	100	1000	−7.6
Na^+	5.0	100	600	−7.1
K^+	5.0	100	600	−6.7
Mg^{2+}	5.0	25	400	−5.5
Ca^{2+}	5.0	4	250	−8.2
Sr^{2+}	5.0	16	400	−6.8
Ba^{2+}	5.0	4	300	−7.6
Mn^{2+}	4.0	4	400	−7.3
Co^{2+}	4.0	2.5	250	−6.8
Ni^{2+}	4.0	2.5	250	−6.8
Cu^{2+}	5.0	1.6	20	−6.6
Pb^{2+}	5.0	1.6	30	−7.1
Zn^{2+}	5.0	1.6	40	−7.1
UO_2^{2+}	3.5	1.60	15	−5.9
Al^{3+}	3.1	0.40	3.5	−6.6
Ce^{3+}	5.0	0.10	2.0	−8.3
La^{3+}	5.0	0.16	1.4	−8.9
Hydrolysed aluminium	5.0	0.16	0.25	0
Hydrolysed thorium	5.0	0.04	0.08	0

4 Discussion

The results in Figure 2 suggest that the formation of tea cream particles can be represented as a phase separation process which resembles that obtained from a binary liquid mixture which becomes immiscible on cooling. In the tea solution, however, the separated phase forms stable colloidal particles rather than a bulk phase, indicating that a mechanism is operative which stabilizes the particles and prevents their coalescence to form a continuous phase. An additional interesting feature is that the morphology of the particles appears to vary according to the initial concentration of the tea cream phase.

The process of particle formation appears to be reproducible and, in general, leads to the formation of spherical particles. However, the rate at which the particles are formed appears to be an important factor, as attempts to form smaller particles by very rapid cooling of the tea infusions led to immediate coagulation of the system. This implies that the molecular entities conferring colloid stability had not had sufficient time to take up interfacial conformations. By slow cooling of the system the requisite morphology was attained, giving stable colloidal particles. This suggested that a certain period of time, of the order of minutes, was essential for the molecules to organize themselves (self-assembly) into a structure appropriate for stability; the order of the time scale suggests that

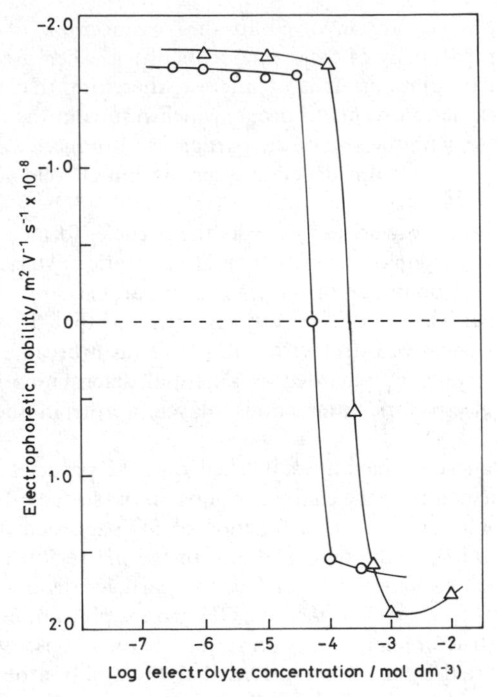

Figure 5 *Electrophoretic mobility of dialysed tea cream particles obtained as a function of electrolyte concentration at pH* 5.0 *and* 25 °C *using* (○) *thorium nitrate solutions and* (△), *aluminium chloride solutions*

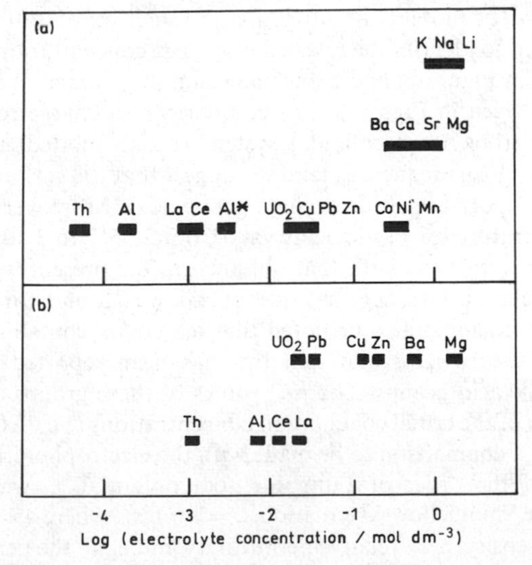

Figure 6 *Reversal of charge concentrations (r.c.c.) plotted in logarithmic form for various cations: (a) for dialysed tea cream particles (see also Table 1); (b) for sodium pectinate.*[11] (Al* at pH = 3)

probably macromolecules are involved in the organization of the particles. Indeed, it is known that many of the components of tea, when used individually, form macromolecular solutions. It is of interest, therefore, that when they are present together association seems to occur, which results in the formation of a stable colloidal phase, with the size of the particles so produced varying between approximately 20 nm and 10 μm depending on the initial solids content of the infusion.

A major question therefore arises as to why the particles formed are colloidally stable and whether the origin of the stability is electrostatic or steric.[15] The results obtained by microelectrophoresis of the tea cream particles as a function of pH indicated that the particles were negatively charged and that the surface charge became increasingly negative with increase in pH. Thus it became apparent that H^+ and OH^- ions could be regarded as potential determining ions for these systems and that electrostatic interactions played a role in the stabilization mechanism.

The next major question, having established that the particles were charged, was to determine the nature of the charged groups on the surface. The form of the curve of electrophoretic mobility as a function of pH suggested the presence of acidic groups with a pK_a in the region of 3.0. In the pH region where phenolic groups were expected to ionize, *ca.* pH 9.0, the particles dissolved. A possible interpretation of this result is that phenolic OH groups play an integral role in a hydrogen-bonded structure and that ionization disrupts this. Moreover, this contention is supported by the observations that the addition of polar organic solvents such as ethanol and acetone and the addition of urea to tea infusions also bring about dissolution of tea cream particles. However, the above observations do not rule out the possibility of hydrophobic interactions and also coulombic interactions playing a role in the cream structure.

The electrophoretic mobility as a function of ionic concentration for a series of ions has been used to identify the reversal of charge concentration for each. These results are given in Figures 4 and 5 and the resultant reversal of charge spectrum for tea cream is given in Figure 6. For comparison, a charge reversal spectrum obtained on a carboxylated colloidal system is also plotted, namely that for sodium pectinate.[11] The sequences tend to suggest that the surface resembles that of a carboxylated polysaccharide such as pectic acid. Moreover, the pK_a values quoted in the literature for pectic acids vary from 2.95[16] to 3.40.[17] These values are in reasonable agreement with that obtained in the present work. Fatty acids tend to have higher values, *e.g.* a value of 4.4 was found for surface pK_a of arachidic acid.[18] It should also be noted that tea cream contains a polyphenolic fraction termed thearubigins and that this has been reported by Roberts[2] to contain carboxylic acid groups. The pK_a values of these groups are not known.

Determination of the critial coagulation concentrations (c.c.c.) for the tea cream particles allowed a comparison to be made with the electrophoretic results. As the particles were of the order of 1 μm, the zeta potential, ζ, was calculated by application of the Smoluchowski equation, $\zeta = u\eta/\varepsilon_r\varepsilon_0$, where u = electrophoretic mobility, η = viscosity, ε_r = relative permittivity and ε_0 = the permittivity of free space. The ζ values at which coagulation occurred are listed in Table 1. It can be seen that in some instances coagulation of the particles occurred at a finite value

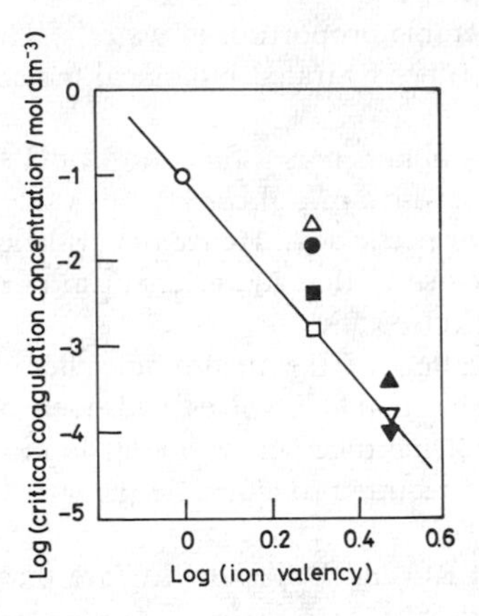

Figure 7 *Logarithm of the critical coagulation concentration (c.c.c.) for dialysed tea cream particles, coagulated with various cations, plotted against the logarithm of the cation valency (see also Table 1). Results at pH* 5.0 *unless stated otherwise:* ○, Li^+, Na^+, K^+; △, Mg^{2+}; ●, Sr^{2+}; ■, Ca^{2+}; □, Cu^{2+}; ▲, Al^{3+} *at pH* 3.1; ▽, La^{3+}; ▼, Ce^{3+}

of ζ, at an average value of −7.1 mV, whereas in other instances coagulation occurred essentially at the charge reversal concentration. The latter situation occurred when thorium and aluminium were used as the cationic species at pH values where these ions are known to form hydrolysed species in solution.[19] The hydrolysed species are known to adsorb strongly to the surface, reducing the charge to zero, and then making it positive. Consequently, it is possible to distinguish between coagulation, with electrostatically stabilized species, which occurs as a consequence of electrical double layer compression where coagulation takes place at finite ζ-potentials, and strong specific adsorption where coagulation occurs in the region of zero ζ.

Coagulation due to electrical double layer compression has been examined in some detail in the theory of stability of lyophobic colloids developed by Derjaguin and Landau[20] and Verwey and Overbeek.[21] For the purposes of the present discussion it can be said that this approach led to a very approximate expression for the critical coagulation concentration of the form c.c.c. = constant/$A^2 v^6$, where A is the composite Hamaker constant[22] and v the valency of the cation used as a coagulant. The implication of this equation is that a plot of log(c.c.c.) against log v should be linear with a slope of −6. A plot of this type is shown in Figure 7 for a number of the cations used; the drawn line is that having a slope of −6. It is not appropriate here to give a more sophisticated interpretation of these results; they are plotted in this way to show the strong inference that tea cream particles behave as lyophobic colloidal dispersions.

Conclusions

The coagulation and electrophoretic behaviour of the tea cream particles indicate that their behaviour is essentially that of lyophobic colloids, even though the particles contain a considerable proportion of water.[10] However, similar behaviour has been found with other hydrated biological materials, *e.g.* polymorphonuclear leucocytes.[13]

The electrophoretic data obtained as a function of pH suggest the presence of acidic groups on the surface with a pK_a of *ca.* 3.0; this value is of the same order as that previously obtained for pectic acid. Hence the evidence from electrophoretic and coagulation studies indicates that the mechanism of colloid stabilization of the particles is primarily electrostatic.

At present the organization of the macromolecules within the particles is unknown. It seems probable that it involves hydrogen bonding between polyphenolic molecules and caffeine together with contributions from other species. Clearly, further studies are required to elucidate the structure.

Acknowledgement: The AFRC and Unilever PLC are thanked for their support for this work.

References

1. A. E. Bradfield and M. Penny, *J. Soc. Chem. Ind.*, 1944, **63**, 306.
2. E. A. H. Roberts, in 'The Chemistry of Flavanoid Compounds,' ed. T. A. Geissman, Pergamon Press, Oxford, 1962, Ch. 15.
3. E. A. H. Roberts, *J. Sci. Food Agric.*, 1963, **14**, 700.
4. R. L. Wickremasinghe and K. P. W. C. Perera, *Tea Q.*, 1966, **37**, 131.
5. R. F. Smith, *J. Sci. Food Agric.*, 1968, **19**, 530.
6. P. D. Collier, R. Mallows, and P. E. Thomas, *Phytochemistry*, 1972, **11**, 867.
7. P. Rutter, *PhD Thesis*, University of Leeds, 1971.
8. P. Rutter and G. Stainsby, *J. Sci, Food Agric.*, 1975, **26**, 455.
9. R. D. Bee, M. J. Izzard, R. S. Harbron, and J. M. Stubbs, *Food Microstruct.*, 1987, **6**, 47.
10. R. S. Harbron, *PhD Thesis*, University of Bristol, 1986.
11. H. G. Bungenberg de Jong, in 'Colloid Science,' ed. H. R. Kruyt, Elsevier, Amsterdam, 1949, Vol. II, Ch. IX.
12. D. B. Keyes and J. H. Hildebrand, *J. Am. Chem. Soc.*, 1917, **39**, 2126.
13. D. J. Wilkins, R. H. Ottewill, and A. D. Bangham, *J. Theor. Biol.*, 1962, **2**, 165, 176.
14. H. G. Bungenberg de Jong, *Arch, Neerl. Physiol.*, 1940–41, **25**, 431.
15. R. Buscall and R. H. Ottewill, in 'Polymer Colloids,' eds. R. Buscall, T. Corner, and J. F. Stageman, Elsevier Applied Science, Barking, 1985.
16. J. T. Davies, D. A. Haydon, and E. K. Rideal, *Proc. R. Soc. London, Ser. B*, 1956, **145**, 375.
17. A. Katchalsky, N. Sharit, and H. Eisenberg, *J. Polym. Sci.*, 1954, **13**, 69.
18. R. H. Ottewill and D. J. Wilkins, *Trans. Faraday Soc.*, 1962, **58**, 608.
19. C. J. Force and E. Matijević, *Kolloid Z. Polym.*, 1968, **225**, 33.
20. B. V. Derjaguin and L. Landau, *Acta Physicochim. URSS*, 1941, **14**, 633.
21. E. J. W. Verwey and J. Th. G. Overbeek, 'Theory of Stability of Lyophobic Colloids,' Elsevier, Amsterdam, 1948.
22. H. C. Hamaker, *Physica*, 1937, **4**, 1058.

Aspects of Stability in Milk and Milk Products

By Douglas G. Dalgleish

HANNAH RESEARCH INSTITUTE, AYR KA6 5HL, UK

1 Introduction

Although milk is often regarded as the archetypal food colloidal system, extensive research has made it clear that milk is something less than ideal when its colloidal properties are studied in detail—or rather that the view taken of milk has been over-simplified. This arises from the fact that the particles in milk, particularly the casein micelles, have structures and properties which change with circumstances and make it difficult to apply to them the detailed principles of colloid chemistry. Hence, although much is known about the structure and properties of the casein micelle at its natural pH and in its natural milieu, milk, the understanding of the behaviour becomes less as the conditions are changed. Processing, of course, causes just such changes in conditions.

In this paper an attempt is made to clarify this assertion by considering the behaviour of a number of dairy products, or rather the particles of which they are composed, with reference to their stability or instability. The stability of unprocessed or simply pasteurized milk is generally assured, except in the worst cases of contamination by microorganisms; it is in the area of dairy products that problems arising from instability become significant. On the other hand, processing is often dependent on controlled instability, such as the renneting of milk to give a cheese curd, and the acidification which leads to the formation of yogurts and cottage cheeses.

2 Milk and its Constituents

Cow's milk contains on average about 40 g l^{-1} of fat and 32 g l^{-1} of protein, of which approximately 25 g l^{-1} is casein and 7 g l^{-1} is serum proteins (predominantly β-lactoglobulin and α-lactalbumin).[1] The latter proteins exist in solution as monomers or dimers, but the caseins form the so-called casein micelles, which are complexes of caseins with calcium phosphate,[2] forming particles of diameter 50–300 nm,[3,4] containing thousands of individual protein molecules. It is on the properties of the caseins and casein micelles that the properties of milk largely depend. Like the caseins, the milk fat is packaged into particles (natural fat globules) with diameters in the range 1–10 μm, which are stabilized by surface layers of phospholipid (the natural fat globule membrane).[5] In the milk as secreted there is little interaction between the major protein classes and the fat because the casein micelles are intact and possess hydrophilic surfaces and the fat globules also have naturally hydrophilic surfaces. In addition, in untreated milk there are no interactions between the casein micelles and the serum proteins to disturb the integrity of the casein micelles.

The structures of the casein micelles have been much debated,[6-9] and agreement on the details of their structures is by no means complete. Four types of casein (α_{s1}, α_{s2}, β, and κ) go to make up the micelles, of which the α_s- and β-caseins combine with calcium phosphate to form the interior of the micelle, while the κ-casein appears to be located on or near to the micellar surface.[10] It is this surface layer of κ-casein which has the major effect of determining the stability of the casein micelles, and its modification will have important effects on micellar behaviour.[11-13] On the other hand, the other caseins of the micelle contribute more to the structure and less to the stability; they may be removed (especially the β-caseins) at least in part without a significant increase in the micellar instability. This type of observation is of some importance, because it implies that there is no unique compositional definition of a casein micelle (except perhaps a functional one). Certainly, the casein micelles cannot be treated as hard-core particles; not only does the presence of the surface layer of κ-casein lead to steric stabilization,[11-13] but the core itself is subject to variation depending on the conditions of the experiment or the way the milk has been treated.

An example of this is the partial dissociation of casein micelles which occurs as a result of cooling and storage at 4 °C; under these conditions about 40% of the β-casein can dissociate from the micelles.[14] Similarly, dissociation of casein occurs when the pH of the milk is decreased from its natural value of about 6.7.[15-17] The former of these changes appears to be reversible, insofar as only small changes in behaviour are found when the milk is rewarmed. However, alteration of pH is a different matter, and it is probable that the pH-induced alterations are not truly reversible. This reflects the fact that the major effect of a decrease in pH is to dissolve quantities of calcium phosphate from the micelles, and that it is difficult to re-form it in its original state as the pH is raised again.

The casein micelle must therefore be viewed as a variable particle which may alter its nature during processing.

3 Processing of Milk

Figures 1 and 2 show simply a number of methods of processing milk, leading to defined products. A number of products (*e.g.* yogurt, cheese, cottage cheese) depend on the controlled development of a measure of instability leading to coagulation, by acidification or renneting (addition of a specific proteolytic enzyme), or both. On the other hand, there are many milk products (UHT milk and cream, homogenized/evaporated milk, and casein-stabilized emulsions) which are intended to have long shelf-lives, and in which instability cannot be allowed to develop.

In both of these classes of product, difficulties can be encountered, either in maintaining stability or in developing controlled instability. In the long-life systems, processes of age-gelation occur, some of which may be explained by

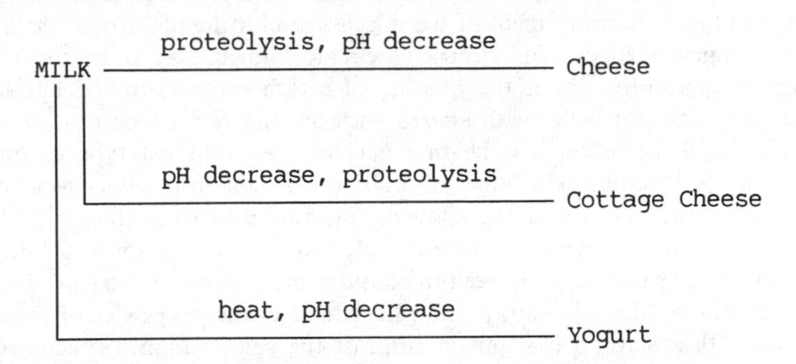

Figure 1

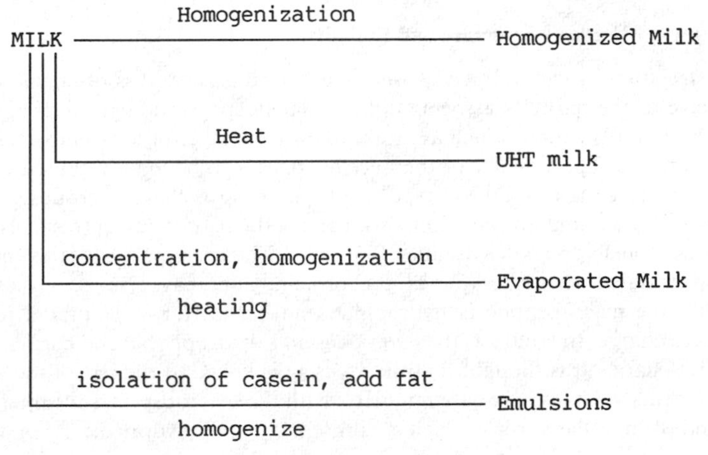

Figure 2

chemical change (*e.g.* proteolysis[18]), and some by physical processes[19] (or in many instances cannot yet be explained). Manufacture of cheeses and yogurts depends on the controlled development of instability, according to well defined operational parameters, and departures from the norm cannot be tolerated. An obvious example of this is in the formation of cheese curd, where the rates of acidification by starter cultures and of the renneting reaction must be approximately the same. If the activity of the starter culture responsible for developing acidity is altered, then the curd will be of inferior quality. The reasons for this are partly connected with the changes in casein micelles induced by acidification and partly arise from the rates and mechanisms of aggregation reactions and their pH dependences.

Processing by its nature involves the imposition of certain reactions on a changing colloidal system, because the colloidal particles in the milk alter their nature and behaviour during processing. For example, changing pH causes disintegration or rearrangement of the micelles, and if the pH drops low enough (about 5.2) new particles of isoelectric casein are formed.[15] Also, heating to high enough temperatures causes the binding of serum proteins to the micellar κ-casein, to produce micelles with altered surfaces and new properties.[20] Combinations of pH alteration and heating may cause a different type of micellar breakdown.[21,22] Homogenization of milk breaks the frameworks of the fat globules and the casein micelles, allowing interaction between them,[23,24] but in doing so alters the properties of the incomplete micelles bound to the fat surfaces. These examples emphasize the need to consider each milk and dairy product as a separate system. Much is known and understood of the properties of the casein micelles at their natural pH, but in contrast the compositions, structures, and properties of the particles formed during processing are by no means fully investigated or understood, and it is to these which research should be directed.

4 Casein Micelles—Structure and Stability

The structures of casein micelles were mentioned earlier. It is often sufficient to conceive of the particles as being approximately spherical aggregates of protein and calcium phosphate which are stabilized by the layer of κ-casein which forms their surfaces. The stability of the micelle appears to be mainly caused by steric factors, rather than by DLVO-type repulsion between charged groups.[25] The κ-casein molecule contains two domains, namely the *para*-κ-casein (residues 1–105) and caseinomacropeptide (residues 106–169).[1] The latter part contains the single phosphorylated serine residue of the protein, and may also be glycosylated: overall, the macropeptide is hydrophilic and carries the bulk of the protein's negative charge. In contrast, the *para*-κ-casein is hydrophobic and carries a slight positive charge. It is thought that the κ-casein is linked to the rest of the micellar components via its *para*-κ-casein moiety, with the macropeptides forming a layer around the micellar surface which is diffuse and highly hydrated,[26,27] in addition to carrying negative charge.[28–30]

The charges on the casein micelles do not, however, appear to be the source of micellar stability.[25] Consequently, studies of micellar stability have more recently centred around steric stabilization, using a number of ways to describe the interactions between the particles.[11,31,32] One, which appears to be self-consistent,[11] suggests that the macropeptides form a layer *ca.* 12 nm thick around the micelle, with each molecule of κ-casein in this layer occupying an area of *ca.* 480 nm^2. This model appears to place only about 12% of the κ-casein on the micellar surface, compared with original estimates of 100%.[10] However, it appears probably that either (i) the micelles are not truly spherical (*i.e.* they have 'bumpy' surfaces) or (ii) the κ-casein may exist as polymeric units.[33] In either case, not all of the κ-casein will be observed by hydrodynamic or electrophoretic measurements. Alternatively, the κ-casein may be mostly near, rather than on, the surface.

Loss of steric stabilization is achieved by removing the macropeptide moieties of the κ-casein molecules. This also has the effect of diminishing the micellar charge.[28-30] The action of rennet, however, is consistent with the removal of the sterically stabilizing layer,[12] although there are arguments on the appropriate way of calculating the interaction energy of particles whose steric stabilization is being destroyed in this way.[31,32] Collapsing the hairy layer by the addition of ethanol leads to micellar aggregation,[13] although the mechanism of this is apparently more complex than a simple loss of steric stabilization. Even when the hairy layer is collapsed, there is still a residual stability, although this is readily destroyed when rennet removes the macropeptide. It is probable, therefore, that although the main mechanism of stabilization of casein micelles is steric, the charges on the macropeptides may also play a part. However, the relative contributions of the two effects have not been established in detail.

Therefore, with respect to aggregation at neutral pH, casein micelles are relatively stable unless the hairy layer is destroyed. However, a second cause of instability results from the dissociation of the micelles, and this can be achieved either by decreasing the pH of milk[15-17] or, to a limited extent, by cooling.[14] In the latter instance, β-casein appears to be the only casein liberated into the solution (suggesting a role for hydrophobic bonding in the manner of incorporation of this casein into the micellar structure), but in the former all types of caseins are dissociated at temperatures of 20 °C and below.[17]

The effect of reducing the pH is to dissociate, partly or wholly, the micellar calcium phosphate (Figure 3a),[15] which acts as a cement binding the caseins together.[34,35] Removal of the calcium phosphate by EDTA treatment at pH 7.0 causes almost complete dissociation of the micelle.[36] However, at lower pH values caused by acidification there is some loss of protein from the micelles but dissociation is by no means complete, especially at 30 °C (Figure 3b).[17] At pH values below 5.2 the caseins begin to precipitate isoelectrically, so that particles begin to re-form.[15,37,38] It is therefore apparent that the structures of casein micelles in milks which have been acidified to pH values below the natural pH contain particles which are not casein micelles (as defined by their properties of composition and stability at pH 6.7). As the pH is increasingly lowered, they cease to behave as casein micelles.[16] The properties of these particles are not well understood, principally because of limited research. However, it is just these properties which are critical in defining the properties of a cheese curd or yogurt. Because this change in properties changes the natures of the particles, it deprives the researcher of one most useful variation (behaviour of a constant particle with changing pH) which might have allowed exploration of the stability and structure of casein micelles in more detail.

5 Homogenized Milk

During the act of homogenization, the natural fat globules are broken down into much smaller units, with diameters of less than 1 μm. The newly created fat surface is coated sparsely with apparently intact casein micelles, and with fragments of casein micelles.[23,24,39] The stability of these synthetic milk fat globules therefore depends largely on the casein, and indeed the particles share some properties with casein micelles.

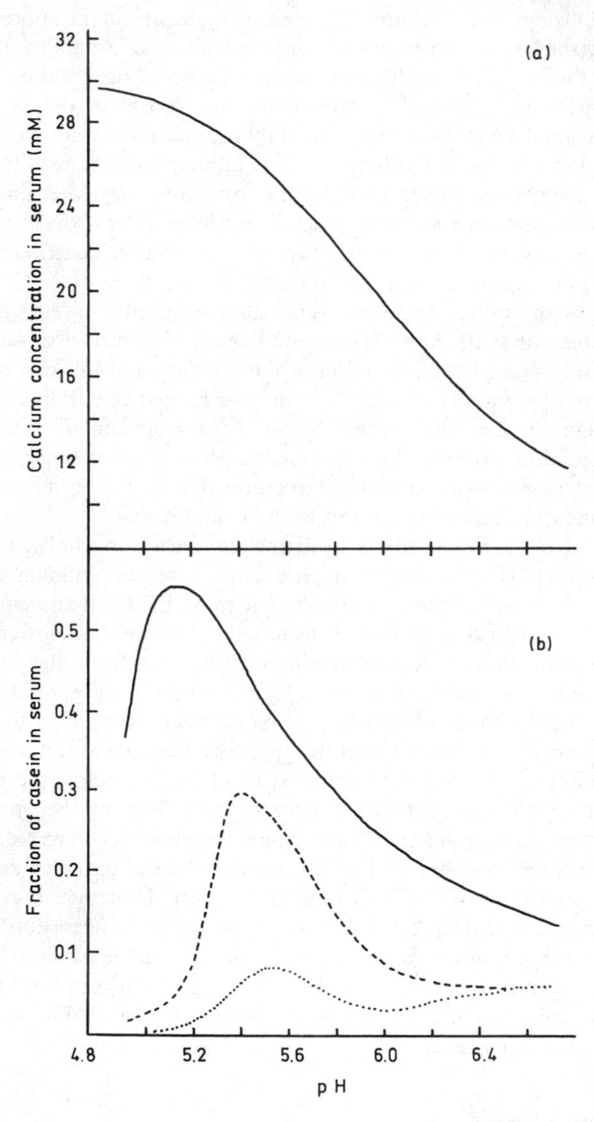

Figure 3 *Dissociation of casein micelles as the pH is lowered. (a) Solubilization of calcium phosphate as the pH of milk is decreased from 6.7. The curve shown is for the amount of calcium in the milk serum and represents a typical experimental curve. (b) Solubilization of casein during the pH adjustment. Results are shown for* (——) 4 °C, (- - -) 20 °C, *and* (···) 30 °C

Notable among these shared properties are the renneting processes. As in the case of intact casein micelles, the rennet attacks the κ-casein of the casein aggregates and casein micelles which are bound to the fat–water interface, and causes destabilization of the fat globules and consequent aggregation.[40-42] The stabilizing agency is therefore still κ-casein, rather than a mixture of the caseins (as in simple caseinate–fat emulsions). It is conceivable that the semi-intact micelles bound to the fat–water interface may perform the role of 'super steric' stabilizers. On the other hand, the breakdown of κ-casein may produce 'hot spots' through which attractive forces act (although the nature of such forces has not yet been investigated). The observation that in many instances the casein micelles are largely intact when they are bound to the surface suggests that synthetic fat globules should behave as large micelles; however, when the coverage of the surface is least, the particles are the least stable to rennet.[42] This creates something of a paradox: in the largest globules the micelles are furthest apart on the fat surface,[23,24] and there must be proportionately large amounts of monomeric or oligomeric caseins adsorbed on the surface between the more intact micelles as it is extremely unlikely that the fat globule surface is bare of protein. However, this protein appears to play little part in stabilizing the system as it would in a simple caseinate–oil emulsion. This may arise partly because the caseins, whether or not they are in micellar form, still retain their association with calcium phosphate. This will in turn partly neutralize the negative charge of the proteins and possibly also bind to the caseins in such a way as to prevent their acting as steric stabilizers. It is therefore probable that it is the large micellar fragments bound to the interface which govern the interactions of the fat globules, even although such fragments are far apart from one another.

It is unlikely that the apparently intact casein micelles on the surfaces of the fat globules are indeed in their native state, as native casein micelles, being thoroughly stabilized as described above, will not bind to hydrophobic surfaces. For example, intact casein micelles do not bind to polystyrene latices (D. G. Dalgleish, unpublished work). Therefore, it is unlikely that micelles rearrange as they approach a fat surface. The necessary disruption to the micellar structure is provided by the act of homogenization, and it has been demonstrated that such disruption and re-formation of casein micelles can occur during homogenization.[44] Hence, the adsorption of the apparently intact casein micelles is likely to be by means of interactions of micellar β- or α_s-caseins with the fat–water interface. This composition of the interfacial layer has not been conclusively demonstrated, and will in fact be difficult to establish, because it is evident that one type of casein can displace another at the interface.[45] Therefore, attempts to dissociate the adsorbed micellar fragments to leave the caseins actually on the interface may only lead to rearrangement of the surface material.

If the micelles are disrupted before adsorption, it might be expected that some κ-casein should be in solution. There is no evidence that this occurs and the location of all the κ-casein is uncertain. More research is needed to ascertain the details of the natures of the surfaces of fat globules, especially those of large size, having low protein-to-fat ratios, in homogenized milks.

If the altered stability of homogenized milk to rennet, compared with that of untreated milk, can be explained to some extent by the distribution of the κ-casein, the mechanisms by which homogenization alters the heat stability of milks, particularly concentrated milks, are virtually unknown. Homogenization generally decreases the coagulation time (*i.e.* increases the potential instability) at temperatures in the range 120–140 °C.[46] The mechanism of heat-induced coagulation even of skim or raw milk is not fully understood,[47] and the relationship between changes in the structures of the particles brought about by the homogenization and their effect on heat stability cannot be described in detail. It has been suggested that the increase in surface area occupied by the caseins (not all of which can be κ-caseins) leads to greater instability, which may be caused by either the precipitation of calcium phosphate or the interaction of denatured whey proteins and caseins, or other reactions. In this context the possibility that the κ-casein is re-distributed during homogenization may be relevant to the final properties of the homogenized milk.

6 Casein-stabilized Emulsions

Caseinate is used as an emulsifier in a number of products, and it is accepted that caseins are very highly surface active.[48,49] It is important to distinguish between, on the one hand, emulsions composed of oil and caseinate and, and on the other, homogenized whole or concentrated milks, because in the former there are no structures such as casein micelles, nor is there any calcium phosphate. Indeed, the presence of calcium is deleterious to the stability of such emulsions,[45] especially if ethanol is present.[50] The stabilizing layer of casein is, if not monomeric, very close to being so.[51] Moreover, the caseins on the surface can be displaced by those in solution, so that the composition of the surface layer can change according to circumstances.[45,51,52] It is therefore at least conceivable that the properties of emulsions stabilized by caseins may differ depending on their composition (*i.e.* the relative amounts of α_s- and β-caseins on the fat–water interface and whether there is sufficient serum casein to allow interchange).

There is, however, little evidence that the properties of the emulsions do change. As β-casein can displace α_s-casein from an interface, an emulsion originally formed using sodium caseinate will equilibrate to maximize the amount of β-casein actually adsorbed on the oil–water interface. The kinetics of this are such that equilibrium can be attained within a few hours at most.[51] Less complex binary mixtures of caseins require apparently much less time to equilibrate.[52] The slower equilibration of the sodium caseinate may result from a number of causes, among which are the possibility that the caseins are not monomeric and the observation that commercial sodium caseinates are altered chemically with respect to their laboratory analogues.

It is the latter factor which appears to be important in determining the stability of some liqueur products. Caseinates from different sources have different stability properties when used as emulsifiers in cream liqueurs,[53] for reasons yet to be fully explained. There are certainly detectable chemical differences shown by the altered elution profiles of the different caseinates from ion-exchange columns,[54] which probably arise from reactions occurring during drying of the product (Figure 4). The precise nature of these chemical changes has yet to be determined. The different caseinates exhibit different sensitivities to calcium ions;

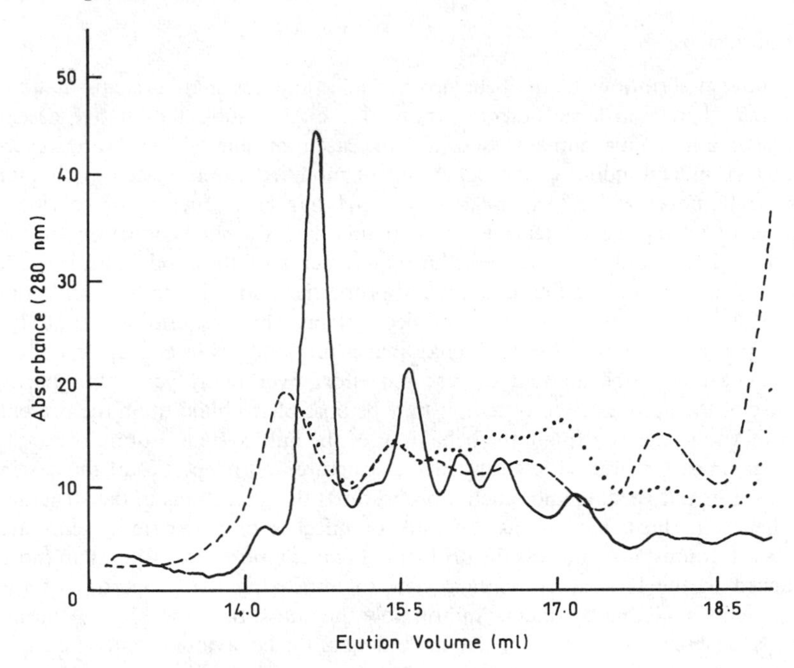

Figure 4 *Elution profiles of caseinates in the κ-casein region. (—) Laboratory freeze-dried caseinate; (- - -) spray-dried caseinate; (· · ·) roller-dried caseinate. Profiles determined using a Pharmacia FPLC system, with Mono-Q column, as described in reference 56*

the most sensitive are those prepared in the laboratory and freeze-dried, followed by the less sensitive spray-dried material, and finally by the very insensitive roller-dried material. The solubility of these caseinates decreased in proportion to their decreased sensitivity to calcium.[54] However, the greater the extent of modification of the casein, the greater was its ability to form a stable cream liqueur,[53] possibly because of the insensitivity to the presence of calcium. In contrast to casein micelles, whose properties are relatively independent of the ionic strength (insofar as the ionic strength can be varied without altering the structure of the particles), caseinate emulsions prepared from isolated α_{s1}- or β-caseins or from sodium caseinate show the behaviour to be expected from a colloidal system, namely reversible aggregation when the ionic strength is raised.[45,55] This indicates that caseinate emulsions depend more on charge stabilization and less on steric stabilization than do either casein micelles or synthetic milk fat globules. The properties of a caseinate-stabilized emulsion therefore depend on the composition of the mixture (proportions of fat to protein), on the homogenization system (which produces a particular droplet size distribution, and in turn defines how much of which caseins bind to the surface), and on the chemical composition of the caseinate. There appears to be no means at present of relating these variables to such factors as the overall stability of the product.

7 Conclusions

The above descriptions of the behaviour of milk and caseinate systems make it clear that there is a large body of knowledge on the subject of native casein systems in a milk environment (especially the casein micelles). There is, however, much less understanding of the behaviour of modified casein systems, whether they be homogenized milks, milks whose pH has been altered, or caseinate systems in which the proteins have been modified during production of the caseinate. If research is to address the requirements of the food industry with respect to understanding the fundamental properties and behaviour of milk and caseins in food systems, it is these modified systems which require further study.

It is perhaps only in respect of milks that a full study is in concept possible. Partly because of the amount of research effort over many years, but partly because of the nature of the system, it may be possible to build upon the present body of knowledge to explain the behaviour of the milk system. Future research, however, must be directed less toward the structure and properties of the casein micelle in its native state and much more towards the alterations in the structure and hence in the behaviour as the milk or micelles or caseinate systems are processed. It must also address the problem of real components, rather than those produced in the laboratory, with a view to improving the properties of the components available by understanding how they have been modified. Alternatively, such a study will allow the understanding of the behaviour of milk proteins, even in their modified forms, in the production of foods.

References

1. H. Swaisgood, in 'Developments in Dairy Chemistry—1,' ed. P. F. Fox. Elsevier Applied Science, Barking, 1982, p. 1:
2. D. G. Schmidt, in 'Developments in Dairy Chemistry—1,' Ed. P. F. Fox, Elsevier Applied Science, Barking, 1982, p. 61.
3. C. Holt, A. M. Kimber, B. Brooker, and J. H. Prentice, *J. Colloid. Interface Sci.*, 1978, **65**, 555.
4. D. G. Schmidt, P. Walstra, and W. Buchheim, *Neth. Milk Dairy J.*, 1973, **27**, 128.
5. H. Mulder and P. Walstra, 'The Milk Fat Globule,' Pudoc, Wageningen, 1974.
6. D. F. Waugh, in 'Milk Proteins—Chemistry and Molecular Biology, II,' ed. H. A. McKenzie, Academic Press, New York, 1971, p. 3.
7. D. Rose, *Dairy Sci. Abstr.*, 1969, **31**, 171.
8. J. Garnier and B. Ribadeau Dumas, *J. Dairy Res.*, 1970, **37**, 493.
9. C. W. Slattery and C. Evard, *Biochim. Biophys. Acta*, 1973, **317**, 529.
10. W. H. Donnelly, G. P. McNeill, W. Buchheim, and T. C. A. McGann, *Biochim. Biophys. Acta*, 1984, **789**, 136.
11. C. Holt and D. G. Dalgleish, *J. Colloid Interface Sci.*, 1986, **114**, 513.
12. P. Walstra, V. A. Bloomfield, G. J. Wei, and R. Jenness, *Biochim. Biophys. Acta*, 1981, **669**, 258.
13. D. S. Horne, *Biopolymers*, 1984, **23**, 989.
14. D. T. Davies and A. J. R. Law, *J. Dairy Res.*, 1983, **50**, 67.
15. A. C. M. van Hooydonk, H. G. Hagedoorn, and J. J. Boerrigter, *Neth. Milk Dairy J.*, 1986, **40**, 281.
16. A. C. M. van Hooydonk, J. J. Boerrigter, and H. G. Hagedoorn, *Neth. Milk Dairy J.*, 1986, **40**, 297.
17. D. G. Dalgleish and A. J. R. Law, *J. Dairy Res.*, 1988, **55**, 529.
18. B. A. Law, A. T. Andrews, and M. E. Sharpe, *J. Dairy Res.*, 1977, **44**, 145.

19. A. T. Andrews, *J. Dairy Res.*, 1875, **42**, 89.
20. P. Smits and J. H. van Brouwershaven, *J. Dairy Res.*, 1980, **47**, 313.
21. H. Singh and P. F. Fox, *J. Dairy Res.*, 1985, **52**, 65.
22. D. G. Dalgleish, Y. Pouliot, and P. Paquin, *J. Dairy Res.*, 1987, **54**, 39.
23. H. Oortwijn, P. Walstra, and H. Mulder, *Neth. Milk Dairy J.*, 1977, **31**, 134.
24. H. Oortwijn and P. Walstra, *Neth. Milk Dairy J.*, 1979, **33**, 134.
25. T. A. J. Payens, *J. Dairy Res.*, 1979, **46**, 291.
26. P. Walstra, *J. Dairy Res.*, 1979, **46**, 317.
27. M. C. A. Griffin and G. C. K. Roberts, *Biochem. J.*, 1985, **228**, 273.
28. M. L. Green and G. Crutchfield, *J. Dairy Res.*, 1971, **38**, 151.
29. K. N. Pearce, *J. Dairy Res.*, 1976, **43**, 27.
30. D. F. Darling and J. Dickson, *J. Dairy Res.*, 1979, **46**, 441.
31. D. F. Darling and A. C. M. van Hooydonk, *J. Dairy Res.*, 1981, **48**, 189.
32. D. G. Dalgleish and C. Holt, *J. Colloid Interface Sci.*, 1988, **123**, 80.
33. J. H. Woychik, E. B. Kalan, and M. E. Noelken, *Biochemistry*, 1966, **5**, 2276.
34. C. Holt, D. T. Davies, and A. J. R. Law, *J. Dairy Res.*, 1986, **53**, 557.
35. L. C. Chaplin, *J. Dairy Res.*, 1984, **51**, 251.
36. C. Holt, *J. Dairy Res.*, 1982, **49**, 29.
37. S. P. F. M. Roefs, P. Walstra, D. G. Dalgleish, and D. S. Horne, *Neth. Milk Dairy J.*, 1985, **39**, 119.
38. D. Rose, *J. Dairy Sci.*, 1968, **51**, 1897.
39. P. Walstra and H. Oortwijn, *Neth. Milk Dairy J.*, 1982, **36**, 103.
40. G. Humbert, A. Driou, J. Guerin, and C. Alais, *Lait*, 1980, **60**, 574.
41. E. W. Robson and D. G. Dalgleish, *J. Dairy Res.*, 1984, **51**, 417.
42. E. W. Robson and D. G. Dalgleish, in 'Food Emulsions and Foams,' ed. E. Dickinson, Special Publication No. 58, Royal Society of Chemistry, London, 1987, p. 65.
43. D. G. Dalgleish and E. W. Robson, *J. Dairy Res.*, 1985, **52**, 539.
44. P. Walstra, *Neth. Milk Dairy J.*, 1980, **34**, 181.
45. E. Dickinson, R. H. Whyman, and D. G. Dalgleish, in 'Food Emulsions and Foams,' ed. E. Dickinson, Special Publication No. 58, Royal Society of Chemistry, London, 1987, p. 40.
46. A. W. M. Sweetsur and D. D. Muir, *J. Dairy Res.*, 1983, **50**, 291.
47. D. D. Muir, *Int. J. Biochem.*, 1985, **17**, 291.
48. E. Dickinson, *Food Hydrocolloids*, 1986, **1**, 3.
49. D. E. Graham and M. C. Phillips, *J. Colloid Interface Sci.*, 1979, **70**, 403, 415, 427.
50. W. Banks, D. D. Muir, and A. G. Wilson, *J. Food Technol.*, 1981, **16**, 587.
51. E. W. Robson and D. G. Dalgleish, *J. Food Sci.*, 1987, **52**, 1694.
52. E. Dickinson, S. E. Rolfe, and D. G. Dalgleish, this volume.
53. D. D. Muir and D. G. Dalgleish, *Milchwissenschaft*, 1987, **42**, 770.
54. D. G. Dalgleish and A. J. R. Law, *J. Soc. Dairy Technol.*, 1988, **41**, 1.
55. S. Bullin, E. Dickinson, S. J. Impey, S. K. Narhan, and G. Stainsby, in 'Gums and Stabilizers for the Food Industry,' eds. G. O. Phillips, D. J. Wedlock and P. A. Williams, Vol. 4, Elsevier Applied Science, Barking, 1987, p. 337.
56. D. T. Davies and A. J. R. Law, *J. Dairy Res.*, 1987, **54**, 369.

Ultrasonic Measurements in Food Emulsions and Dispersions

By Malcolm J. W. Povey and David J. McClements

PROCTER DEPARTMENT OF FOOD SCIENCE, UNIVERSITY OF LEEDS, LEEDS LS2 9JT, UK

1 Ultrasonic Velocity Measurement

The measurement of ultrasonic velocity is relatively straightforward and can often be conducted with very high accuracy (*ca.* 0.1%) and with relatively cheap (*ca.* £6k) and robust equipment. The probes are normally non-invasive and cheap (*ca.* £200 for a pair).

Unlike the high-power ($>10\,\mathrm{kW\,m^{-2}}$) applications of ultrasonics generally found in the laboratory for cleaning, cell disintegration, and emulsification purposes, a 'pitch and catch' technique is used for velocity measurements. This is a low-power ($<100\,\mathrm{mW^{-2}}$) non-destructive technique operating in the pulsed mode rather than the continuous operation of high-power ultrasonics. Although any frequency above about 16 kHz is considered to be 'ultrasonic,' the frequency range 100 kHz–10 MHz is generally used and a pulsed technique is employed. The power levels are also much lower; average power levels are less than 100 mW, compared with $10\,\mathrm{kW\,m^{-2}}$ for the continuous ultrasonic irradiation needed in high-power applications which also operate at much lower frequencies (*ca.* 20–50 kHz).

There have been no reports of hazards associated with the use of the ultrasonic pulse echo technique in process industries and overall the technique can be regarded as non-invasive, non-destructive, and safe. Information can be obtained from the ultrasonic signal about the physical properties of foods which is unavailable by other means. For these reasons, ultrasonics is being taken seriously as a sensor technology for the food industry.

The pulse echo technique is depicted in Figure 1. A piezoelectric transducer is used to convert an electrical signal into a mechanical signal, which is transmitted across a bond between the transducer and the sample. The transducer is excited by a pulsed electrical field of a few nanoseconds duration. The field can be many hundreds of volts in magnitude. Once inside the sample, the pulse travels in a manner analogous to that of a shout in a canyon. It is trapped inside because of the acoustical mismatch between the cell walls and the sample, only small amounts of energy escaping at each reflection. It is these small amounts of energy leaking out of the sample which are detected in a reverse manner to that by which

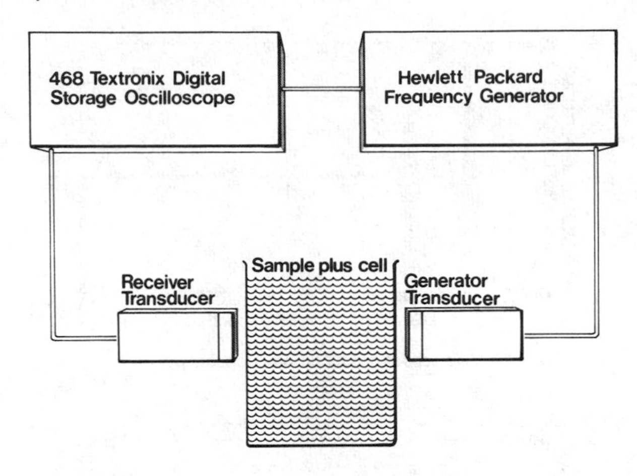

Figure 1 *Block diagram of an ultrasonic velocity measurement apparatus*

they were generated. Either a second transducer may be used or the original, generating transducer may be used as the detector. The two-probe system has the advantage of being more sensitive.

An 'echo train' is generated, which can be viewed on an oscilloscope (Figure 2) or used to trigger a timer–counter. The time between successive echoes, is then measured. If the distance travelled by the pulse is known, then the velocity of sound can be calculated. Using this technique, a precision of $0.1\ m\ s^{-1}$ in $1500\ m\ s^{-1}$ can be routinely obtained and apparatus is on sale which claims to improve on this figure 10-fold. Accuracies of $1\ m\ s^{-1}$ are easily obtained.

The pulse actually consists of a number of cycles of mechanical compressions and rarefractions, whose frequency is the ultrasonic frequency. This frequency is normally determined by the material and the dimensions of the transducer. To all intents and purposes it is determined by the manufacturer of the transducer, although sometimes the specification can be unreliable. This frequency can make a large difference to the results obtained, as will be seen later.

In addition, the acoustical matching of the probe needs to be checked. Assuming that only 'off-the-shelf' probes are used, then there are only two types of matching to be considered: either the probes are matched acoustically to steel, or they are matched to water; which is more appropriate depends on the application. For emulsion work, the water-matched probes will be more suitable. For solid dispersions, the steel-matched probes may be better, although not necessarily so. The 'matching factor' is called the acoustical impedance and can be calculated from the product of density and the velocity of ultrasound.

As the path length and the beam profile of the ultrasound need to be well defined, this will normally mean that a special cell must be designed for a given application. Such cells can usually be built to fit in-line, however, and need provide no obstruction to flow in continuous flow applications. Manufacturers

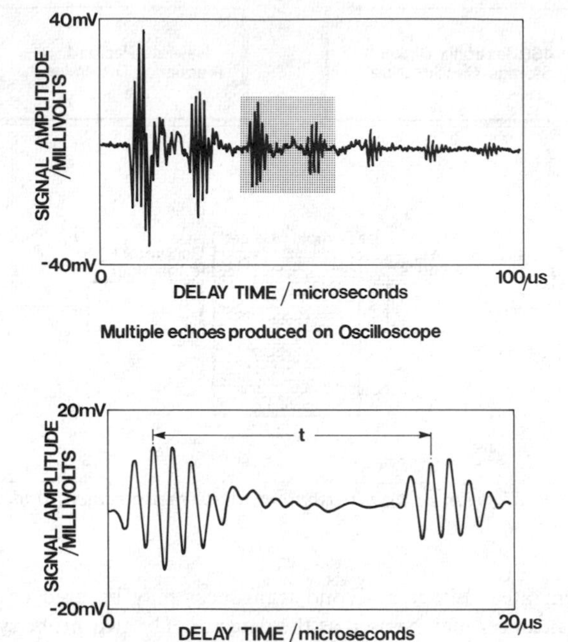

Figure 2 *Pulse echo train from the apparatus in Figure 1, detected by an oscilloscope*

such as Paar Scientific already make such cells which are capable of being cleaned in place and of operating at pressures up to 5 bar.

The pulse echo technique described above will normally be restricted to working at one frequency. In many applications this could be a disadvantage and efforts are now being directed to the design of ultrasonic cells capable of providing information about ultrasonic velocity over a wide range of frequencies. However, such equipment is in its infancy.

The pulse echo train represented in Figure 2 is also capable of providing information about the absorption of the ultrasound in the sample. However, this can only be done at much lower levels of accuracies, of the order of 5% at best. The technique involves fitting an exponential to the peaks of successive echoes and determining its exponent. This, when related to the pulse path length, gives the attenuation coefficient for ultrasonics in the sample of interest. Because of the difficulties involved in making attenuation measurements, little attention will be given to this subject in this paper.

2 The Wood Equation

Introduction—The simplest formulation for the velocity of ultrasound in dispersions is given by the Wood equation. It involves assuming that both density and

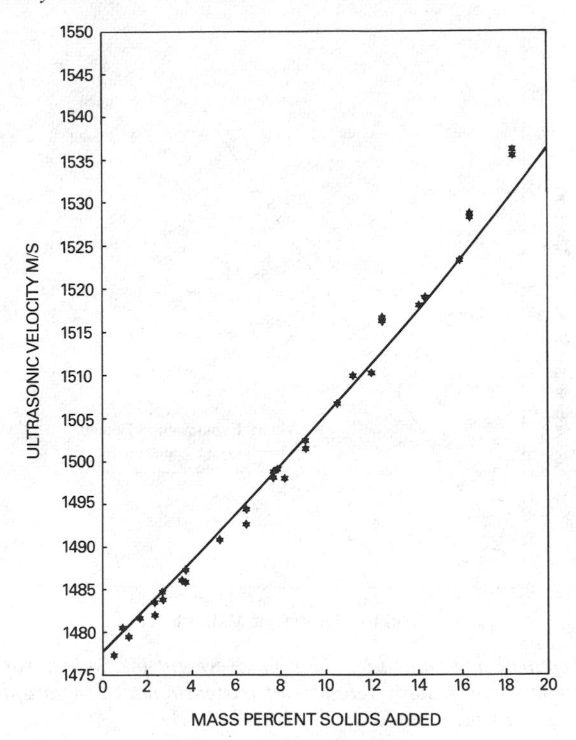

Figure 3 *Experimental data for ultrasonic velocity against solid content for tristearin added to paraffin oil. The measurements were made at* 25 °C *using* 1 MHz *ultrasonics*

adiabatic compressibility are volume averages in the dispersion. These volume averages are then substituted into the equation for the velocity of a compressional ultrasonic wave as follows:

$$\left.\begin{aligned} \rho &= \rho_1\varphi + \rho_2(1-\varphi) \\ \beta &= \beta_1\varphi + \beta_2(1-\varphi) \\ \mathrm{v} &= (\rho\beta)^{-\frac{1}{2}} \end{aligned}\right\} \qquad (1)$$

where ρ is the density of the dispersion/emulsion, β is the adiabatic compressibility, φ is the volume fraction of the dispersed phase, the subscripts 1 and 2 refer to the dispersed and continuous phases, respectively, and v is the velocity of compressional ultrasound.

Until recently, it has been assumed that the Wood equation had a very wide application.[2] Such an assumption certainly simplifies the problem of data reduction in ultrasonics. Equation (1) gives a parabolic fit between velocity and volume fraction. As the density and compressibility of the two constituent phases can be determined from ultrasonic measurements of the components in isolation, then the volume fraction of the dispersed phase can be obtained. This is the basis of the determination of solid content in crystallizing fats.[3]

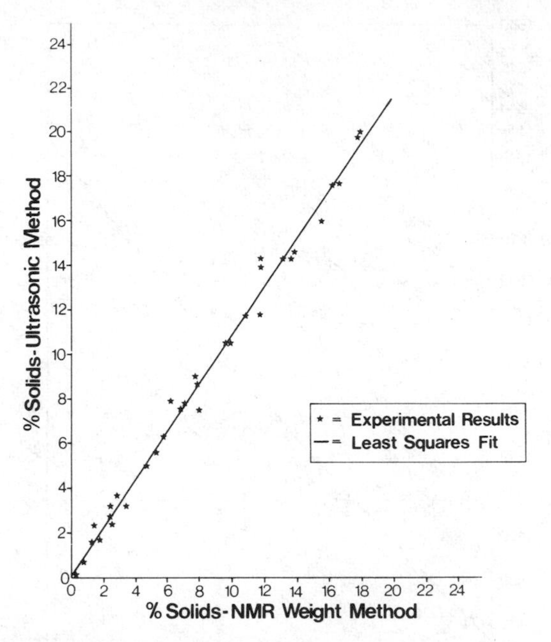

Figure 4 *Solid content determined by the pNMR weight method* vs. *solid content determined by ultrasonic velocity for tristearin added to paraffin oil. The measurements were made at* 18 °C *and* 1 MHz.

However, recent evidence, which will be discussed later, suggests that the Wood equation only has limited application for water-in-oil and oil-in-water emulsions. Nevertheless, we have shown that it works very well for mixtures of triglycerides containing dispersions of solid fat crystals.[4] We will examine this work first as it gives a good idea of how ultrasonics may be applied to a specific food system.

Crystallizing Fats—Using the technique described above, together with the Wood equation, the mass fraction of triglyceride crystal suspended in paraffin oil was determined ultrasonically.[3] For a triglyceride such as tristearin, which has negligible solubility in paraffin oil, an extremely good fit between velocity and mass fraction of crystal was obtained (Figure 3). A comparison with the pNMR technique is also shown (Figure 4). If the triglyceride was soluble in the paraffin oil, then this could be detected (Figure 5). To a first approximation, the ultrasonic velocity was found to be independent of the type of solid triglyceride present.

Empirical relationships between ultrasound velocity, the carbon number, and the degree of unsaturation of liquid triglycerides have also been established,[4] permitting either the prediction of ultrasound velocity from the molecular formulae of a known mixture of triglycerides or, alternatively, using the technique as a means of characterizing liquid oils.

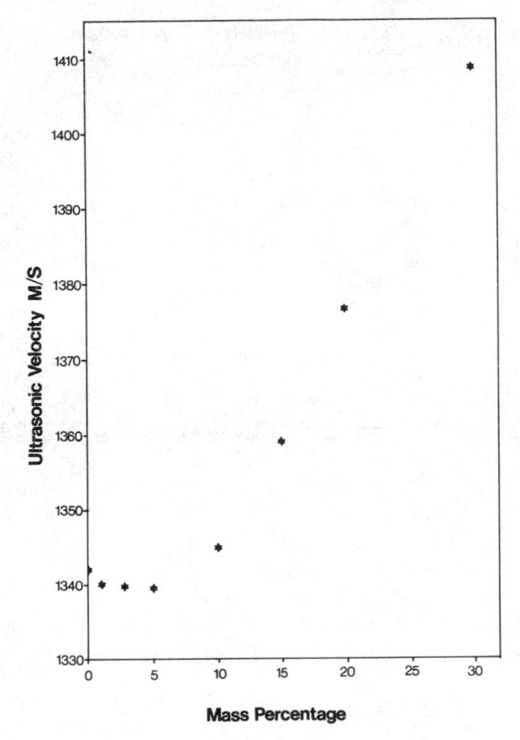

Figure 5 *Plot of ultrasonic velocity* vs. *mass of tristearin added to paraffin oil at* 56°C. *Initially, the tristearin dissolves in the paraffin oil, depressing the velocity. Once the solution becomes saturated and the tristearin enters as solid, the velocity then rises, following the same slope as other triglycerides*

3 Scattering Theory

Weak Scattering—We are now convinced that the Wood equation cannot, except fortuitously, work with emulsions consisting of two phases which are significantly different in either density or thermal diffusivity. On the face of it this would appear to rule out ultrasonics as a useful technique for characterizing emulsions. However, we shall show that, on the contrary, more useful information can be obtained, provided that the frequency can be varied. To understand why this is so, it is necessary to consider scattering theory.

In Figure 6, the ultrasonic field scattered by an oil droplet suspended in water is represented. The scattering is assumed to be weak, that is, the intensity of the field scattered by the particle is much lower than that of the incident field. For this to be the case it is necessary for the particle to be acoustically reasonably well matched to its surroundings and for the scattering to be non-resonant. Resonant scattering will be dealt with later. Two types of weak scattering are of importance

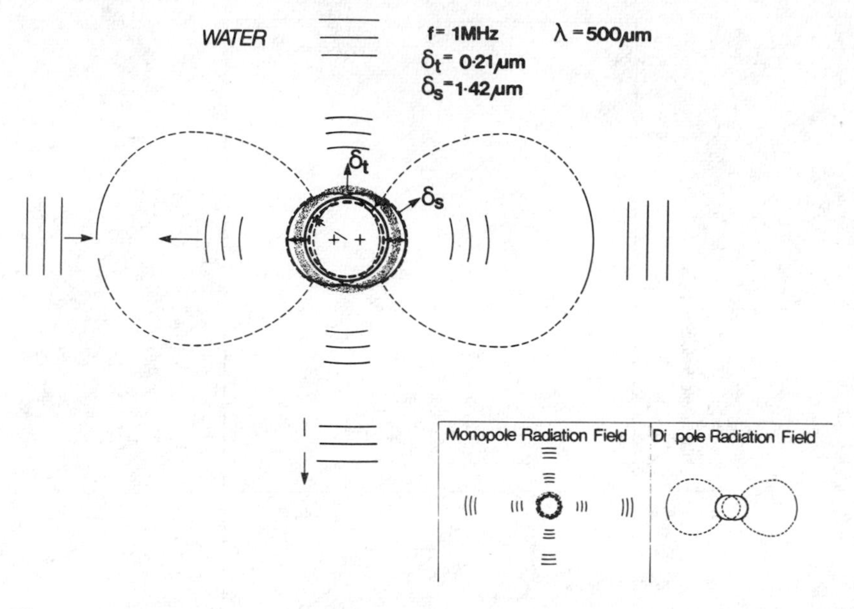

Figure 6 *Representation of the ultrasonic field created by scattering from an oil droplet of radius* r *suspended in water at* 20 °C. *A frequency of* 1 MHz *is used.*

in emulsions, thermal and visco-inertial. These two types of scattered ultrasound both remove energy from the forward beam, which would normally be detected by the receiver. It will appear, therefore, as if the ultrasound has been adsorbed from the beam by the particles. In reality, part of the beam has been redirected in another direction. Hence the signal will appear to have been attenuated and this signal reduction can be measured and related to the properties of the dispersion. The velocity is also affected.

Scattering therefore acts in a different way to ultrasonic absorption, in which the ultrasound is attenuated by direct conversion into heat by a variety of molecular relaxation processes, although its effect cannot be distinguished by a transducer sampling the forward beam.

Thermal Scattering—The frequency chosen for the illustration, 1 MHz, is typical of many ultrasonic measurements. The incoming ultrasonic wave will cause the particle to pulsate in response to the successive compressions and rarefractions associated with the wave. Associated with these pulsations is a rise and fall in temperature. However, within the particle, because of its differing compressibility and heat capacity, the temperature fluctuation will be different from its surroundings. Heat conduction then becomes possible. This will cause the temperature fluctuations to move out of phase with the incident ultrasound. The temperature fluctuations can themselves become a source of ultrasound, this time out of phase

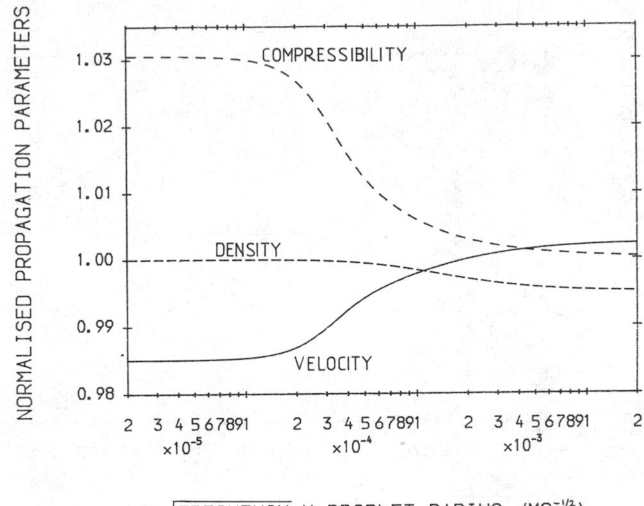

Figure 7 *Effective density, effective compressibility, and velocity plotted against* $f^{\frac{1}{2}}r$ *for a hexadecane-in-water emulsion at room temperature. The emulsion contains a volume fraction of* 0.1 *hexadecane*

with the incident wave. It is in this way that the thermal properties of the dispersion can become a source of ultrasonic scattering. The penetration depth of the thermal wave scattered by the particle is very small, 0.21 μm in our illustration. The scattered ultrasound will be spherically symmetrical, in contrast to the incident plane wave. This type of radiation is termed 'monopole.'

Visco-inertial Scattering—A second type of weak scattering is also illustrated in Figure 6. If the particle has a density which is different from its surroundings, it will move relative to the continuum in response to the incoming ultrasound because it will have a different inertia. However, this relative motion will be damped if the continuum is visco-elastic, viscous drag reducing the amplitude of the motion. The motion of the particle relative to its surroundings gives rise to two different scattered fields, a shear wave and a compressional wave. In an emulsion, the continuum is liquid and a shear wave will be highly damped. In our illustration it will have a penetration depth of 1.42 μm. The scattered ultrasound will have a $\cos\theta$ dependence on angle, having a form termed 'dipolar.'

In general, the effects of visco-inertial and thermal scattering are intermingled. However, when the ultrasonic wavelength is much greater than the droplet size, which is the case for most ultrasonic measurements in emulsions, they can be considered to act independently. Thermal scattering causes excess attenuation and leads to an effective compressibility of the emulsion which is greater than the volume average value [equation (1)] by an amount which depends on the particle size, the frequency used and the difference between the thermal properties of the

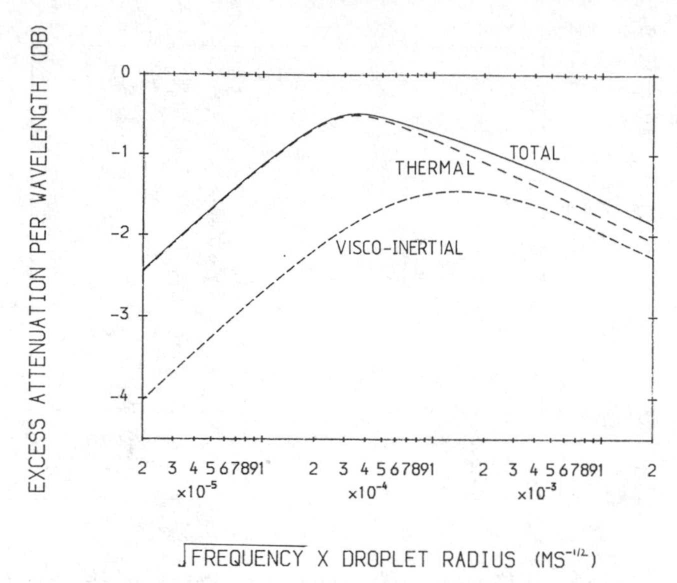

Figure 8 *Contributions of the visco-inertial and thermal attenuations per wavelength to the overall attenuation in the hexadecane-in-water emulsion described in Figure 7*

component phases. Viscous scattering also causes excess attenuation but leads to an effective density which is lower than the volume average value [equation (1)] by an amount which depends on the particle size, the frequency used, and the density difference between the phases. The overall attenuation in an emulsion is the sum of the viscous and thermal attenuations, whereas the velocity is determined by the magnitude of the effective density and compressibility [equation (1)].

Figure 7 illustrates the dependence of the effective density, the effective compressibility, and the velocity on particle size and frequency for a hexadecane-in-water emulsion, where both visco-inertial and thermal scattering are important. At low frequencies and particle sizes the velocity has a lower limit which is determined by the difference in thermal properties between the two phases. At high frequencies and particle sizes the velocity has an upper limit which depends on the density difference between the two phases. In the intermediate region the velocity increases with increasing particle size and frequency. The contribution of the visco-inertial and thermal attenuations per wavelength to the overall attenuation figure, $\alpha\lambda$, is shown in Figure 8. It can be seen that $\alpha\lambda$ has a maximum value in the intermediate range but tends to zero in the high and low $f^{\frac{1}{2}}r$ limits, thermal scattering being the predominant source of scattering in this emulsion over most of the frequency range considered.

The dependence of the ultrasonic velocity and attenuation on particle size and frequency means that it should be possible to determine the particle size in emulsions by ultrasonic measurements.

Multiple Scattering and Self-consistency—Weak scattering is complicated by one other phenomenon and a mathematical poser. The phenomenon of multiple scattering is relatively easy to understand but very difficult to analyse. All it involves is the fact that the scattered wave will itself be scattered by other particles and may recombine to form part of the 'incoming' wave incident upon the particle. Without going into further detail we have used Lloyd and Berry's[5] treatment of the subject to incorporate multiple scattering into our formulations. We present for the first time in this paper a comparison between experiment and multiple scattering theory in the area of ultrasound propagation in emulsions.

The mathematical poser is the problem of 'self-consistency.' In practical terms, this means that a theory should give the correct results for each component of the emulsion/dispersion, when the other is absent. None of the theories that we use is self-consistent in this sense and some error is clearly to be expected as a result. However, because the theories are formulated to give the appropriate limit for the continuum and as in emulsions concentrations above 60% usually result in phase inversion, this does not present a great problem. Nevertheless, errors can be expected to be significant at high concentrations owing to lack of self-consistency. In the sunflower oil emulsions studied by us, it seems that the errors due to lack of self-consistency are greater than the differences between various treatments of the multiple scattering problem.

Strong and Resonant Scattering—Before detailing in specific terms a theoretical treatment of our results, the question of strong and resonant scattering needs to be addressed. Strong scattering occurs when the scattered field intensity is of the same order as that of the incident field. This can result in the disappearance of the direct signal altogether, although the ultrasonic energy will still be present in the system, bouncing around between individual scatterers. Under these circumstances, the scattered wave becomes incoherent, phase correlation between different parts of the wave having been lost altogether. Under these circumstances the ultrasonic energy, whilst still present in the system, becomes undetectable by normal means. In addition, strong scattering problems are much more difficult to deal with theoretically and we shall not consider them further here.

Although resonant scattering can also be regarded as a form of strong scattering, information can be extracted from this phenomenon. In particular, Gaunaurd and Uberall[6] have shown that, in the case of air bubbles suspended in water, scattering will be most intense at the Minnaert frequency, w_m:

$$w_m = [3\rho_b v_b/(\rho_f a^2)]^{\frac{1}{2}}$$

where ρ_b is the bubble density, ρ_f is the fluid continuum density, a is the bubble diameter, and v_b is the velocity of sound in the bubble. Below resonance, the sound velocity is reduced below its continuum value and is given by the Wood equation. It is frequency dependent in the region of the resonance and independent of frequency at high frequencies and has the continuum value (see Figure 9). Note the large effect on attenuation. Small bubbles of the order of tens of micrometres in diameter can have a large effect at low volume fractions on ultrasonic propagation at a few megahertz. Consequently, care needs to be taken

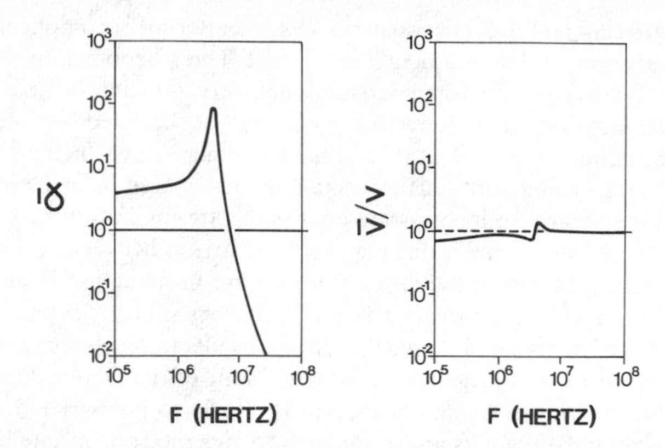

Figure 9 *Dependence of ultrasonic velocity and attenuation on frequency for bubbles of diameter* 1 μm *and a volume fraction of* 0.00001
(Adapted from *J. Acoust. Soc. Am.*, 1981, **69**, 362)

to remove any air from an emulsion before it is studied ultrasonically. Alternatively, this resonance effect could be used to characterize and detect the presence of air in an emulsion.

Calculation of Ultrasonic Velocity and Attenuation in Emulsions—We have used the approach adopted by Epstein and Carhart[7] to obtain the scattering coefficients. These coefficients are, in effect, the amplitudes of the scattered waves expanded as monopole, dipole, quadrupole, *etc.*, fields. The expansion is carried out as a set of spherical harmonics which are then fitted across the boundary between the particle and the continuum. For sunflower oil-in-water emulsions, only two terms in the expansion were needed. The approach of Lloyd and Berry[5] was used to combine the contributions from many particles into a dispersion. Formally, this approach is to be preferred to that of the more widely used method of Waterman and Truell[8] because it is more precise. The Waterman and Truell approach has been shown to be identical with that of Urick and Ament,[9] which we used in earlier work.[2] The method of Epstein and Carhart[7] differs from another well known treatment of this subject by Allegra and Hawley[10] in that their treatment applies to liquid particles whereas Allegra and Hawley's is appropriate to solid particles. There are errors in Epstein and Carhart's paper and the detailed formulation of the equations on which our computer calculations are based appear in a paper submitted elsewhere.[11]

Experimental Measurement of Ultrasonic Velocity in Emulsions

Variation with Particle Size and Frequency—Experimental measurements of the ultrasonic velocity in a series of sunflower oil-in-water emulsions were compared with the computer calculations and are presented in Figure 10. These values were calculated from measurements of the velocity in emulsions of varying particle size

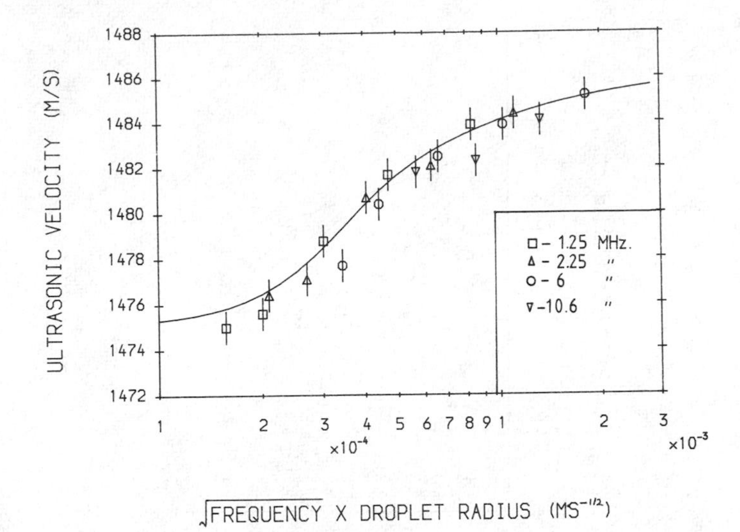

Figure 10 *Dependence of ultrasonic velocity,* v, *on* $f^{\frac{1}{2}}r$ *for a* 10% w/w *mass fraction of sunflower oil-in-water emulsion at* 20 °C. 2% w/w *Tween* 20 *was used as the emulsifier. The data were taken at frequencies of* 1.25, 2.25, 6.0, *and* 10.6 MHz

(0.14–0.74 μm) over a range of frequencies and were plotted against $f^{\frac{1}{2}}r$ as this term is proportional to the particle size relative to the viscous and thermal skin depths, *i.e.* r/δ_v and r/δ_t.

The effects of thermal scattering on ultrasonic velocity are clearly illustrated. For low frequencies and particle sizes the velocity falls appreciably below that predicted by the Wood equation (1486.2 m s^{-1}). As the frequency or droplet size increases the velocity increases, tending towards its upper limit for high values of these parameters. For sunflower oil-in-water this upper limit is close to the Wood prediction as viscous scattering is negligible in this system ($\rho/\rho' = 0.92$).

Figure 10 shows that there is excellent agreement between the theory and experiments. It is important to note that if the technique is to be used to determine particle sizes in emulsions, there is an upper and lower limit of particle sizes which can be detected, because the velocity has an upper and a lower limit. For example, if we assume that the practical limits for ultrasonic measurement are 0.1 and 100 MHz and if we take the upper and lower limits of velocity dispersion in Figure 10 as $f^{\frac{1}{2}}r \approx 2 \times 10^{-4}$–$2 \times 10^{-3}$, then the lower limit for particle sizing will be 20 nm at 100 MHz and the upper limit will be 6 μm at 0.1 MHz. These limits will depend on the emulsion and will be extended to larger sizes if visco-inertial scattering is significant.

Variation with Particle Concentration—The variation of the ultrasonic velocity with the mass fraction of oil droplets is shown in Figure 11. It can be seen that the

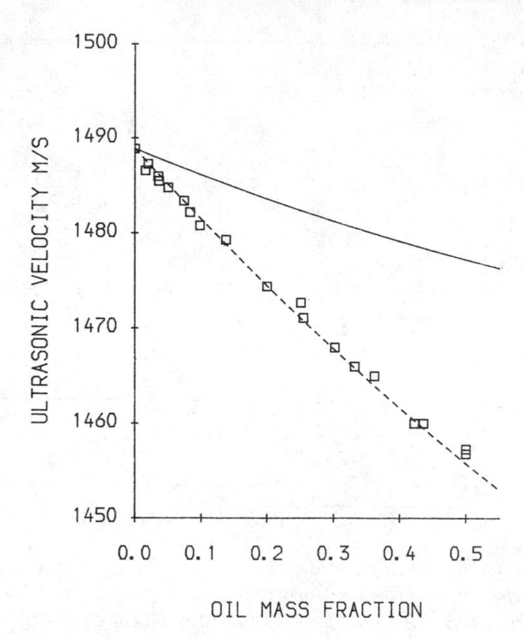

Figure 11 *Variation of ultrasonic velocity with the mass fraction of sunflower oil droplets at* 1 MHz *and* 20 °C. *The solid line is the Wood equation and the dotted line is the prediction of the calculation outlined in the last part of Section 3. The squares are experimental data*

difference between the velocity predicted by the Wood equation and that by scattering theory increases as the mass fraction of droplets increases and is very significant at high mass fractions (*ca.* 30 m s^{-1}). There is very good agreement between the theory and experiments at mass fractions up to about 0.3. However, for larger mass fractions the experimental results lie above the predictions, probably owing to the lack of self-consistency in the multiple scattering formulation described earlier. The consequences of these results are two-fold. First, the Wood equation does not give good predictions for these systems, particularly at lower frequencies and particle sizes. Second, the ultrasonic technique may be used to determine volume fractions in emulsions if the particle size is known or *vice versa*, even at volume fractions as high as 40%.

Multiple scattering contributes significantly to the velocity at high volume fractions. At low volume fractions ($\varphi < 0.1$), the effect of multiple scattering is negligible. However, at volume fractions of about 0.5 the contribution from multiple scattering has risen to 2.5 m s^{-1}, which is about 10% of the overall velocity dispersion.

Discussion—*Particle Sizing.* We conclude that over a restricted but nevertheless valuable particle size range the ultrasonic technique can be used to obtain particle sizes. Particle size distributions could also be fitted if frequency were scanned over the range 0.1–100 MHz. Further information on particle size distribution would be available if it were possible to make accurate attenuation measurements. These results must be extended to emulsions other than sunflower oil-in-water by carrying out model calculations. The particle size ranges quoted above will necessarily be different. If the particle size distribution is already known, then the particle compressibility, phase volume fraction, or the viscosity of the fluid boundary layer to the particle may be obtained in this way.

These measurements can be carried out in systems in which it is normally very difficult to measure particle size, using relatively inexpensive equipment, on-line and non-invasively.

Effects of Adsorbed Layers. The technique outlined above could be used to monitor adsorption on the surface of particles as this would change the particles' effective size, compressibility, and thermal properties. Sayers[12] has considered the effect of an adsorbed layer on a solid particle in a solid dispersion and shown that scattering will increase as a result. Frequency scanning of the velocity of sound, combined if possible with attenuation measurements will give the best results. Even more information could be obtained, for instance separation of the contributions of the thermal and visco-inertial terms, if the angular dependence of the scattering could be measured.

Frequency Scanning and Creaming. We conclude that it is essential to be able to vary the frequency in any ultrasonic apparatus being used to study a variety of emulsions. If the apparatus operates at a single frequency, it will be unable to distinguish particle size effects from volume fraction effects. In the study of creaming of sunflower oil-in-water, for instance, larger particles will have a disproportionate effect on the ultrasonic velocity. Therefore, rather than giving a true measure of volume fraction, the results will be weighted by the presence of the larger particles which cream more quickly. The ultrasonic velocity is fitted best by the Sauter diameter (d_{32}).[13]

Phase Inversion. On phase inversion in an emulsion, only phase volume need be conserved. Consequently, a change in ultrasonic velocity can be expected, particularly if the particle size alters.

Flowing Systems. These measurements can be carried out in flowing systems and in this case the Doppler frequency shift of the scattered signal can give further information about the flow-rate of the emulsion, in addition to all the information detailed above. The signal can be time gated to separate the scattered signal from various points in the flow, permitting characterization of the flow profile. To do this it is only necessary to adapt the electronic system, and the ultrasonic cell can remain unchanged.

Flocculation and Ostwald Ripening. Ultrasonics may be of use in studying these phenomena. In Ostwald ripening, the velocity should be affected directly as the dispersed phase volume will remain constant whilst the particle size changes. However, this will only be the case where the velocity is frequency dependent.

Flocculation is more complicated because its impact on velocity will depend on a number of factors. As a floc amounts to a locally concentrated system, some of the comments in Section 5 apply. Once the particles begin to 'touch,' in any of the senses described in Section 5, then flocculation will begin to impact on velocity.

Conclusion. The measurements detailed above need to be combined with accurate measurement of density and close control over temperature. The temperature coefficient of ultrasonic velocity in water at room temperature is $3\,\mathrm{m\,s^{-1}\,K^{-1}}$. Hence temperature control to within 0.1 K is necessary to control the velocity in the continuum of an oil-in-water emulsion to within $0.3\,\mathrm{m\,s^{-1}}$. This is generally an appropriate level in the laboratory, as velocity can be measured with a precision of $0.1\,\mathrm{m\,s^{-1}}$. Density should also be measured to a similar precision, *i.e.* 0.1%, and this can be done with oscillatory densitometers. These measurements would normally need to be made on the individual components making up the emulsion, as well as on the emulsion itself.

Given the appropriate information about the physical properties of the individual components which make up the emulsion, then one or more of the following pieces of information may be available from measurements of ultrasonic velocity and attenuation as a function of frequency and angle: dispersed phase volume fraction, particle size, particle compressibility, particle viscosity, thermal diffusivity of the particle boundary layer, and shear viscosity of the particle boundary layer. In addition, creaming and sedimentation, Ostwald ripening, phase inversion, and floculation may also come within the ambit of ultrasonic measurement.

5 Ultrasonic Velocity in Concentrated Emulsions

The approach outlined above works in oil-in-water and water-in-oil emulsions and a wide variety of other emulsions[2] at concentrations of up to 40%. It will break down under the following conditions: (a) the ultrasonic wavelength becomes of the order of or less than the particle size; (b) the shear wave and/or thermal damping length becomes of the order of or greater than particle separation; (c) the particles touch. Under these circumstances, the semi-phenomenological theory of Biot[14,15] may be all that can be used. Certainly this is the case if the two phases become interpenetrating. At high concentrations, phase inversion can occur. Once this has happened, it may be possible to employ the theory outlined in Section 4 again.

6 Dispersions of Solid Particles in a Liquid Continuum

The treatment outlined in Section 4 applies to these systems, together with the comments in Section 5 on concentrated systems. However, the appropriate scattering model is that of Allegra and Hawley,[10] rather than that of Epstein and Carhart.[7]

Sugar in Fat—The reservations noted above about strong scattering are very important. In our experience, sugar crystals are capable of strongly scattering ultrasonic fields of 1 MHz when suspended in liquid fats. Whilst the acoustic mismatch between the sugar and the fat is considerable, we suspect that an equally important factor is that the particle size approaches that of the ultrasonic wavelength, which is of the order of 1 mm. For particle sizes of 100 μm upwards in fats, we would expect Rayleigh scattering to be an increasingly important factor.

7 Solid Dispersions

Sayers[12] has treated the subject of a dispersion of solid particles suspended in a solid matrix along the lines discussed in Section 4. Again, it must be emphasized that this treatment applies only to weakly scattered systems. With chocolate, for example, the sugar particles scatter strongly and at present we are in no position to deal with this system adequately.

Solid foams are an interesting area to study as the bubble size can be arranged so that it varies from weakly scattering through resonant scattering to a situation where the bubble size is much greater than the ultrasonic wavelength. In this final situation, the systems that we have studied appear to follow the Wood equation. This is an area which will repay further study.

8 Conclusion

Generally, it is a mistake to approach ultrasonic data in an empirical manner. It may provide the researcher with useful insights into systems but it is a hazardous course for process measurement. If the factors that determine velocity are not understood, then a fluctuation in process parameters out of the control of the sensor system can completely invalidate the results and the users of the system will be none the wiser as to why this is so.

The key to the successful application of ultrasonics is a good understanding of its propagation through the systems of interest. We have shown that even within the class of systems termed food colloids, ultrasonic behaviour is very rich and varied and that a good understanding of the phenomena contained therein can provide valuable information about those systems, which is otherwise unavailable.

To date, ultrasonics has proven to be at least as good as pNMR for SFC determinations in crystallizing fats.[3] In contrast to pNMR, the technique can be used without any 'fudge factors' and the ultrasonic technique is not as sensitive to the interfering effect of water.

Once the fats have been incorporated into an emulsion, a more sophisticated approach is generally needed, involving measurements at at least two frequencies, one of which must be in a region in which the ultrasonic velocity is independent of particle size.

For the technique to realize its full potential as a method of determining not just dispersed phase volume fraction but also particle size, particle size distribution, particle compressibility, boundary layer properties, flow-rate and flow profile, flocculation and Ostwald ripening, the following problems need to be solved.

The problem of strong scattering systems needs to be addressed. Careful theoretical and experimental studies of such systems would be immensely valuable, as a large number of food dispersions are at present unavailable to ultrasonic probing owing to their strong scattering. Chocolate is an example of such a system in which the sugar crystals dominate the scattering, swamping information about, for example, solid fat content. New ultrasonic probes may need to be designed for such studies but there is no reason, in principle, why these systems should not prove tractable. New theoretical techniques, derived from other disciplines such as semiconductor theory, may be necessary.

New ultrasonic cells, capable of delivering information on Doppler-shifted scattering and the frequency dependence of velocity, need to be designed. This problem is soluble in principle, but the practical questions of building cells which can be used in-line in a factory are non-trivial. In practice, a simple route may be first to develop the technique as an analytical tool for use in the laboratory.

A study of resonant scattering would be valuable as a means of characterizing aerated food systems. Many of the theoretical problems have already been solved but the practical application of these ideas is necessary.

We are already studying the problems outlined above, with a view to providing a basic scientific underpinning to what we believe will be a very valuable technique in the colloid scientist's armoury. However, studies of the angular dependence of the ultrasonic scattering could prove exceedingly valuable in elucidating the properties of adsorbed boundary layers.

References

1. M. J. W. Povey, in 'Food Emulsions and Dispersions,' ed. E. Dickinson and G. Stainsby, Elsevier Applied Science, Barking, 1988.
2. D. J. McClements and M. J. W. Povey, *Adv. Colloid Interface Sci.*, 1987, **27**, 285.
3. D. J. McClements and M. J. W. Povey, *Int. J. Food Sci. Technol.*, 1988, **22**, 419.
4. D. J. McClements and M. J. W. Povey, *J. Am. Oil Chem. Soc.*, 1988, **65**, 1787.
5. P. Lloyd and M. V. Berry, *J. Phys. D*, 1967, **91**, 678.
6. G. C. Gaunaurd and H. Uberall, *J. Acoust. Soc. Am.*, 1981, **69**, 362.
7. P. S. Epstein and R. R. Carhart, *J. Acoust. Soc. Am.*, 1953, **25**(3), 553.
8. P. C. Waterman and R. Truell, *J. Math. Phys.*, 1961, **2**, 512.
9. R. J. Urick and W. S. Ament, *J. Acoust. Soc. Am.*, 1949, **21**, 115.
10. J. R. Allegra and S. A. Hawley, *J. Acoust. Soc. Am.*, 1972, **51**, 1545.
11. D. J. McClements and M. J. W. Povey, *J. Phys. D*, 1989, **22**, 38.
12. C. M. Sayers, *Wave Motion*, 1985, **7**, 95.
13. T. Oshawa, *Jpn. J. Appl. Phys.*, 1969, **7**, 795.
14. M. A. Biot. *J. Acoust. Soc. Am.*, 1962, **34**, 1254.
15. P. R. Ogushwitz, *J. Acoust. Soc. Am.*, 1985, **77**, 441.

The Packing and Movement of Particles

By S. F. Edwards

CAVENDISH LABORATORY, UNIVERSITY OF CAMBRIDGE, CAMBRIDGE
CB3 0HE, UK

1 Packing as a Problem in Theoretical Physics

It is well known that powders can, in appropriate circumstances, flow like liquids and can have variable density, *e.g.* a freshly sieved flow has a noticeably lower density than a compressed flow. The results are intuitive and reproducible, having a marked similarity to (but with notably greater richness and diversity) the thermal equilibrium properties of say a liquid. A thermal system is ergodic, *i.e.* it will get into every conceivable situation it can subject to the conservation of energy. From this fact (or possibly hypothesis), one can build up thermodynamics. A powder or floc does not have this property; the grains of a powder are not in constant motion seeking out comfortable positions, only to be continually jogged into new positions. How then can one build up a theory of their properties? Because fracture is much easier, powders and flocs flow in a more complex way than liquids. Whereas fracture can always be avoided in a (visco-elastic) liquid by moving sufficiently slowly, this will not work for a powder which, as it is not in a state of agitation, has no characteristic time scale to be slower than. Hence powders and flocs provide a real challenge to theoretical physics, offering not only new phenomena but also the challenge of constructing a new formalism to describe a familiar but, from the point of view of basic physics, uncharted form of matter.

2 How Can One Describe an Assembly of Spheres?

First let us recapitulate the basis of normal statistical mechanics. Suppose there are N particles whose energy written in terms of their positions and velocities is H. The value of H in an isolated system is conserved and called E. The ergodic postulate is that all configurations that can occur, do occur, subject only to $H = E$. In mathematical terms this is written as a probability, $\delta(E - H)$.
The total number of configurations is

$$\int \delta(E - H)\, \mathrm{d}\,(\text{all}) \tag{1}$$

and is called $\exp(S/k)$, where S is the entropy and k Boltzmann's constant. S is

$S(E,V,N)$. Gibbs then showed that if one defines the temperature T by $T = \partial E/\partial S$, the free energy of the system is given by

$$\exp(-F/kT) = \int \exp(-H/kT)\,\mathrm{d(all)} \tag{2}$$

and E is given in terms of F and T by

$$E = F - T\frac{\partial F}{\partial T} \tag{3}$$

or

$$F = E - TS \tag{4}$$

for now F is $F(V,N,T)$ and

$$S = -\frac{\partial F}{\partial T} = S(V,N,T) \tag{5}$$

A liquid can be characterized by the temperature and density and if we impose a definite fixed pressure, the density or temperature alone is sufficient. A powder is rather like this latter case. Let us suppose that a powder is characterized by its density. As there is no dynamic motion, a Maxwell demon let loose on a powder could make all sorts of different arrangements which had the same density. Clearly the simplest assumption that we can make would be to say that if we shake up a box of spheres arriving at a certain density it will be much the same as any other box of spheres shaken up to give the same density. Hence we argue that V is now the analogue of E and that there will be a function W which is the analogue of H. The normalized distribution

$$\exp(-S/k)\,\delta(E-H) \tag{6}$$

is replaced by

$$\exp(-S/\lambda)\,\delta(V-W) \tag{7}$$

where S is still the entropy, but it is natural now to measure it in terms of a length unit, λ, rather than an energy unit k. The transition from a microcanonical ensemble to a canonical ensemble comes by defining X (the analogue of T):

$$\mathrm{X} = \frac{\partial V}{\partial S} \tag{8}$$

and Y (the analogue of F):

$$\exp(-Y/\lambda X) = \int \mathrm{d(all)}\exp(-W/\lambda X) \tag{9}$$

with

$$V = Y - X\frac{\partial Y}{\partial X} \tag{10}$$

Hence the 'fluffiness' of the powder is given by X. Our problem is now to discover W and evaluate V as a function of X.

3 The Packaging of Spheres

To illustrate the solution, let us take the simplest case where we describe the powder in particulate terms by ascribing a specific volume surrounding each particle according to its co-ordination, c. Then, if there are n_c particles of co-ordination c,

$$\sum_c n_c = N \tag{11}$$

and we take the linear approximation

$$W = \sum_c n_c v_c \tag{12}$$

The value of c can vary from 4 (the minimum) to 12 (almost 13). For the sake of this model we take v_c to vary, $v_0 < v_c < v_i$, with a uniform distribution, *i.e.*

$$\sum_c F(v_c) \to \varepsilon^{-1}\int_{v_i}^{v_i} \mathrm{d}v f(v)$$

ε having the dimension of volume. Therefore, we need to evaluate

$$Y = -\lambda X \log \int \Pi_c \mathrm{d}n_c \exp\left(-(\Sigma' n_c v_c / (\lambda X)\right) \delta\left(\Sigma n_c - N\right) \tag{13}$$

Some mathematics now leads to the result

$$Y = \frac{N}{2}\left[(v_i + v_0) + v_0 - v_i)\coth\left(\frac{\varepsilon}{2\lambda X}\right)\right] \tag{14}$$

Thus $X = 0$, the closest packed powder, has

$$Y = Nv_0 \tag{15}$$

and

$$V = Nv_0 \tag{16}$$

whereas $X = \infty$, the least close packed powder, has

$$Y = \frac{N(v_0 + v_i)}{2} + \frac{\lambda NX(v_0 - v_i)}{\varepsilon} \tag{17}$$

and

$$V = \frac{N(v_0 + v_i)}{2} \tag{18}$$

this result corresponding to all configurations being equally likely. Using equation (10), an extrapolation from the densest to the fluffiest powder is given by

$$V = \frac{N}{2}\left[v_i + v_0 + (v_0 - v_i)\coth\left(\frac{\varepsilon}{2\lambda X}\right) - \frac{\varepsilon}{2\lambda X}(v_0 - v_i)\operatorname{cosech}^2\left(\frac{\varepsilon}{2\lambda X}\right)\right] \tag{19}$$

Although this model is crude, it shows how a start can be made on the problem, and suggests that W can be expanded in the form

$$W = \Sigma n_c v_c + \Sigma n_c n_d v_{cd} + \dots \tag{20}$$

4 The Packing of Irregular Shapes

Implicit in Section 3 is that although there is no order, there is one way one can label the particles and, as they do not move, this can be taken to be any labelling, in particular, say, a simple cubic labelling. It is important to realize that this is simply a convenience. Thus our irregular particle is now said to be of type a, has position i (*i.e.* a deformation would bring it to a simple point r_i) and an orientation (three Euler angles). We see at this point an extra subtlety compared with spheres. When the powder settles, in addition to its co-ordination, it can respond to its surroundings and will orient itself so as to minimize its contribution to the volume, to whatever extent it can, that is, it will adjust the degree of freedom which is degenerate for a sphere (the reader might envisage the packing of spheroids to see this point). Now it will be seen that the type of particle at r_i cannot change; the other degrees of freedom can. Hence we have to work out Y for a *given* distribution of types a, but adjustable orientation and co-ordination, and *then* average over the distribution of particles. This is a problem known in theoretical physics as a 'spin glass.'

The order of the averaging really does matter, and in certain problems is radically different; for example the dilute magnetic alloy has a phase change in the 'glass' (or 'quenched' in current usage), but not in 'liquid' (or 'annealed' in current usage) average. A polymer in the presence of a disordered background collapses in the 'annealed' case, but reaches a finite but reduced size in the 'quenched' case. This promises to be an exciting problem, but the solution is not yet resolved.

Let us now consider soft particles.

5 The Packing and Flow of Soft Particles

The difficulty with the flow of hard particles is that in general they cannot flow at all. There has to be an elaborate co-operative motion along particular pathways in phase space. Soft particles can distort themselves to make flow possible, and so appear to be much more amenable. Nevertheless, they pose problems which have defied analytical progress. To see what the difficulty is, we first ask the question of how one can describe a series of surfaces in space. This question is being asked in, of all places, the theory of elementary particles, in a development called 'string theory'. Here a particle has a trajectory which instead of being a line in space time is a surface. Our problem is worse, because it is a set of closed surfaces. Nevertheless, we can imagine a set of surfaces defined by an equation

$$\varphi(r) = 0 \tag{21}$$

The surfaces move and at a time t become

$$\varphi(r,t) = 0 \tag{22}$$

Clearly, the elastic tension that is built up on a grain of gel in motion is not described by so simple a form, but this problem is already complicated enough. The question now is to ask whether an equation can be derived for the probability of finding a $\varphi(r)$ at the time t. It can, by using the equation

$$\mu \frac{\partial \varphi}{\partial t} = -\lambda \nabla \frac{\nabla \varphi}{|\nabla \varphi|} \delta(\varphi) \nabla \varphi$$

give the velocity

$$v = \frac{1}{4\pi} \int \frac{1}{|v - v'|} \cdot \frac{\lambda}{\nu m \rho} \left(\bar{\nu} \frac{\nabla \varphi}{|\nabla \varphi|} \right) \delta(\varphi) \nabla \varphi \; \mathrm{d}^3 r - r' + v_{\mathrm{ext}} \tag{23}$$

where λ, μ, ν, and m are constants specifying the system and v_{ext} is the velocity at infinity. This equation is more difficult to solve than even its complex form suggests, for the following reason. The equation has the property that no part of the surface can cross any other part. Thus if we start equation (23) representing a set of spheres, these can distort as a result of the motion, but they cannot interpenetrate. The disaster is that any approximation to the equation does allow interpenetration. A similar problem arises with a polymer. Suppose we have a polymer with equation $R(s,t)$ and ask for the probability that this is $r(s)$ at t, the motion being Brownian with a random coil being immersed in a liquid. The equation of motion is

$$\left\{ \frac{\partial}{\partial t} - \iint \mathrm{d}s \mathrm{d}s' \frac{\partial}{\partial r(s)} G[r(s) - r(s')] \left[\frac{\partial}{\partial r(s')} + \frac{3}{l} \cdot \frac{\partial^2 r}{\partial s'^2} \right] \right\} p = 0 \tag{24}$$

where G is called the Oseen tensor and is given roughly by

$$G = |r(s) - r(s')|^{-1} \tag{25}$$

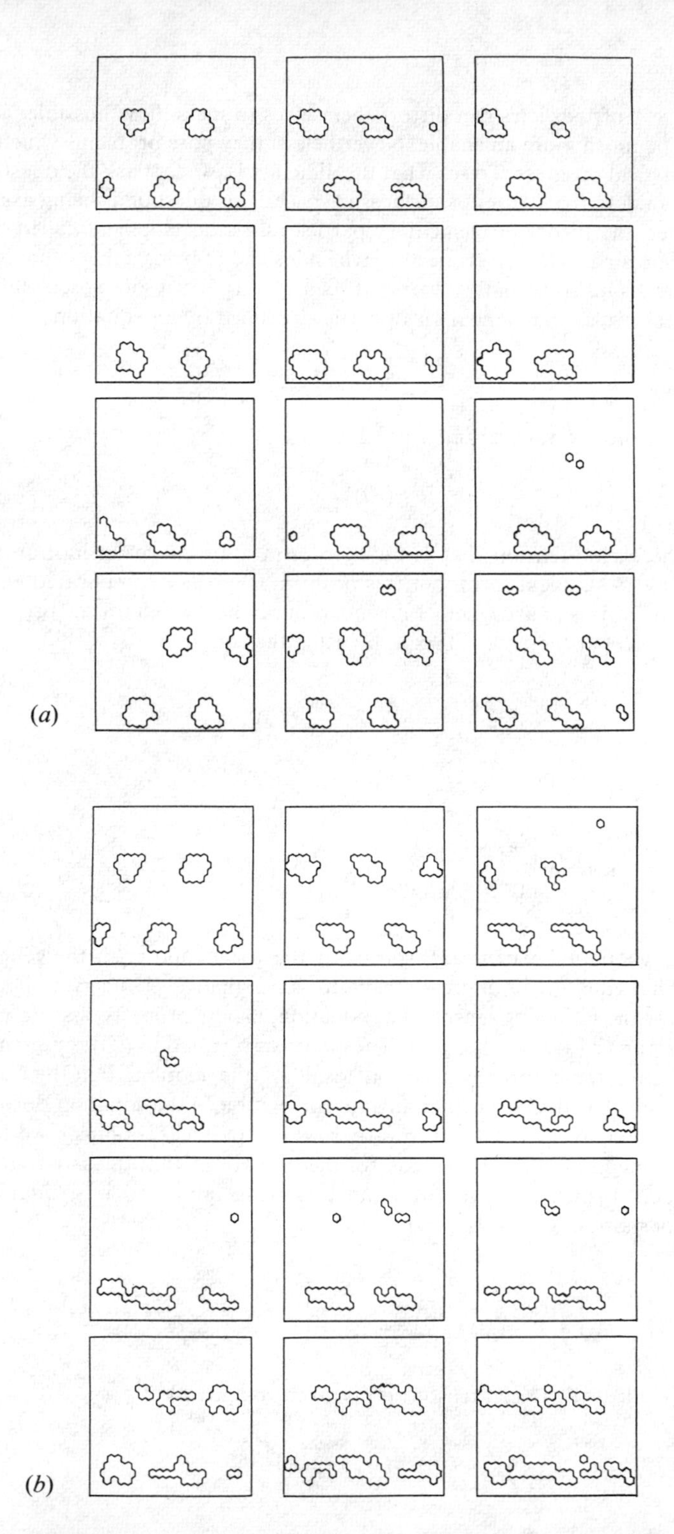
(a)
(b)

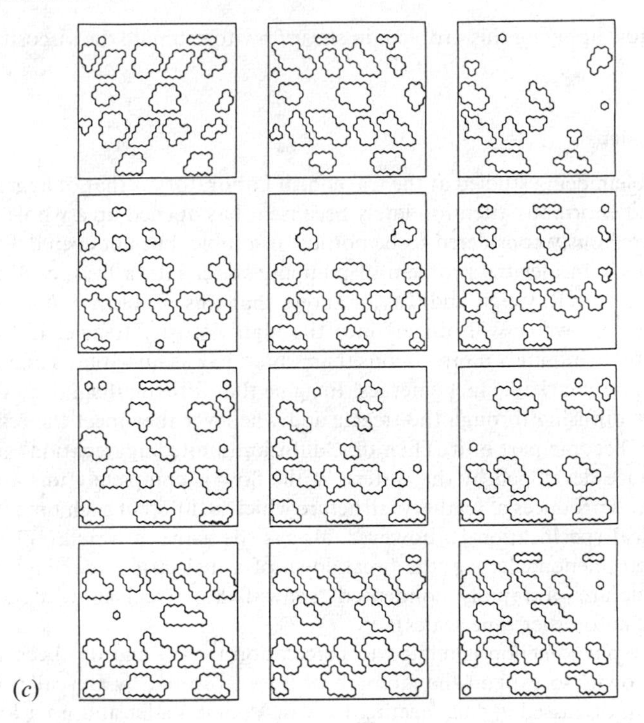

Figure 1 *(a) A single particle being sheared at the minimum shear rate used. The plot is of a section through the periodically repeated cell, which is why several images are seen in each frame. The frames are taken after one cycle of shear = 1.22. The shear moves the top of the frame to the left. (b) A single particle being sheared at the maximum shear rate used. (c) Five particles being sheared at the minimum shear rate*

It is easy to show that equation (24) describes a polymer which never cuts itself. However, I have been unable to solve it. If we approximate G by, say, its average, we obtain the 'pre-average' approximation of Zimm, which does allow the unphysical passage of the molecule through itself.

After this difficult analytical discussion, let us now try to simulate the globule on a computer.

6 Simulation of Soft Particle Flow

This turns out to be possible. One models the particles by an assembly of packed cubes, and moves them in Brownian motion by taking a cube from one point on the surface and putting it on another. The change in surface tension energy is calculated and a Monte Carlo technique called the Metropolis algorithm is used to ensure that the particle likes to keep its spherical shape. Some colleagues have obtained convincing pictures of how the globules flow in Brownian motion, and some are shown in Figure 1.

We are now applying this problem in shear flow to estimate the viscosity of the system.

7 Aggregation

A final problem being studied at the Cavendish Laboratory is that of aggregation. This has had enormous attention lately because it has opened up a whole class of problems previously considered to be not just insoluble, but not even defined. To illustrate this, consider the problem of pumping water into a treacle. A conventional analysis by Rayleigh and Taylor shows that this problem is unstable and the low-viscosity water will 'finger' into the high-viscosity treacle. If treacle is considered to be infinitely more viscous than water, it is as unstable as one can get.

However, a remarkable fact emerges. Imagine that little mathematical droplets of water are diffusing through the treacle and whenever they meet the real water surface they become part of it. Then this 'diffusion-limited aggregation' gives the correct surface developed by the water. As the flow is completely unstable, it is statistical, and produces a 'feathery' structure which is different each time it is run. The statistical specification is, however, always the same, a 'fractal.' There are many related phenomena, *e.g.* the breakdown of a dielectric under high voltage or, the structure of rapidly condensed 'snow flakes' (real snow flakes have symmetry due to other time scales).

This same phenomenon is related to percolation theory and has been used by Ball and Brown[1] to explain the strength of flocs as found by Buscall.[2] Related problems are discussed at this meeting by Van Voorst Vader and go a long way towards obtaining an understanding of the strength of Bingham solids such as fats and paints.

8 Sieving

The real subtlety of sieving lies in the difference between balls and feathers. Balls will fall to give the (lower) close packed density as described in Section 2. One can readily simulate this on a computer and also do reasonable analytical work. Spherical particles cannot overhang and, in terms of Section 2, there is a minimum co-ordination number. However, feathers can overhang, so that if one describes the surface by $R(s_1,s_2)$ or in two dimensions by $R(s)$, the arc length s is constantly increasing, but periodically the curve touches itself, and a region is enclosed, arc length is lost, and a void is included. Thus a fascinating mathematical problem arises as to the difference in density of an assembly of deposited spheres and of elongated particles. If one pours sugar, grain for grain it will have a higher density than tea leaves. Can one calculate it?

Equations for the process have been derived, and attempts are currently being made to solve them.

References

1. W. D. Brown, PhD Thesis, University of Cambridge, 1986; W. D. Brown and R. C. Ball, *J. Phys.* A **18** L517.
2. R. Buscall, *J. Non-Newtonian Fluids*; R. Buscall, *Coll. and Surf.* 1982, **5**, 269.

Discussions

Paper by Lillford and Judge

Dr. D. R. Haisman (Consultant, Bedford): Cake structures can be stabilized after baking by 'dropping the cake on the floor,' thus fracturing the bubble structure and converting a foam to a sponge. Is the reason why it is bad to slam the oven door during baking that if the bubble structure is fractured at this stage, while the matrix is still in a plastic state, the structure collapses?

Dr. P. J. Lillford (Unilever Research, Colworth): That is probably correct, although direct evidence is not available. Furthermore, it implies that prodding the centre with a fork to check the completion of cooking can also be overdone.

Dr. S. C. Warburton (Cambridge University): Ashby and Gibson contain in their theory the explanation of why Dr. Lillford obtains parallel lines of approximate gradient 2. He plots bulk foam failure stress *versus* bulk foam density. The full theory takes into account the wall properties (*i.e.* the foam plot should be bulk foam failure stress/cell wall failure stress *vs.*bulk foam density/cell wall density). As the water content varies, the wall properties will alter. Not only will ignoring the cell wall properties shift the curves, but it may also cause them to deviate from the gradient 2 (if the cell wall properties are not constant for different density foams).

Dr. P. J. Lillford: I fully concur. The point I wished to make was that one must *expect* dramatic effects of the action of water on biological materials.

Paper by Princen

Prof. Dr. A. Prins (Agricultural University, Wageningen): The analysis of the behaviour of a foam or an emulsion under deformation is carried out under the assumption that the surface tension is constant and uniform. If it is accepted that by pulling or pushing films out or into the Plateau borders the surface tension of both film and plateau border will deviate from the equilibrium value, what is the effect on the analysis presented?

Dr. H. M. Princen (General Foods, USA): Indeed, our analysis in reference 19 assumes that the surface tension in the films remains constant, so that the angles between the films remain at 120°. As indicated by Prof. Prins, the tensions of the stretching and shrinking films will, in reality, deviate in a positive or negative sense from the equilibrium value, which will affect the instantaneous angles between the films in a rather complex fashion. The change in film tension has in fact, been evaluated also by Frankel *et al.*[20] and was found to be proportional to $C_a{}^{2/3}$. It is therefore possible, in principle, to take this effect into account by coupling the individual film velocities, film tensions, and film angles, and by evaluating the viscous energy dissipated in the transition regions in this modified system. This is a complex task, and I believe that it would lead, at most, to a second-order correction.

Recently, Kraynik and Reinelt (personal communication; to be published) have used a 'force or film tension approach,' instead of our 'viscous energy dissipation approach.' They specifically take into account the continuously changing film tensions and angles. When the results of their analysis are averaged over a cycle, they appear to arrive at a viscous contribution that is identical with ours. This gives us added confidence that our predicted dependence of the viscous terms on C_a is unaffected.

Paper by Weaire

Prof. Dr. P. Walstra (Agricultural University, Wageningen): During coarsening of a foam the surface tension will generally alter. As long as this is only a lowering of surface tension due to the decrease in surface area, this will only slow down the coarsening. But if the surface has visco-elastic properties (dilational surface properties), the surface tension may vary from place to place. To what extent can this affect the course of the coarsening?

Prof. D. L. Weaire (Trinity College, Dublin): My own concern has primarily been the difficult problem of identifying the behaviour of the idealized system in which surface energy is constant with time, and from place to place. One may well question this assumption in relation to real systems, and also that of constant cell wall thickness. I cannot easily comment on these problems *in general*, but it would seem that for simple 2D soap froth systems the assumptions of the idealized model are not grossly in error. A relevant investigation, conducted by ourselves and also by Glazzier, is the direct experimental test of von Neumann's law. We found it to hold *on average*, and not to change with time. However, individual cells did not conform to the law. This should be pursued in more detail.

Prof. Dr. A. Prins (Agricultural University, Wageningen): In discussing what is going on during the deformation of a polyhedric foam, it is perhaps wise to distinguish clearly between the processes of rearrangement of films and disproportionation of bubbles. Usually disproportionation of bubbles describes the growth of bigger bubbles at the expense of smaller ones and rearrangement describes the response of the foam structure when more than three films meet in one Plateau border. Can you comment on this?

Prof. D. L. Weaire: Our two-dimensional simulation includes both types of topological change. Although the growth of some cells at the expense of others is easily understood, its overall effect upon the cell statistics is not, because of the involvement of the other type of rearrangement. Incidentally, a word of caution should be added regarding the simple law, due to Von Neumann, which has been used to determine the growth of individual cells. There is some experimental evidence that it holds only *on average*, but we do not understand why.

Paper by Ronteltap and Prins

Dr. D. R. Haisman (Consultant, Bedford): What is the lower size limit of bubbles that your optical fibre technique can detect? If the response of the instrument depends on the tip detecting a change in refractive index of the medium, could it also be used to measure solid or liquid 'particles' in suspension?

Dr. A. D. Ronteltap (Agricultural University, Wageningen): The minimum diameter of the fibre tip is 20 μm. The lower size limit of gas length detected is 50 μm. An observed gas length of 50 μm might correspond to a bubble diameter of 50 μm or to a bigger bubble, depending on whether the bubble is pierced diametrically or not. This introduces an up until now unsolved statistical error in the left tail of the distribution, giving a smaller number of small bubbles than are actually present. However, the error will be rather small.

Whether the apparatus can be used to measure liquid 'particles' in suspension depends on the difference in refractive index of the two phases and the size of the suspended particles. Furthermore, these suspensions must be concentrated to avoid 'particles' being pushed aside. Solid 'particles' cannot be measured because they cannot be pierced.

Dr. D. C. Clark (AFRC Institute of Food Research, Norwich): (a) During your talk you stated that coalescence which involves film rupture and implies film instability was film thickness dependent. Many extremely stable foams are generated which drain to very thin common black films. Can you comment on this?

(b) The vertical orientation of your fibre-optic probe would be conducive to foam drainage down the fibre and possible pendant drop formation at the tip. Could you give some indication of the influence of such effects on the measurements of bubble size, given that droplet growth would produce a constantly changing radius of curvature at the tip?

Dr. A. D. Ronteltap: (a) Unfortunately, little is known about the actual process of film rupture in aqueous foams, although many mechanisms have been proposed. Depending on the mechanisms of coalescence the film thickness is of more or lesser importance. In beer foam, fat enhances coalescence. The mechanisms that might explain this phenomenon, such as spreading of fat substances on the film surface giving an unstable configuration or the piercing of the film by hydrophobic particles, all predict that coalescence occurs preferably when the film thickness is low. Of course, these mechanisms depend also on other physical parameters, for example surface viscosity. Common black films generally lose their stability when they are subjected to evaporation.

(b) The probe is run through the foam at a speed of about 10 cm s^{-1}. The measurement therefore takes less than 1 s. Drainage is not likely to influence observations within this short period. Furthermore, it is not very likely that a droplet grows at the end of the tip owing to drainage because surface tension will be more important than gravity for such a small dimension of the tip. A thin liquid layer at the tip will not essentially change the optics of the glass-fibre probe.

Dr. A. Howe (AFRC Institute of Food Research, Norwich): Does the insertion of the probe perturb the position of the droplets in the foam?

Dr. A. D. Ronteltap: From statistics it follows that for a line through the foam the sum of the gas lengths observed divided by the height of the foam must be equal to the gas fraction. For our probe in the experiment only 70% of the actual gas fraction is found. Probably this is partly due to the fact that the probe is not infinitesimal and partly because some bubbles are pushed aside.

Dr. L. R. Fisher (CSIRO, Australia): Presumably such a probe will burst the bubbles through which it passes. Given this, do you have any evidence that the bubble size distribution measured reflects the real bubble size distribution? For example, is the size distribution measured dependent on the velocity of the probe tip?

Dr. A. D. Ronteltap: Piercing the probe through series of larger beer films stretched horizontally in a vertical glass tube did not cause film rupture. I therefore think that films in the foam do not burst either. Furthermore, moving the probe upwards through the foam after the experiment gives a very similar looking result as is found moving the probe downwards. No effect of the velocity of the glass-fibre probe on the size distribution is measured.

Mr. E. Pelan (Unilever Research, Sharnbrook): The method depends on drawing a fibre-optic probe through the foam to measure bubble size using the reflected signal. How do different bubble sizes and different points of entry on the bubble surface affect the measurement?

Dr. A. D. Ronteltap: Different bubble sizes and different points of entry on the bubble surface affect the measurement. The direct measured gas lengths do not correspond directly with the actual bubble radii. However, the actual bubble size distribution can be calculated with a statistical method described by Weibel.[14]

Paper by Bamforth

Dr. D. Clark (AFRC Institute of Food Research, Norwich): Hydrophobic peptides are associated with bitter tastes. Does the addition of hydrophobic peptides at the end of processing significantly alter taste?

Dr. C. W. Bamforth (Bass plc): Extensive tasting trials reveal this not to be the case. Beers are, of course, inherently bitter.

Dr. D. Clark The foaming of proteins can be highly pH dependent. During the generation of beer foams and their drainage and disproportionation, CO_2 comes out of solution and potentially could cause some pH changes. Has this been investigated and could this account for some of the differences observed between CO_2 and N_2 blown foams?

Dr. C. W. Bamforth: I am unaware of any research done on pH in relation to foaming.

Dr. J. Mingins (AFRC Institute of Food Research, Norwich): Have you analysed the foams for their composition to assess which species are involved in foaming? You may recognize the competition of different polypeptides.

Dr. C. W. Bamforth: Our approach has been to look at liquid beer *per se* and the foaming properties of isolated fractions therefrom. There is scope for looking at the foam itself—provided it is generated in proportions such that the materials which enter it are representative of those which determine foam stability in the trade.

Dr. L. Fisher (CSIRO, Australia): Egg white is well known to cause allergic reactions in some people. Would you expect this to be a problem in the commercial use of egg albumin to stabilize beer foams?

Dr. C. W. Bamforth: We have looked into this matter extensively. The albumen preparation incurs rigorous hydrolysis and pH and heat treatment, all of which have been shown to reduce its allergenic potency greatly. Furthermore, beers containing hydrolysed albumen caused absolutely no symptoms in highly egg-sensitive individuals.

Dr. L. Fisher (CSIRO, Australia): As mean bubble sizes increase with time in a foam, the area of the foam will decrease and the adsorbed protein films will become compressed. At a critical fractional area reduction, the protein will undergo interfacial coagulation to produce a mechanically strong film. It would appear that such films are good candidates for long-term stabilization of foams to produce 'lacing'. Do you have any evidence for or against such a mechanism?

Dr. C. W. Bamforth: The mechanism for foam 'solidification' during ageing is uncertain. Some work has been done to indicate that hop resins are essential, through their role in 'precipitating out' polypeptides. It is relevant, however, that at very high N_2 levels you can get such stable foams that they are dragged down with the beer and will not cling. It is known that N_2 yields much smaller bubbles.

Prof. Dr. A. Prins (Agricultural University, Wageningen): From a physical point of view, is it correct to speak about materials which are positive or negative with respect to the foam stability (of beer) when it is observed that a particular material in one case acts as a foam stabilizer (whatever that may mean) and in another case as a foam destabilizer?

Dr. C. W. Bamforth: Our findings are that constituents of beer are generally either 'foam-positive' (*e.g.* amphipathic polypeptides) or 'foam-negative' (*e.g.* most lipids). Only a relatively few materials (*e.g.* ethanol) can act in both respects. This is probably through different mechanisms. Relatively low concentrations of ethanol probably act beneficially by lowering surface tension. Higher concentrations may interfere with interface structure.

Dr. T. Galliard (RHM Research and Engineering Ltd, High Wycombe): I recall some earlier work from Dr. J. J. Wren's laboratory which showed that some polar lipids, and especially lysophosphatidylcholine, enhanced head retention in beer, *i.e.* have a positive effect in your terminology. However, you stated that all lipids have a negative effect on head retention. Could you enlighten me?

Dr. C. W. Bamforth: Not *all* lipids are 'foam-negative.' We have found, for example, that digalactosyl diglyceride at low concentrations can enhance beer foams; phospholipids however are particularly deleterious to foam (see, for example, *J. Inst. Brew.*, 1968, **74**, 257). Their presence during processing can be beneficial by suppressing process foaming and thereby enabling more 'foam-positives' to survive into the finished product. It is essential, however, that the lipids are removed in the final beer clarification stages.

Dr. A. Howe (AFRC Institute of Food Research, Norwich): Why is lacing not uniform over the height of the glass when the rate of sipping is constant?

Dr. C. W. Bamforth: The extent of lacing will depend on (a) the rate at which foam 'solidification' occurs and (b) the rate at which foam decays. The relative rates of each will determine to what extent foam clings throughout 'sipping.'

Dr. J. Mitchell (University of Nottingham): The use of Shreeve's Winkle gives a small bubble size and good head retention to ordinary bitter. In view of this, is the distinctive nature of the foam on Guinness due to the presence of N_2 with its low water solubility compared with CO_2, resulting in a low degree of bubble disproportionation, or is it due to the dispersion method?

Dr. C. W. Bamforth: Beers containing N_2 do not appear to display superior foams, although it has never been convincingly established why. Thus a beer may contain in excess of 4000 p.p.m. CO_2 but only 10–20 p.p.m. N_2.

Dr. B. Wedzicha (University of Leeds): Sulphur dioxide is sometimes present in beer as a residue or from fermentation. Does it have any effect on the head properties of beer? If so, is there any relationship between sulphur dioxide content and effect, and what is the explanation?

Dr. C. W. Bamforth: It has been suggested that sulphur dioxide added in the boiling stage of wort production can lower foam stability by causing the cleavage of disulphide bonds in polypeptides, thereby yielding lower molecular weight, less foam-stabilizing moieties.

Paper by Smith

Miss J. F. Heathcock (Cadbury Schweppes plc, Reading): Extrusion often leads to distortion of bubbles at the extreme-elongated features. How do you take account of this in your calculations?

Dr. A. C. Smith (AFRC Institute of Food Research, Norwich): The Ashby treatment deals with foams comprising identical symmetrical pores. Any treatment of a foam in which there is a distribution of pore sizes requires that either the Ashby treatment is either applied locally where the bubble size is constant or in an averaged way over the bulk. Local mechanical testing would explicitly reveal the effect of elongated pores. The Ashby approach relates the relative density to the pore dimension: $\rho/\rho_w \propto (t/l)^2$ for an open cell and $\rho/\rho_w \propto t/l$ for a closed cell. The treatment of mechanical properties is two-dimensional and hence the relative mechanical property is inversely proportional to a power of the pore dimension in the plane of testing. For example, if elongation resulted in smaller pores, the strength and stiffness would increase when the foam is tested so as to deform these pores.

Prof. P. Richmond (AFRC Institute of Food Research, Norwich): You said that by studying the extraction you can eliminate the effect of debris accumulation of your mechanical measure during penetration. Is it obvious that the effect of debris is the same during penetration and removal?

Dr. A. C. Smith: The passage of the probe into the foam may be accompanied by the accumulation of debris around and in front of the probe. This results in a ramped base line of the force–displacement response. The removal of the probe isolates the force contribution owing to the movement of the probe through accumulated material and offers a correction to the magnitude of the force peaks after penetration some distance into the foam bulk. In cases where the probe is used to obtain information of the first pore wall or the one-dimensional structure, this correction is not necessary.

Prof. Dr. P. Walstra (Agricultural University, Wageningen): The inhomogeneity of the material used in making extruded foam-like foods may significantly affect the rheological and fracture properties of such foods. One aspect may be that the formation of the gas cells needs nucleation, and nucleation will then occur at certain 'weak' sites in the solid material. Do you have any idea to what extent such complications may modify the validity of the Ashby theory?

Dr. A. C. Smith: The Ashby treatment deals with identical foams which represent only an approximation to many solid food foams. Relating mechanical properties to density does not require a knowledge of processing conditions or the mechanism of foam formation. Inhomogeneities in the extruded material enter the Ashby treatment through the properties of the foam walls which may themselves be seen as complex composite solids. The origin and manipulation of the foam structure and density, however, remain a significant problem in understanding the extrusion cooking process. The rheological and thermal properties of these materials appropriate to the foam process are yet to be fully studied.

Paper by Dickinson and Woskett

Dr. J. Mitchell (University of Nottingham): As you say, the lack of dependence on protein type to displacement by surfactants is surprising. de Feijter's work just compares β-casein and β-lactoglobulin. If you looked at a wider range of proteins, would you expect this still to hold? I am particularly thinking of proteins such as ovalbumen which show extensive surface coagulation and therefore form a continuous probably irreversible network at the interface.

Dr. E. Dickinson (University of Leeds): The lack of sensitivity to protein type may simply be a reflection of the large difference between the surfactant molecule surface binding energy (say 10–15 kT) and the average protein segment binding energy (say 2–3 kT), as compared with the small difference between binding energies for various proteins. In the absence of any specific protein–surfactant interactions, I would expect the ease of protein displacement to correlate with the surface pressure at constant bulk protein concentration. It would, as you say, be interesting to have data for more proteins, especially those which show extensive surface coagulation.

Dr. D. G. Dalgleish (Hannah Research Institute, Ayr): The interaction of SDS with proteins is imagined to be relatively non-specific (hence the use of SDS–PAGE methodology). Your illustrations show a difference between proteins in SDS mixtures: is this not unexpected in view of the non-specific and dominant effect of SDS?

Dr. E. Dickinson: Whether the interaction between protein and anionic surfactant is primarily specific or non-specific depends on the relative concentrations of the two components in the system. The first part of the binding isotherm (low surfactant bulk concentration) is thought to be mainly specific, involving electrostatic interaction between the ionic head group and positive charged groups on the protein, and also some additional hydrophobic interaction between the surfactant tail and neighbouring non-polar regions of the protein. At high surfactant concentrations, co-operative binding occurs, and this is non-specific, being mainly driven by the same sort of associative interactions which leads to surfactant micelle formation in the absence of protein. To my knowledge, the use of SDS in protein analysis methodology generally involves a substantial excess of surfactant over protein—and hence generally involves non-specific SDS–protein interactions. Differences between proteins would be expected to be significant, however, at SDS concentrations well below the critical micelle concentration (cmc), but not at concentrations approaching the cmc, or above the cmc. In this latter case, what we are really looking at is partitioning of protein into surfactant micelles, and this one would expect to be relatively non-specific.

Dr. L. Fisher (CSIRO, Australia): Is there any effect of the age of the adsorbed protein film on the ability of small molecule surfactants to displace the protein?

Dr. E. Dickinson: Though it has not been studied in detail, the experimental information which does exist seems to indicate that protein displacement by small molecule surfactants is insensitive to the age of the adsorbed protein film. Indeed, it does not appear to matter whether the surfactant is present in the premix prior to homogenization, or whether it is added afterwards (J.A. de Feijter, J. Benjamins and M. Tamboer, *Colloids Surf.*, 1987, **27**, 243). Where the displacing agent is itself a macromolecule, the position may, however, be somewhat different. Certainly, we have found (E. Dickinson, A. Murray, B. S. Murray, and G. Stainsby, in 'Food Emulsions and Foams,' ed. E. Dickinson, Royal Society of Chemistry, London, 1987, p. 86) that, in emulsions made with gelatin, it is possible to displace essentially all the adsorbed gelatin by casein in the fresh emulsion, but only a small fraction of it in the 24-h-old emulsion. Unfortunately, we have not repeated the experiment with casein replaced by a small molecule surfactant, so I cannot give a direct comparison. It seems intuitively reasonable, however, to suggest that it may be more difficult to displace a highly visco-elastic gel-like adsorbed protein layer in penetrating the network structure of the adsorbed film. With the gelatin + casein system, there is evidence from surface rheology (E. Dickinson, A. Murray, B. S. Murray, and G. Stainsby, in 'Food Emulsions and Foams,' ed. E. Dickinson, Royal Society of Chemistry, London, 1987, p. 86) that casein does displace gelatin from the primary interface, but that it still can remain associated with the interface in the form of secondary layers. There is no reason to believe that this same phenomenon cannot occur also with protein + surfactant systems, given the experimental evidence for strong protein–surfactant interactions in many systems.

Dr. H. Kerr (Flour Milling and Baking Research Association, Chorleywood): On one of the figures showing the effect of small molecule surfactants on protein desorption, you commented that the surfactant did not differentiate between the two proteins which are structurally very different and that this seemed surprising. Is it not true that the area of surface occupied by each protein's attachment point is similar? Could this in part be one reason that these small molecule surfactants do not 'notice' that the proteins are different (given that these two proteins are attached by different numbers of points and would need different total energy inputs to desorb them fully)?

Dr. E. Dickinson: I would expect that the proportion of segments at the interface (in trains) would be larger for a disordered protein such as β-casein than for a globular protein such as β-lactoglobulin. So the total adsorption energies for the two proteins would also be expected to be different. What may be more important is the average adsorption energy per segment; this I would not expect to vary much from protein to protein. So long as the surfactant concentration is sufficiently high, and the surfactant–surface binding energy sufficiently favourable, it will not be much more difficult to displace 100 segments than to displace 50 for a flexible polymer. Where the protein structure is very rigid, however, one might expect a substantial difference due to the low configurational entropy. But this is just speculation at the present time; I have no direct experimental evidence.

Dr. A. Howe (AFRC Institute of Food Research, Norwich): Addition of small molecule surfactants to cream liqueur systems leads to stabilization to creaming by gel formation. Why does the gelation occur?

Dr. E. Dickinson: It is possible that the weak gel formation in these emulsions may be the result of interactions between casein and surfactant at the oil–water interface and in the bulk aqueous phase, with the likely involvement of liquid crystalline mesophases at the surface of the emulsion droplets at the higher surfactant concentrations. Casein alone does not produce any signs of gelation at concentrations several times those used in these simulated cream liqueurs.

Paper by Clark, Mackie, Smith, and Wilson

Dr. D. S. Horne (Hannah Research Institute, Ayr): You are proposing a charge neutralization mechanism for your aggregation process; might I suggest that your results could be unified and your scheme confirmed by attempting to relate the kinetics of increase in turbidity to the charge of your protein complex? Your results as a function of pH, ionic strength, *etc*., might then be expected to fall on a universal curve provided proper account was made of the influence of these parameters on the protein charge.

Dr. D. C. Clark (AFRC Institue of Food Research, Norwich): I think it likely that your proposed scheme is valid; however, given the interdependence of all the solution conditions that we have examined plus the added complication of a time-dependent change in aggregation and rearrangement of aggregate composition, I wonder how fruitful such an approach would be.

Dr. L. R. Fisher (CSIRO, Australia): In your measurements of film thickness, how do the results compare with considering theoretical balanced forces? In particular, at high salt concentrations the films seem to be rather thick relative to the Debye length. We have found (see L. R. Fisher and E. E. Mitchell, this volume) that at the highest salt concentration studied in the system triglyceride–10^{-3}% lysozyme + NaCl–triglyceride, in aqueous solution, the aqueous film thickness is greater than that predicted from double-layer repulsion alone, and have speculated that steric repulsion is also involved. It would be interesting to know whether you are getting a similar effect at high salt concentrations.

Dr. D. C. Clark: Yes, apart from the electrolyte-free system, all the film thicknesses are much greater than twice the effective double layer thicknesses. Simple charged soap films also give thicknesses beyond the effective range of normal double layer repulsion (M. N. Jones and G. Ibbotson, *Trans. Faraday Soc.*, 1970, **68**, 2396), but with the protein films here the thicknesses are even greater. The salt dependence runs counter to classical screening but with adsorbed proteins this is not too surprising. It is interesting to speculate on the role of the protein chains, but more quantitative measurements of adsorbed layer thicknesses and interfacial conformations are needed.

Dr. A. Howe (AFRC Institute of Food Research, Norwich): Can you comment on the trimodal size distribution of protein aggregates formed at short times (Figure 5)? Have you used any other methods to investigate the dimensions of the aggregates?

Dr. D. C. Clark: It is possible that the large particles (19 and 40 μm) may be an artefact from the Malvern analysis due to the high turbidity and resulting multiple scattering. However, care was taken to keep turbidity to within the recommended range for the instrument. Also, data were recorded in the presence of 0.025 M NaCl and gave similar particle distributions, although the protein concentration had to be raised significantly to increase the scatter to a measurable level. Particles corresponding to the 2, 5 and 19 μm species were observed in thin films (Figure 8) by microscopy; 40 μm particles were not observed but these may have been expelled from the film during formation.

Dr. T. Galliard (RHM Research and Engineering Ltd, High Wycombe): Your turbidity method would not distinguish between interaction/aggregation between the different proteins on the one hand and precipitation of one component on the other; for example, when measuring pH response, is this not important in the interpretation of your results?

Dr. D. C. Clark: BSA and lysozyme are highly soluble proteins and in separate experiments the isolated components did not precipitate under the range of solution conditions described. At the comparatively low protein concentrations addressed in this study, we believe that the turbidimetric assay is a good measure of aggregate formation which was found to be reversible in dissociation experiments induced by dilution or increasing ionic strength.

Dr. J. Mitchell (University of Nottingham): Is there a correlation between the extent of aggregation between BSA and lysozyme in bulk, and foam stability, *e.g.* does your effect of pH on aggregation mirror the effect of pH on foam stability found by Poole *et al.*[9] I was worried that your large aggregates would not diffuse to the interface or would only have time to do so, in your bubbling test you used but not in the whipping method used by Poole *et al.*

Dr. D. C. Clark: The influence of pH on the extent of aggregation of BSA and lysozyme as judged by turbidity is shown in Figure 6 and is in good agreement with the data of Poole *et al.*[9] on foam stability. This is perhaps surprising since the concentrations used in the experiments are different. I am not sure of the adsorption properties of the large aggregates. Dilution experiments suggest that they are too stable to dissociate prior to adsorption and their size and hence slow diffusion may initially preclude their adsorption due to population of the interface by faster diffusing low molecular weight aggregates and non-aggregated protein.

Paper by Fisher and Mitchell

Dr. B. S. Murray (University of Leeds): I think you said that even in the case where the two oil surfaces with adsorbed β-casein 'stick' together, coalescence eventually does take place. Firstly, I do not understand how film rupture could be propagated if the β-casein is truly bridging the two surfaces, and secondly, are you in a position to say how the 'average' coalescence time for the β-casein films compares with that for the lysozyme films?

Dr. L. R. Fisher (CSIRO, Australia): To answer your second question first, the time between adhesion and coalescence does indeed seem to be shorter with β-casein films of any age compared with lysozyme films of age greater than 3 h, say. I should also mention that the behaviour of β-casein appears to be rather sensitive to the source of the material. The material for which we report results here was of very high purity. We are currently seeking reasons for the variation between samples; with some samples (tested since the submission of our paper), we find no adhesion at all!

In answer to your first question, we simply have no idea how coalescence is initiated. Perhaps β-casein aggregates at the interface, eventually leaving sufficiently large 'bare patches' to initiate coalescence. It is very hard to devise a good experimental test of this and other speculations.

Dr. A. M. Howe (AFRC Institute of Food Research, Norwich): You conjectured that the destabilization of films with adsorbed lysozyme was a result of interfacial compressive effects arising from continued adsorption of the lysozyme on the droplet surface but not in the film. Can you test or have you tested this hypothesis by bringing together droplets with adsorbed lysozyme in a medium free of protein (*e.g.* remove the aqueous lysozyme before bringing the droplets together)?

Dr. L. R. Fisher: That is a neat idea which looks to be worth following up.

Dr. D. C. Clark (AFRC Institute of Food Research, Norwich): You postulated that one explanation of the appearance of strains in the draining film of lysozyme between the oil droplets could be additional protein adsorption in the non contact zone. Graham and Phillips (*J. Colloid Interface Sci.*, 1979, **70**, 403) showed with radiolabelled lysozyme that the surface concentration of protein rapidly reached a plateau value and no further adsorption was observed in their systems. Can you comment on this?

Dr. L. R. Fisher: The Graham and Phillips results were for adsorption at the air–water interface, although there is no reason to doubt that in this respect adsorption at the oil–water interface would follow a similar pattern. Nevertheless, it may need only a small increase in the amount of adsorbed protein to produce quite a large increase in surface pressure. We find, in fact, that for lysozyme adsorbing at the MCT 810–water interface from 10^{-3}% w/w aqueous solution, the interfacial tension continues to decrease (*i.e.* the surface pressure increases) over at least 24 h.

Dr. E. Dickinson (University of Leeds): In the coalescence experiments in which β-casein was aged in solution prior to bringing the droplets together, was any difference observed between the cases in which the protein solution was aged in the absence of the droplet surface and that in which there was ageing of the adsorbed protein film?

Dr. L. R. Fisher: For solutions older than *ca.* 3 h, the solutions were always aged in the absence of the droplets, so I am afraid that I cannot give a definite answer to your question.

Dr. F. van Voorst Vader (Unilever Research, Vlaardingen): Droplet adhesion was also observed between PO surfaces covered with PVA and PO droplets (F. van Voorst Vader and H. Dekker, *J. Colloid Interface Sci.*, 1981, **83**, 377). Is this effect similar?

Unpublished measurements confirm the occurrence of droplet adhesion when using sodium caseinate solutions.

Did you evaluate the adhesion energy from film contact angles?

Dr. L. R. Fisher: If the PO droplets were not initially coated with PVA, then polymer bridging is clearly the most likely mechanism for adhesion in your experiments.

We are not able to measure the contact angles between the droplets in our experiments with sufficient precision to have sufficient confidence in them to calculate adhesion energies. It is possible in principle, though, to calculate adhesion energies from the change in contact area on adhesion, combined with the relevant geometric parameters for the droplets, and we are now in the process of doing this.

Paper by Rickayzen

Dr. J. Mingins (AFRC Institute of Food Research, Norwich): As a theoretician, what information would you like experimental scientists to provide to help in this field?

Prof. G. Rickayzen (University of Kent): At the present time, the theories are best at predicting the density profiles of fluids and solvation forces between solid bodies immersed in fluids. Accordingly, measurements of these properties would provide the most direct comparison with theory. The work by Israelachvili and co-workers has been invaluable in this regard.

If one looks more to experiments on colloids, then more reliable measurements on critical concentrations for coagulation and the Schultz–Hardy rule would be invaluable. Looking to the future, more data on kinetic processes such as electrophoresis and electro-osmosis would be helpful.

Dr. H. Kerr (Flour Milling and Baking Research Association, Chorleywood): The models you have presented use hard spheres to represent molecules. How would the model's predictions be affected if other shapes were used to represent molecules?

Prof. G. Rickayzen: Studies of the profiles of fluids with non-spherical molecules already exist. For example, a fluid of dumb-bell molecules comprising two hard spheres with centres separated by a fixed distance has been studied by Sullivan, Barker, Gray, Streett, and Gubbins (*Mol. Phys.*, 1981, **44**, 597). The main features of the profile are similar to those presented in this paper. However, there is also a cusp in the profile resulting from the fixed separation of the centres. The results also suggest that in this case, where the wall is hard and otherwise structureless, the molecules tend to align parallel with the wall rather than perpendicular to it. This preference becomes more marked the more elongated the molecule becomes and the greater the bulk density of the fluid. In general, one would expect the structure of the profile near the wall to reflect the geometry of the individual molecules and their short-range order.

Dr. M. Povey (University of Leeds): Can thermodynamics provide an adequate description of the behaviour of interfaces? The question arises because the scale of many interfacial processes is such that the kinetics of particle motion, for example for small particles for which Brownian motion is important, means that relaxation times for molecular interaction have to be taken into account. This is surely the case when, for many emulsions, they are not truly in a state of thermodynamic equilibrium.

Prof. G. Rickayzen: This paper has been concerned only with the *structure* of fluids at fluid–solid interfaces and not with other behaviour. As long as the relaxation time for the fluid is short compared with other times, and this is commonly the case, the fluid can be taken to be in thermodynamic equilibrium and the thermodynamic approach to structure can be used. Even when the colloid as a whole is not in equilibrium, the fluid may be locally in equilibrium with its neighbouring particle and the interface is still determined by thermodynamics.

Dr. K. Hammond (Unilever Research, Port Sunlight): In your introductory remarks you mentioned the theory relating to electrophoresis, although you did not address the matter in detail in the main body of your presentation. My question relates to the calculation of zeta potentials via electrophoretic mobilities, and Henry's (1931) treatment to account for discrepancies between the Smoluchowski and Debye–Hückel theories.

Henry's calculation assumed the external field could be superimposed on the particle field, and that the latter could be described by a linear Poisson–Boltzmann equation. Such a treatment is only really valid for particles of low potential ($\zeta < 25$ mV). Moreover, the approach does not take into account the consequences of particle movement, such as field distortion, 'double-layer' rearrangement, *etc.* I am aware of efforts to address these deficiencies; notably the early works of Overbeek (1943) and Booth (1950) and subsequent approaches by Wiersema *et al.* (1969) and O'Brien and White (1978). However, to those of us less versed in the inherent complexities involved, such treatments appear rather cumbersome and user unfriendly. Is it, in your opinion, likely that a readily applicable 'solution' or correction method of a generic nature is likely to emerge in the near future?

Prof. G. Rickayzen: I mentioned the zeta potential in the paper to show that, although it is a useful and much-used concept, it is not easily identified in the molecular theory of the interface. In fact, whereas the classical theory of the interface leads to an electrical potential which varies monotonically with distance from the surface, molecular theory gives rise to an oscillation (even rapidly oscillating) potential just at the distances usually used in defining the zeta potential. There is thus no simple estimate of the potential near the surface.

To try to answer the question, then, one will need to add dynamics to the picture described in the paper. In principle this can be done, but in practice I believe the results are not likely to emerge in the near future. Whether the result or treatment will be user friendly will no doubt depend on the eye of the user.

Paper by Clark, Lips, and Hart

Dr. L. R. Fisher (CSIRO, Australia): Did you encounter problems with your systems affecting the performance of the electrodes?

Dr. A. Lips (Unilever Research, Colworth): No. All the electrodes performed reproducibly. We did, of course, avoid liquid junction effects.

Dr. D. G. Dalgleish (Hannah Research Institute, Ayr): In the plot of caseinate voluminosity against [Ca^{2+}] the apparent voluminosity drops at high [Ca^{2+}]. Is this evidence for a change in the voluminosity, or is it simply an expression of experimental error?

Dr. A. Lips: Although our analysis could suggest a change in voluminosity at high [Ca^{2+}], this cannot be concluded as the surface potentials become net positive and the assumption of indifference of the Cl^- ion (co-ion) becomes questionable.

Dr. J. Mingins (AFRC Institute of Food Research, Norwich): 1. Have you checked your assumption that co-ions are indifferent, *e.g.* by doing experiments in chloride and nitrate?

2. Does the correspondence between the Poisson–Boltzmann and Donnan arguments hold for asymmetric electrolytes?

Dr. A. Lips: 1. We have not so far experimented with a range of co-ions. This would be worthwhile.

2. Our approach is intended for systems in which electric fields and potentials are small. In this case the correspondence between Poisson–Boltzmann and Donnan arguments holds for any type of electrolyte with the inverse Debye screening length defined by a summation, in the usual way, over all ionic species.

Paper by Mackie, Hey, and Mitchell

Prof. Dr. P. Walstra (Agricultural University, Wageningen): The viscosity of melted chocolate is greatly increased by small amounts of water. This suggests that tiny water droplets may form necks between hydrophilic particles, thereby aggregating them. The action of lecithin may be, at least partly, to 'bind' the water by formation of reversed micelles. Have you any evidence about the quantity of water in the dodecane dispersions and its possible effect on the aggregation of the particles?

Dr. J. R. Mitchell (University of Nottingham): I would agree with your explanation for the effect of water on the viscosity of molten chocolate. In our systems we were careful to ensure that no water was present. In chocolate, lecithin lowers with viscosity even in the absence of water, so we consider our interpretation to be reasonable. Would you expect lecithin to replace water at a hydrophilic surface?

Dr. J. Mingins (AFRC Institute of Food Research, Norwich): Lecithins have the possibility of forming microemulsions or reverse micelles in non-polar solvents.

Dr. J. R. Mitchell: Then perhaps this is an additional mechanism when water is present.

Dr. D. T. Coxon, (AFRC Institute of Food Research, Reading): The addition of lecithin to molten chocolate produces a viscosity reduction over the range of addition of 0.1–0.5% w/w. At higher levels of lecithin addition there is an increase in the chocolate viscosity. Can you explain this phenomenon from your model system study?

Dr. J. R. Mitchell: There seem two possibilities here: either at high levels of addition, lecithin in the bulk solvent contributes to the rheology, or alternatively at high levels of lecithin there is an adsorbed bi-layer and the surface becomes polar again.

Miss J. F. Heathcock (Cadbury Schweppes plc, Reading): Chocolate systems, in general, have a wide particle size distribution ranging from colloidal particles up to tens of microns. What effect do you think this has on lecithin functionality?

Dr. J. R. Mitchell: The adsorption behaviour of lecithin should be independent of particle size. We therefore consider our proposed mechanism to be also relevant to more polydisperse suspensions of larger particle size. The rheological behaviour would be dominated by the smaller particles.

Dr. D. R. Haisman (Consultant, Bedford): In your investigation of the opposite rheological properties of lecithin-coated silicas in polar and non-polar systems, have you considered comparing behaviour through a homologous series of alcohols, when you might perhaps detect a definite transition point?

Dr. J. R. Mitchell: It would be very interesting to do this. Thank you for the suggestion.

Paper by Baines and Morris

Dr. J. R. Mitchell (University of Nottingham): Your data show very nicely the superior flavour release properties of xanthan compared with other hydrocolloids. It is well known that xanthan can exist in an ordered conformation or a denatured/random coil conformation depending on salt levels and temperature. Is it possible that it only shows its unusual flavour release properties when in the ordered conformation?

Miss Z. V. Baines (Cranfield Institute of Technology): As all the disordered polysaccharides we investigated behaved in the same way, it would be reasonable to expect the same behaviour for disordered xanthan. Unfortunately, it is not really possible to test this experimentally as the counter ions to the polymer contribute sufficient ionic strength to stabilize the ordered form at room temperature, except at very low polymer concentrations where we would not expect to see significant flavour or taste suppression anyway. The alternative of presenting the solutions at 70 or 80 °C might not be too popular with our panellists!

Dr. L. R. Fisher (CSIRO, Australia): You suggested, I believe, that one mechanism of suppression is actual restriction of diffusion, and hence release, of flavour molecules. Could you not check this by measuring the concentration of flavour component remaining in the gel after a given length of mastication? This would not necessarily give evidence of restricted diffusion (unlikely to be important at these polymer concentrations, I should think) but would at least help to isolate concentration of *released* flavour *versus* other factors in the overall perception.

Miss Z. V. Baines: This is an interesting suggestion, which we might well try out. There would be considerable difficulty and experimental error in analysing for residual flavour volatiles, but accurate analysis for sugar should be straightforward and as in all our studies flavour and taste were suppressed in exactly the same way, that should be sufficient.

Dr. A. Lips (Unilever Research, Sharnbrook): Can you comment on whether the observed dependence of flavour perception with polymer concentration in the case of the more random coil systems is similar to that expected for the diffusion coefficient of the flavour molecules.

Miss Z. V. Baines: All our results suggest that hindered diffusion is not a significant factor in flavour and taste suppression in polysaccharide solutions. Some years ago Preston and his group at Monash University in Australia carried out a careful study of diffusion of low molecular weight solutes in polymer solutions, using dextrans with molecular weight ranging from about 10^4 to about 2×10^6, and found that the diffusion coefficient decreased with increasing size of the solute molecules and with increasing polymer concentration, but was almost totally independent of the molecular weight of the polymer.[12] In our own work the perceived intensity of flavour and taste is strongly dependent on molecular weight, but scales with degree of coil overlap, in the same way as objective (and perceived) solution viscosity.

Dr. D. F. Darling (Glace Bolaget AB, Sweden): The concentrations of polysaccharides being used in this study are such that diffusion of small-molecule solutes is unlikely to be inhibited and therefore molecular diffusion as a flavour/taste release mechanism in these systems seems improbable. A more probable interpretation of the effect of polysaccharide concentration on taste perception is that the transport mechanism is convection controlled and that the polysaccharide solution viscosity will then influence the convective and mass-transport process. This will be particularly significant above c^* and for the stabilizers where gelation of the system becomes effective.

Miss Z. V. Baines: This is exactly our own view: the dominant process in flavour/taste suppression in thickened systems is restricted mixing between the bulk of the solution and the surface.

Paper by Wedzicha and Zeb

J. Mingins (AFRC Institute of Food Research, Norwich): Do you consider that phase-transfer reagents play any role in the distribution of components in food systems?

Dr. B. L. Wedzicha (University of Leeds): Phase-transfer reagents are normally charged, non-polar molecules which are capable of forming ion pairs with oppositely charged ions in aqueous solution, thereby allowing the extraction into an organic solvent of species which are normally insoluble in the organic phase. Thus, for example, 8-hydroxyquinolinium cations associate with a variety of inorganic anions to form ion pairs which are extractable into butanol and 3-methylbutanol (M.-S. Kuo and H. A. Mottola, *Anal. Chim. Acta*, 1980, **120**, 255). Food systems are likely to contain suitable phase-transfer reagents in the form of the many naturally occurring bases, although none has yet been identified in this role. The potential presence of such reagents is certainly another factor which needs to be considered in providing a proper description of solute partitioning in foods.

Paper by van Voorst Vader and Groeneweg

Mr. J. Madsen (Grindsted Products A/S, Denmark): 1. (a) Can monoglycerides be used in the system? How much?

2. Can you work with the system with water droplet size about 3 μm.

Dr. F. van Voorst Vader (Unilever Research, Vlaardingen): 1. Yes, about 0.5–1% GMO for paraffin oil. For polarizable solvents (CCl_4, benzene) higher concentrations are needed.

2. If an elastic phase is formed spontaneously at the interface (see the example of isostearic acid glycerol ester in water–paraffin) this should be possible.

Dr. A. Lips (Unilever Research, Sharnbrook): You refer to the sedimentation of concentrated emulsions and the possible relevance of theories based on diffusion-limited aggregation and cluser–cluster aggregation models which have implicit fractal type representations of aggregate structures. Can one expect these ideas to represent the formation of aggregates in concentrated dispersions?

Dr. F. van Voorst Vader: The essential point that I take from the theory of elastic percolation is that the weak spots that determine the deformation properties of a sediment can be described by the deformation properties of randomly oriented chains as analysed by Kantor and Webman[8] and Brown.[9] This does not necessarily imply that the whole sediment geometry can be adequately described as a fractal.

Available data on the relationship between storage modulus and volume fraction can be described by equations (6) and (7) up to high values of the volume fractions φ. At present it is not clear whether the phenomenological parameters φ_c, $G'(\varphi)$ and τ can be assigned to a certain simulation model or that they only contribute 'effective' parameters in the chemical engineering sense.

Paper by Barker and Grimson

Prof. Dr. P. Walstra (Agricultural University, Wageningen): During slow sedimentation, Brownian motion of the particles can considerably affect the sedimentation rate, either slowing down sedimentation of individual particles or enhancing the rate owing to sedimentation of groups of particles. The latter effect depends on Peclet number and the volume fraction. Can such refinements be introduced into your theory?

Dr. G. C. Barker (AFRC Institute of Food Research, Norwich): It can be shown that the addition of a stochastic term to Burger's equation leads to the renormalization of the constant ε. The inclusion of noise leads to larger values of ε and therefore to broader shocks. However, as pointed out by Prof. Edwards, Burger's equation is one-dimensional so that simple noise is not an accurate representation of Brownian motion. The details of Brownian motion must be included at the microscopic level and, therefore, in this model they appear in a rigorous derivation of a velocity function, $v(\varphi)$. This is only possible in a few cases where Brownian effects are not too strong so that, at this stage, we should restrict applications to the region of Peclet numbers where gravitational effects dominate, *i.e.* large particles. This regime is relevant for many sedimenting food systems.

Dr. E. Dickinson (University of Leeds): You have used a Monte Carlo technique to simulate the kinetics of sedimentation. Normally, the Monte Carlo approach invokes no concept of time, but you have reported the velocity of sedimentation from your computations. Could you explain this, please?

Dr. G. C. Barker: In Figure 5, time is used as a convenient label for the number of Monte Carlo steps per particle. The configurations generated during the simulation are time-ordered but, as in all Monte Carlo calculations, the quantitative association of steps per particle with time is problematic.

Paper by Purslow

Miss J. F. Heathcock (Cadbury Schweppes plc, Reading): Are there any general observations on the effect of temperature on the fracture mechanisms of these meat systems?

Dr. P. Purslow (AFRC Institute of Food Research, Bristol): The effect of temperature on the fracture of frozen meat has already been discussed; there is a transition from essentially elastic to elastic–plastic fracture with increasing temperature. In terms of the fracture of cooked meat and reformed meat, there is obviously a strong effect of cooking temperature, because of the different structural changes caused at different cooking temperatures. For example, beef cooked to 80°C has a higher breaking stress along the muscle fibre direction than at 60 °C (P. E. Bouton, P. V. Harris and W. R. Shorthose, *J. Texture Stud.*, 1975, **6**, 297). Lastly, the temperature at which the fracture test is carried out has an effect. Meat cooked to 80 °C is less tough when subsequently tested at 70 °C than tested at room temperature. This is because some of the heat-induced denaturation of the collagen is reversible (D. A. Ledward and R. A. Lawrie, *J. Sci. Food Agric.*, 1975, **26**, 691).

Dr. M. L. Green (AFRC Institute of Food Research, Reading): Can you describe how fat affects fracture processes in meat?

Dr. P. Purslow: We have tested the effect of small inclusions of comminuted adipose tissue on the adhesive strength of junctions between meat pieces bound together with myosin gel (Donnelly and Purslow, unpublished observations). These results show that fat inclusions depress the adhesive strength of the meat–myosin gel junctions. This diminution in strength is slightly greater than can be accounted for on the basis of the fat inclusions 'diluting' the protein concentration of the myosin gel.

Dr. J. R. Mitchell (University of Nottingham): The two examples of fracture you gave for meat pieces held together by a protein gel showed (a) fracture at the interface between the gel and the meat, and (b) fracture in the meat piece. Does fracture ever occur in the protein gel itself, and if not does that mean that studies on isolated myosin gels are irrelevant to the strength of reformed meats?

Dr. P. Purslow: In reformed meats where the pieces of meat in the product are quite small, the gel phase is in effect a continuous phase and, for a whole reformed steak to be broken into two pieces, some amount of fracture of this continuous gel phase must inevitably occur. In more traditional products such as sectioned and formed hams (*i.e.* traditional hams), which still constitute the major share of reformed meat products, because the pieces of meat stuck together are large it is quite likely that a bite-sized specimen will contain just one junction, and that the gel is *not* continuous phase. In this case, the strength of a specimen cut from the product may indeed have little to do with the intrinsic strength of the myosin gel, if these products fail by adhesive, interfacial separation, or cohesive failure in the meat pieces, *i.e.* if the cohesive strength of the protein gels is high. Many of the studies of the 'binding strength' of protein gels that appear in the literature have used junctions such as that described in this paper, (J. J. MacFarlane, G. R. Schmidt, and R. H. Turner, *J. Food Sci.*, 1977, **42**, 1603; D. G. Siegel and G. R. Schmidt, *J. Food Sci.*, 1979, **44**, 1129; D. G. Siegel and G. R. Schmidt, *J. Food Sci.*, **44**, 1686); others have argued that the strength and stiffness of isolated gels have a bearing on binding function (T. Nakayama and Y. Sato, *J. Texture Stud.*, 1979, **2**, 75; M. Ishioroshi, K. Samejima, and T. Yasui, *J. Food Sci.*, 1979, **44**, 1280). Careful comparison of these two types of investigation show that factors (such as actin-to-myosin ratio, the level of added salt in the gel, its pH, and temperature of cooking), which are said to affect binding do have a different pattern of effects in the strength of isolated gels than in the strength of meat–gel junctions (P. D. Jolley and P. P. Purslow, in 'Food Structure–its Creation and Evaluation,' eds. J. M. V. Blanshard and J. R. Mitchell, Butterworths, London, 1988, pp. 231–264). This again argues that the mechanism of failure of these model meat–gel junctions is *not* cohesive-in-adhesive failure—because if it were, then the effects of these factors would be the same in both kinds of test. This distinction has helped clear up a deal of confusion over the effects of certain factors on binding strength.

Paper by Borchert

Mr. D. Kochakji (T. J. Lipton Inc., USA): If you break a wiener emulsion during processing, is there any method to 're-form' the emulsion and still have an acceptable end product?

Dr. L. Borchert (Oscar Mayer Foods Corp., USA): No, there is no method to re-form the broken emulsion.

Paper by Madsen

Dr. R. D. Bee (Unilever Research, Sharnbrook): Is there a critical size at which water drops produce spattering in margarines or spreads?

Mr. J. Madsen (Grindsted Products A/S, Denmark): Not really, even margarine with water droplets as small as 1–2 μm can give heavy spattering if the droplets coalesce, and margarine with water droplets up to 5–10 μm can give minimal spattering. In spreads (70% fat) there is heavy spattering in all products during open-pan frying.

Paper by Campbell

Dr. F. van Voorst Vader (Unilever Research, Vlaardingen): According to the

Curie principle, growth of those crystal planes minimize the surface free energy of the crystal which predominates. Moreover, the molecules in the surface layers of amphiphile crystals assume the most suitable orientation to this purpose. For instance, monoglyceride crystals that grow in contact with water orient their polar groups towards the aqueous phase and thus are hydrophilic, while the same crystals when grown in contact with oil orient their hydrophobic tails outwards and thus are hydrophobic. Experiments performed by Dr. D. A. Bargeman indicated that films of stearic acid solidified in contact with air were initially hydrophobic, but became gradually hydrophilic when left in contact with water, indicating a reorientation of their surface layer. My question is whether your results support the above statements?

Mr. I. J. Campbell (Unilever Research, Sharnbrook): Experiments performed with fatty acids clearly demonstrated that crystals grown in contact with water are more hydrophilic than those crystals grown in contact with air. Similar but less significant differences were found for monoglyceride crystals. These results would appear to be in agreement with the work of Dr. Bargeman. The effect of prolonged exposure to water of crystals, solidified in contact with air, was not examined. However, the observation that the surface hydrophilicity increases with time seems feasible, particularly if one considers the dynamic equilibrium of solubilization and recrystallization occurring in fat crystal systems.

Paper by Harbron, Ottewill, and Bee

Dr. B. L. Wedzicha (University of Leeds): What are the resolubilization characteristics of tea cream?

Dr. R. S. Harbron (University of Birmingham): It can be seen from the 'phase diagram' shown in Figure 2 that the solubilization/reformation of tea cream is a reversible process. As the tea infusion is heated, the cream particles begin to shrink and apparently dissolve in the supporting medium. Eventually a critical temperature is reached at which shrinkage no longer occurs and this is taken to be the 'cream point.' However, the particles are not completely dissolved as a sub-microscopic residue remains even at temperatures of 100 °C. This is presumed to be further polymerized high molecular weight polyphenolics. The re-formation of the cream particles is the reverse of the above, provided that the cooling rate is slow (*i.e.* cooling under ambient conditions). The tea cream particles apparently grow from the insoluble residue mentioned above. Hysteresis was not observed to any significant degree in the heating–cooling cycle.

Dr. H. Kerr (Flour Milling and Baking Research Association, Chorleywood): With regard to the data relating concentration to morphology of particles in the tea infusions, where does a normal cup of tea come on this scale, and would the 'cream' be forming where milk is present?

Dr. R. S. Harbron: Beverage strength is approximately 0.35% w/w and cream is indeed visible in an infusion of this strength which has been allowed to cool. Milk proteins interact strongly with many components of tea, particularly the polyphenols, and interfere with the aggregation process we have described, probably by reducing the secondary aggregation.

Dr. M. A. Tung (Technical University of Nova Scotia, Canada): You have shown light micrographs, scanning electron micrographs and transmission electron micrographs of shadowing the particles in black tea, all of which show surface features of the particles. Have you carried out thin-section preparation and transmission electron microscopic studies to show *internal* structure of the particles? If so, what are the fine structure characteristics of the particles?

Dr. R. S. Harbron: Electron micrographs of the sections of the cream particles have been made and the variation in morphology with concentration has been described in detail (R. D. Bee, M. J. Izzard, R. S. Harbron, and N. M. Stubbs, *Food Microstruct.*, 1987, **6**, 47.

Paper by Dr. S. M. Gaud
(Presented at the meeting, but not reproduced in this volume)

Mr. D. Kochakji (T. J. Lipton Inc., USA): You mentioned enzymatic activity in distilled vinegar; could you be more specific as to the enzyme found in distilled vinegar?

Dr. S. M. Gaud (Kraft Inc., USA): Amylase, but not very often.

Paper by Dalgleish

Miss J. F. Heathcock (Cadbury Schweppes plc): What is your current thinking of the effect of heating using direct steam injection?

Dr. D. G. Dalgleish (Hannah Research Institute, Ayr): Inasmuch as we have not applied our techniques to caseins isolated from UHT products, it is not possible to give a detailed answer in terms of the chemical damage caused to milk or to casein by either direct or indirect heating of milk.

Prof. Dr. P. Walstra (Agricultural University, Wageningen): Direct steam injection appears to cause aggregation.

Dr. D. G. Dalgleish: This is true, but we do not understand the chemistry in detail. In the paper, I was referring specifically to the observed effects of different drying regimes on the final product. We have yet to extend our measurements to milks heated in various ways, which will be a much more difficult problem.

Dr. D. R. Haisman (Consultant, Bedford): Do the differences in stabilizing efficiency and calcium sensitivity between roller-dried and spray-dried caseinates extend to their stabilities in acid systems?

Dr. D. G. Dalgleish: We have not studied this aspect; certainly all of the casein was precipitable at pH 4.6 in all of our samples. We have no information on the acid sensitivity at, say, pH 5.0–5.5, where caseins generally exhibit some instability.

Dr. D. C. Clark (AFRC Institute of Food Research, Norwich): Do the differences in the chromatographic profiles obtained with your casein samples compared with industrial preparations reflect differences in states of aggregation?

Dr. D. G. Dalgleish: The chromatography uses high concentrations of urea, so I do not think that aggregation of the α_s- and β-caseins is important. In our procedure the κ-casein is alkylated, so that aggregation *via* disulphide formation is also present. So our chromatography does not reflect aggregation. However, it is clear that in the original caseinates there is aggregation, as estimated by amounts of insoluble material, after attempting to dissolve the caseinate in hot water. As might be expected, roller-dried material was the worst, but some spray-dried caseinates also gave quite large amounts of insoluble material.

Dr. B. B. Gupta (St. Ivel Ltd., Bradford-on-Avon, Wilts): 1. The quality of skim milk from which the sodium caseinate was prepared commercially is likely to affect the elution profile which you obtained on separation by FPLC.

2. Have your prepared sodium caseinate from a poor-quality skim milk and separated by FPLC? If so, what type of profile did you obtain?

Dr. D. G. Dalgleish: My own caseinates were made from our own Institute milk, which is high quality in respect of bacteriological contamination. We have not attempted to make caseins from specific low-quality milks, and to study their properties. However, from what I have seen of differently prepared caseins, the alterations in elution profiles which I see in commercial caseinates are not those which I would expect to find in milks which had been proteolytically attacked.

Paper by Povey

Dr. P. D. Fletcher (University of Hull): Is there any effect on the ultrasonic wave propagation due to chemical relaxation processes (as in ultrasonic relaxation investigations of the kinetics of micelle dynamics, cation hydration, *etc.*)?

Dr. M. Povey (University of Leeds): Any effects due to chemical relaxation processes in the component phases will be accounted for since the velocity and attenuation of the component phases are measured and used in the theoretical calculations.

Miss J. F. Heathcock (Cadbury Schweppes plc, Reading): What is the lower limit of resolution for particle size determinations by this technique and can it be used for sizing asymmetric particles?

Dr. M. Povey: For the sunflower oil-in-water emulsions investigated in this work the lower limit of the particle size that can be determined is about 20 μm (see text). The technique could be used for sizing asymmetric particles, however more complex theoretical equations would be needed, *e.g.* those of C. C. Habeyer (*J. Acoust. Soc. Am.*, 1982, **72**, 870).

Mrs. K. Boode (Agricultural University, Wageningen): In your paper you did not deal with the shape of the emulsion droplets. I assume, however, that the theory you applied applies to spheres. Is that correct and, if so, can this technique be applied to emulsions with irregularly shaped droplets?

Dr. M. Povey: The technique could be applied to emulsions with irregularly shaped droplets, but more complex theoretical equations are required. A number of workers have included asymmetric shaped particles in their fomulations, *e.g.* Habeyer (see above).

Dr. L. R. Fisher (CSIRO, Australia): The problem of working with real systems containing both air bubbles and oil droplets seems to be a very important one. Do you see prospects for distinguishing between the two by using, say, the ultrasonic scattering properties of the system?

Dr. M. Povey: Ultrasonic scattering due to air bubbles are very different to that of oil droplets or water droplets suspended in water or oil, respectively, owing to thermal scattering and acoustic impedance differences. Thus it may be possible to distinguish between the different scattering phenomena by measuring the dependence of ultrasonic velocity and attenuation on frequency. Alternatively, the air could be removed, or its particle size varied by compressing the food.

Studies on the Aggregation of Casein Micelles

By David S. Horne

HANNAH RESEARCH INSTITUTE, AYR KA6 5HL, UK

For some time we have been studying the aggregation of casein micelles induced by either the addition of ethanol or by rennet proteolysis. Our studies have used both static, principally turbidimetric, and dynamic light-scattering techniques. This paper concentrates on a particular aspect of this study, the fractal nature of the aggregates.

Recent computer simulations have stimulated interest in defining the structure of colloidal aggregates in terms of the concept of the fractal dimension.[1] Fractal objects show dilation symmetry: their essential geometric factors are invariant to scale change. Such aggregate structures or clusters can be quantitatively characterized by their fractal dimension, D, a measure of the cluster packing density, such that

$$N(r) = (r/r_0)^D \tag{1}$$

where $N(r)$ is the number of seed particles of radius r_0 contained inside a sphere of radius r within the aggregate or cluster. For a growth process we can take r to be some characteristic size of the cluster (*e.g.* its radius of gyration) and interpret the above equation as a law relating mass to size at various stages of growth. For the trivial case of a solid object, the fractal dimension D is equal to d, the Euclidean dimension of space. However, for all cases of interest here, $D < d$, and is typically non-integral. Typical values for the fractal dimension of structures resulting from aggregation processes fall in the range 1.7–2.5 whether obtained in simulation calculations using various reaction models or by experimental measurement in a variety of reactions.[2]

From a scattering viewpoint, all fractals show a power-law dependence of the scattered intensity, I, on the wave vector, q, as given by

$$I \propto q^{-D} \tag{2}$$

when $R^{-1} \ll q \ll r_0^{-1}$, r_0 being the monomer radius, R some characteristic size of the aggregate, and D the fractal dimension. Normally, for large length scales, angular dependence of laser light scattering is employed to measure the fractal dimension. However, the wave vector is also a function of the wavelength of the light employed and it was recently demonstrated that the fractal dimension of an aggregate is similarly accessible through measurement of the wavelength dependence of the turbidity of the suspension.[3] All fractal dimensions reported in this study were measured using this turbidimetric technique.

To limit effects of micellar polydispersity, as found in a normal sample of skimmed milk, the aggregation studies were carried out on a defined micellar size fraction, designated pellet 4, prepared by differential centrifugation and resuspension in milk ultrafiltrate.[4]

Initial rate studies of ethanol-induced aggregation of casein micelles under the same reaction conditions of pH, ionic strength, and calcium level showed the reaction to become diffusion-limited at ethanol concentrations greater than 29% v/v. In the experiments described below data were obtained as a function of micelle concentration at an ethanol concentration of 30% v/v.

First, we consider the time dependence of the wavelength exponent, *i.e.* the gradient of the double logarithmic plot of turbidity against wavelength for one particular micelle concentration. It is readily shown[3] that this gradient is equal to $4.2 - \beta$ and that β, effectively the wavelength exponent of the structure factor, approaches the fractal dimension as the aggregate size increases. The experimentally observed behaviour is shown in Figure 1. The error bars on each point are those obtained from the linear regression analyses of log τ on log λ. For this particular example, measurements were also carried out on the angular dependence of the light scattering and on the kinetics of growth of the particle radius with time. Both methods yielded values for the fractal dimension close to the asymptotic value of 2.3, obtained using the turbidimetric technique by averaging the last six points in Figure 1 at the longer times.

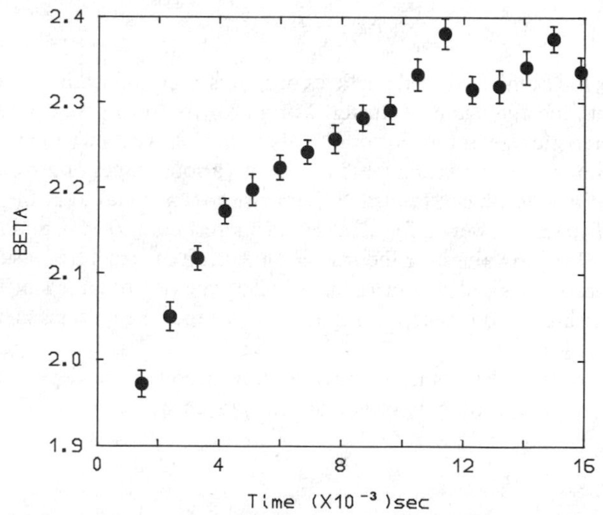

Figure 1 *The turbidity exponent,* β, i.e. *the wavelength exponent of the cluster structure factor, as a function of reaction time for an aggregating casein micelle system (pellet 4) in* 30% v/v *ethanol solution, illustrating its approach to an asymptotic value, the fractal dimension, as the cluster grows*

This value of 2.3 is higher than the value of 1.7 obtained in simulations of diffusion-limited cluster–cluster aggregation[5,6] or even reaction-limited cluster–cluster aggregation at 2.0,[7] but it is lower than that obtained in the other major reaction model, diffusion-limited particle–cluster aggregation, which yields a value of 2.5.[8,9]

Experimentally, results have been obtained for several, mainly mineral, systems. The transition from diffusion-limited to reaction-limited regimes and the consequent change in the aggregate fractal dimension from 1.7 to 2.0 has been observed in aggregates of colloidal gold.[10] Aubert and Cannell[11] have shown that slow aggregation of colloidal silica yielded clusters with $D = 2.08 \pm 0.05$ but that rapid aggregation could give clusters with $D = 1.75$ or 2.08, although the clusters with the lower fractal dimension always restructured to give $D = 2.08$.

Our experimental value for this particular set of reaction conditions is larger than those encountered so far in either simulation studies or in experimental measurement of the fractal dimension. From an analysis of the rate of change of β with time, it was suggested[3] that this higher value was due to an annealing and restructuring process concurrent with cluster formation, a mechanism not fully realized so far in simulation studies. Such attempts[12–14] as have been made, however, on fully developed clusters have increased the fractal dimension only slightly.

Simulation studies are, however, carried out necessarily at low concentration. Precisely, they mimic infinite dilution whereas our experiments are at a low but finite concentration. Kolb *et al.*[15] have suggested that higher fractal dimensions

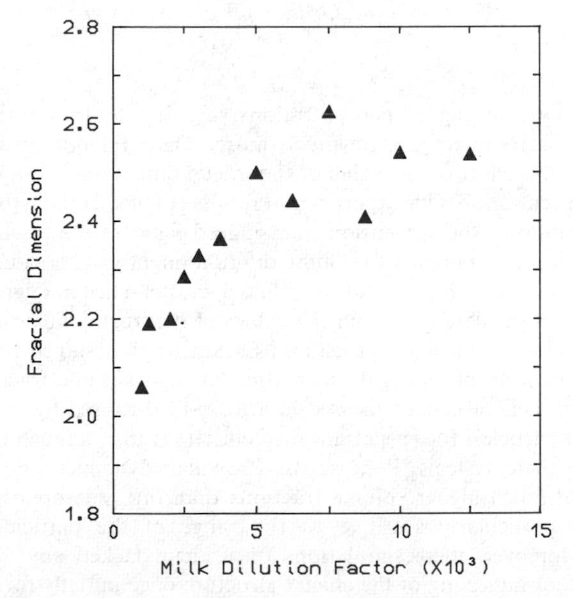

Figure 2 *Fractal dimensions obtained by turbidimetric techniques for a range of micelle concentrations. The ethanol concentration was* 30% v/v *in all instances*

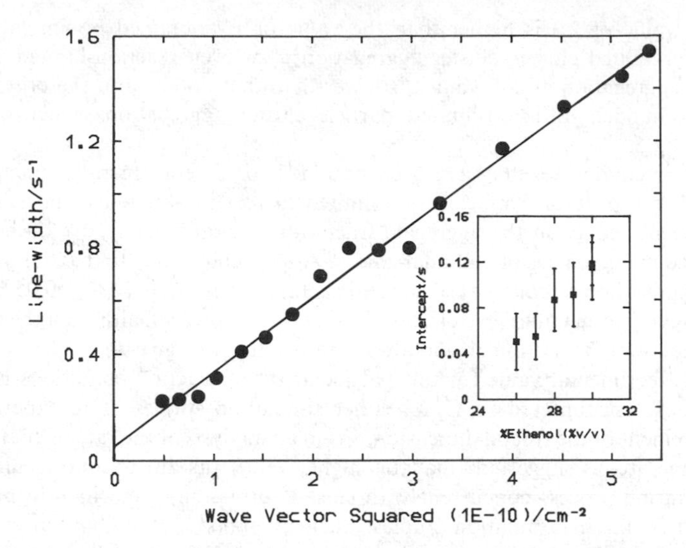

Figure 3 *Dependence of relaxation rate or line width of the autocorrelation function obtained for a micellar aggregate as a function of the square of the wave vector,* q. $q = 4\pi n/\lambda(\sin\theta/2)$, *where* θ *is the scattering angle,* λ *the laser light wavelength, and* n *the refractive index of the medium. The solid line is a linear regression fit to the data points shown. The intercept yields the rotational diffusion coefficient, the inset showing how this varies with ethanol concentration for aggregates grown on a similar time scale*

should be found at higher concentrations, arguing that in this regime the developing clusters interpenetrate much more. They did not speculate on the magnitude of this effect on the value of the fractal dimension or on how it would vary with concentration. Our experimental results (Figure 2) show that increasing the concentration of the suspension does indeed increase the fractal dimension and suggest that extrapolation to infinite dilution might yield a value close to the simulation result of 1.7 for diffusion-limited cluster–cluster aggregation. The general shape of the data curve and the values of the fractal dimension shown in Figure 2 are close to those of Ansell and Dickinson,[16] obtained in a Brownian dynamics simulation of aggregate formation at high volume fraction. In their study, each particle or cluster diffused in a force-field caused by the presence of neighbouring particles. Interpenetration of clusters is thus a much more important factor in these systems. Because the Brownian dynamics simulations were carried out at much higher volume fractions than our measurements, they are likely to be particularly sensitive to the range of the particle interactions employed. Moreover, these simulations once again lacked any allowance for consolidation or annealing of the cluster structure once initially formed.

One further aspect of structural interest is the anisotropy of the clusters formed in ethanol-induced aggregation of casein micelles. Based initially on an analysis of the hydrodynamic behaviour of clusters generated in simulation studies, Lindsay *et al.*[17] concluded that the fractal structure of their gold sols was anisotropic, allowing the effects of their rotational diffusion to map through into the decay of the autocorrelation function. Chu[18] has shown that when $qR_g \gg 1$, the first cumulant Γ, *i.e.* the logarithmic derivative of the autocorrelation function when $t \rightarrow 0$, can be written as a linear function in q^2 with a positive intercept, the rotational diffusion coefficient. It is possible that the gold sol clusters were too small to meet the criterion $qR_g \gg 1$, since their plot of Γ *versus* q^2 showed no intercept, passing directly through the origin, the gradient being the translational diffusion coefficient. Our data for the much larger casein micelles do, however, produce intercepts (as in Figure 3), indicating a substantial rotational contribution. Measuring this intercept as a function of ethanol concentration, we find that the anisotropy increases with increasing ethanol concentration over the range 26–30% v/v (inset, Figure 3). This is precisely the range over which this reaction passes from reaction-linked to diffusion-limited kinetics. It is therefore possible that more compact spherical aggregates are being produced at the lower alcohol levels and, if not more open ones, at least aggregates with higher aspect ratios at higher ethanol concentrations. The latter varying behaviour is not reflected in the measured fractal dimension, however, this remaining at a uniform value of 2.2–2.3 across the alcohol concentration range. These results are, however, preliminary and at a single casein micelle concentration. They are therefore to be subjected to more extensive time and concentration dependence studies.

References

1. B. B. Mandelbrot, 'The Fractal Geometry of Nature,' Freeman, San Francisco, 1982.
2. See, for example, F. Family and D. P. Landau (eds.), 'Kinetics of Aggregation and Gelation,' North-Holland, Amsterdam, 1984; H. E. Stanley and N. Ostrowsky (eds.), 'On Growth and Form,' Nijhoff, Dordrecht, 1986; L. Pietronero and E. Tosatti (eds.), 'Fractals in Physics,' Elsevier, Amsterdam, 1986.
3. D. S. Horne, *Faraday Discuss. Chem. Soc.*, 1987, **83**, 259.
4. D. S. Horne and D. G. Dalgleish, *Eur. Biophys. J.*, 1985, **11**, 249.
5. P. Meakin, *Phys. Rev. Lett.*, 1983, **51**, 1119.
6. M. Kolb, R. Botet, and R. Jullien, *Phys. Rev. Lett.*, 1983, **51**, 1123.
7. M. Kolb and R. Jullien, *J. Phys. (Paris) Lett.*, 1984, **45**, L977.
8. T. A. Witten and L. M. Sander, *Phys. Rev. Lett.*, 1981, **47**, 1400.
9. P. Meakin, *Phys. Rev. A*, 1983, **27**, 604.
10. D. A. Weitz, M. Y. Lin, and C. J. Sandroff, *Surf. Sci.*, 1985, **158**, 147.
11. C. Aubert and D. S. Cannell, *Phys. Rev. Lett.*, 1986, **56**, 738.
12. R. Botet and R. Jullien, *Phys. Rev. Lett.*, 1985, **55**, 1943.
13. P. Meakin, *J. Colloid Interface Sci.*, 1986, **112**, 187.
14. P. Meakin, *J. Chem. Phys.*, 1985, **83**, 3645.
15. M. Kolb, R. Jullien, and R. Botet, in 'Scaling Phenomena in Disordered Systems,' eds. R. Pynn and A. Skjeltrop, Plenum Press, New York, 1985, p. 71.
16. G. C. Ansell and E. Dickinson, *Faraday Discuss. Chem. Soc.*, 1987, **83**, 167.
17. H. M. Lindsay, M. Y. Lin, D. A. Weitz, P. Sheng, Z. Chen, R. Klein, and P. Meakin, *Faraday Discuss. Chem. Soc.*, 1987, **83**, 153.
18. B. Chu, 'Laser Light Scattering,' Academic Press, New York, 1974.

Kinetics of the Partial Coalescence Process in Oil-in-Water Emulsions

By K. Boode and P. Walstra

DEPARTMENT OF FOOD SCIENCE, WAGENINGEN AGRICULTURAL UNIVERSITY, WAGENINGEN, THE NETHERLANDS

It was shown by van Boekel and Walstra[1] that the presence of crystals in the oil phase can induce considerable coalescence in oil-in-water (O/W) emulsions, especially in a velocity gradient or in a cream layer. This is, however, not true coalescence, as irregular clumps of fat particles are formed, which disturb the energetically favoured spherical shape. Therefore, the process is called partial

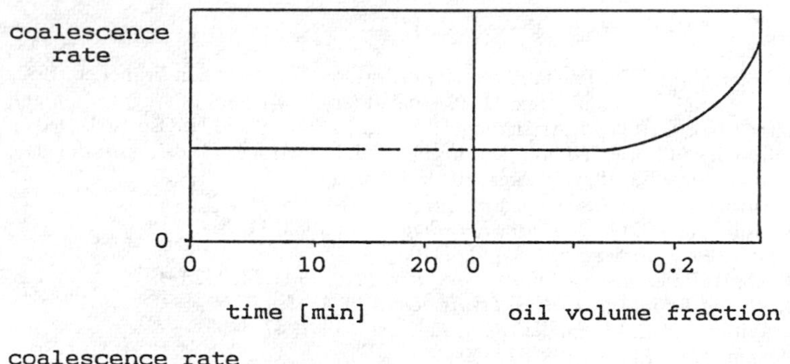

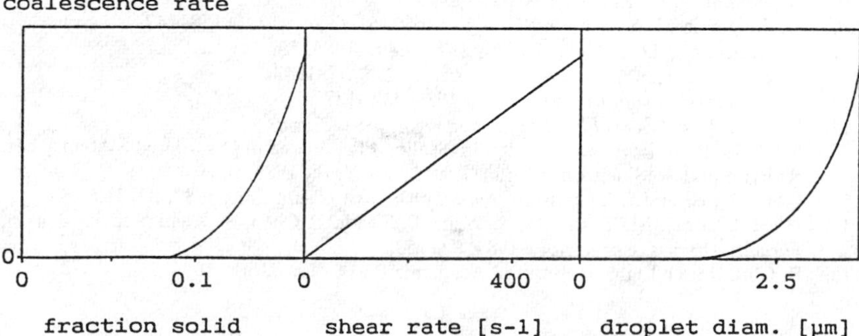

Figure 1 *Coalescence rate (*e.g. *expressed as rate of increase of average particle size) as a function of different variables*

coalescence. It occurred only if fat crystals were observed at the O/W boundary, hence a mechanism based on piercing of the thin film between approaching droplets by slightly protruding crystals was postulated. It was observed that the rate of the partial coalescence process was (initially) independent of time, almost independent of oil volume fraction, and increased strongly with increasing shear rate, average droplet size, and proportion of solid fat: it also depended on the emulsifiers used. In Figure 1 some of these relations are shown schematically.

Since then it has become clear that partial coalescence also may occur in other ways. Figures 2 and 3 give examples of the various types of change in droplet size

The experiments were all carried out with emulsions with some spread in droplet size, although the distribution was narrow enough to allow the use of a mean diameter.

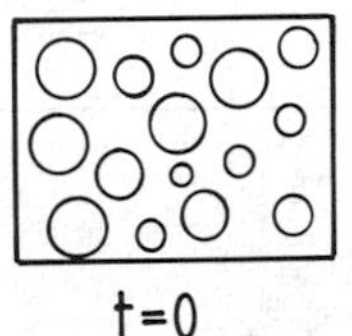

Type A: The fat droplets that participate in partial coalescence form irregularly shaped clumps. On warming, an emulsion is obtained with droplets larger than at the beginning of the experiment.[1]
Example: paraffin mixture in PVA solutions.

t = 30

Type B: In some instances the clumps become so large that they cream out of the emulsion when it is warmed. The remaining fat globules are larger than the original ones.[1,2]
Example: natural cream.

t = 30

Type C: Some fat droplets participate in rapid partial coalescence, leading to clumps creaming out of the emulsion, while the remaining fat globules show an unaltered size distribution.[2,3]
Example: milk fat in whey protein solutions.

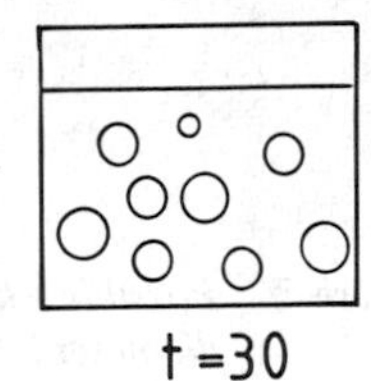

Type D: As in types B and C, large clumps are formed that cream out of the emulsion on heating. This time, however, the remaining emulsion does show an altered size distribution with smaller globules.[2,3]
Example: saturated triglycerides in SDS solutions.

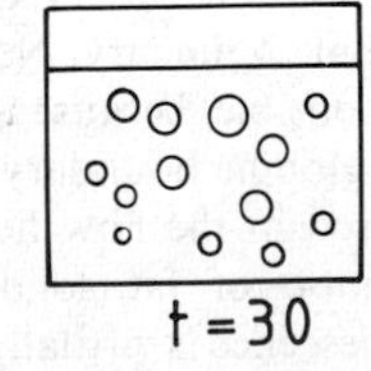

Figure 2 *Various types of partial coalescence (highly schematic). In some instances type A changes into type B in the course of the process, and similarly types C and D*

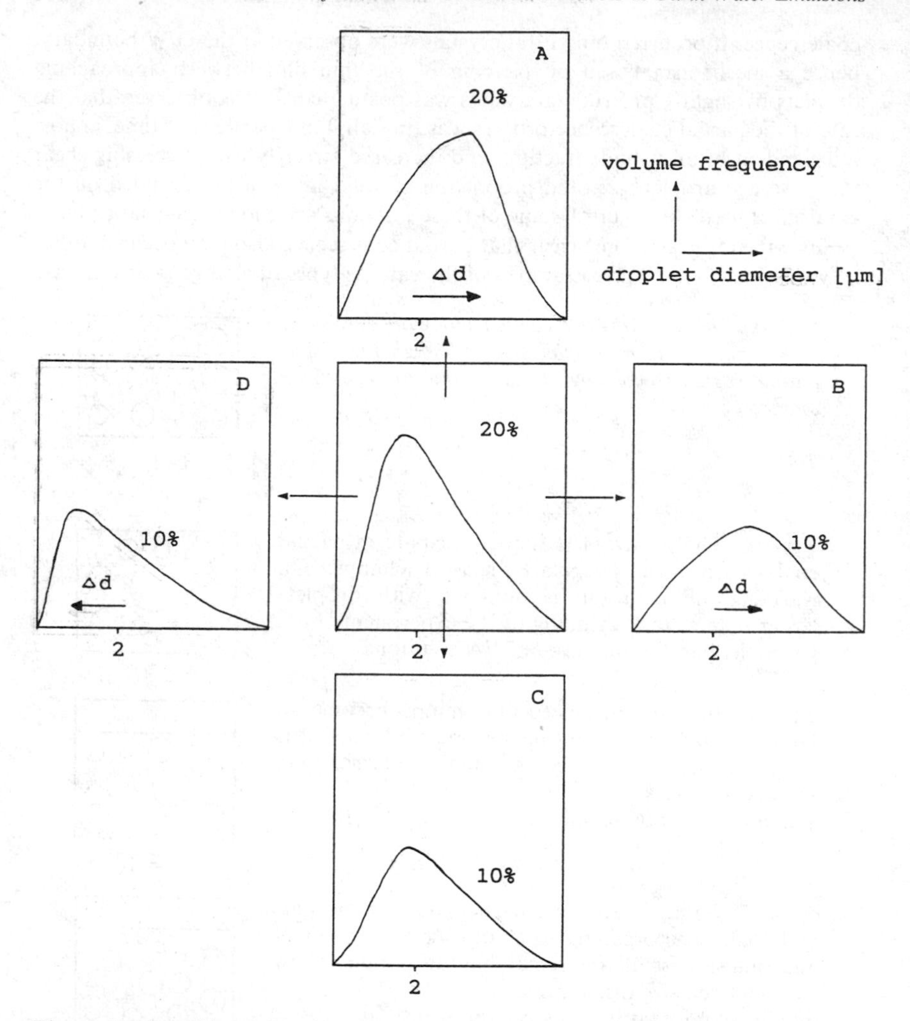

Figure 3 *Possible changes in droplet size distribution (volume frequency* versus *diameter)*

distribution observed.[2,3] In several emulsions, crystals were not seen at the globule boundary. Nevertheless, the same mechanism of piercing crystals may be responsible, because in the clumps obtained there always seemed to be crystals at the globule boundary. Possibly, a change in crystal orientation can somehow be caused in the flow field. The dependence of the partial coalescence rate on the fraction of fat solid was found to be variable; in some emulsions partial coalescence is virtually zero if over half of the fat is solid,[4] whereas in others much more of the fat can be crystallized. It appears as if the kind of fat present in the droplets has the greatest effect on the type of partial coalescence that occurs. Presumably the amount, size, shape, and arrangement of the crystals are important. Moreover, the contact angle (oil, crystal, water) may play a part.[5] Up to now, no clear pattern has manifested itself and consideration of accepted theories of coagulation kinetics, even if various types of colloidal interaction are taken into account, does not lead to an explanation. Further studies are in progress.

References

1. M. A. J. S. van Boekel and P. Walstra, *Colloids Surf.*, 1981, **3**, 109.
2. J. P. Melsen, *PhD Thesis* Wageningen Agricultural University, 1987.
3. K. Boode and P. Walstra, unpublished work.
4. P. Walstra, *Lipids*, 1976, **2**, 411.
5. D. Darling, *J. Dairy Res.*, 1982, **49**, 695.

The α-Gel Phase of Glycerol Lactopalmitate in Whipped Emulsions

By J. M. M. Westerbeek and A. Prins

AGRICULTURAL UNIVERSITY, DEPARTMENT OF DAIRYING AND FOOD PHYSICS, BOMENWEG 2, 6703 HD WAGENINGEN, THE NETHERLANDS

and

K. Kussendrager

D.M.V. CAMPINA BV, NCB-LAAN 80, 5460 BA VEGHEL, THE NETHERLANDS

1 Introduction

In whippable emulsions, both proteins and α-tending emulsifiers are used as surface-active components. The α-tending emulsifiers are added to the product because they promote extensive structure formation.[1,2] This structure consists of agglomerates of flocculated fat particles which form a continuous network throughout the whipped product. It has been suggested that the crystallization behaviour and the structure-inducing properties of these emulsifiers are related to each other, but no satisfactory explanation for this dependence has yet been given.

Buchheim *et al.*[3] showed by means of transmission electron microscopy that partial coalescence may probably be the binding mechanism of the fat particles, but this explanation is not completely satisfactory. For instance, the mechanism does not explain why the product does not churn like natural cream during whipping. Further, we found that these aggregated structures can easily be redispersed by the addition of an anionic surfactant such as sodium dodecyl sulphate (SDS). This would be impossible in the case of partial coalescence.

It seems more likely that some kind of reversible flocculation process occurs in these whippable emulsions. The occurrence of mesomorphic phases between emulsion droplets may lead to reversible structure formation,[4] but it is generally accepted that α-tending emulsifiers do not form mesomorphic phases.[1] Further, these emulsifiers are used below the melting point of their hydrocarbon chains in whippable products.

It is well known that amphiphiles such as monoglycerides[5] or phospholipids[6] are able to form lamellar α-gel phases from the liquid crystalline phases below their melting points. These phases consist of bilipid layers of emulsifier molecules with crystalline hydrocarbon chains, separated by thin water layers. If an α-tending emulsifier was able to form a stable α-gel phase, then a reasonable explanation may be given for the reversible gel properties of the flocculated fat particle network. The particles are flocculated in a deep minimum as a result of attractive van der Waals forces. Repulsive hydration forces protect the flocculated particles against coalescence or partial coalescence. The SDS effect may be explained by the fact that gel phases swell under the influence of anionic amphiphiles.[5]

Evidence for the existence of an α-gel phase may be obtained from both differential scanning calorimetric (DSC) experiments and X-ray diffraction. With DSC it should be possible to detect changes in the melting point of the emulsifier molecules as a result of hydration phenomena. The swelling properties of a lipid bilayer as a result of hydration can be studied by means of X-ray diffraction. The experiments were performed with glycerol lactopalmitate (GLP), an example of an α-tending emulsifier.

2 Materials and Methods

Sample Preparation—The GLP powder was prepared by a method analogous to that of Buchheim *et al.*[3] by spray-drying and an aqueous emulsion containing 30% GLP, 5% sodium caseinate and 12% glucose. This emulsion was obtained by homogenization at 70 °C in a Rannie piston homogenizer at a pressure of 100 atm. This emulsion was spray-dried with an A/S NIRO atomizer and the powder was stored at room temperature.

Differential Scanning Calorimetry—The measurements were performed with a Mettler TA-3000 system. The heating curves were not corrected for differences in the amount of sample in the DSC cup.

X-ray Diffraction—The measurements were performed with a Kratky camera equipped with a Braun one-dimensional position-sensitive detector, which was connected to a Braun multi-channel analyser. Each channel of the multi-channel analyser corresponded to a certain diffraction angle. Corrections were made for background noise and the curves were desmeared.

3 Results and Discussion

The experiments were performed with GLP dispersions, free from added fat, for two reasons. GLP is a very complex mixture of mono-, di-, and triglycerides. As this type of emulsifier is always used in emulsions with large specific surface areas, homogenization of GLP is of great importance. A second reason is that interpretation of the results in the case of emulsions containing relatively large amounts of fat is too difficult because of the occurrence of fat crystallization.

In Figure 1 the difference in heating curves for a dry and a wetted powder sample is demonstrated. The powder sample was brought into contact with water at room temperature, *i.e.* about 20 °C below the melting point of GLP. The addition of water clearly affects the transition temperature of the GLP sample.

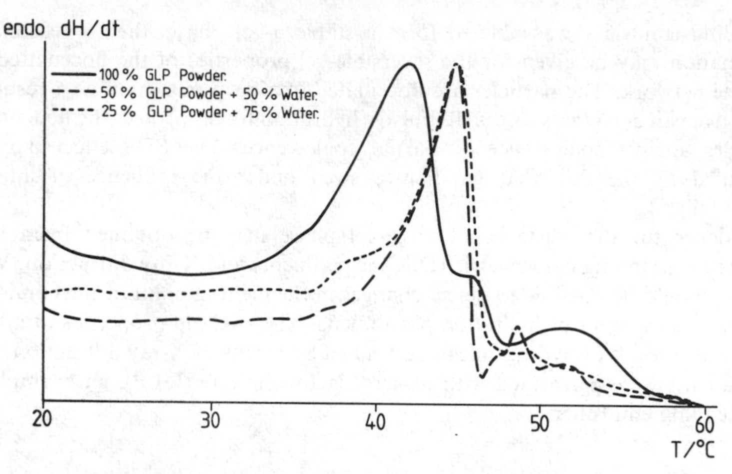

Figure 1 *Heating curve of a spray-dried GLP emulsion in the absence and presence of water. Heating rate:* 1 °C min^{-1}

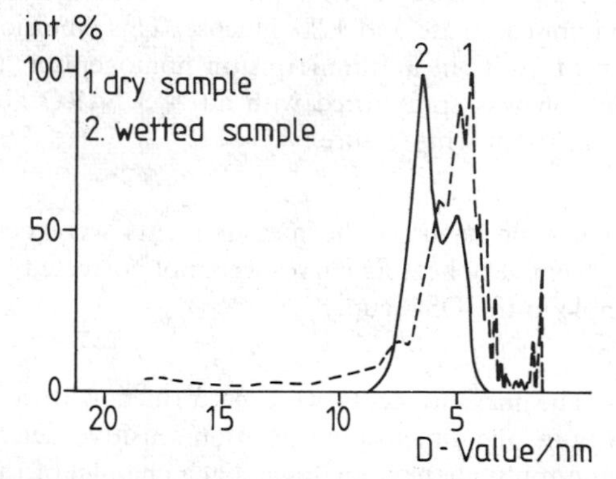

Figure 2 *Desmeared X-ray diffraction patterns for the GLP sample in the absence and presence of water at* 20 °C. *The hydrated sample contained* 67% *water*

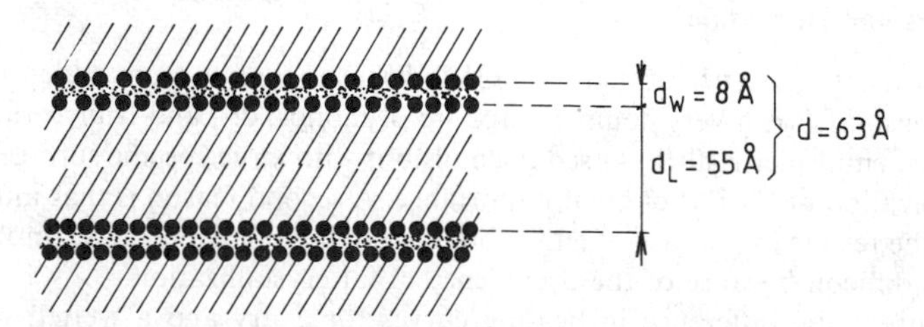

Figure 3 *Schematic molecular model of the structure of the α-gel phase of the GLP at maximum swelling;* d *is the long spacing and* d_w *and* d_L *are the thicknesses of the water and hydrocarbon layers, respectively*

This can only be explained if some specific interaction occurred between water and GLP. The same hydration effect has been found by de Vringer[7] for the melting temperature of a mixture of equal amounts of cetyl and stearyl alcohols.

Figure 2 shows the effect of water on the value of the long spacing of the GLP samples. The presence of water leads to a shift of the long spacing towards a value of about 63 Å. The value of the long spacing for the dry sample is less accurate than that for the wet sample because a broad and scattered signal is obtained. The smeared value for the dry sample is 54 Å, whereas the desmeared value is about 50 Å.

Our main conclusion is that GLP forms a stable α-gel phase in the presence of water. Although the exact values of the length of the dry lipid bilayer, the angle of tilt, and possible changes in this angle caused by hydration are not yet known, we present a hypothetical model for the structure of the hydrated bilipid layer of GLP in Figure 3. In the near future we hope to obtain more precise values of the structural data for this gel phase by neutron diffraction studies.

Acknowledgement. The contribution of Mr. P. Aarts of Unilever Research Laboratories, Vlaardingen, to the X-ray diffraction measurements is greatly appreciated.

References

1. N. Krog, *J. Am. Oil Chem. Soc.*, 1977, **54**, 124.
2. J. Andreasen, paper presented at the International Symposium on Emulsions and Foams in Food Technology, 1973.
3. W. Buchheim, N. M. Barfod, and N. Krog, *Food Microstruct.*, 1985, **4**, 221.
4. S. E. Friberg and C. Solans, *Langmuir*, 1986, **2**, 121.
5. N. Krog and A. P. Borup, *J. Sci. Food Agric.*, 1973, **24**, 691.
6. R. M. Williams and D. Chapman, in 'Progress in the Chemistry of Fats and Other Lipids,' ed. R. T. Holman, Pergamon Press, Oxford, 1970, p. 1.
7. T. de Vringer, *Thesis*, Technical University, Leiden, 1987.

Importance of Milk Proteins to the Whipping Properties of 38% Fat Cream

By E. C. Needs

AFRC INSTITUTE OF FOOD RESEARCH, READING LABORATORY, (UNIVERSITY OF READING), SHINFIELD, READING, BERKS. RG2 9AT, UK

1 Introduction

Whipping transforms cream from an oil-in-water emulsion into a three-phase system as air bubbles become trapped in a network of fat globules. During the initial stages of whipping, milk serum proteins are adsorbed at the clean air–serum interface of newly formed air bubbles. Not until cream has been whipped for 20 s do fat globules become evident at the air interface.[1] Milk-fat globule membrane (MFGM) is removed from the globule surface in contact with the air forming a fat–air interface.[2] The final stages of whipping involve coalescence between fat globules so that bubbles become connected and held by aggregations of coalesced fat globules.[1] In fully whipped cream pockets of the protein-stabilized air–water interface remain interspersed between the adsorbed fat globules.

It is probable that the protein at the interface of whipped cream has a similar composition to that at the air–serum interface of skim-milk foams.[3] When skim-milk foams collapse, bubble ghosts remain.[4] These consist of the interfacial material, which may be harvested, and are found to be composed mainly of β-casein, β-lactoglobulin, and α-lactalbumin.[1]

How necessary are milk proteins in the formation of whipped cream and how much do they contribute to the final structure?

2 Experimental

Soluble milk proteins were removed by diluting separated cream 1:2 with simulated milk ultrafiltrate (SMUF) and reseparating. The process was repeated until the cream had been washed three times. An untreated cream and a control cream washed in skim milk were also prepared. All creams were standardized to 38% fat. Whey protein or casein was added to portions of the washed cream (final concentration 2%).

Total nitrogen was measured by Kjeldahl digestion and fat content by the Gerber method.[5] The whipper was based on the design of Mohr and Keonen.[6] Cream stiffness was measured with an Instron food tester.

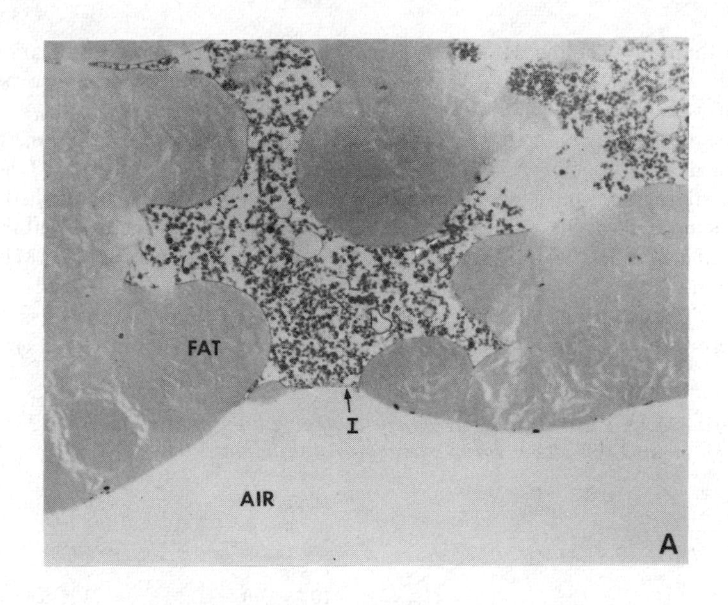

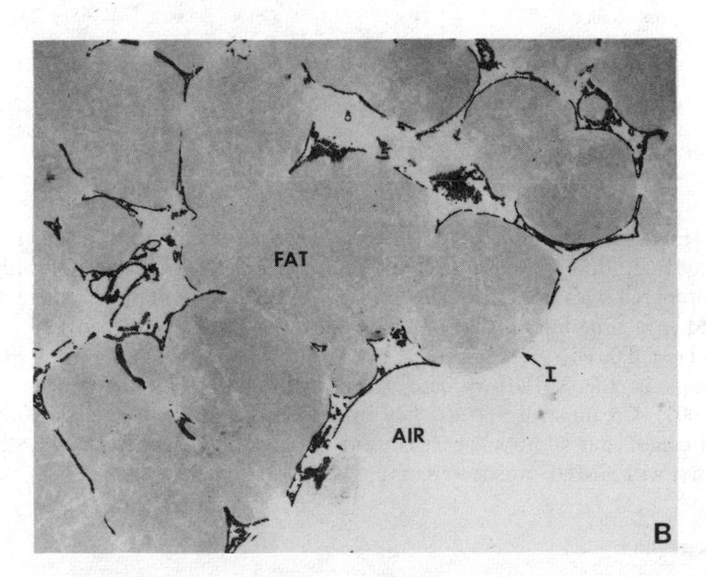

Figure 1 *Structure of whipped cream examined by TEM. (A) Control cream showing coalesced fat globules adsorbed at the interface (I), and serum phase (S) containing numerous casein micelles. (B) Cream washed with SMUF, fat structure similar to A; serum phase showing little evidence of micellar casein. Material adsorbed at air–serum interface may be remnants of primary MFGM*

3 Results

More than 99% of the soluble protein was removed by washing cream with SMUF. It was estimated that if all the residual soluble protein were adsorbed at the air–water interface, a protein loading of 0.4 mg m^{-2} would result. A single monolayer coverage requires[7] between 2.5 and 4 mg m^{-2}. In addition, there remained an average of 285 mg of protein (MFGM) per 100 g of cream.

The micrographs in Figure 1 show the whipped structure of cream washed with either skim milk or SMUF. The control cream had typical structural features of whipped cream. In SMUF-washed cream the casein micelles were absent, MFGM occupying the majority of the air–water interface.

The whipping times and stiffnesses of the five types of cream are shown in Table 1. There was no significant difference between the control and untreated creams.

Table 1 *Means and standard errors for whipping time and stiffness of 38% fat creams containing various amounts of different proteins*

Cream type	*n*	*Whipping time (mean ± SE[b])*	*Stiffness (g) (mean ± SE[b])*
Untreated	11	101 ± 3.0	71 ± 6.4
Control, skim washed	11	98 ± 3.4	65 ± 7.0
SMUF-washed[a]	11	47 ± 3.0	30 ± 6.4
Added protein—whey	11	52 ± 3.9	63 ± 7.1
Added protein—casein	8	81 ± 3.6	38 ± 7.7

[a] SMUF = simulated milk ultrafiltrate.
[b] Standard error of the mean.

SMUF-washed cream whipped in less than half the time of control cream ($p < 0.001$). Addition of whey protein did not significantly increase the whipping time compared with SMUF-washed cream, but cream containing added casein whipped in a time intermediate between those for SMUF-washed and control creams ($p < 0.001$).

Stiffness in the SMUF-washed cream was 54% less than control cream ($p < 0.001$). Cream with added whey protein had a stiffness similar to that of control cream, but addition of casein did not significantly increase the stiffness compared with SMUF-washed cream.

4 Discussion

The differences in whipping times and stiffnesses of the different creams may be related to structural differences in the proteins. β-Casein is a flexible, random-coil protein which will rapidly reduce surface tension by adsorption at a newly formed air–water interface. Whey proteins are highly ordered globular molecules, resistant to surface denaturation.

Brooker[4] has shown that β-lactoglobulin forms stable bubble ghosts able to withstand a temperature of 55 °C for 1.5 h. This suggests that the molecules at the interface are aggregated in some way. Ghost material can be easily disrupted by increasing the pH or addition of urea, and most of the β-lactoglobulin regains its natural conformational state.[8] Thus molecules are aggregated at the interface by hydrophobic and ionic bonding. When cleavage of the disulphide bond occurs oligomers may form between adjacent denatured molecules. These interfacial structures formed by globular whey proteins may confer some rigidity to the whipped cream.

β-Casein does not form stable bubble ghosts, contains no disulphide bonds, has no fixed quaternary structure, and may therefore contribute little to the final stiffness. However, because β-casein is highly surface active once it has been adsorbed at the interface, the surface tension may be lowered to such a level as to reduce the rate of adsorption of fat globule and so increase the whipping time.

References

1. B. E. Brooker, M. Anderson, and A. T. Andrews, *Food Microstruct.*, 1986, **5**, 277.
2. W. Buchheim, *Gordian*, 1978, **78**, 184.
3. M. Anderson, B. E. Brooker, and E. C. Needs, in 'Food Emulsions and Foams,' ed. E. Dickinson, Special Publication No. 58, Royal Society of Chemistry, London, 1987, pp. 100–109.
4. B. E. Brooker, *Food Microstruct.*, 1985, **4**, 289.
5. British Standard No. 696, Part II, British Standards Institution, London, 1969, p. 7.
6. W. Mohr and K. Keonen, *Dtsch. Molk. Ztg.*, 1953, **74**, 468.
7. G. Stainsby, in 'Functional Properties of Food Macromolecules,' eds. J. R. Mitchell and D. A. Ledward, Elsevier Applied Science, Barking, 1986, pp. 315–353.
8. A. T. Andrews, 'Ghost Busters,' AFRC Institute of Food Research, Reading Laboratory Newsletter, Summer 1987.

Low Molecular Weight Surfactants and the Stability of Cream Liqueurs

By Eric Dickinson, Sunit K. Narhan and George Stainsby

PROCTER DEPARTMENT OF FOOD SCIENCE, UNIVERSITY OF LEEDS, LEEDS LS2 9JT, UK

1 Introduction

Many oil-in-water food emulsions contain a mixture of protein and low molecular weight surfactants (usually fatty acid esters). Long-term stability against flocculation and coalescence is attributed mainly to protein adsorbed at the emulsion droplet surface; the role of the surfactant, however, is less clear. Previous studies[1–3] have suggested the formation of interfacial protein–surfactant complexes at low surfactant concentrations, with complete displacement of protein from the interface at high concentrations.

We are concerned here with the role of fatty acid esters [glyceryl monostearate (GMS) and sodium stearoyl-2-lactylate (SSL)] in simulated cream liqueurs (dairy emulsions containing alcohol). Ethanol in the continuous phase of cream liqueurs affects the emulsion properties through its influence on the adsorbed protein.[4,5] Ethanol is a potential displacer of adsorbed protein;[6] finer droplets, and hence more stable emulsions, are produced[7] when ethanol is present (at concentrations up to 20 wt.-%) during the homogenization of casein-stabilized emulsions.

2 Experimental

The method for preparing the liqueurs (see Figure 1) is based on those used earlier.[8,9] The emulsion composition falls within the normal range for cream liqueurs,[10] with trisodium citrate added to reduce sensitivity to calcium ions.[11] The alcohol content is 14 wt.-%.

The GMS was of a commercial grade supplied in the form of micro-beads melting at 55–60 °C; 65% of the esterified fatty acids were stearic and *ca.* 40% of the sample was the monoester. A commercial grade of SSL (m.p. 55 °C), together with colours and flavours, were gifts from Gilbeys of Ireland Ltd.

The extent of normal gravity creaming was quantified by determining visually the position of the sharp cream boundary after storage for 12 weeks at ambient temperature. Analysis for protein was accomplished using a biuret procedure after severe centrifugation of the emulsion to remove fat.[12] Emulsion rheology

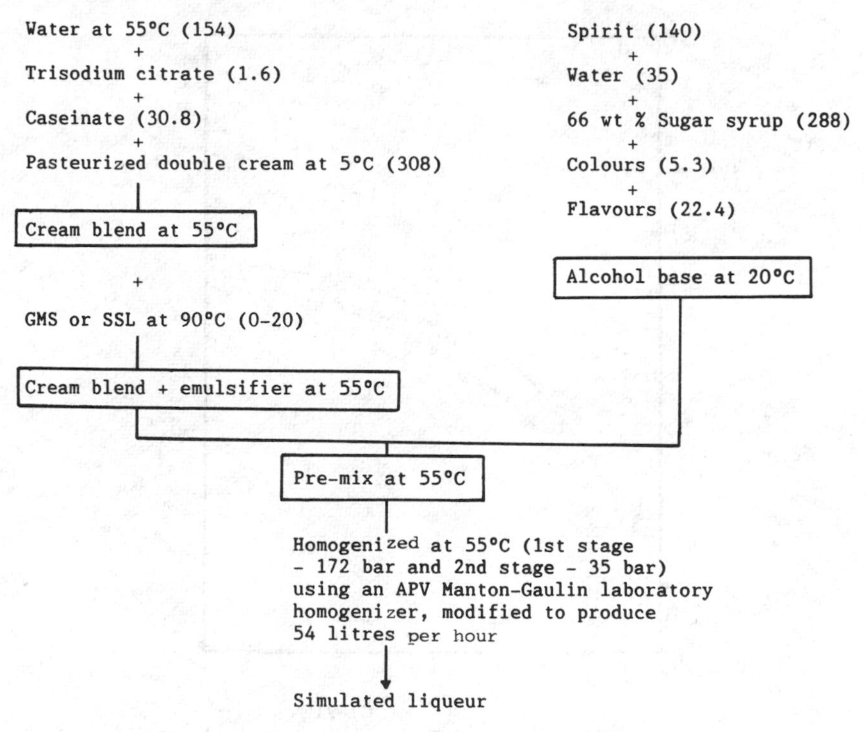

Figure 1 *Outline of method for making* 1 kg *of simulated cream liqueur. Numbers in parentheses refer to the weight (in* g*) of each ingredient used. Pure water is substituted for surfactant as appropriate*

was examined at 20 °C using a Carri-med constant-stress rheometer at shear rates in the range 0–10 s^{-1}.

3 Results and Discussion

The presence of GMS or SSL at concentrations up to 0.5 wt.-% was found to produce no change in droplet-size distribution as measured by a Coulter counter. At the highest levels of surfactant studied (2 wt.-%), however, some increase in the number of large droplets was observed,[12] especially when GMS was present. As 14 wt.-% ethanol reduces the most probable droplet diameter from 1 to 0.6 μm in caseinate-stabilized n-tetradecane-in-water emulsions (no GMS or SSL),[7] it seems reasonable to infer that droplet sizes in our simulated liqueurs are primarily governed by the large proportion of alcohol, rather than by the much lower proportion of surfactant.

The addition of 0.5 wt.-% of surfactant (GMS or SSL) led to a reduction in the amount of cream layer from 2.5 vol.-% to zero. There was no discernible creaming at higher surfactant concentrations. This improved stability seems to be associated with a change in the rheological properties. Emulsions containing 0–0.34

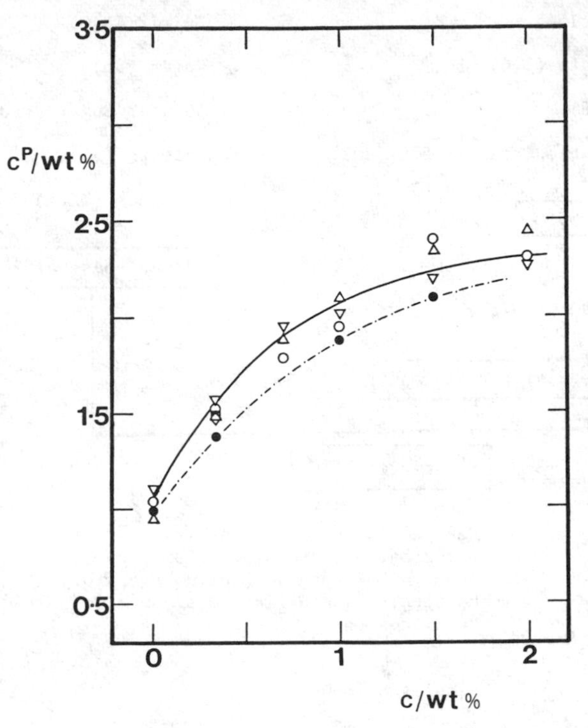

Figure 2 *Displacement of caseinate from the fat droplet surface by GMS or SSL. Protein concentration,* c^P, *measured in the aqueous phase after centrifugation is plotted against surfactant concentration,* c, *in the whole system:* △, ▽, ○, *GMS;* ●, *SSL. Best-fit lines, (—) and (-··-), are drawn through GMS and SSL data points, respectively. The emulsion contains a total concentration of* 3.08 wt.-% *added caseinate, together with* 0.43 wt.-% *casein and* 0.10 wt.-% *whey protein contributed by the cream.*[12] *It is assumed that no milk-fat globule membrane protein is present in the clear solutions analysed, all this protein, whether at the interface or displaced, having been removed by centrifugation*

wt.-% of surfactant were Newtonian down to shear stresses below 0.1 $N m^{-2}$, but those containing at least 0.5 wt.-% of surfactant were found to possess a clear yield stress of a magnitude (*ca.* 10 $N m^{-2}$) sufficient to inhibit creaming completely.[13] An emulsion containing 2 wt.-% of surfactant is semi-solid, and it pours only on shaking.

The results in Figure 2 show that GMS and SSL displace protein from the interface to similar extents, although displacement is by no means complete at the levels used. Although the primary interfacial layer probably contains only ethanol and surfactant, it seems likely that some protein remains associated with the interface in a secondary layer, and therefore provides some steric stabilization. The semi-solid emulsions produced at the highest surfactant concentrations

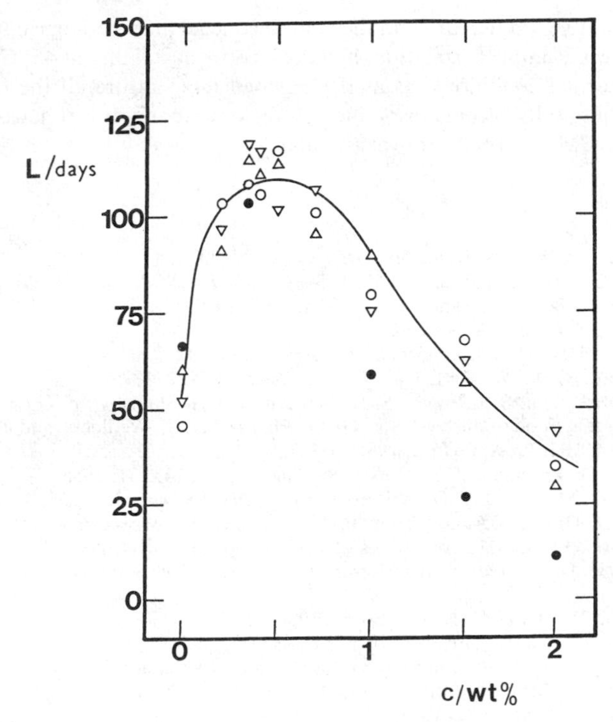

Figure 3 *Shelf-life of simulated cream liqueurs on storage at* 45 °C. *The time,* L, *for serum separation first to become visible is plotted against the surfactant concentration,* c: △, ▽, ○, *GMS*; ●, *SSL. A best-fit line is drawn through the GMS data points*

resemble certain cosmetic and pharmaceutical preparations in which the mechanism whereby the emulsifier controls consistency has been termed 'self-bodying.'[14] Structure formation through self-bodying in our emulsions is probably due to interactions between casein and surfactant at the interface[15] and in the bulk aqueous phase. Analysis of the latter, after removing the cream by centrifugation, showed an increase in glyceride content (expressed as GMS equivalent) from 0.6 mg l^{-1} in the absence of surfactant to 1.2 mg l^{-1} in the presence of 2 wt.-% of GMS[12] (milk fat provides soluble glycerides, whether GMS is added or not). These data provide evidence for complexation in the aqueous phase between surfactant and protein, although the exact numerical values should be treated with caution as some loss of protein, and bound GMS, may have occurred during the severe centrifuging required to produce solutions of sufficient clarity for spectrophotometric analysis.

A commercial cream liqueur must not be perceived as visco-elastic by the consumer. Together with legal constraints, this suggests an upper limit of *ca.* 0.4 wt.-% of added surfactant in order to achieve optimal stability. When storage at 45 °C is used as a guide to shelf-life,[16] the optimal surfactant level comes out to be (see Figure 3) not very different from the combined legal and rheological limit. One does not expect a simple correlation, however, between stability at 45 °C and shelf-life under normal conditions, as at the elevated temperature all the fat is liquefied and calcium salts become insoluble. On the other hand, it is noteworthy that the effects of GMS and SSL are very similar.

References

1. H. Oortwijn and P. Walstra, *Neth. Milk Dairy J.*, 1979, **33**, 134.
2. R. Wüstneck, G. Kretzschmar, and L. Zastrow, *Colloid J. USSR*, 1987, **49**, 207.
3. J. A. de Feijter, J. Benjamins, and M. Tamboer, *Colloids Surf.*, 1987, **27**, 243.
4. W. J. Donnelly, *J. Food Sci.*, 1987, **52**, 389.
5. D. D. Muir and D. G. Dalgleish, *Milchwissenschaft*, 1987, **42**, 770.
6. E. Dickinson and C. M. Woskett, *Food Hydrocolloids*, 1988, **2**, 187.
7. S. Bullin, E. Dickinson, S. J. Impey, S. K. Narhan, and G. Stainsby, in 'Gums and Stabilisers for the Food Industry,' eds. G. O. Phillips, D. J. Wedlock, and P. A. Williams, Vol. 4, IRL Press, Oxford, 1988, p. 337.
8. C. C. Widmar, D. Tripp, and V. G. Ficca, *Br. Pat. Appl.*, 2 145 111, 1985.
9. W. Banks, D. D. Muir, and A. G. Wilson, *J. Soc. Dairy Technol.*, 1982, **35**, 41.
10. W. Banks, D. D. Muir, and A. G. Wilson, in 'Physico-chemical Aspects of Dehydrated Protein-rich Milk Products,' Proceedings of IDF Symposium, Helsingør, Denmark, May 17–19, 1983; Danish Government Research Institute for Dairy Industry, *Hillerød*, 1983, p. 331.
11. D. D. Muir and W. Banks, *J. Food Technol.*, 1986, **21**, 229.
12. S. K. Narhan, *PhD Thesis*, University of Leeds, 1987.
13. E. Dickinson, in 'Food Structure—Its Creation and Evaluation,' eds. J. M. V. Blanshard and J. R. Mitchell, Butterworths, London, 1988, Ch. 4.
14. B. W. Barry, *Rheol. Acta*, 1971, **10**, 96.
15. G. Doxastakis and P. Sherman, *Colloid Polym. Sci.*, 1986, **264**, 254.
16. W. Banks, D. D. Muir, and A G. Wilson, *J. Food Technol.*, 1981, **16**, 587.

Interfacial Competition Between α_{s1}-Casein and β-Casein in Oil-in-Water Emulsions

By Eric Dickinson and Susan E. Rolfe

PROCTER DEPARTMENT OF FOOD SCIENCE, UNIVERSITY OF LEEDS, LEEDS LS2 9JT, UK

and

Douglas G. Dalgleish

HANNAH RESEARCH INSTITUTE, AYR KA6 5HL, UK

1 Introduction

The competitive adsorption between the different components of casein in dairy emulsions has been recognized by several workers.[1-4] Of the two major monomeric components, β-casein is more surface-active than α_{s1}-casein at air–water and oil–water interfaces,[5-7] and so there is a tendency for β-casein to predominate over α_{s1}-casein at fluid interfaces, *e.g.* in the proteinaceous membrane of homogenized milk.

In this paper, we report results of experiments designed to quantify the extent to which α_{s1}-casein or β-casein will displace the other protein from the oil–water interface. An adsorbed film of α_{s1}-casein (or β-casein) is first established at the surface of emulsion droplets, or at a planar oil–water interface, and then the other protein is introduced into the aqueous phase. Competitive adsorption is monitored from the change in protein composition of the serum phase in the case of the emulsions, or the change in surface viscosity in the case of the planar interface.

2 Experimental

Oil-in-water emulsions (20 wt.-% n-tetradecane, 0.5 wt.-% protein, 20 mM imidazole buffer, pH 7) were made at room temperature using a small-scale valve homogenizer at an operating pressure of 300 bar. In exchange experiments, an emulsion made with α_{s1}-casein (or β-casein) was washed free from unadsorbed protein by centrifuging at 10^4 *g* for 15 min, redispersing the cream in buffer, and then repeating the procedure. The amount of unadsorbed protein in the aqueous phase was determined by fast protein liquid chromatography (FPLC) using a Pharmacia Mono-Q ion-exchange column with a linear salt gradient. To the washed α_{s1}-casein (β-casein) emulsion was added an equal volume of a buffered solution of β-casein (α_{s1}-casein). The mixed system was stirred vigorously at 20 °C for a continuous period of 24 h, during which time samples were taken. These were centrifuged at 10^4 *g* for 15 min, and protein concentrations in the aqueous phase were determined by FPLC.

Surface viscosities were measured at the n-tetradecane–water interface (pH 7, 25 °C) using the Couette-type torsion-wire surface rheometer described previously.[8] In exchange experiments, a film adsorbed from a 10^{-3} wt.-% solution of α_{s1}-casein was aged for 24 h, then β-casein was introduced into the aqueous sub-phase at a concentration of 10^{-3} wt.-%.

3 Results and Discussion

Droplet size distributions of the emulsions, as measured with a Coulter counter, were all very similar, irrespective of whether they were made with α_{s1}-casein, β-casein, or a mixture of the two. In emulsions made with α_{s1}-casein + β-casein (0.5 wt.-% protein), the calculated total surface concentration ($\Gamma \approx 3.0$ mg m^{-2}) is rather insensitive to emulsifier composition (see Figure 1). A low total protein content is chosen in order to ensure that most of the protein used to make the emulsion does, in fact, end up at the interface. Analysis of the aqueous phases after washing indicated that *ca.* 18–20% of the protein emulsifier was not adsorbed during homogenization; this figure does not change with the emulsifier composition.

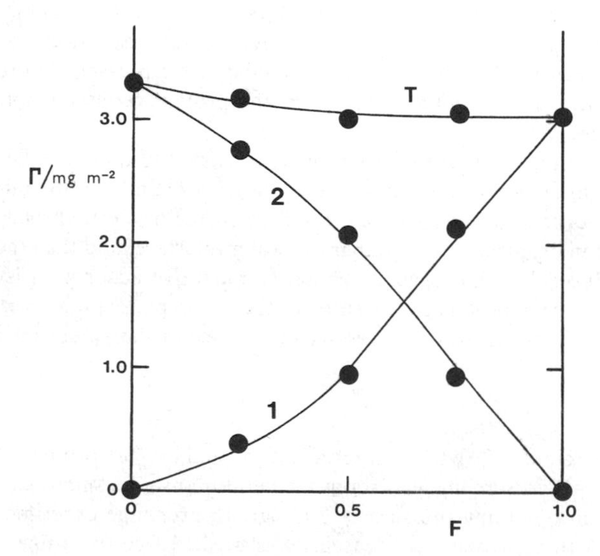

Figure 1 *Interfacial composition of freshly made emulsions* (20 wt.-% *oil*, 0.5 wt.-% *protein*). *Surface concentration*, Γ, *is plotted against weight fraction*, F, *of* α_{s1}-*casein in protein emulsifier:* 1, α_{s1}-*casein;* 2, β-*casein;* T, *total protein*

Figure 1 shows that β-casein is preferentially adsorbed at the oil–water interface during or shortly after homogenization. The emulsion made with the 1:1 mixture of α_{s1}-casein + β-casein contains about twice as much of the latter as the former at the droplet surface.

Figure 2 shows results from the exchange experiments. The data refer to average values over a period of 6–24 h after initial mixing, although steady-state compositions were rapidly attained under the vigorous mixing conditions employed here. When samples are not stirred after initial mixing, there is a much slower approach to equilibrium.[9] It is clear from Figure 2 that it is much easier to displace α_{s1}-casein by β-casein than the other way round, in agreement with our preliminary findings.[9] It is not possible to displace all the adsorbed α_{s1}-casein from the interface in a single mixing experiment because of solubility problems, but this has been achieved in a three-stage process ($\Gamma < 0.1$ mg m^{-2}) using a 10-fold excess of β-casein in solution at each stage. Therefore, it appears that none of the α_{s1}-casein at the droplet surface is irreversibly adsorbed. It is seen in Figure 2

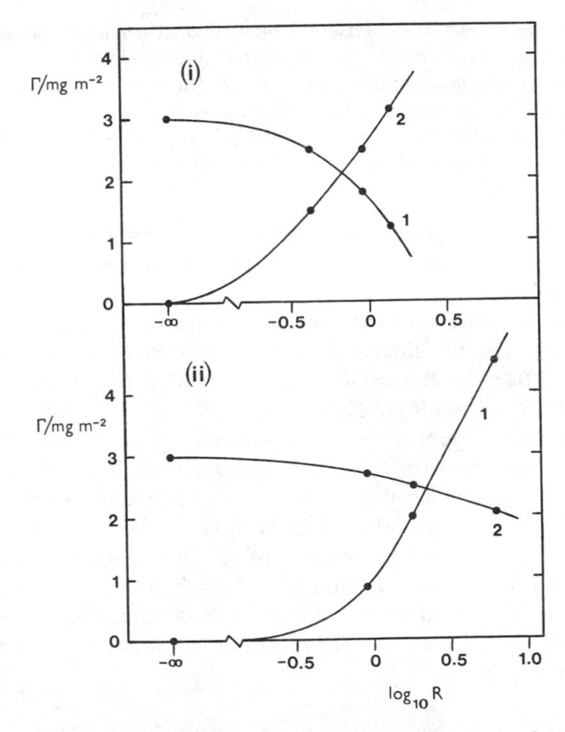

Figure 2 *Interfacial composition in emulsions after exchange of (i) α_{s1}-casein at droplet surface with β-casein in bulk and (ii) β-casein at droplet surface with α_{s1}-casein in bulk. Surface concentration, Γ, is plotted against log where* R *is the concentration of protein in the added solution (expressed as* mg ml^{-1} *of emulsion) divided by the concentration of protein in the original emulsion* (4.7 mg ml^{-1} *of emulsion*): 1, α_{s1}-*casein;* 2, β-*casein*

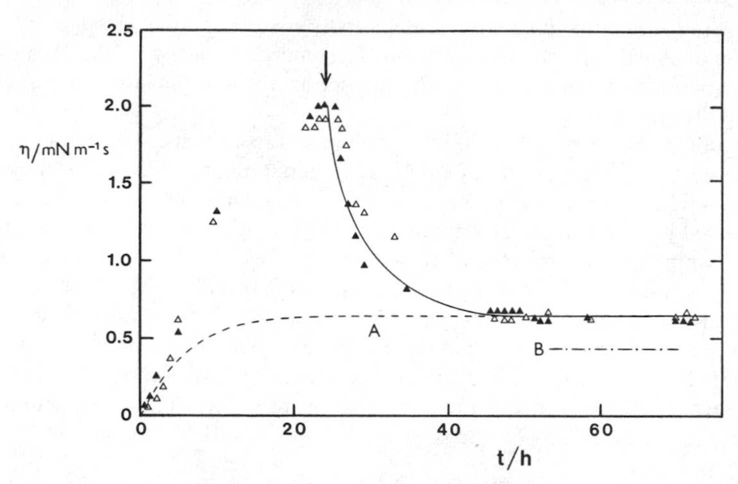

Figure 3 *Displacement of α_{s1}-casein by β-casein at a planar oil–water interface. Surface shear viscosity, η, is plotted against time, t. Points (△) and (▲) denote two independent experiments. The arrow denotes when β-casein is added to aqueous sub-phase after aging α_{s1}-casein film for 24 h. Curve A denotes what happens when the β-casein is present from time t = 0. Curve B denotes the limiting surface viscosity in the absence of α_{s1}-casein*

that more protein is adsorbed from solution than is displaced from the interface. This may be a reflection of accumulation of protein at the interface in the form of secondary layers.

Surface viscosity measurements are sensitive to the composition of adsorbed protein films,[7,8] and are therefore useful in the study of protein competitive adsorption.[10] After 24 h, the surface viscosities of films adsorbed from 10^{-3} wt.-% solutions of the two caseins are substantially different: 0.4 mN m^{-1} s for β-casein as compared with 2.0 mN m^{-1} s for α_{s1}-casein. Figure 3 shows what happens when an adsorbed film of α_{s1}-casein is first aged for 24 h and then β-casein is introduced into the aqueous phase: over a further period of 24 h there is a gradual fall in viscosity to a steady value of 0.65 mN m^{-1} s, which is identical with that reached after 48 h when a mixed solution of 10^{-3} wt.-% α_{s1}-casein + 10^{-3} wt.-% β-casein is adsorbed at the same interface. This shows that the final protein film obtained in the displacement experiment is predominantly but not exclusively β-casein, which is consistent with the results of the emulsion displacement experiments (see Figure 2).

4 Conclusions

Analysis of the aqueous phase after centrifugation of fresh emulsions made with a mixture of α_{s1}-casein + β-casein has shown that β-casein predominates at the interface. Exchange experiments indicate that β-casein will rapidly displace α_{s1}-casein from the emulsion droplet surface; none of the α_{s1}-casein appears to be irreversibly adsorbed. Consistent with a thermodynamic equilibrium between interface and bulk, α_{s1}-casein will also displace β-casein, but to a much lesser extent. Confirmation of these findings is provided from the change in surface viscosity as β-casein displaces α_{s1}-casein from the planar oil–water interface.

Acknowledgement. S. E. R. acknowledges the receipt of an AFRC Co-operative Studentship in conjunction with the Hannah Research Institute.

References

1. H. Mulder and P. Walstra, 'The Milk Fat Globule,' Pudoc, Wageningen, 1974.
2. D. F. Darling and D. W. Butcher, *J. Dairy Res.*, 1978, **45**, 197.
3. A. V. McPherson, M. C. Dash, and B. J. Kitchen, *J. Dairy Res.*, 1984, **51**, 289.
4. E. W. Robson and D. G. Dalgleish, *J. Food Sci.*, 1987, **52**, 1694.
5. J. Benjamins, J. A. de Feijter, M. T. A. Evans, D. E. Graham, and M. C. Phillips, *Discuss. Faraday Soc.*, 1975, **59**, 218.
6. E. Dickinson, D. J. Pogson, E. W. Robson, and G. Stainsby, *Colloids Surf.*, 1985, **14**, 135.
7. J. Castle, E. Dickinson, B. S. Murray, and G. Stainsby, *ACS Symp. Ser.*, 1987, No. 343, 118.
8. E. Dickinson, B. S. Murray, and G. Stainsby, *J. Colloid Interface Sci.*, 1985, **106**, 259.
9. E. Dickinson, R. H. Whyman, and D. G. Dalgleish, in 'Food Emulsions and Foams' ed. E. Dickinson, Royal Society of Chemistry, London, 1987, p. 40.
10. E. Dickinson, *Food Hydrocolloids*, 1986, **1**, 3.

Non-intrusive Determination of Droplet Size Distribution in a Concentrated Oil-in-Water Emulsion from Sedimentation Profiles

By Andrew M. Howe* and Margaret M. Robins

AFRC INSTITUTE OF FOOD RESEARCH, NORWICH LABORATORY, NORWICH NR4 7UA, UK

1 Introduction

Liquid-based food dispersions often contain particles with mean diameters of the order of micrometres, but with a wide distribution of sizes (typically the diameter d ranges from 0.1 to 10 μm). Determination of the full size distribution in such concentrated, opaque, polydisperse systems is generally not possible with conventional particle-sizing techniques. On dilution of the dispersion the full size distribution may be found, usually by means of a combination of techniques. We present a full size distribution determined from the concentration profiles in gravitational creaming of a 20% oil-in-water emulsion. The emulsion was of n-alkane (heptane–hexadecane, 9:1 v/v) droplets stabilized by the non-ionic surfactant Brij 35 in water.[1]

2 Results

Concentration profiles of the emulsion were determined non-intrusively (from a measure of the velocity of ultrasound through the emulsion at a series of heights[2] over a period of 63 days (Figure 1). The diffuse meniscus is characteristic of the creaming of individual polydisperse droplets which rise at velocities U given by Stokes' law:

$$U = \frac{d^2 g \Delta\rho}{18\eta} \tag{1}$$

where g is the acceleration due to gravity and $\Delta\rho$ is the density difference between disperse and continuous phases. The local medium viscosity η is calculated[3] from the local volume fraction φ and the continuous phase viscosity η_0:

$$\eta = \eta_0(1 - \varphi/\varphi_m)^{-2.5\varphi_m} \tag{2}$$

where φ_m is the maximum packing fraction, taken to be 0.70 (the concentration in the non-compacted cream).

* Present Address: Surface Science Group, Research Division, Kodak Limited, Harrow, Middlesex HA1 4TY.

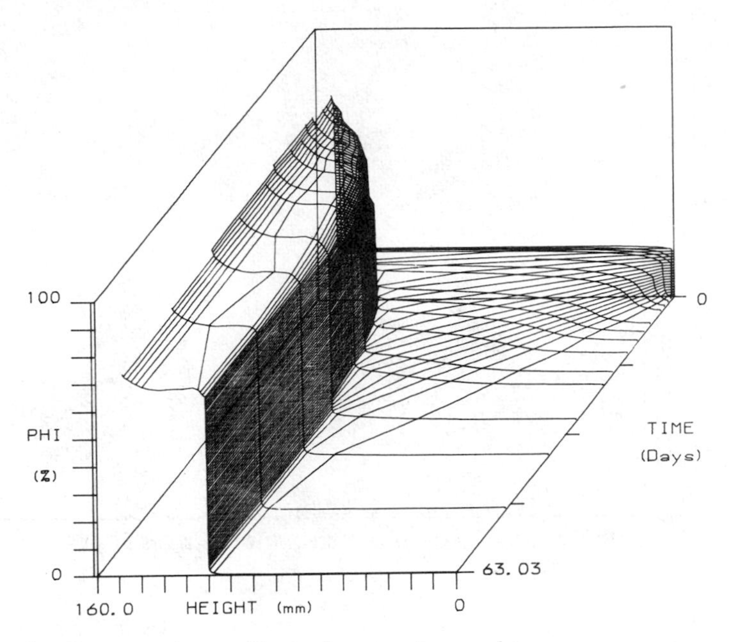

Figure 1 *Concentration profiles in the creaming emulsion*

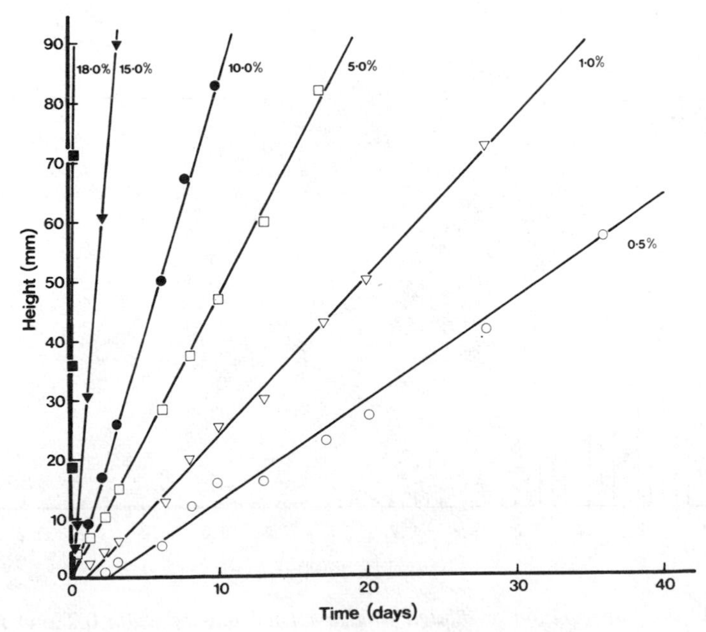

Figure 2 *Height of contours of constant φ during the creaming process*

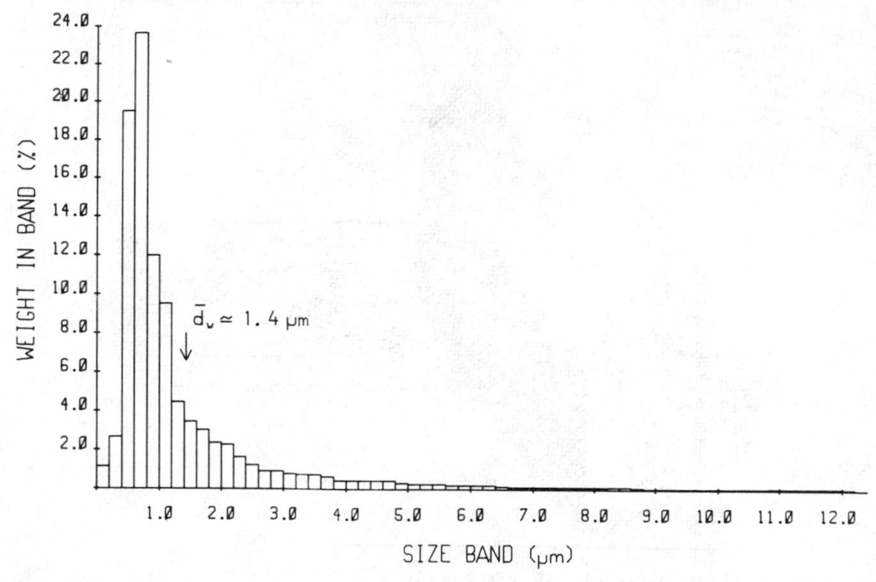

Figure 3 *Size distribution by weight (in bands of width* 0.2 μm*) from concentration profiles*

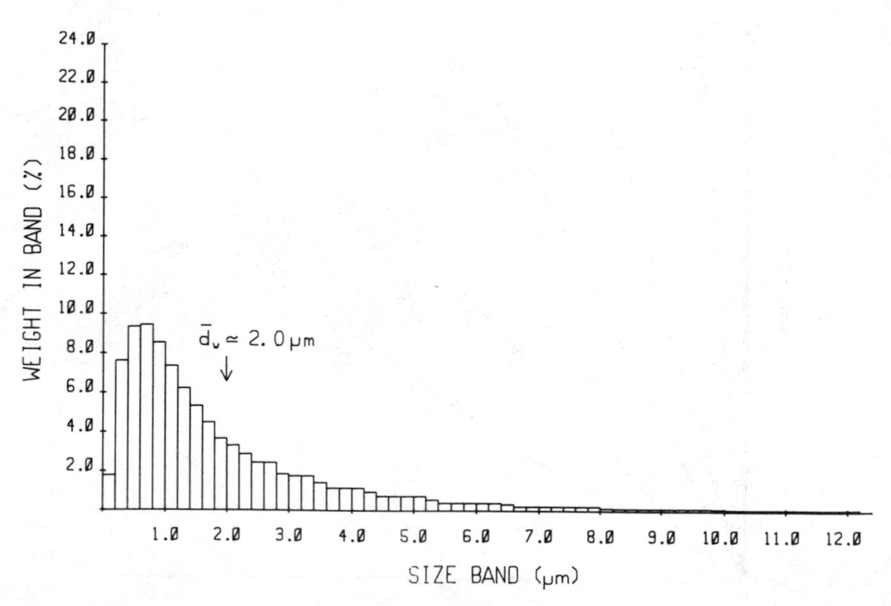

Figure 4 *Size distribution of diluted emulsion (in bands of width* 0.2 μm*) from light scattering*

In Figure 2, the heights of contours of constant φ are plotted during the creaming process. The velocity (gradient) of each contour is the Stokes' velocity of the largest droplet ($d = d_m$) present in that contour, and hence increases with φ. There is a delay in the creaming of the small droplets (for contours of $\varphi < 1\%$). The delay may arise from diffusive, convective, or filtration processes but does not affect the rise velocities.[4]

From the contour velocities and equation (1), a series of droplet diameters d may be calculated. The cumulative size distribution (by weight) is obtained from the velocities of contours at a series of concentrations from $\varphi = 0.28$ to 18% (Figure 3).

A droplet size distribution was obtained by light scattering (using a Malvern Mastersizer) from a diluted sample of the emulsion. The results are shown in Figure 4 and are in good agreement with those from the sedimentation analysis. Photon correlation spectroscopic measurements, on slowly creaming sections taken from the base of the emulsion after 9 days, were in excellent agreement with predictions based on the above results and analysis.[4]

3 Conclusion

The droplet size distribution was calculated from the rise velocities of contours of constant oil concentration. The sizes ranged from <200 nm to 12 μm. The calculated size distribution compared well with that determined from light-scattering measurements on the diluted emulsion. A delay in the creaming of the small droplets (diameter <0.5 μm) was observed, which would lead to errors in the calculated size distribution if a single concentration profile were to be used.

Acknowledgements. We are grateful to David Hibberd and Alan Mackie for helpful discussions and skilled measurements. This work was funded by the Ministry of Agriculture, Fisheries and Food and the Department of Education and Science.

References

1. P. A. Gunning, D. J. Hibberd, A. M. Howe, and M. M. Robins, *Food Hydrocolloids*, 1988, **2**, 119.
2. A. M. Howe, A. R. Mackie, and M. M. Robins, *J. Dispersion Sci. Technol.*, 1986, **7**, 231.
3. R. C. Ball and P. Richmond, *Chem. Phys. Liq.*, 1980, **9**, 99.
4. C. Carter, D. J. Hibberd, A. M. Howe, A. R. Mackie, and M. M. Robins, *Prog. Colloid Polym. Sci.*, 1988, **76**, 37.

Effect of Heat on the Emulsifying Properties of Gum Arabic

By R. C. Randall, G. O. Phillips, and P. A. Williams

FACULTY OF RESEARCH AND INNOVATION, NORTH EAST WALES INSTITUTE, CONNAH'S QUAY, DEESIDE, CLWYD CH5 4BR, UK

1 Introduction

Gum arabic, a naturally occurring polysaccharide, is widely used in the food industry in, for example, the stabilization and encapsulation of citrus oils and other beverage flavour emulsions.[1] The gum is a highly branched polysaccharide consisting of D-galactose, L-arabinose, L-rhamnose, D-glucuronic acid, and 4-*O*-methylglucuronic acid and contains *ca.* 2% of proteinaceous material as an integral part of the structure. Most of the protein is associated with a high molecular mass component, probably as an arabino-galactan protein complex.[2,3] Recent studies[4] have indicated that the structure of the gum can be interpreted in terms of a 'Wattle Blossom Model,' where arabino-galactan blocks of molecular mass 200 000 are linked to a main polypeptide chain.

Gum arabic is often subjected to prolonged heat during processing, and Anderson and McDougall[5] have recently shown that this can lead to auto-hydrolysis with precipitation of protein-rich material. As we have recently shown that the proteinaceous fraction plays a functional role in the emulsification process,[6] our studies have now been extended to clarify the effect of heat on the emulsification action and stability of the gum.

2 Materials

Spray-dried Kordofan gum arabic was kindly supplied by Agrisales (Corby, UK). It was in the mixed salt form and contained 1.1% Ca^{2+}, 0.84% K^+ and 0.019% Na^+ on a dry weight basis. A 15% w/w solution had a natural pH of 4.3. Its specific rotation was $-32.5°$. Sweet oil orange oil was a gift from Concorde Flavours (Corby, UK).

3 Methods

Gel Permeation Chromatography (GPC)—Gum arabic solution (1 cm^3, 1.5% w/w) in 0.5 mol dm^{-3} sodium chloride solution was passed through a 3 μm Millipore filter and injected on to a K15 (90 × 1.5 cm i.d.) jacketed column containing Sephacryl S500 gel (Pharmacia) and maintained at 25 °C. The eluent flow-rate was 19.8 $cm^3 h^{-1}$ and detection was made using a Cecil 202 UV spectrophotometer at a wavelength of 218 nm. V_0 and V_t were determined using DNA (molecular mass 1.8×10^7) and sodium azide, respectively.

Preparation of Emulsions—Orange oil (100 cm^3, 20% w/w) emulsions containing gum arabic (19% w/w) were prepared using a Silverson laboratory emulsifier. The emulsions were maintained at 25 °C and mixing was continued for precisely 5 min in order to ensure reproducibility.[6]

Emulsions were also prepared using gum arabic solutions that had been refluxed for various times at 100 ± 2 °C.

Emulsification Efficiency—The efficiency of emulsification was assessed from turbidity measurements as reported previously.[6,7] Freshly prepared emulsion (1 cm^3) was diluted to 1 dm^3 with distilled water and the absorbance measured at 650 nm using a Perkin-Elmer diode-array spectrophotometer. The absorbance was shown to increase as the droplet size decreased; two typical histograms illustrating droplet size distributions obtained for emulsions having high (0.84) and low (0.25) absorbance readings using a ZM Model Coulter Counter equipped with a 50-μm orifice electrode are shown in Figure 1.

Adsorption of Gum Arabic at the Oil–Water Interface—The adsorption of gum arabic at the oil–water interface was followed by monitoring the GPC profile

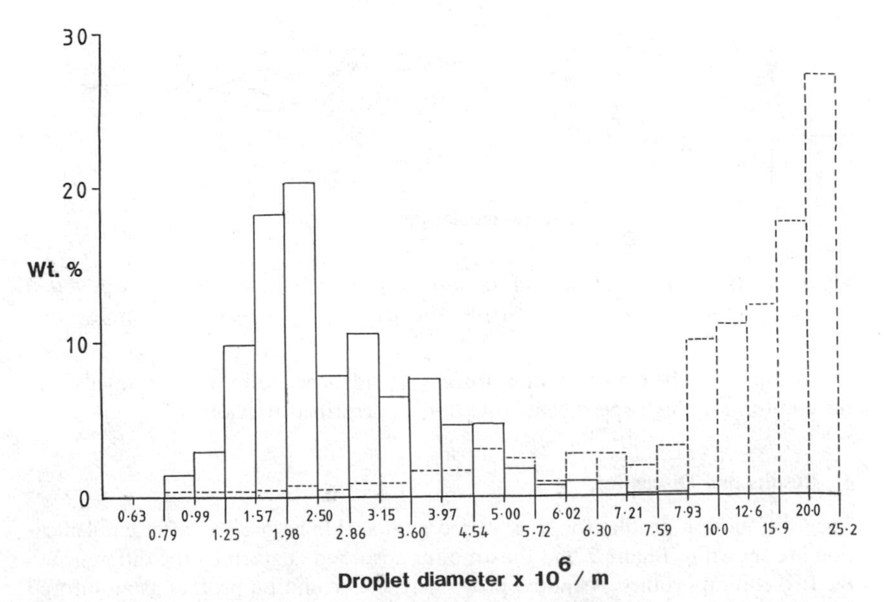

Figure 1 *Histogram showing droplet size distributions for emulsions.* ——, *Absorbance at* 650 nm = 0.84; - - - , *Absorbance at* 650 nm = 0.25

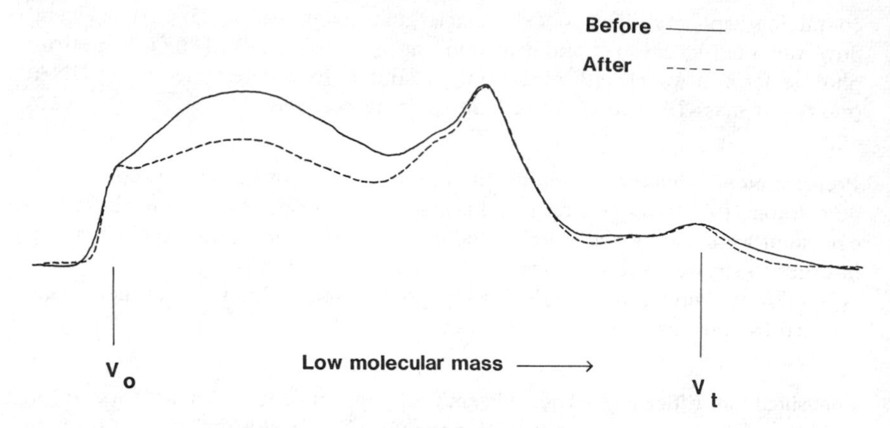

Figure 2 *Gel permeation chromotograms of a solution of spray-dried gum arabic before and after emulsification*

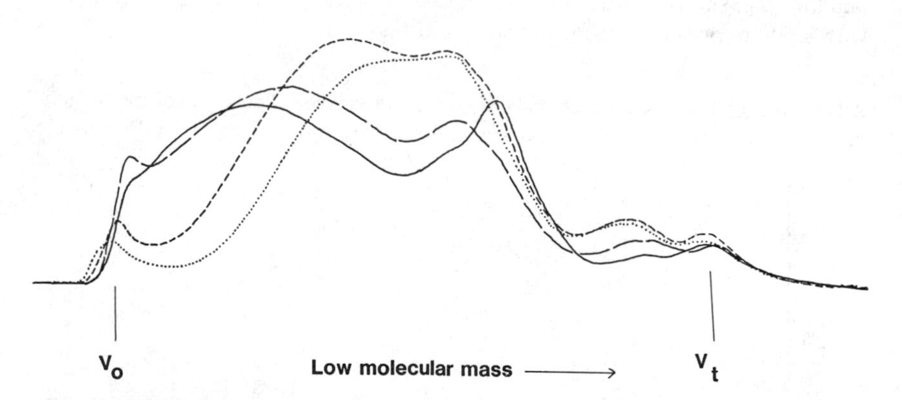

Figure 3 *Gel permeation chromatograms of spray-dried gum arabic as a function of heating time. Time* (h):———, 0; — — —, 1; -----, 3; . . ., 6.5

before and after the emulsification process. Unadsorbed gum was separated from the emulsion by high-speed centrifugation as described previously.[6]

4 Results and Discussion

The GPC elution profiles for spray-dried gum arabic before and after emulsification are shown in Figure 2 and the amount adsorbed is given by the difference in the two elution profiles. As noted previously,[2,6] the elution profiles as monitored by UV spectrophotometry do not give a quantitative assessment of the molecular mass distribution, owing to variations in the molar absorptivities of the different

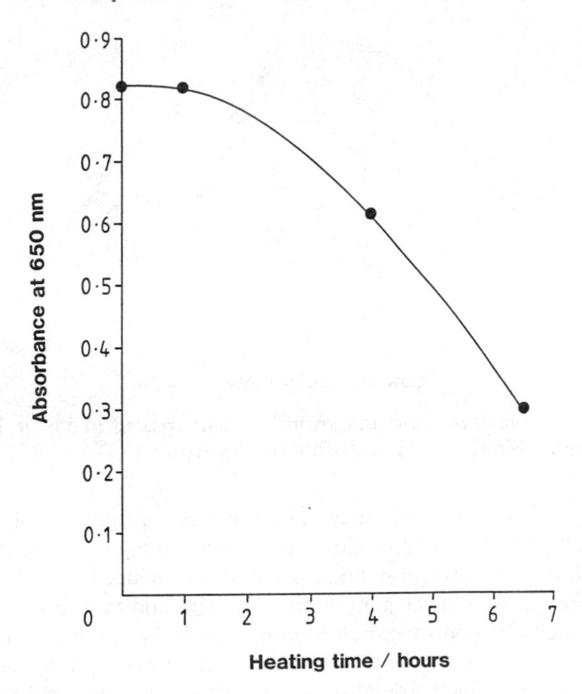

Figure 4 *Emulsification efficiency of heat-treated gum arabic as monitored by the absorbance at* 650 nm

chemical species present. The high molecular mass peak has a high molar absorptivity, but nevertheless represents only a very small percentage of the total gum. During emulsification, high molecular mass material adsorbs preferentially, and this is in accord with our previous work using unprocessed gum arabic samples,[6] which also showed that only a small proportion of the total gum (1–2%) was able to adsorb and that this is the fraction rich in protein.

Figure 3 illustrates the effect of heat on the GPC profiles of the gum. During heating at 100 °C for up to 3 h, there is a decrease in the intensity of the high molecular mass peak with a corresponding increase in the intensities of the lower molecular mass peaks. Continued heating leads to further loss of the high molecular mass fraction, which on standing precipitates out of solution in the form of a brown solid. This is accompanied by a decrease in the area under the GPC curve after refluxing for 6.5 h. Anderson and McDougall[5] showed that the insoluble material was only a very small percentage of the total gum (0.2%), but that this was very rich in protein (71.9%). The optical rotation of the dissolved gum remained unaltered following refluxing, indicating that the main polysaccharide macrostructure remained unaffected.

The emulsification efficiency of the heated gum arabic solutions was assessed from turbidity measurements as shown in Figure 4. Turbidity decreases with the time of refluxing, indicating an increase in the average droplet size and hence a

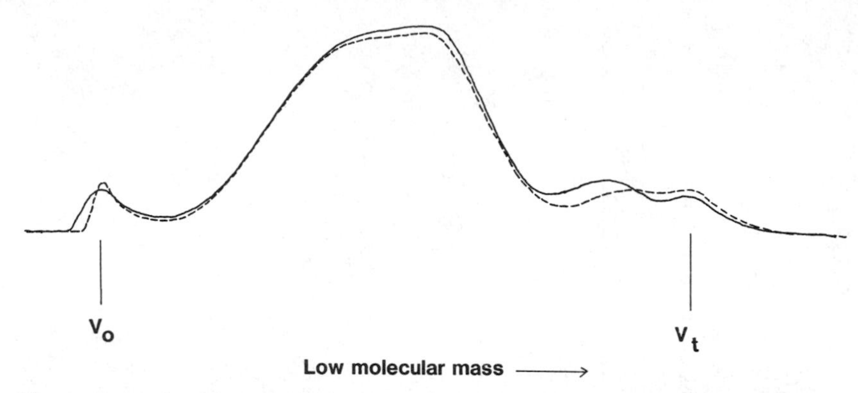

Figure 5 *Gel permeation chromatograms of heat-treated* (6.5 h *at* 100 °C) *gum arabic before* (——) *and after emulsification* (- - -)

reduction in emulsification efficiency. The GPC elution profiles of heat-treated gum arabic before and after emulsification are given in Figure 5, and it is apparent that very little, if any, adsorption takes place. It is concluded, therefore, that only a very small proportion of the gum is surface active and that this corresponds to the high molecular mass protein-rich fraction which is precipitated from solution on refluxing. The surface activity is probably due to hydrophobic groups in the polypeptide chains, which are able to adsorb at the interface with the oil. Coalescence is then prevented by the interaction of the hydrophilic carbohydrate residues protruding away from the droplet surfaces into solution. Such characteristics account for the versatility of gum arabic in promoting emulsification and stabilization in the wide range of food systems.

Acknowledgement. The authors are indebted to Agrisales Ltd. for their advice and enthusiastic support during the course of this work.

References

1. M. Glicksman, in 'Food Hydrocolloids,' ed. M, Glicksman. Vol II, CRC Press, Boca Raton, FL, 1983.
2. M. C. Vandevelde and J. C. Fenyo, *Carbohydr. Polym.*, 1985, **5**, 251.
3. Y. Akiyama, S. Eda, and K. Kato, *Agric. Biol. Chem.*, 1984, **48**, 235.
4. S. Connolly, J. C. Fenyo, and M. C. Vandevelde, *Food Hydrocolloids*, 1987, **5/6**, 477.
5. D. M. W. Anderson and F. J. McDougall, *Food Additives Contam.*, 1987, **4**, 247.
6. R. C. Randall, G. O. Phillips, and P. A. Williams, *Food Hydrocolloids*, 1988, **2**, 131.
7. K. N. Pearce and J. E. Kinsella, *Agric. Food Chem.*, 1978, **26**, 716.

Gel Formation After Heating of Oil-in-Water Emulsions Stabilized by Whey Protein Concentrates

By Gérard Masson and Rolf Jost

NESTLE RESEARCH CENTRE, NESTEC LTD., VERS-CHEZ-LES-BLANC,
CH-1000 LAUSANNE 26, SWITZERLAND

1 Introduction

Proteins stabilize oil-in-water (O/W) emulsions by coating the surface of the dispersed oil droplets and providing surface charges which generate repulsive forces between droplets.[1] Gelation of O/W emulsions stabilized by whey proteins can occur after heat treatment of the fluid emulsions. The emulsifying capacity and the gelation ability of the whey proteins have received particular attention.[2-4] It has been demonstrated that the emulsifying capacity of whey proteins can be affected by the protein concentration, the oil volume fraction, the pH, and the ionic strength of the aqueous phase and also by the homogenization efficiency. The gelation capacity of O/W emulsions after heat treatment has been correlated with the physico-chemical properties of the fluid emulsions.

2 Materials and Methods

Materials—Whey protein concentrates (WPC) were obtained by ultrafiltration of sweet Gruyère-type cheese whey and then spray-dried. Grape seed oil, soyabean oil, and sunflower oil were used as vegetable oils in the oil phase. The aqueous phase consisted of WPC dispersed in water at defined pH and ionic strength.

Homogenization of Emulsions—WPC dispersions and vegetable oil were heated separately at 50 °C, then mixed, stirred, and homogenized with a valve homogenizer (Büchi Model 196) thermostated at 50 °C. The homogenization step was repeated 15 times.

Characterization of Emulsions—A Malvern PC 100 photon correlation spectrometer was used to measure the mean size and the size distribution of oil droplets. The electrophoretic mobility of the dispersed oil droplets was measured using a Mark II particle micro-electrophoresis apparatus (Rank Brothers).

The most important physico-chemical properties of WPC-stabilized O/W emulsions at different pH are shown in Table 1.

Table 1 *Physico-chemical properties of WPC-stabilized O/W emulsions as a function of pH. Temperature of measurements, 25 °C; viscosity of the dilution buffer,* 0.89 mPa. s

Property	*pH of O/W emulsion* 3.5	5.5	7	9
Final pH after dilution	3.50	5.55	7.03	8.96
Ionic strength (mol. l^{-1})	0.0035	0.0166	0.0088	0.0055
Viscosity (mPa. s)	1.37	2.42	1.36	1.46
Mean particle diameter (nm)	379	1348	361	363
Polydispersity index	0.204	0.214	0.205	0.206
Debye–Hückel parameter (nm^{-1})	0.1944	0.4238	0.3082	0.2438
Electrophoretic mobility (10^{-8} m^2 s^{-1} V^{-1})	1.598	−1.497	−2.747	−3.597
Zeta potential (mV)	21.94	−19.41	−36.93	−48.90
Surface charge density (10^{-3} C. m^{-2})	2.829	−5.324	−7.576	−8.30
Protein coverage surface for 100 ml emulsion (m^2)	158	45	166	165
Maximum possible protein coverage (mg m^{-2})	5.3	18.7	5.1	5.1

Heat Gelation of Emulsions—After homogenization and cooling to 20 °C, the fluid emulsions were poured into glass beakers and incubated for 30 min in a water-bath at 90 °C. The gels formed were left for 12 h at room temperature.

Gel Firmness—Depending on their appearance, the gels formed were classified. The gels classified 'self-supporting' were submitted to compression tests on an Instron Model 1112 Texture Analyzer.

3 Results

Correlations Between Emulsion Fineness and Gelation Ability—The influence of the number of passages of the emulsion in the homogenizer on the emulsion fineness and on the gel rigidity is shown in Figure 1.

Effect of Variation of Protein and Oil Content—Gel firmness is influenced by changes in both the whey protein content and the oil content of the initial emulsion. This effect is shown in Figure 2.

Effect of pH on Heat Gelation Capacity—The pH range within which WPC-stabilized O/W emulsions can be gelled extends from 3.5 to 8. In the area of the isoelectric pH (4.5–5.5), firm and smooth gels were obtained, provided that the WPC used were undenatured.

4 Conclusions

High oil volume fraction can be incorporated in a gel matrix made of whey proteins. Stable O/W emulsions must be produced before subjection to heat

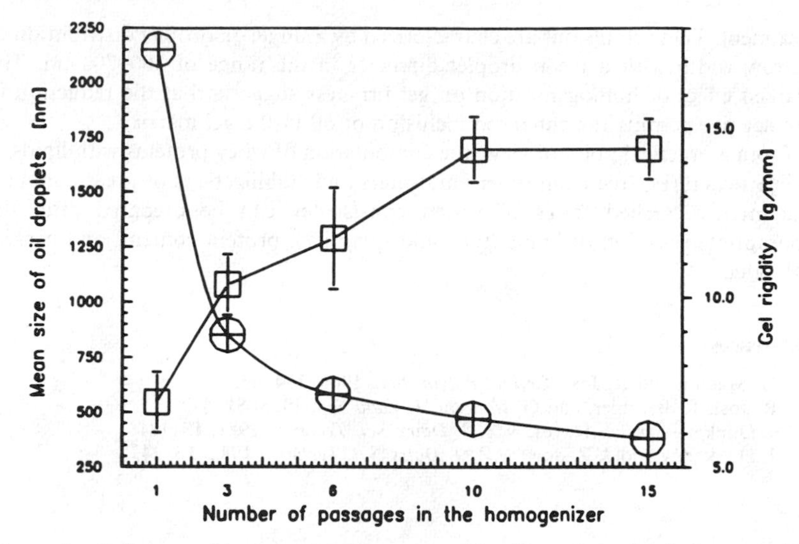

Figure 1 *Influence of homogenization on emulsion fineness and gelation capacity. The emulsion tested contained* 6% *WPC and* 15% *vegetable oil. Homogenization was performed on a Büchi Model B*-196 *homogenizer at* 50 °C. (⊕) *Mean size of oil droplets as measured by light scattering;* (□) *rigidity of the gels formed from the emulsions*

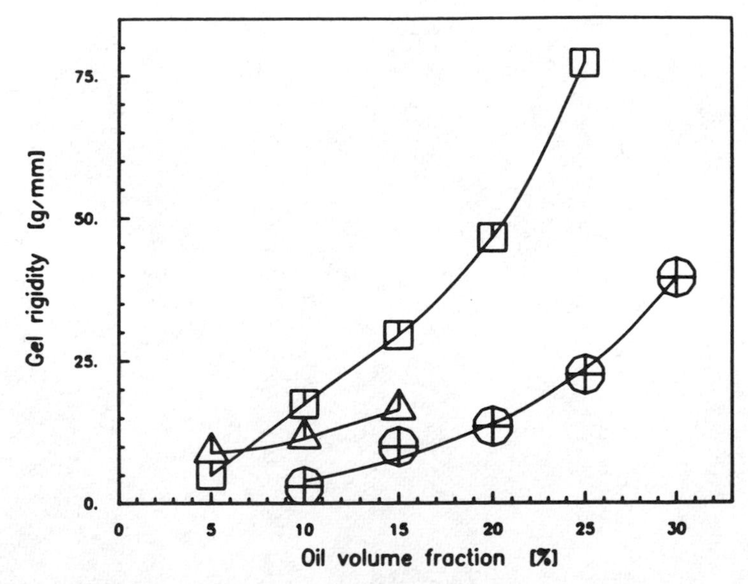

Figure 2 *Rigidity of the gel as a function of oil volume fraction and of protein concentration.* (⊕)5 vol.-% *WPC;* (□) 10 vol.-% *WPC;* (△) 5 vol.-% *egg white proteins*

treatment. These emulsions are characterized by a single oil droplet distribution of narrow width with a mean droplet diameter in the range of 300–700 nm. The marked effect of homogenization on gel firmness suggests that the reduction in droplet size permits the enhanced inclusion of oil in the gel matrix.

From a practical point of view, the combination of whey proteins with lipids in edible structures, free from other emulsifiers and stabilizers, is of great commercial interest. Gelled foods of varied consistency can be prepared after the appropriate selection of lipids (type and quantity), protein content, and correct pH value.

References

1. G. Masson and R. Jost, *Colloid Polym. Sci.*, 1986, **264**, 631.
2. R. Jost, R. Baechler, and G. Masson, *J. Food Sci.*, 1986, **51**, 440.
3. J. Dunkerley and J. Hayes, *N.Z. J. Dairy Sci. Technol.*, 1981, **15**, 181.
4. J. Dunkerley and J. Zadow, *N.Z. J. Dairy Sci. Technol.*, 1982, **16**, 243.

Rheological Study of Interactions Among Wheat Flour, Milk Proteins, and Lipids of Béchamel Sauce

By L. Patricia Martinez-Padilla and Joël Hardy

LABORATOIRE DE PHYSICOCHIMIE ET GÉNIE ALIMENTAIRES, E.N.S.A.I.A., INSTITUT NATIONAL POLYTECHNIQUE DE LORRAINE, 2 AVENUE DE LA FORÊT DE HAYE, 54500 VANDOEUVRE LES NANCY, FRANCE

1 Introduction

The highly thixotropic behaviour of Béchamel sauce was evaluated using two models. One was a rheological–kinetic model proposed by Tiu and Boger,[1] adapted in previous work for constant shear stress experiments. This model includes the following:

(a) the phenomenological relationship of Herschel-Bulkley modified by a structural parameter (λ):

$$\tau - \tau_{yo} = \lambda(K_0 + \dot{\gamma}^{n0}) \tag{1}$$

where K_o = consistency coefficient at time zero (Pa s^{no}), n_o = flow behaviour index at time zero (dimensionless), λ = structural parameter, equal to unity at time zero and λ_e at equilibrium (dimensionless), and τ_{yo} = yield stress at time zero (Pa);

(b) a second-order equation to describe the decay of the structure:

$$\mathrm{d}\lambda/\mathrm{d}t = -K_{b_1}(\lambda - \lambda_e)^2 \tag{2}$$

where $\mathrm{d}\lambda/\mathrm{d}t$ is the change in structural parameter with respect to time (s^{-1}) and K_{b_1} is a rate constant which is a function of the shear rate (s^{-1}).

The other model is an empirical one,[2] obtained by multiple logarithmic regression of the three variables $\dot{\gamma}$, $\tau - \tau_{yo}$ and t:

$$\tau - \tau_{yo} = K' \dot{\gamma}^{n'} t^{-p'} \tag{3}$$

where K' = time consistency index (Pa $s^{n'+p'}$), n' = time flow behaviour index (dimensionless) and p' = thixotropic index (dimensionless).

The aim of this study was to show that these parameters could be used to explain and quantify the effects of interactions among wheat flour, milk, and butter on the consistency and thixotropy of Béchamel sauce with the help of several constants given by the two rheological thixotropic models.

2 Experimental

Béchamel powder, wheat flour, spray-dried milk powder, UHT milk (1.5% fat), and butter (82% fat) were obtained from commercial suppliers. The moisture contents were 5.6%, 9.4%, 3.7%, 90.1%, and 16% respectively.

The investigated suspensions were as follows. BECH = 9.1% Béchamel sauce powder in UHT milk (commercial sauce); FI = 4.2% wheat flour in distilled water; FM = 4.2% wheat flour + 10.4% powder milk in distilled water; and FMB = 4.2% wheat flour + 10.4% powder milk + 5.3% butter in distilled water (laboratory Béchamel sauce).

Preparation of Gelatinized Suspensions—Béchamel sauce and suspensions were prepared as described by Evans and Haisman.[3] A rotary evaporator, with a thermocouple, was used to stir and heat the suspensions in a thermostatically controlled oil-bath. All the suspensions were rapidly heated (from 40 to 96 °C in 5 min), held at 96 °C for 25 min (the time necessary for completion of gelatinization), and then cooled in 10 min to 60 °C. Samples (4.5 ml) were placed on the rheometer plate, where their temperature was allowed to stabilize at 20 °C for 2 min.

Measurement of Rheological Properties—A controlled stress, thermostatically controlled cone (3.0 cm radius, 4°) and plate rheometer (Carrimed C.S.) was used to measure shear rate ($\dot{\gamma}$) as a function of time (t). The rheometer was controlled by a microcomputer (Apple IIe). Three measurements were made at each applied shear stress, a fresh sample being used for each one. A light paraffin oil was applied to the edge of the sample to prevent excessive drying. The measurements were made at 20 $\pm$ 0.1 °C. The sample was subjected to a sudden, instantaneous constant shear stress that was maintained for 40 min. The interval of shear rate studied was 0–650 s^{-1}. For the initial yield stress (τ_{y0}) of the sample, determinations were made just until the sample began to flow, using the more sensitive angular displacement system (cone 2 cm radius, 1°). Sugar solutions and motor oil (SAE 10) were used to calibrate the rheometer.

3 Results and Discussion

Figure 1 shows a typical rheogram for Béchamel sauce. When τ is applied, $\dot{\gamma}$ increases over a period of time, and hence the apparent viscosity decreases until a steady value is reached ($\dot{\gamma}$). The acceleration of shear rate is more important on the first part of the curve (initial time of shear) than on the final part. The results show that as the shear rate increases the rate of structural breakdown becomes greater. Rheological constants were evaluated using the procedure reported in a previous paper[2] and are plotted in Figures 2 and 3 for each model. In both models, the correlation coefficients for the constants evaluated were greater than 0.9.

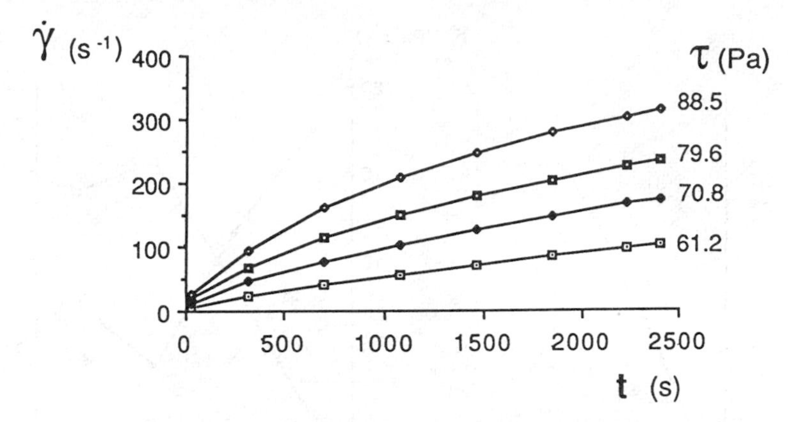

Figure 1 *Typical plot of shear rate* ($\dot{\gamma}$) *as a function of shear stress* (τ) *and time* (t). *Béchamel sauce; temperature* 20 °C

The simplest suspension (FI) shows a larger value for the equilibrium structural parameter (λ_e), a small rate of breakdown (K_{b_2}) and a small thixotropic index (p'), indicating that the fluid is more resistant to structural breakdown than the more complex suspensions (FM, FMB, and BECH), in which the rate of structural decay is increased and the structure is significantly broken (λ_e).

At time zero, the yield stress increases exponentially with concentration. This strong concentration dependence is similar to that reported for carbohydrate suspensions,[4] but the suspensions studied here are much more complex.

The flow behaviour indices (n and n') decrease as the suspensions become more complex, indicating the high non-Newtonian behaviour.

The consistency indices (K_0 and K') increase with increasing total solids. On the other hand, the consistency index K' (empirical model) could be used to quantify the textural consistency of the samples and this value is affected directly by shearing and time of shearing.

If τ_{y0} is considered to be the shear stress necessary to destroy the network structure,[5] the higher τ_{y0} observed as the system becomes more complex could be attributed to the intermolecular hydrogen bonds and molecular entanglements produced when the starch suspensions are heated[6] or when protein solutions are concentrated.[7]

With milk, changes in proteins during heating result in cross-linking and binding with carbohydrates and lipids.[8] κ-Casein has been shown to be the most reactive of the milk casein proteins, showing multi-stranded structures in solution,[9] and perhaps plays an improtant role in cross-linking and binding in gel structures.

On heating the milk and wheat flour in Béchamel sauce, it is probable that intermolecular interactions occur between the κ-casein and the free amylose, amylopectin, and starch granules liberated from the wheat flour.[10] This type of interaction has been seen with carrageenan–casein mixtures[11] and results in thixotropic structures. These interactions could cause strong binding, resulting in the higher values of τ_{y0} and K'.

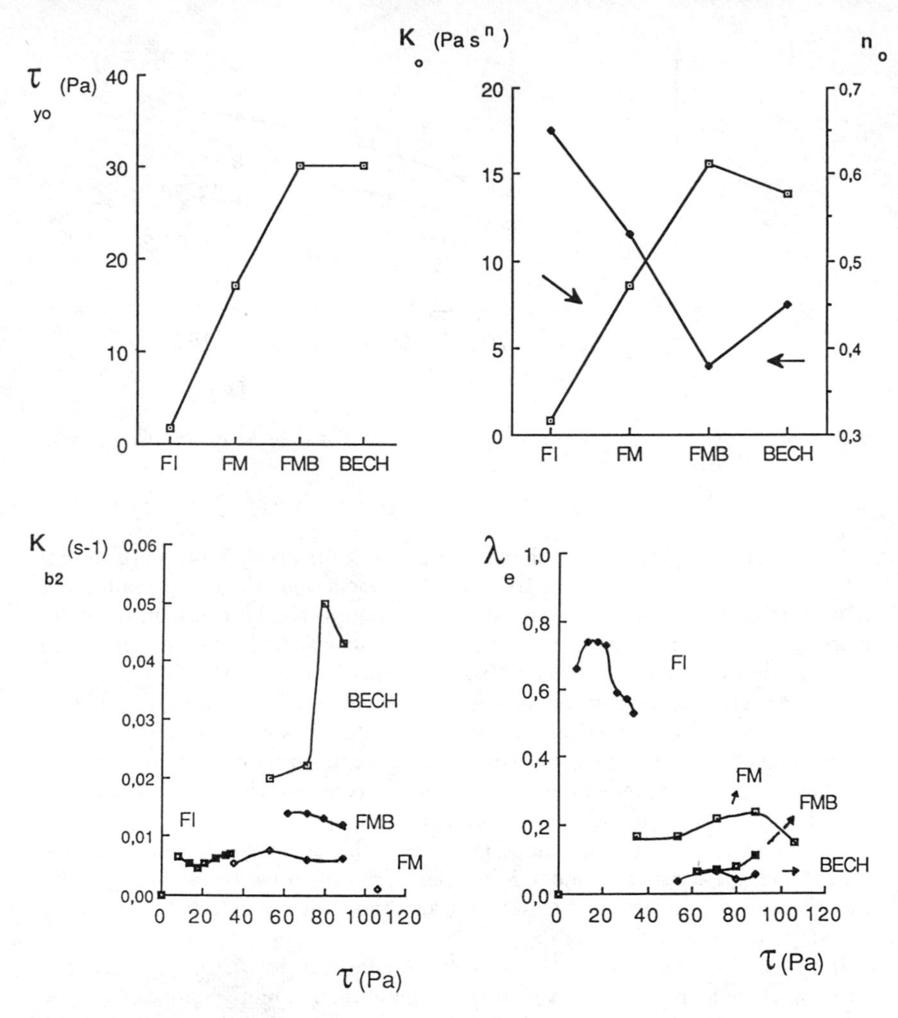

Figure 2 *Parameters of phenomenological model. Rate constant* (K_{b_2}) *and structural parameter* (λ_e) *as a function of shear stress* (τ) *and Herschel-Bulkley constants* (τ_{y0}, K_0, *and* n_0) *at time zero. BECH = commercial sauce; FMB = laboratory sauce, flour-milk-butter suspensions; FM = flour–milk suspension; FL = flour suspension*

It is important to note that the τ_{y0} values, are more than ten times higher for the flour–milk suspensions than for the flour–water suspensions, supporting the idea of intermolecular interactions between casein, amylose, and amylopectin. However, they are not very strong and are destroyed by low shearing, thus explaining the thixotropic behaviour. Otherwise, when lipids are added to the four–milk suspensions, the consistency (K') is increased more than the thixotropy (p')

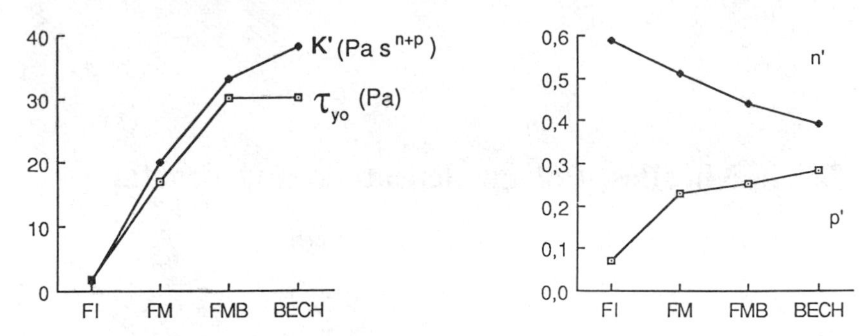

Figure 3 *Parameters of empirical model for* $0 \leqslant \dot{\gamma} \leqslant 600\ s^{-1}$ *and* $0 \leqslant t \leqslant 2400$ s. *Abbreviations as in Figure 2*

observed in the empirical model. This could imply that the interactions of lipids are stronger than the interactions of the casein with the starch.

4 Conclusion

The rheological thixotropic parameters obtained could be used to explain the interactions between macromolecules which are used to modify food consistency. These constants provide evidence for and indirectly quantify the casein–starch–milk lipid interactions in Béchamel sauce. Work is in progress to study these interactions.

References

1. C. Tiu and D. V. Boger, *J. Texture Stud.*, 1974, **5**, 329.
2. L. P. Martinez-Padilla and J. Hardy, paper presented at the 22nd Colloque Annuel du Groupe Français de Rhéologie, Toulouse, 16–18 November, 1987.
3. D. I. Evans and D. R. Haisman, *J. Texture Stud.*, 1979, **10**, 347.
4. R. E. Lang and C. Rha, *J. Texture Stud.*, 1981, **12**, 47.
5. L. Duran and E. Costell, *J. Texture Stud.*, 1982, **13**, 43.
6. M. Glicksmann, 'Gum Technology in the Food Industry,' Academic Press, New York, 1969.
7. M. A. Tung, *J. Texture Stud.*, 1978, **9**, 3.
8. E. W. Evans, in 'Interactions of Food Components,' eds. G. G. Birch and M. G. Lindley, Elsevier Applied Science, Barking, 1986.
9. T. H. M. Snoeren, P. Both, and D. G. Schmidt, *Neth. Milk Dairy J.*, 1976, **30**, 132.
10. A. C. Eliasson, *J. Texture Stud.*, 1982, **13**, 43.
11. T. A. J. Payens, *J. Dairy Sci.*, 1972, **55**, 141.

Casein Micelles, Polycondensation, and Fractals

By D. S. Horne, T. G. Parker, and D. G. Dalgleish

HANNAH RESEARCH INSTITUTE, AYR KA6 5HL, UK

1 Introduction

The caseinate proteins (α_{s1}, β, κ, and α_{s2}) of bovine milk exist as large, roughly spherical aggregates, historically termed micelles. Because so many technological properties of milk depend on the properties of these particles, it is essential to understand their structure and stability. This demands a workable model for the structure of the casein micelle, which will be consistent with the relatively sparse knowledge of its genesis in the lactating mammary gland and which will also satisfy all that is known of the physical properties of the caseins and their micellar aggregates.

2 Micellar Properties

Casein micelles are approximately spherical but vary in size. Electron microscopy shows a spread of diameters from 20 to 300 nm[1] or even larger.[2] The electron micrographs reveal a raspberry-like structure and this, combined with the association tendencies of the caseins, has led to the proposal that casein micelles possess a submicellar structure. Although the subunits are considered to be fairly uniform in size, they must vary in composition (with the average composition being that of the original milk), to allow for the uneven distribution of κ-casein required to permit a surface location of this protein. The latter requirement follows from the observations that small micelles contain a greater proportion of κ-casein than large micelles,[3-5] that rennet attacks micellar κ-casein and destabilizes the micellar system to form a coagulum,[6] and that rennet action decreases the hydrodynamic radius.[7] Direct evidence from electron microscopy for the location of the different caseins in micelles is conflicting, with some studies on native micelles suggesting that κ-casein is predominantly concentrated either on the periphery of at least the larger micelles or in a bridging position.[8-11] However, one study suggests a uniform distribution of κ-casein throughout the micelle but allows a decreasing content of κ-casein as micelle size is increased.[12]

Casein micelles also possess a high voluminosity of 4–5 ml g^{-1}, which has been ascribed mainly to a loose sponge-like structure with much interstitial water.[3] Further evidence for an open structure follows from the extent to which micellar components react with a number of micelle-penetrating reagents, some up to a few nanometres in size.[13]

3 Micellar Models

Schmidt[14] has classified models of micelles into three categories: coat-core models,[15] internal structure models,[16-19] and sub-micellar models.[14,20-23] The internal structure models are branching-chain polymer models involving bi- and trifunctional monomeric units. Slattery[24] used Flory's analyses[25] of the weight distribution of polymers produced in such systems, and concluded that the polycondensation theory for a trifunctional system gave results inconsistent with observed micellar weight distributions: this was also true of the structural scheme of Garnier and Ribadeau Dumas.[18] The most important criticism of these chain-polymer models is that a reaction of bi- and trifunctional monomers cannot be stopped, so that infinite rather than limited aggregates (casein micelles) would be formed.

We have considered a number of possible applications of polyfunctional condensation mechanisms to milk proteins, for example the aggregation of individual caseins[26] and the precipitation of protein when milk is heated.[27] The criticisms of such mechanisms expressed by Slattery[24] can be met by considering the more complex condensation model introduced below.

In this, we assign to κ-casein the role of monofunctional chain terminator. We consider that α_s-caseins act as trifunctional monomers, providing branch points, and that β-casein acts as a bifunctional unit, propagating chains by linear growth during the formation of casein micelles by random polymerization of the proteins. Following the work of Gordon and co-workers,[28-30] it is possible to calculate number- and weight-average molecular weights in such mixtures after defining a statistically derived probability generating function to describe the polymer growth. Alternatively, we can calculate the weight distribution, not easily obtained by the statistical method, from a kinetic approach using diffusion-controlled reaction of functionalities.[31] Both calculations assume that the functionalities react via second-order kinetics, that no ring formation occurs, and that functionalities are not buried by steric hindrance as the cluster grows. Such factors will be important with monomers as large as casein proteins, but are difficult to include in either of the approaches. It is possible, however, to develop a computer simulation of this polyfunctional condensation reaction.

4 Computer Simulation Model

For simplicity, the cluster is grown on a square, two-dimensional lattice, the lattice spacing being equivalent to the monomer size. The simulation is a variant of one of the earliest models of colloidal aggregation where particles are added one at a time to the growing cluster using linear trajectories.[32,33] The process begins by placing a single particle at the centre of the lattice. The functionality of this particle is defined by generating a random number between 0 and 1, the size of this number being used to determine the particle's functionality. These preset limits thus also allow the proportions of κ-, β- and α_s-casein to be fixed. The next particle is chosen and its functionality similarly defined. The particle trajectory, a straight line parallel to $\pm x$ or $\pm y$ axes, is picked at random from all possible trajectories which can bring the particle into contact with the seed (growth site) or the growing cluster. The particle is considered to have contacted the cluster when it occupies a lattice site which is a nearest neighbour to one of the sites occupied by a member of the cluster. The incoming particle is permitted to stick to the cluster if the impact site has a free functionality. The functionalities of the two interacting units are each decremented by one, the new particle is fixed at the site, and the record of cluster geometry is updated.

If the nearest neighbour occupied site or sites have no free functionalities, the particle attempts to progress deeper into the cluster along its original trajectory until it meets a head-on failure to join the cluster or until it completes a fly-by of the cluster. Following this total failure, a new particle trajectory is picked at random for this particle, and this is continued until the particle is accepted into the cluster. The number of such failures reflects the rate of growth of the cluster or the reactivity of the cluster surface and will be investigated in future studies.

Once a new particle has been added, the number of accessible 'active' sites around the cluster is summed. If this number is zero, a closed or stable cluster has been generated, incapable of further accretion. If the number is non-zero, cluster growth continues. No reorganization of the cluster is permitted at any stage. Occasionally, the contact site is nearest neighbour to two or more possible growth points. If the incoming particle has sufficient functionalities, it reacts with the maximum possible, *i.e.* ring formation is permitted.

5 Results and Discussion

Figure 1 shows a cluster of 800 particles generated following the above protocol. The random accretion process has resulted in a highly ramified and porous structure. This results only from the varying functionality or reactivity of the accepting site. If every site accepts particles on every contact, such linear or ballistic trajectories produce a compact, space-filling aggregate.[34] Instead, by randomly assigning site reactivity we have generated a cluster which resembles those generated by a diffusion-limited aggregation model where each single incoming particle follows a random walk trajectory.[35,36]

Such diffusion-limited aggregates were shown to be fractal, *i.e.* they were found to be invariant to a change in scale of the picture. Quantitatively, for such structures, a power-law relationship between cluster mass and radius of gyration exists:

$$M \sim R_g{}^D \tag{1}$$

The exponent, D, also called the fractal dimension, was found to have a value of about 1.7 using a variety of diffusion-limited aggregation models in two dimensions.[37] For models with linear, ballistic trajectories, however, compact structures were produced and on a two-dimensional lattice, D was found to be equal to 2, the topological dimension of the surface.[38] This has led to the belief that a Brownian particle trajectory is necessary to produce a fractal structure. The cluster in Figure 1 was generated using ballistic trajectories and a distribution of functionalities. At each stage of growth, as each particle was added, the radius of gyration of the cluster was computed. The double logarithmic plot of cluster mass *versus* radius

```
                                      1 9  84
                                  144777144 78
                                     184  474  5
                                    44444  44   4
                                    87741  8 9474
                                   1771     4 749
                                44  47      844
                                47  54184 4771
                           84  178  44 4787
                           8744478887784  41
                              7414  47
                              9  4787744
                           14      4 4  4
                            8     4747 871
                            48  47  848  41
                             8  41  448847
                           587848   784  741          41
                            74 174474 4  814         17 1
                            1   1  4  48 8 84        877
                                  48   787 441        441  1
                                  1   84 8 774    171414477  7441
                      11          984 4 17445    1 8 744 774774
                     1774          4 8   74      44747   84 4 4
                       17         787  4774    477  1     1 8 4
           11          147        544178 45 44748  41     4744
          177             471 5     4  1  4  4 774444      74
            4              4787     587   147  8474 77     84
           17              14 8   14744  1 47484744 771   1
           17        1       5 8 1   1    4 8    88448   84
           17       44   44777747        1747    47844477
            4      5 8  1 5 4478   1   44474     81 1  41
          474     4448      97    4   44 71 1    71 74 8741
          4 8     744      1774 17  184 7144     784471 71
          778 9   48       98879 481477 4871   14   1  4
          4771745 44       4718 447  44 177747744    844
          598 75 54    1411748788 7484    1 17444  17471
          147 87187      7787     4874      4144     788
         48 4  84478 14  44  4    47444     1 4 148848 49
         574774    7944  4 944   471  1      1       977
          4 484474777874774      487874                81
          4787917474 44 44       417 41
           1   4741     77    5  8 4 441
               74       1744 4747717148
               1        17 114  174  17875
                          4   17474  8 41
                         44    4448447471
                    9 44       141 77411
                    4 4        474471
                   5774          44
                     8   5
                     78848
                     4
                     4
                     74471
                   874  9
                   44
```

Figure 1 *A typical two-dimensional cluster of* 800 *polyfunctional particles grown on a square lattice using ballistic trajectories. The cluster consists of* 16% κ*-casein* (f = 1), 42% β*-casein* (f = 2) *and* 42% α_s*-casein* (f = 3). *The numerals reflect the nature and residual functionality of the monomers at this stage of cluster growth. Thus,* 7, 8 *and* 9 *are* f = 3 *monomers with zero, one and two functionalities remaining, respectively;* 4 *and* 5 *are* f = 2 *with zero and one functionality;* 1 *indicates a reacted* f = 1 *monomer*

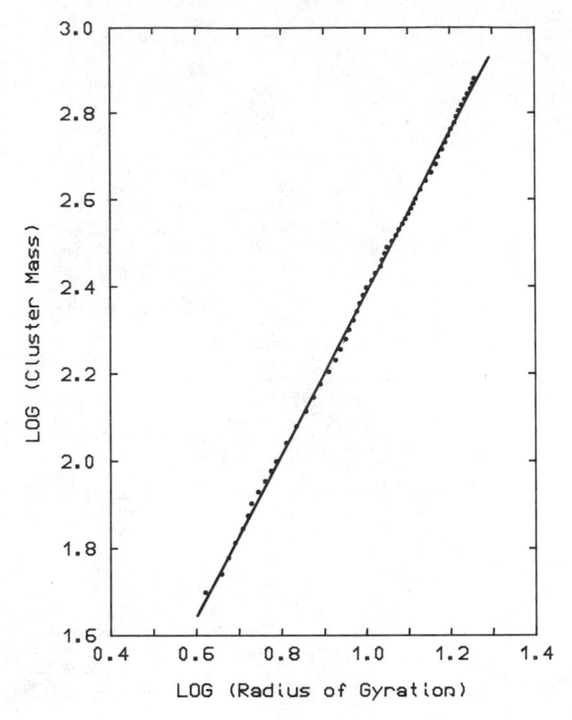

Figure 2 *Log–log plot of cluster mass* versus *radius of gyration for the cluster in Figure* 1 *calculated during cluster growth after the addition of every tenth particle. The solid line is the linear regression fit to the data and has slope* D = *fractal dimension* = 1.86 (r = 0.998)

of gyration is shown in Figure 2. The plot is linear and the gradient is 1.85, close to the fractal dimension of 1.7 recorded for the diffusion-limited aggregate. It is anticipated in this model, however, that the fractal dimension will be influenced by the functionality distribution. This awaits further study.

The deep fjords in the structure will allow the penetration of small reagents to the very core of the micelle. Despite this, the water trapped in the internal open areas will contribute to a high voluminosity. There are also areas of high local density around perhaps a particularly reactive accumulation of particles. When projected into three dimensions, such local variations could give the appearance of a sub-unit structure. It may therefore be unnecessary to postulate the existence of sub-micelles when the required inhomogeneous structure is provided by the polyfunctional mechanism.

The monofunctional units (κ-casein) terminate chains, but the growing cluster can circumvent the blocked site. Hence, although the majority of the κ-casein is found on the cluster perimeter, a significant fraction (*ca.* 38%) remains buried. Limited studies suggest that the higher the κ-casein content, the easier it is to generate small clusters. Reducing the κ-casein content allows greater growth.

Because growth is a truly random process in this simulation, a range of stable cluster sizes can be created using the same functionality distribution. This occurs because chain terminators are surrounded on a random basis and, conversely, active functionalities are randomly shielded and sterically hindered by other outer growing regions of the cluster.

Our simulation model produces structures which qualitatively reproduce the salient properties of the casein micelle: a highly porous structure, a wide size distribution, a surface location for κ-casein but with a significant fraction buried, an inhomogeneous internal structure which side-steps the requirement for sub-units, and the probability that the κ-casein content will vary with micelle size. The model now needs to be refined and quantified and, perhaps, extended to three dimensions. As to the nature of the functionalities, is this the role of colloidal calcium phosphate, as appears to be implicit in the results of Holt *et al.*[39] on the dissociation of caseins when micellar calcium phosphate is dissolved in a controlled way?

References

1. D. G. Schmidt, P. Walstra, and W. Buchheim, *Neth. Milk Dairy J.*, 1973, **27**, 128.
2. B. E. Brooker and C. Holt, *J. Dairy Res.*, 1978, **45**, 355.
3. D. G. Schmidt and T. A. J. Payens, *Surf. Colloid Sci.*, 1976, **9**, 165.
4. D. G. Schmidt, *Neth. Milk Dairy J.*, 1980, **34**, 42.
5. T. C. A. McGann, W. J. Donnelly, R. D. Kearney, and W. Buchheim, *Biochim. Biophys. Acta*, 1980, **630**, 261.
6. D. G. Dalgleish, in 'Developments in Dairy Chemistry, 1,' ed. P. F. Fox, Elsevier Applied Science, Barking, 1982, p. 157.
7. P. Walstra, V. A. Bloomfield, G. J. Wei, and R. Jenness, *Biochim. Biophys. Acta*, 1981, **669**, 258.
8. R. M. Parry and R. J. Carroll, *Biochim. Biophys. Acta*, 1969, **194**, 138.
9. R. J Carroll and H. M. Farrell, *J. Dairy Sci.*, 1983, **66**, 679.
10. M. Horisberger and M. Vonlanthen, *J. Dairy Res.*, 1980, **47**, 185.
11. S. Kudo, S. Iwata, and M. Mada, *J. Dairy Sci.*, 1979, **62**, 916.
12. M. Horisberger and M. Vauthey, *Histochemistry*, 1984, **80**, 9.
13. B. Ribadeau Dumas and J. Garnier, *J. Dairy Res.*, 1970, **37**, 269.
14. D. G. Schmidt, in 'Developments in Dairy Chemistry, 1,' ed. P. F. Fox, Elsevier Applied Science, Barking, 1982, p. 61.
15. D. F. Waugh, in 'Milk Proteins, Chemistry and Molecular Biology, II,' ed. H. A. McKenzie, Academic Press, New York, 1971, p. 3.
16. D. Rose, *Dairy Sci. Abstr.*, 1969, **31**, 171.
17. T. A. J. Payens, *J. Dairy Sci.*, 1966, **49**, 1317.
18. J. Garnier and B. Ribadeau Dumas, *J. Dairy Res.*, 1970, **37**, 493.
19. J. Garnier, *Neth. Milk Dairy J.*, 1973, **27**, 240.
20. C. V. Morr, *J. Dairy Sci.*, 1967, **50**, 1744.
21. T. A. J. Payens, *J. Dairy Res.*, 1979, **46**, 291.
22. C. W. Slattery and R. Evard, *Biochim. Biophys. Acta*, 1973, **317**, 529.
23. C. W. Slattery, *Biophys. Chem.*, 1977, **6**, 59.
24. C. W. Slattery, *J. Dairy Sci.*, 1976, **59**, 1547.
25. P. J. Flory, 'Principles of Polymer Chemistry,' Cornell University Press, Ithaca, NY, 1953.
26. D. G. Dalgleish and T. G. Parker, *J. Dairy Res.*, 1979, **46**, 259.
27. T. G. Parker, D. S. Horne, and D. G. Dalgleish, *J. Dairy Res.*, 1979, **46**, 377.
28. M. Gordon, *Proc. R. Soc. London, Ser. A*, 1962, **268**, 240.
29. D. S. Butler, M. Gordon, and G. N. Malcolm, *Proc. R. Soc. London, A*, 1966, **295**, 29.

30. M. Gordon, T. G. Parker, and W. B. Temple, *J. Comb. Theory*, 1971, **11**, 142.
31. D. G. Dalgleish and T. G. Parker, unpublished calculations.
32. M. J. Vold, *J. Colloid Sci.*, 1963, **18**, 684.
33. D. N. Sutherland, *J. Colloid Interface Sci.*, 1966, **22**, 300.
34. P. Meakin, *Phys. Rev. B*, 1983, **28**, 5221.
35. T. Witten and L. Sander, *Phys. Rev. Lett.*, 1981, **47**, 1400.
36. P. Meakin, *Phys. Rev. A*, 1983, **27**, 604.
37. P. Meakin, *Phys. Rev. A*, 1986, **33**, 3371.
38. P. Meakin, *J. Colloid Interface Sci.*, 1985, **105**, 240.
39. C. Holt, D. T. Davies, and A. J. R. Law, *J. Dairy Res.*, 1986, **53**, 557.